Current Genome Science

Advances in Genome Science: Changing Views on Living Organisms (Volume 1)

Editor

Christian Neri

Research Director at INSERM
Laboratory of Neuronal Cell Biology & Pathology
Psychiatry and Neuroscience Center U894
2 ter rue d'Alesia 75014 Paris
France

Bentham Science Publishers
Executive Suite Y - 2
PO Box 7917, Saif Zone
Sharjah, U.A.E.
subscriptions@benthamscience.org

Bentham Science Publishers
P.O. Box 446
Oak Park, IL 60301-0446
USA
subscriptions@benthamscience.org

Bentham Science Publishers
P.O. Box 294
1400 AG Bussum
THE NETHERLANDS
subscriptions@benthamscience.org

CONTENTS

Foreword *i*

Preface *ii*

List of Contributors *iii*

CHAPTERS

1. **Endometrial Cancer as a Familial Tumor: Pathology and Molecular Carcinogenesis** **3**
   Kouji Banno, Megumi Yanokura, Kennta Masuda, Arisa Ueki, Iori Kisu, Yusuke Kobayashi, Wataru Yamagami, Nobuyuki Susumu and Daisuke Aoki

2. **Replication Origins and Timing of Temporal Replication in Budding Yeast: How to Solve the Conundrum?** **15**
   Matteo Barberis, Thomas W. Spiesser and Edda Klipp

3. **The Role of Androgen Receptor Mutations in Progression of Prostate Cancer** **45**
   G.N. Brooke, J. Waxman and C. L.Bevan

4. ***MUTYH* Associated Polyposis (MAP)** **66**
   M.L.M. Binderup and M.L. Bisgaard

5. **Gene Clusters, Molecular Evolution and Disease: a Speculation** **101**
   Alireza Baradaran-Heravi, Leah I. Elizondo and Cornelius F. Boerkoel

6. **Expression and Function of Kruppel Like-Factors (KLF) in Carcinogenesis** **146**
   Diab Thoria, Bureau Christophe,Hanoun Naima, Torrisani Jérôme, Vinel Jean-Pierre, Buscail Louis and Cordelier Pierre

7. **Episcopic 3D Imaging Methods: Tools for Researching Gene Function** 169
Wolfgang J Weninger and Stefan H Geyer

8. ***Drosophila Lethal Giant Larvae* Neoplastic Mutant as a Genetic Tool for Cancer Modelling** 189
Francesca Froldi, Marcello Ziosi, Annalisa Pession and Daniela Grifoni

9. **Genetic Insights Through Genome Wide Association Studies in Type 2 Diabetes Mellitus will Lead to New Therapeutics** 210
Marcel G. M. Wolfs, Naishi Li, Jingyuan Fu, Cisca Wijmenga, Timon W. van Haeften, Marten H. Hofker

10. **Indispensable Roles of Plastids in Arabidopsis Thaliana Embryogenesis – Update** 241
Mehdi Nafati and Kentaro Inoue

11. **Mass Spectrometry-Based Approaches Toward Absolute Quantitative Proteomics** 261
Keiji Kito and Takashi Ito

12. **Regulation of Interactions with Sliding Clamps During DNA Replication and Repair** 293
Francisco J. López de Saro

13. **Revealing Gene Function and Genetic Diversity in Plants and Animals via TILLING and EcoTILLING** 316
Barkley NA and Wang ML

14. **Variations Around the Arrestin Fold** 352
Laurence Aubry, Dorian Guetta and Gérard Klein

15. **Role of Calcium Regulated Kinases: Expressional and Functional Analysis in Abiotic Stress Signaling** 377
Ritika Das, Amita Pandey and Girdhar K. Pandey

16. Multiple Hsp70 Isoforms in the Eukaryotic Cytosol: Evidence for Both Redundant and Specialized Functions **426**
Mehdi Kabani and Céline N. Martineau

17. Mitochondrial and Nuclear Genes of Mitochondrial Components in Cancer **452**
E. Kirches

Index **491**

FOREWORD

Genome science has become essential in advancing knowledge in the wide field of biological and clinical research. The eBook 'Advances in Genome Science' is intended to provide the community with a multi-disciplinary and accessible view of some of the latest developments in the field, allowing readers to capture the relevance and diversity of genomics.

Emerging in the 1980s with the availability of exciting new techniques that allowed larger systematic studies of more than one gene, gaining speed in the 1990s with the genome projects and today, going through a period of conceptual and technological refinement and differentiation. Genomics is at the brink of yielding the scientific and medical progress the public and the research communities in basic, applied and clinical science have been hoping for and expecting.

The challenges that still lie ahead are great and the best use of the possibilities of genomics will be necessary to tackle them and provide solutions to urgent medical and environmental problems. While the substantial investment in large scale genomics studies has repeatedly been discussed in critical terms, genomics researchers as a result can now focus on the evaluation of the valuable, comprehensive datasets that have been generated. Theoretical approaches have started to yield testable models of biological processes, hypothesis-driven investigation into disease mechanisms or signaling pathways has been provided with exiting new leads, identification of new therapy targets has a broader, more detailed knowledge base to start from.

'Advances in Genome Science' allows a look at the cutting-edge genomics approaches in use today that bridge the gap to translational science and personalized medicine and are certain to change the way we think about genetic information even more.

Erich Wanker

PI Neuroproteomics and Molecular Mechanisms of Neurodegeneration at
Max Delbrück Center for Molecular Medicine, Berlin-Buch
Full Professor of Molecular Medicine at Charité University Medicine Berlin
Germany

PREFACE

'Genome science has contributed to understanding how biological processes may develop in living organisms on a global-scale level, resulting in significant progresses in various fields such as molecular evolution, development, plant genetics, aging and disease research. This eBook brings together reviews from different aspects of genome science research and selected topics provide an updated introduction to the diversity of impact in the field. The eBook is a valuable reference and handbook for researchers and clinicians, as well as students and medical professionals interested in the impact of genome science on our understanding of normal and pathological biological processes.'

Christian Neri

Research Director at INSERM
Laboratory of Neuronal Cell Biology & Pathology
Psychiatry and Neuroscience Center U894
2 ter rue d'Alesia 75014 Paris
France

List of Contributors

Kouji Banno

Department of Obstetrics and Gynecology, Keio University School of Medicine, Tokyo, Japan

Megumi Yanokura

Department of Obstetrics and Gynecology, Keio University School of Medicine, Tokyo, Japan

Kennta Masuda

Department of Obstetrics and Gynecology, Keio University School of Medicine, Tokyo, Japan

Arisa Ueki

Department of Obstetrics and Gynecology, Keio University School of Medicine, Tokyo, Japan

Iori Kisu

Department of Obstetrics and Gynecology, Keio University School of Medicine, Tokyo, Japan

Yusuke Kobayashi

Department of Obstetrics and Gynecology, Keio University School of Medicine, Tokyo, Japan

Wataru Yamagami

Department of Obstetrics and Gynecology, Keio University School of Medicine, Tokyo, Japan

Nobuyuki Susumu

Department of Obstetrics and Gynecology, Keio University School of Medicine, Tokyo, Japan

Daisuke Aoki

Department of Obstetrics and Gynecology, Keio University School of Medicine, Tokyo, Japan

Matteo Barberis

Institute for Biology, Theoretical Biophysics, Humboldt University Berlin, Invalidenstraβe 42, 10115 Berlin, Germany

Thomas W. Spiesser

Institute for Biology, Theoretical Biophysics, Humboldt University Berlin, Invalidenstraβe 42, 10115 Berlin, Germany

Edda Klipp

Institute for Biology, Theoretical Biophysics, Humboldt University Berlin, Invalidenstraβe 42, 10115 Berlin, Germany

Greg N. Brooke

Department of Surgery and Cancer,Androgen Signalling Laboratory, Imperial College London, London W12 0NN, UK

Jonathan Waxman

Department of Surgery and Cancer,Androgen Signalling Laboratory, Imperial College London, London W12 0NN, UK

Charlotte. L.Bevan

Department of Surgery and Cancer,Androgen Signalling Laboratory, Imperial College London, London W12 0NN, UK

Marie Louise M. Binderup

Department of Cellular and Molecular Medicine, University of Copenhagen, Panum Institute 24.4, Blegdamsvej 3, 2200 Copenhagen N, Denmark

Marie L. Bisgaard

Department of Cellular and Molecular Medicine, University of Copenhagen, Panum Institute 24.4, Blegdamsvej 3, 2200 Copenhagen N, Denmark

Alireza Baradaran-Heravi

Department of Medical Genetics, Child and Family Research Institute, University of British Columbia, Vancouver, British Columbia, Canada

Leah I. Elizondo

Centre for Molecular Medicine and Therapeutics, Child and Family Research Institute, Department of Medical Genetics, University of British Columbia, Vancouver, Canada

Cornelius F. Boerkoel

Department of Medical Genetics, Child and Family Research Institute, University of British Columbia, Vancouver, British Columbia, Canada

Diab Thoria

INSERM U1037, CRCT Team 10, 1, avenue Jean Poulhes, BP 84225, 31432 Toulouse Cedex 04, France

Bureau Christophe

INSERM U1037, CRCT Team 10, 1, avenue Jean Poulhes, BP 84225, 31432 Toulouse Cedex 04, France

Hanoun Naima

INSERM U1037, CRCT Team 10, 1, avenue Jean Poulhes, BP 84225, 31432 Toulouse Cedex 04, France

Torrisani Jérôme

INSERM U1037, CRCT Team 10, 1, avenue Jean Poulhes, BP 84225, 31432 Toulouse Cedex 04, France

Vinel Jean-Pierre

INSERM U1037, CRCT Team 10, 1, avenue Jean Poulhes, BP 84225, 31432 Toulouse Cedex 04, France

Buscail Louis

INSERM U1037, CRCT Team 10, 1, avenue Jean Poulhes, BP 84225, 31432 Toulouse Cedex 04, France

Cordelier Pierre

INSERM U1037, CRCT Team 10, 1, avenue Jean Poulhes, BP 84225, 31432 Toulouse Cedex 04, France

Wolfgang J. Weninger

IMG, Centre for Anatomy and Cell Biology, Medical University of Vienna, Waehringer Str. 13, A-1090 Vienna, Austria

Stefan H. Geyer

IMG, Centre for Anatomy and Cell Biology, Medical University of Vienna, Waehringer Str. 13, A-1090 Vienna, Austria

Francesca Froldi,

Cancer Modelling in Drosophila, Alma Mater Studiorum, Departments of Biologia Evoluzionistica Sperimentale, and Patologia Sperimentale, Bologna, Italy

Marcello Ziosi,

Cancer Modelling in Drosophila, Alma Mater Studiorum, Departments of Biologia Evoluzionistica Sperimentale, and Patologia Sperimentale, Bologna, Italy

Annalisa Pession

Cancer Modelling in Drosophila, Alma Mater Studiorum, Departments of Biologia Evoluzionistica Sperimentale, and Patologia Sperimentale, Bologna, Italy

Daniela Grifoni

Cancer Modelling in Drosophila, Alma Mater Studiorum, Departments of Biologia Evoluzionistica Sperimentale, and Patologia Sperimentale, Bologna, Italy

Marcel G. M. Wolfs

Department of Pathology and Medical Biology, Medical Biology Section, Molecular Genetics, University Medical Center Groningen, University of Groningen, Hanzeplein 1, 9713 GZ Groningen, the Netherlands

Naishi Li

Department of Endocrinology, Key Laboratory of Endocrinology of Ministry of Health, Peking Union Medical College Hospital, Peking Union Medical College, Chinese Academy of Medical Science, 100730, Beijing, China

Jingyuan Fu

Department of Genetics, University Medical Center Groningen, University of Groningen, PO Box 30001, 9700 RB Groningen, the Netherlands

Cisca Wijmenga

Department of Genetics, University Medical Center Groningen, University of Groningen, PO Box 30001, 9700 RB Groningen, the Netherlands

Timon W. van Haeften

Department of Internal Medicine, G 02-228, University Medical Center Utrecht, PO Box 85500, 3508 GA Utrecht, the Netherlands

Marten H. Hofker

Department of Pathology and Medical Biology, Medical Biology Section, Molecular Genetics, University Medical Center Groningen, University of Groningen, Hanzeplein 1, 9713 GZ Groningen, the Netherlands

Mehdi Nafati

Department of Plant Sciences, University of California, Davis, CA 95616, USA

Kentaro Inoue

Department of Plant Sciences, University of California, Davis, CA 95616, USA

Keiji Kito

Department of Life Science, School of Agriculture, Meiji University, Kawasaki 214-8571, Japan

Takashi Ito

Department of Biophysics and Biochemistry, Graduate School of Science, University of Tokyo, Tokyo 113-0032, Japan

Francisco J. López de Saro

Department of Molecular Evolution, Centro de Astrobiología (CSIC-INTA), 28850 Torrejón de Ardoz, Madrid, Spain

Noelle A. Barkley

USDA-ARS, Plant Genetic Resources Conservation Unit (PGRCU), 1109 Experiment Street, Griffin, GA 30223, USA

Ming L. Wang

USDA-ARS, Plant Genetic Resources Conservation Unit (PGRCU), 1109 Experiment Street, Griffin, GA 30223, USA

Laurence Aubry

CEA, DSV, iRTSV, Laboratoire Biologie à Grande Echelle, F-38054 Grenoble, France

Dorian Guetta

INSERM, U1038, F-38054 Grenoble, France

Gérard Klein

BGE/OdyCell, Institut de Recherches en Technologies et Sciences pour le Vivant, CEA-Grenoble, 17 avenue des Martyrs, 38054 Grenoble Cedex 9, France

Ritika Das

Department of Biology, University of Rochester, Rochester, New York 14627, USA

Amita Pandey

Department of Plant Molecular Biology, University of Delhi South Campus, Benito Juarez Road, Dhaula Kuan, New Delhi-110021, India

Girdhar K. Pandey

Department of Plant Molecular Biology, University of Delhi South Campus, Benito Juarez Road, Dhaula Kuan, New Delhi-110021, India

Mehdi Kabani

Laboratoire d'Enzymologie et Biochimie Structurales (LEBS), CNRS, Bât. 34, Avenue de la Terrasse, 91198 Gif-sur-Yvette, France

Céline N. Martineau

Laboratoire d'Enzymologie et Biochimie Structurales (LEBS), CNRS, Bât. 34, Avenue de la Terrasse, 91198 Gif-sur-Yvette, France

Elmar Kirches

Department of Neuropathology, Otto-von-Guericke University, Magdeburg, Gemany

Send Orders of Reprints at reprints@benthamscince.net
Advances in Genome Science, Vol. 1, 2013, 3-14

CHAPTER 1

Endometrial Cancer as a Familial Tumor: Pathology and Molecular Carcinogenesis

Kouji Banno*, Megumi Yanokura, Kennta Masuda, Arisa Ueki, Iori Kisu, Yusuke Kobayashi, Wataru Yamagami, Nobuyuki Susumu and Daisuke Aoki

Department of Obstetrics and Gynecology, School of Medicine, Keio Unversity, Tokyo, Japan

Abstract: Some cases of endometrial cancer are associated with a familial tumor and are referred to as hereditary nonpolyposis colorectal cancer (HNPCC or Lynch syndrome). Such tumors are thought to be induced by germline mutation of the DNA mismatch repair (MMR) gene, but many aspects of the pathology of familial endometrial cancer are unclear and no effective screening method has been established. However, the pathology of endometrial cancer with familial tumor has been progressively clarified in recent studies. At present, about 0.5% of all cases of endometrial cancers meet the clinical diagnostic criteria for HNPCC. A recent analysis of the three MMR genes (*hMLH1, hMSH2* and *hMSH6*) revealed germline mutations in 18 of 120 cases (15.0%) of endometrial cancer with familial accumulation of cancer or double cancer, with a frameshift mutation of the *hMSH6* gene being the most common. Many cases with mutation did not meet the current clinical diagnostic criteria for HNPCC, indicating that familial endometrial cancer is often not diagnosed as HNPCC. The results suggest that the *hMSH6* gene mutation may be important in carcinogenesis in endometrial cancer and germline mutations of the MMR gene may be more prevalent in cases associated with familial accumulation of cancer. An international large-scale muticenter study is required to obtain further information about the pathology of endometrial cancer as a familial tumor.

Keywords: HNPCC, Endometrial cancer, DNA mismatch repair gene, *hMLH1*, *hMSH6*.

INTRODUCTION

The incidence of endometrial cancer among malignant gynecological tumors has increased with lifestyle and environmental changes, and has reached almost half of all cases of uterine cancer in Western countries. The absolute number and prevalence

***Address correspondence to Kouji Banno:** Department of Obstetrics and Gynecology, School of Medicine, Keio University, Shinanomachi 35 Shinjuku-ku, Tokyo 160-8582, Japan; Tel:+81-3-3353-1211; Fax:+81-3-3226-1667; E-mail: kbanno@sc.itc.keio.ac.jp

Christian Neri (Ed)

of cases of endometrial cancer have increased worldwide and control of this cancer is an urgent task. However, many aspects of the mechanism of carcinogenesis and pattern of advancement are unclear. Environmental factors such as obesity and a high estrogen level are thought to play important carcinogenic roles, but a close association with hereditary disposition has also been suggested, since double cancer and an increased incidence of cancer in relatives are common in patients with endometrial cancer.

A high incidence of concomitant endometrial cancer has been shown in female patients with hereditary nonpolyposis colorectal cancer (HNPCC), a typical familial tumor that is inherited in an autosomal dominant manner, indicating the presence of HNPCC-associated familial endometrial cancer. HNPCC has a high risk of development of colorectal cancer, and accumulation of colorectal cancer in a family line was initially reported by Wartin *et al.* in 1913. Subsequent analysis of this family led Lynch to propose the disease concept of cancer family syndrome in 1971 [1,2]. In 1990, the International Collaborative Group (ICG)-HNPCC established the following clinical diagnostic criteria for HNPCC, which are referred to as the classic Amsterdam Criteria: 1) HNPCC is diagnosed when 3 or more patients with histologically confirmed colorectal cancer are present in a family line and one is a first relative of the other two, 2) colorectal cancer develops over two generations, and 3) one case is diagnosed at younger than 50 years old [3].

CLINICAL DIAGNOSTIC CRITERIA FOR HNPCC

Since the Amsterdam Criteria for HNPCC were proposed in 1991, several other diagnostic criteria, including the Japanese Criteria and Bethesda Criteria, have been published. The confusion caused by the different criteria was resolved by revision of the Amsterdam Criteria by the ICG-HNPCC in 1999, to give the new Amsterdam Criteria [4] (Table **1**). These criteria address endometrial cancer, small intestinal cancer, urethral cancer, and kidney cancer, in addition to the colorectal cancer included in the classic criteria. Cases not meeting the classic Amsterdam Criteria may meet the new Amsterdam Criteria, and this has resulted in an increased number of cases diagnosed as HNPCC. In addition, discovery of HNPCC has become possible through investigation of familial medical histories of endometrial cancer patients. The revision also recognized the importance of cooperation among gynecologists for identification of HNPCC. One concern with the new criteria is the omission of ovarian, breast and stomach cancer, which may also be associated with HNPCC.

Table 1: Clinical Diagnostic Criteria for HNPCC

Amsterdam Minimum Criteria (1990)
1. At least 3 cases of colorectal cancer in relatives (verified pathologically)
2. One is a first degree relative of the other two
3. At least two successive generations should be affected
4. One case of colorectal cancer diagnosed before the age of 50 years old
5. FAP should be excluded
Revised Amsterdam Criteria II (1999)
1. At least 3 relatives with an HNPCC-associated cancer (cancer of the colorectum, endometrium, small bowel, ureter or renal pelvis)

2-5: As for the minimum criteria

MECHANISM OF HNPCC CARCINOGENESIS AND ASSOCIATION WITH ENDOMETRIAL CANCER

Six variants of the MMR gene, the causative gene in HNPCC, have been cloned: *hMSH2, hMLH1, hMSH3, hMSH6, hPMS1* and *hPMS2*. An aberration in one of these genes prevents accurate repair of base mismatches produced during DNA replication, resulting in production of a DNA chain of altered length. This phenomenon is called microsatellite instability (MSI) and can lead to an increased frequency of errors in target genes involved in carcinogenesis, resulting in cancerization of the cell. Among the MMR genes, germline mutations of *hMLH1* on chromosome 3 and *hMSH2* on chromosome 2 are thought to cause most cases of HNPCC, but it is unclear if these mutations are also causative for HNPCC-associated endometrial cancer. Mutation of *hMSH6* has also been proposed to be important for development of HNPCC-associated endometrial cancer, but the details are unclear (Fig. **1**) [5, 6].

The TGF-β type II receptor, which is involved in inhibition of cell proliferation, and BAX, which is associated with apoptosis induction, are candidate target genes in HNPCC [7,8]. The *E2F* [9], *β-catenin* [10] and *TCF-4* [11] genes are also candidates, but the frequencies of aberration of these genes in endometrial cancer are low. Variation of the target gene involved in carcinogenesis in different organs has also been suggested. *PTEN* is also a commonly reported target gene in endometrial cancer [12], but its function remains unclear. Endometrial cancer develops through a mechanism similar to that of HNPCC, but may have unique biological characteristics. An analysis of endometrial cancer as a familial tumor has not been performed and many aspects of the cancer are unclear, including its genetic penetrance.

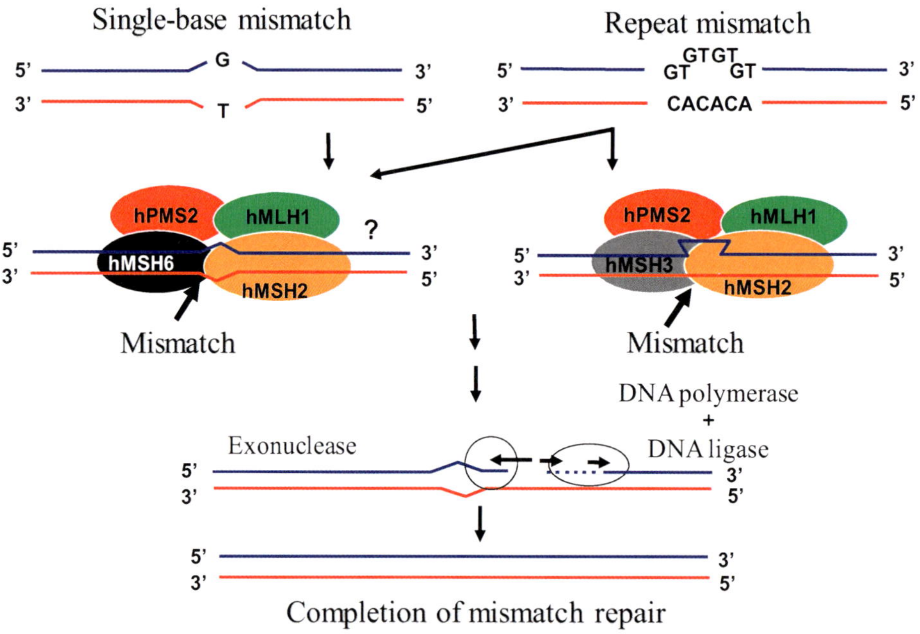

Figure 1: The DNA mismatch repair (MMR) mechanism in humans.

CLINICAL CHARACTERISTICS OF HNPCC

The frequency of HNPCC is unclear due to variation among reports, but HNPCC is thought to account for about 5% of all colorectal cancer cases [13-18] and to have the following clinicopathological characteristics: autosomal dominant inheritance; genetic penetrance of about 85% at 80 years of age; development at a young age; frequent development in the right colon; often poorly differentiated mucinous adenocarcinoma; diploid on cytometric analysis; marked intratumor infiltration of lymphocytes; microsatellite instability; high risk of development of endometrial cancer, urinary tract cancer, and small intestinal cancer; and a favorable prognosis [19]. The reason for the favorable prognosis despite many cases being poorly differentiated mucinous adenocarcinoma is unclear, but may be due to lymphocyte infiltration in the tumor and a low rate of lymph node metastasis. An association between aberration of MMR, the causative gene for HNPCC, and reduced sensitivity to anticancer drugs such as cisplatin has also been reported (Fig. **2**) [20].

What is hereditary non-polyposis colorectal cancer (HNPCC)?

1) Familial tumor that exhibits an autosomal dominant pattern of inheritance
2) Caused by mutation of DNA mismatch repair (MMR) genes in germ cells
3) Found in approximately 5% of cases of colorectal cancer
4) Develops at a young age (mean: 42 years old)
5) Appears commonly in the right colon
6) Frequently poorly differentiated mucinous adenocarcinoma
7) Microsatellite instability (MSI)-positive
8) Associated with high risks of endometrial, ureter, renal pelvis, and small intestinal cancers

Figure 2: Clinical characteristics of HNPCC.

GYNECOLOGICAL CANCERS ASSOCIATED WITH HNPCC

The close association of endometrial cancer with HNPCC is apparent in the new Amsterdam Criteria. However, other gynecological cancers have also been reported in the first relatives of patients with HNPCC, with the incidence of endometrial cancer being highest at 9-19%, followed by stomach cancer and ovarian cancer at 6-14% and 5-7%, respectively [21]. An association of ovarian cancer with HNPCC has been suggested, but may not be as close as that of endometrial cancer, and a link between breast cancer and HNPCC has also attracted recent attention. Muir-Torre syndrome is an autosomal dominant hereditary disease of concomitant sebaceous gland tumors (adenoma, epithelioma and carcinoma) and malignant visceral tumors, and the incidences of colorectal and urogenital cancers are high among malignant visceral tumors, followed by breast cancer and malignant blood diseases. The lifetime incidence of breast cancer is about 12%. Mutation of the *hMSH2* MMR gene has been found in 2 cases of Muir-Torre syndrome [22,23] and this disease is assumed to have the pathology of hereditary breast cancer and to be a subtype of HNPCC.

ENDOMETRIAL CANCER AND MICOSATELLITE INSTABILITY (MSI)

To investigate the status and clinicopathological characteristics of familial endometrial cancer, Banno *et al.* [24] surveyed the familial and medical histories of 385 patients who underwent treatment for endometrial cancer. MSI analysis was performed in 38 of these patients. The familial medical histories showed that 2 of the 385 cases met the new Amsterdam Criteria for HNPCC, giving a rate of HNPCC of about 0.5%. Investigation of familial accumulation of cancer in 890 relatives (439 men and 451 women) of the 38 endometrial cancer patients who underwent MSI analysis revealed high incidences of endometrial cancer, colorectal cancer and ovarian cancer, suggesting that a hereditary factor common to HNPCC is also involved in endometrial cancer. MSI analysis detected the presence of at least one of 5 microsatellite markers (D2S123, D3S1284, D5S404, D9S162, and hMSH2 intron 12) in 12 of the 38 cases (31.6%). This rate is very high compared to MSI in cancers of other organs, demonstrating that abnormal DNA mismatch repair plays an important role in endometrial cancer. The patients with MSI showed a tendency to have double cancer (such as ovarian cancer) compared with patients with microsatellite stability (MSS), although the difference was not significant (27% *vs.* 15%). Regarding prognosis, none of the MSI-positive cases were fatal (0/11, 0%), while 5 MSI-negative (MSS) cases were fatal (5/27, 19%). The difference was not significant, but this tendency is similar to that for HNPCC-associated colorectal cancer. The incidences of moderately differentiated adenocarcinoma G2 (36%) and poorly differentiated adenocarcinoma G3 (18%) tended to be higher in MSI-positive endometrial cancer, although the difference was not significant. These findings appear contradictory with the favorable prognosis, but interestingly they may reflect the biological characteristics of endometrial cancer induced by abnormal DNA mismatch repair [25].

GERMLINE MUTATIONS IN ENDOMETRIAL CANCER PATIENTS WITH FAMILIAL ACCUMULATION OF CANCER

Hirai *et al.* [26] have recently performed an analysis of germline mutation of three MMR genes (*hMLH1, hMSH2* and *hMSH6*) in 120 patients with endometrial cancer with strongly hereditary familial cancer accumulation or double cancer. Cases in which two or more first relatives had HNPCC-associated tumors (colorectal, endometrial, small intestinal, urethral, and renal pelvic cancers) and stomach, ovarian and breast cancers were designated as Group A, and cases with synchronous or metachronous double cancer of HNPCC-associated tumors (colorectal, endometrial, small intestinal, urethral and renal pelvic cancers) with stomach, ovarian or breast cancer were designated as Group B. Of the 120 cases, 57 were in Group A and 48 in Group B, with 15 in both groups. White blood cell-derived DNA was obtained from the patients and germline mutations were analyzed by direct sequencing of PCR products amplified using primers for all exons of *hMLH1, hMSH2* and *hMSH6.* Germline mutation of the MMR genes was detected in 18 of the 120 cases (15.0%), including 9 of the 57 cases (15.8%) in Group A and 4 of the 48 cases (8.3%) in Group B. A high rate of germline mutation of MMR genes (5/15, 33.3%) was found in cases meeting the criteria for both groups. Mutations were detected in the *hMSH6* gene in 9 cases (50%), the *hMLH1* gene in 5, and the *hMSH2* gene in 4. Mutations of the *hMSH6* gene were concentrated in regions near exons 4-5 and were frameshifts in most cases. The mutations in 7 of the 18 cases had already been reported in the literature and registered in a mutation database, but those in the other 11 cases had not been reported previously [26].

SCREENING OF ENDOMETRIAL CANCER AS HNPCC

The clinical characteristics of HNPCC in colorectal cancer have gradually been clarified and HNPCC is thought to account for about 5% of all colorectal cancer cases. In contrast, there are only a few reports on endometrial cancer, despite its similar association with MMR gene aberration. An interview of families of 385 patients with endometrial cancer designed to investigate the familial medical histories of cancer revealed 2 patients who met the clinical diagnostic criteria for

HNPCC [24]. This indicates that only 0.5% of all endometrial cancer cases are diagnosed as HNPCC, and this rate is markedly lower than that in colorectal cancer.

Microsatellite instability caused by MMR gene aberration is detectable by PCR using microsatellite markers. In screening for HNPCC, use of 5 microsatellite markers (BAT25, BAT26, D5S346, D2S123 and D17S250) is recommended. However, a germline mutation of the MMR gene was detected in only 1 of 12 patients with MSI-positive endometrial cancer (8.3%) and this mutation was a nonsense mutation at codon 100 of the *hMLH1* gene. This mutation was already registered as a functional gene mutation in the Human Gene Mutation Database (HGMD), but this case did not meet the new Amsterdam Criteria for HNPCC [26]. This suggests that current clinical diagnostic criteria and MSI analysis have limited value in screening for HNPCC-associated endometrial cancer.

In Hirai *et al.* [26], germline mutation of the MMR gene was present in 15.8% of patients with endometrial cancer with familial accumulation of cancer, but not meeting the clinical diagnostic criteria for HNPCC. This suggests that MMR gene aberration should be kept in mind in management of patients with a tendency for familial cancer, regardless of the clinical assessment. In particular, familial development of breast, ovarian and stomach cancer, which are not recognized as HNPCC-associated cancers, should be investigated in interviews on familial medical histories. No germline mutation of the MMR gene was present in cases with double cancer alone without familial cancer accumulation, but all double cancers developed earlier than endometrial cancer, which is very interesting from the perspective of screening for endometrial cancer.

The *hMSH6* gene was the most commonly mutated MMR gene found by Hirai *et al.* [26], rather than *hMLH1* or *hMSH2*, which are frequently mutated in colorectal cancer. This finding indicates the importance of *hMSH6* in development of endometrial cancer in HNPCC, and suggests an association between the MMR genotype and the cancer phenotype. Most interestingly, the mutations of *hMSH6* were concentrated in exons 4-5 and were frameshifts. One of the *hMSH6* mutations was a reported functional mutation, but the concentration in exons 4-5 differed from reports in Western patients. Thus, it is possible that this gene mutation is characteristic for HNPCC-associated endometrial cancer [27,28].

CARCINOMA OF THE LOWER UTERINE SEGMENT (LUS) AND LYNCH SYNDROME

Westin *et al.* [29] diagnosed Lynch syndrome meeting the Amsterdam criteria II in 14.2% (5/35) of cases with carcinoma of the LUS. A *hMSH2* gene mutation was present in all 5 cases, with MSI and a reduced hMSH2 protein level on immunostaining. Four further cases showed reduced hMSH2 and hMSH6 levels on immunostaining and high MSI, strongly suggesting *hMSH2* gene mutation, but did not meet the Amsterdam criteria II; and one case showed a reduced hMLH1 level despite the absence of aberrant DNA hypermethylation. All 10 patients (29%) were regarded to have Lynch syndrome [29]. The frequency of Lynch syndrome in carcinoma of the LUS was 14.2% even if limited to the 5 cases with definite Lynch syndrome, which is very high compared to the frequency of 1-2% in general endometrial cancer.

In an immunohistological study of expression of mismatch repair gene-encoded proteins (hMLH1, hMSH2, hMSH6, and PMS2), Garg *et al.* found carcinoma of the LUS in 5 of 32 patients with reduced protein expression, but only in 1 of 39 with normal protein expression, indicating an association between carcinoma of the LUS and Lynch syndrome [30]. On the other hand, Watanabe *et al.* found that 24.2% (8/33) of non-LUS endometrial cancer cases were MSI-H compared to 0% (0/13) of carcinoma of the LUS cases in MSI analysis [31], suggesting a possible difference in genetic background between US and Japanese patients [29]. Only Westin *et al.* have previously proposed an association between carcinoma of the LUS and Lynch syndrome, and a further investigation of this association is required.

FUTURE DIRECTIONS AND CONCLUSION

Many HNPCC-associated cases of endometrial cancer caused by MMR gene aberration do not meet the current clinical diagnostic criteria for HNPCC. Gene analysis focusing on the *hMSH6* gene may allow efficient identification of familial endometrial cancer in cases of endometrial cancer with a tendency of familial accumulation of cancer. The findings that some cases of endometrial cancer are familial tumors and that an abnormal DNA mismatch repair gene is linked to development of the cancer are important for understanding the biological characteristics of endometrial cancer. Thus, it is important to identify and analyze patients with this type of endometrial cancer. However, cases of familial endometrial cancer that meet the current diagnostic criteria account for only 0.5% of all cases, and a close examination of family history is essential in screening for this cancer. An improved understanding of familial tumors among gynecologists and an international large-scale muticenter study are required for further progress, with establishment of a system of genetic tests and counseling for patients. There are currently many aspects of familial endometrial cancer that are unclear, but clarification of the pathology and development of a surveillance system and genetic tests are likely to produce new diagnostic and therapeutic methods for various types of endometrial cancer.

CONFLICT OF INTEREST

The author(s) confirm that this chapter content has no conflict of interest.

ACKNOWLEDGEMENTS

Declared none.

DISCLOSURE

This is an updated report of the earlier review article by K. Banno " Endometrial cancer as a familial tumor: pathology and molecular carcinogenesis (review) ". Curr. Genomics, 2009, 10: 127-132.

REFERENCES

[1] Lynch, H.T.; Krush, A.J. The cancer family syndrome and cancer control. *Surg. Gynecol. Obstet.*, **1971**, *132*, 247-250.

[2] Lynch, H.T.; Lynch, J.F. Hereditary nonpolyposis colorectal cancer. *Semin. Surg. Oncol.*, **2000**, *18*, 305-313.

[3] Vasen, H.F.; Mecklin, J.P.; Khan, P.M.; Lynch, H.T. The International Collaborative Group on Hereditary Nonpolyposis Colorectal Cancer. *Dis. Colon Rectum.*, **1991**, *34*, 424-425.

[4] Vasen, H.F.; Watson, P.; Mecklin, J.P.; Lynch, H.T. New clinical criteria for hereditary nonpolyposis colorectal cancer (HNPCC, Lynch syndrome) proposed by the International Colaboratory Group on HNPCC. *Gastroenterology*, **1999**, *116*, 1453-1456.

[5] Miyaki, M.; Konishi, M.; Tanaka, K.; Kikuchi-Yanoshita, R.; Muraoka, M.; Yasuno, M.; Igari, T.; Koike, M.; Chiba, M.; Mori, T. Germline mutation of MSH6 as the cause of hereditary nonpolyposis colorectal cancer. *Nat. Genet.*, **1997**, *17*, 271-272.

[6] Akiyama, Y.; Sato, H.; Yamada, T.; Nagasaki, H.; Tsuchiya, A.; Abe, R.; Yuasa, Y. Germline mutation of hMSH6/GTBP gene in an atypical hereditary nonpolyposis colorectal cancer kindred. *Cancer Res.*, **1997**, *57*, 3920-3923.

[7] Markowitz, S.; Wang, J.; Myeroff, L.; Parsons, R.; Sun, L.; Lutterbaugh, J.; Fan, R.S.; Zborowska, E.; Kinzler, K.W.; Vogelstein, B.; et al. Inactivation of the Type II TGF-βreceptor in colon cancer cells with microsatellite instability. *Science*, **1995**, *268*, 1336-1338.

[8] Rompino, N; Yamamoto, H.; Ionov, Y.; Li, Y.; Sawai, H.; Reed, J.C. Perucho M. Somatic frameshift mutation in the BAX gene in colon cancers of the microsatellite mutator phenotype. *Science*, **1997**, *275*, 967-969.

[9] Ikeda, M.; Orimo, H.; Moriyama, H.; Nakajima, E.; Matsubara, N.; Mibu, R.; Tanaka, N.; Shimada, T.; Kimura, A.; Shimizu, K. Close correlation between mutation of E2F and hMSH3 genes in colorectal cancers with microsatellite instability. *Cancer Res.*, **1998**, *58*, 594-598.

[10] Miyaki, M.; Iijima, T.; Kimura, J.; Yasuno, M.; Mori, T.; Hayashi, Y.; Koike, M.; Shitara, N.; Iwama, T.; Kuroki, T. Frequent mutation of beta-catenin and APC gene in primary colorectal tumors from patients with hereditary nonpolyposis colorectal cancer. *Cancer Res.*, **1999**, *59*, 4506-4509.

[11] Planck, M.; Wenngern, E.; Borg, A.; Olsson, H.; Nibert, M. Somatic frameshift alterations in mononucleotide repeat-coating genes in different tumor types from an HNPCC family with germline MSH2 mutation. *Genes Chromosomes Cancer*, **2000**, *29*, 33-39.

[12] Tashiro, H.; Blazes, M.S.; Wu, R.; Cho, K.R.; Bose, S.; Wang, S.I.; Li, J.; Parsons, R.; Ellenson, L.H. Mutation in PTEN are frequent in endometrial cancer but rare in other common gynecological malignancies. *Cancer Res.*, **1997**, *57*, 3953-3940.

[13] Lynch, H.T.; Smyrk, T.C.; Watson, P.; Lanspa, S.J.; Lynch, J.F.; Lynch, P.M.; Cavalieri, R.J.; Boland, C.R. Genetics, natural history, tumor spectrum, and pathology of hereditary nonpolyposis colorectal cancer. *Gastroenterology*, **1993**, *104*, 1535-1549.

[14] Moslein, G.; Tester, D.J.; Lindor, N.M.; Honchel, R.; Cunningham, J.M.; French, A.J.; Halling, K.C.; Schwab, M.; Goretzki, P.; Thibodeau, S.N. Microsatellite instability and mutation analysis of hMSH2 and hMLH1 in patients with sporadic, familial and hereditary colorectal cancer. *Hum. Mol. Genet.*, **1996**, *5*, 1245-1252.

[15] Wu, Y.; Nyström-Lahti, M.; Osinga, J.; Looman, M.W.; Peltomäki, P.; Aaltonen, L.A.; de la Chapelle, A.; Hofstra, R.M.; Buys, C.H. MSH2 and MLH1 mutations in sporadic replication error-positive colorectal carcinoma as assessed by two-dimensional DNA electrophoresis. *Genes Chromosomes Cancer*, **1997**, *18*, 269-278.

[16] Herfarth, K.; Kodner, I.J.; Whelan, A.J.; Ivanovich, J.L.; Bracamontes, J.R.; Wells Sa Jr.; Goodfellow, P.J. Mutations in MLH1 are more frequent than in MSH2 in sporadic colorectal cancers with microsatellite instability. *Genes Chromosomes Cancer*, **1997**, *18*, 42-49.

[17] Genuardi, M.; Anti, M.; Capozzi, E.; Leonardi, F.; Fornasarig, M.; Novella, E.; Bellacosa, A.; Valenti, A.; Gasbarrini, G.B.; Roncucci, L.; Benatti, P.; Percesepe, A.; Ponz de Leòn, M.; Coco, C.; de Paoli, A.; Valentini, M.; Boiocchi, M.; Neri, G.; Viel, A. MLH1 and MSH2 constitutional mutations in colorectal cancer families not meeting the standard criteria for hereditary nonpolyposis colorectal cancer. *Int. J. Cancer.*, **1998**, *75*, 835-839.

[18] Nomura, S.; Sugano, K.; Kashiwabara, H.; Taniguchi, T.; Fukayama, N.; Fujita, S.; Akasu, T.; Moriya, Y.; Ohhigashi, S.; Kakizoe, T.; Sekiya, T. Enhanced detection of deleterious and other germline mutations of hMSH2 and hMLH1 in Japanese hereditary nonpolyposis colorectal kindreds. *Biochem Biophys Res Commun.*, **2000**, *271*, 120-129.

[19] Horii, A.; Han, H.J.; Shimada, M.; Yanagisawa, A.; Kato, Y.; Ohta, H.; Yasui, W.; Tahara, E.; Nakamura, Y. Frequent replication errors at microsatellite loci in tumors of patients with multiple primary cancers. *Cancer Res.*, **1994**, *54*, 3373-3375.

[20] Fink, D.; Nebel, S.; Aebi, S.; Zheng, H.; Cenni, B.; Nehmé, A.; Christen, R.D.; Howell, S.B. The role on DNA mismatch repair in platinum drug resistance. *Cancer Res.*, **1996**, *56*, 4881-4886.

[21] Watoson, P.; Lynch, H.T. The tumor spectrum in HNPCC. *Anticancer Res.*, **1994**, *14*, 1653-1640.

[22] Kurse, R.; Rütten, A.; Lamberti, C.; Hosseiny-Malayeri, H.R.; Wang, Y.; Ruelfs, C.; Jungck, M.; Mathiak, M.; Ruzicka, T.; Hartschuh, W.; Bisceglia, M.; Friedl, W.; Propping, P. Muir-Torre phenotype has a frequency of DNA mismatch-repair gene mutations similar to that in hereditary nonplyposis colorectal cancer families defined by the Amsterdam Criteria. *Am J Hum Genet.*, **1998**, *63*, 63-70.

[23] Bapat, B.; Xia, L.; Mandlensky, L.; Mitri, A.; Tonin, P.; Narod, S.A.; Gallinger, S. The genetic basis of Muir-Torre syndrome includes hMLH1 locus. *Am J Hum Genet.*, **1996**, *59*, 763-769.

[24] Banno, K.; Susumu, N.; Hirao, T.; Yanokura, M.; Hirasawa, A.; Aoki, D.; Udagawa. Y.; Sugano, K.; Nozawa, S. Two Japanese kindreds occurring endometrial cancer meeting new clinical criteria for hereditary non-polyposis colorectal cancer (HNPCC), Amsterdam Criteria II. *J Obstet Gynaecol Res.*, **2004**, *30*, 287-292.

[25] Banno, K.; Susumu, N.; Yanokura, M.; Hirao, T.; Iwata, T.; Hirasawa, A.; Aoki, D.; Sugano, K.; Nozawa, S. Association of HNPCC and endometrial cancer. *Int J Clin Oncol.*, **2004**, *9*, 262-269.

[26] Hirai, Y.; Banno, K.; Suzuki, M.; Ichikawa, Y.; Udagawa, Y.; Sugano, K.; Miki, Y. Molecular epidemiological and mutation analysis of DNA mismatch repair (MMR) genes in endometrial cancer patients with HNPCC-associated familial predisposition to cancer. *Cancer Sci.*, **2008**, *99*, 1715-1719.

[27] Berends, M.J.; Wu, Y.; Sijimons, R.H.; Mensink, R.G.; van der Sluis, T.; Hordijik-Hos, J.M.; de Vries, E.G.; Hollema, H.; Karrenbeld, A.; Buys, C.H.; van der Zee, A.G.; Hofstra, R.M.; Kleibeuker, J.H. Molecular and clinical characteristics of MSH6 variants: an analysis of 25 index carriers of germline variant. *Am J Hum Genet.*, **2002**, *70*, 26-37.

[28] Hampel, H.; Frankel, W.; Panescu, J.; Lockman, J.; Sotamaa, K.; Fix, D.; Comeras, I.; La Jeunesse, J.; Nakagawa, H.; Westman, J.A.; Prior, T.W.; Clendenning, M.; Penzone, P.; Lombardi, J.; Dunn, P.; Cohn, D.E.; Copeland, L.; Eaton, L.; Fowler, J.; Lewandowski, G.; Vaccarello, L.; Bell, J.; Reid, G.; de la Chaelle, A. Screening for Lynch syndrome (hereditary nonpolyposis colorectal cancer) among endometrial cancer. *Cancer Res.*, **2006**, *66*, 7810-7817.

[29] Westin, S.; Lacour, R.; Urbauer, D.; Luthra, R.; Bodurka, D.; Lu, K.; Broaddus, R. Carcinoma of the lower uterine segment: A newly described association with Lynch syndrome. *J Clin Oncol.*, **2008**, *26*, 5965-5971.

[30] Garg, K.; Leitao, M.; Kauff, D.; Hansen, J.; Kosarin, K.; Shia, J.; Soslow, R.A. Selection of endometrial carcinomas for DNA mismatch repair protein immunohistochemistry using patient age and tumor morphology enhances detection of mismatch repair abnormalities. *Am J Surg Pathol.*, **2009**, *33*, 925-933.

[31] Watanabe, Y.; Nakajima, H.; Nozaki, K.; Ueda, H.; Obata, K.; Hoshiai, H.; Noda, K. Clinicopathologic and immunohistochemical features and microsatellite status of endometrial cancer of the uterine isthmus. *Int J Gynecol Pathol.*, **2001**, *20*, 368-373.

Send Orders of Reprints at reprints@benthamscince.net
Advances in Genome Science, Vol. 1, 2013, 15-44

CHAPTER 2

Replication Origins and Timing of Temporal Replication in Budding Yeast: How to Solve the Conundrum?

Matteo Barberis[1,2,*], Thomas W. Spiesser[1] and Edda Klipp[1,*]

[1]Institute for Biology, Theoretical Biophysics, Humboldt University Berlin, Invalidenstraße 42, 10115 Berlin, Germany and [2]Max Planck Institute for Molecular Genetics, Ihnestraße 73, 14195 Berlin, Germany

Abstract: Similarly to metazoans, the budding yeast *Saccharomyces cereviasiae* replicates its genome with a defined timing. In this organism, well-defined, site-specific origins, are efficient and fire in almost every round of DNA replication. However, this strategy is neither conserved in the fission yeast *Saccharomyces pombe*, nor in *Xenopus* or *Drosophila* embryos, nor in higher eukaryotes, in which DNA replication initiates asynchronously throughout S phase at random sites. Temporal and spatial controls can contribute to the timing of replication such as Cdk activity, origin localization, epigenetic status or gene expression. However, a debate is going on to answer the question how individual origins are selected to fire in budding yeast. Two opposing theories were proposed: the "replicon paradigm" or "temporal program" *vs.* the "stochastic firing". Recent data support the temporal regulation of origin activation, clustering origins into temporal blocks of early and late replication. Contrarily, strong evidences suggest that stochastic processes acting on origins can generate the observed kinetics of replication without requiring a temporal order. In mammalian cells, a spatiotemporal model that accounts for a partially deterministic and partially stochastic order of DNA replication has been proposed. Is this strategy the solution to reconcile the conundrum of having both organized replication timing and stochastic origin firing also for budding yeast? In this review we discuss this possibility in the light of our recent study on the origin activation, suggesting that there might be a stochastic component in the temporal activation of the replication origins, especially under perturbed conditions.

Keywords: Budding yeast, DNA replication, origins of replication, temporal program, stochastic firing, genomic instability, Clb5, Sic1.

INTRODUCTION

In eukaryotic cells, DNA replication is restricted to a specific time window, called S phase. Successful progression through S phase requires that replication be properly regulated to ensure that the entire genome is duplicated exactly once, without errors,

***Address correspondence to Matteo Barberis and Edda Klipp:** Humboldt University Berlin, Institute for Biology - Theoretical Biophysics, Invalidenstraβe 42, 10115 Berlin, Germany; Tel: +49 30 2093 8383; Fax: +49 30 2093 8813; E-mails: matteo.barberis@biologie.hu-berlin.de and edda.klipp@rz.hu-berlin.de

Christian Neri (Ed)

in a timely fashion, as any errors might generate chromosomal defects which can lead to genomic instability and death [1, 2]. As a result, DNA replication has evolved into a tightly regulated process involving the coordinated action of numerous factors that function in all phases of the cell cycle.

Eukaryotic replication starts from multiple locations referred to as replication origins throughout each chromosome. These sequences direct the formation of a number of protein complexes leading to the assembly of two bidirectional DNA replication forks at each origin. Research in the past years has identified many of the protein components of these complexes and the time during the cell cycle they assemble at the origin [1]. The density of active replication origins determines S phase dynamics and chromosome stability during mitosis [3], and in the budding yeast *Saccharomyces cerevisiae* a direct correlation between the length of S phase and the number of the replication origins has been demonstrated [4]. Experimental and computational studies have identified and mapped over 700 potential origin function target sites on the budding yeast genome [5-10] which are highly redundant. Despite the contribution that multiple origins per chromosome may give to an efficient genome duplication, direct experimental evidence indicates that yeast chromosomes have many more replication origins than they need for their timely replication during the S phase. In fact, several origins on chromosome III can be deleted without substantially affecting the ability to inherit this chromosome faithfully during cell division [11, 12]. Therefore, for the purpose of genome duplication, it is a challenge to investigate the relation between the number of active origins and the replication time, in particular whether a minimal origin set could be sufficient to complete DNA replication in the correct timing of the S phase.

Based on our recent study [13], we discuss in this review how origin firing could be globally coordinated to ensure an even distribution of replication initiation sites across the genome, highlighting the fact that in budding yeast, as for all eukaryotes, activation of the temporal program of origin activation can be driven by stochastic events.

SPATIAL AND TEMPORAL REGULATION OF ORIGIN FIRING

Eukaryotic cells initiate DNA synthesis from hundreds of replication origins distributed over the chromosomes that constitute the genome. The temporal program of origin activation is imposed by well-defined cis-acting sequence features. The chromosomal sequences required for an origin of replication vary significantly between different eukaryotic organisms. In the unicellular eukaryote *S. cerevisiae*, three to four DNA sequences of 10–15 base pairs (bp) spread over 100–150 bp are sufficient to act as an origin and include the highly conserved and essential **A**utonomously **R**eplicating **S**equence (ARS) [14].

Efficient replication is guaranteed by multiple replication origins which are activated, or fired, per chromosome. For example, yeast chromosome III uses 11 origins for its duplication while chromosome X uses about 20 [5, 6, 15]. The budding yeast has a genome of 13.5 million base pairs (Mbp) distributed over 16 chromosomes, and each single chromosome is considerably smaller than the *Escherichia coli* genome of 4.6 Mbp. Yet yeast replication origins occur on average every 20-40 kb, a hundred times more densely distributed than one would predict by comparison to the *E. coli* genome. The difference in fork migration rates may explain in part the need for multiple replication origins per eukaryotic chromosome. DNA replication forks migrate at rates about 30 times slower in yeast (about 3 kb/min) compared to *E. coli* (about 100 kb/min) [5, 16]. The use of multiple initiation events per chromosome probably compensates for slower fork migration rates in maintaining an efficient rate of genome duplication and S phase progression in eukaryotic cells. Based on this reasoning, budding yeast would need about 100 replication origins to duplicate its genome at a rate sufficient to accommodate its S phase, about four times less than the current estimation [5, 6].

In reality, budding yeast has many more replication origins than necessary to complete the replication program. In fact, only a small number of origins seem to represent classic replicators, while the majority corresponds to zones of inefficient, closely spaced start sites none of which are indispensable for origin activity. The fact that potentially only a subset of them is required to correctly replicate DNA according to a precise timing characteristic of the S phase could be related to several independent mechanisms that may affect the efficiency in origin firing or the timing of origin activation, including DNA sequence features intrinsic to the specific origin. Moreover, for many origins it is clear that temporal and spatial controls can contribute to the timing of replication modulating origin activity. The time during S phase when a chromosomal region is replicated correlates strikingly with chromatin structure and gene expression in many organisms, and transcriptionally inactive genes and/or heterochromatin are often the last portions of the genome to be duplicated [17-19]. A wide number of studies have established a role for cis-factors including chromosome context and chromatin structure in controlling the efficiency and timing of origin activation during S phase in *S. cerevisiae*, describing specialized chromatin-binding proteins, chromatin-modifying enzymes and chromatin remodelling proteins in controlling origin activity [20, 21]. Recently, a role of transcriptional activity on the replication timing has been discussed [22-24]. Moreover, different studies highlight the link between origin activity and origin localization [25, 26] and epigenetic factors regulating the time of firing [27-29].

ORIGIN ACTIVATION AND REPLICATION INITIATION

DNA replication is regulated through interaction between cis-acting sequences and trans-acting initiation factors. Specifically, replication origins are determined by the binding, in late mitosis and early G1 phase, of initiator proteins which assembly into well-defined complexes called pre-replicative complexes (pre-RC). This process is referred to "licensing", and the pre-RC recognizes the origins and assembles to prepare them for firing. Formation of the pre-RC requires minimally the six subunit origin recognition complex (ORC), Cdc6, Cdt1 and Mcm2-7 [1, 30]. Subsequently, in early S phase, the pre-RC is converted into an active pre-initiation complex (pre-IC), with the inclusion of Cdc45, GINS (complex containing the four subunits Sld5, Psf1, Psf2 and Psf3, essential for both initiation and progression of DNA replication) and DNA polymerases, that facilitates unwinding the origins and starting the replication process [1, 30]. The process is summarized in Fig. (**1**). The activation of the replication machinery has still to be highlighted in many of its regulatory events, but a critical step is the availability of specific kinase activities at the G1/S transition. CDK (Cdk1-Clb5,6) and DDK (Cdc7-Dbf4) complexes phosphorylate components of the pre-RC setting the correct timing of S phase progression [31-39]. As shown in Fig. (**1**), in particular Cdk1-Clb5,6 activity exploits an essential role in the process, due to the fact that, remaining high from S to G2/M phases, it phosphorylates and, thus, negatively regulates ORC [40], Cdc6 [41-45], Cdt1 and Mcm2-7 [46-49]. The effect of these phosphorylations is to prevent the assembly of new pre-RCs until the next G1 phase [50] and differs from protein to protein, as ORC is inactivated while Mcm2-7 and Cdt1 are exported from the nucleus, and Cdc6 is degraded. In addition, the physical interaction between Clb5 and ORC helps to inhibit re-initiation [51]. Cdk1-Clb phosphorylation of the initiation factors Sld2 and Sld3 is also a critical step for the initiation of DNA replication [34-37].

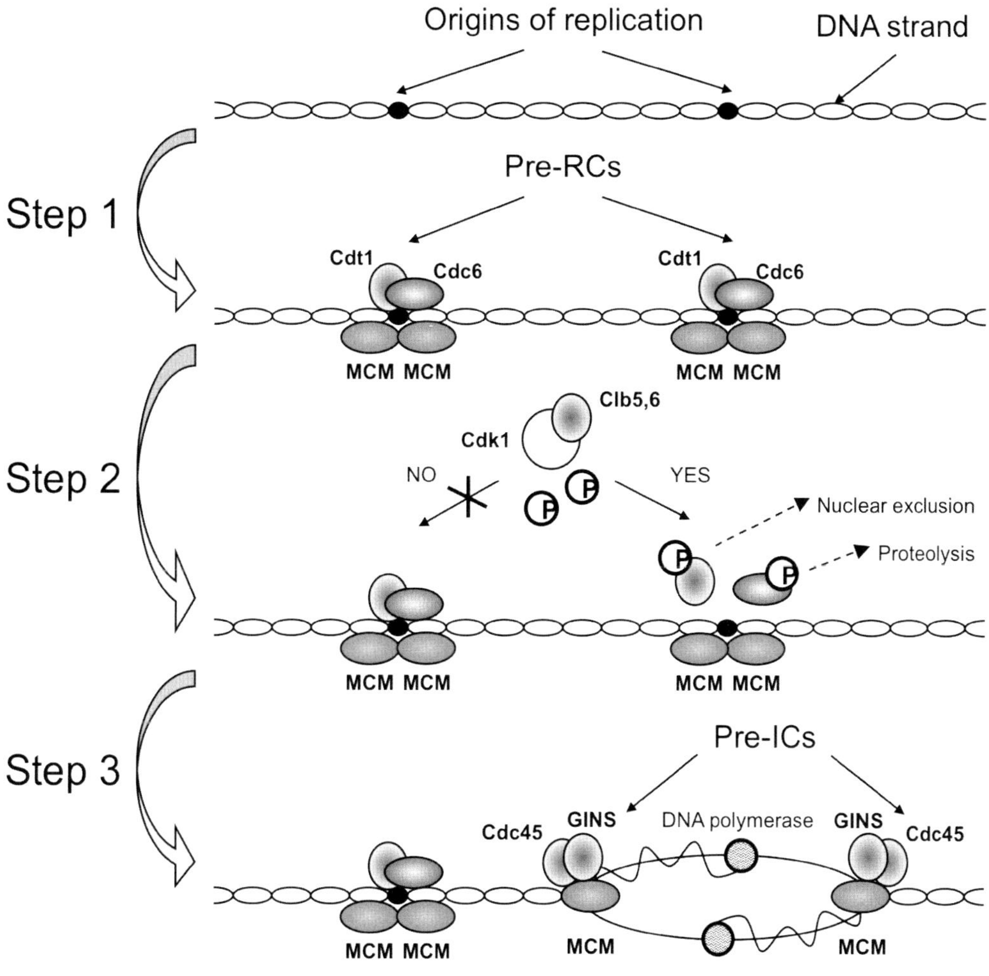

Figure 1: Schematic representation of the DNA replication process. Pre-RC components are recruited to the replication origins in a stepwise manner during late M and early G1 phases of the cell cycle. At the origins, ORC recruits Cdc6 and Cdt1, which in turn are required for the subsequent loading of the Mcm2–7 complex. After pre-RC has assembled, its activation during S phase involves the assembly of a second set of factors like Cdc45 and GINS as well as the activity of Cdk1-Clb5,6, which phosphorylates and determines the fate of specific components of the replication machinery. Origin firing and DNA replication then start bidirectionally (modified from Barberis and Klipp, *Genome Inform.* 2007).

ORIGIN ACTIVATION BY CLB-DEPENDENT CDK ACTIVITY

Several studies have suggested that yeast chromosomes contain early and late replicating domains and exhibit replication timing profiles that are consistent with a highly regulated chronological program [7, 9, 52], favoring the argument that, differently from other eukaryotes, in budding yeast the origins of replication fire according to a defined temporal program [53]. Each individual origin is characterized by both a specific reproducible time of activation and a firing efficiency during the replication process in the S phase of cell cycle [54]. Generally, early origins are positioned toward the central portion of chromosomes, while late origins are positioned near the telomeres such that the central portion of chromosomes replicate before the ends [55, 56]. Detailed data are available for yeast chromosome VI [57, 58]. However, there is not a consistent relationship between replication origin efficiency and the time at which an origin fires during S phase, being some late firing origins efficient and others not (data for chromosomes VI and XIV are reported [58, 59]). Some late origins are inefficient because they are located near an earlier firing origin, which replicates the later origin before it has a chance to fire [60, 61]. Thus, competition between two closely spaced origins can influence origin efficiency and, consequently, the replication profile of origin activation [62].

Regulation of the S phase program occurs at several stages, affecting origin firing, replication fork elongation, fork velocity and fork stability, all dependent on S phase-promoting kinase activities. Origin selection is in fact synchronised with cell cycle progression by coupling origins that are activated preferentially during early or late S phase to the expression of specific Cdk-cyclin complexes. In mammalian cells, initiation of early origin firing is regulated by the trans-acting factors Cdk2-cyclin A or Cdk2-cyclin E kinase activities. In addition, Cdc7 is also reported to be involved in the regulation of entry into S phase. These two kinases are involved in the loading on origins of Cdc45 by phosphorylating subunits of Mcm2-7 complex [63]. Trans-factors that regulate late origin firing have been suggested as well, where Cdk1-cyclin A2 has been proposed to have this essential role [64]. Early or late firing of the origins during S phase accounts for much of the temporal programming of replication in budding yeast, and specific Cdk-cyclin complexes play a dominant role in defining the origins that are active during early and late S phase. Cdk1-Clb6 plays the major role in early origin activation, before the degradation of Clb6 during S phase [65]. Cdk1-Clb5 is known to be indispensable for activation of late replication origins, showing that *clb5Δ* cells failed to activate these origins and subsequently showed an extended S phase [52, 66]. Interestingly, this late origin firing defect is suppressed although entry into S phase and is significantly delayed in *clb5Δ clb6Δ* cells, suggesting that other B-type cyclins promote firing of both early and late replication origins.

There is a strong interest in understanding mechanisms of origin selection for the proper replication timing. This temporal program of replication refers to the propensity for different regions of the genome to be replicated at different times during S phase, resulting in a temporally regulated origin activation [67]. However, despite the fact that quite some details controlling the activation of the replication origins have been elucidated – especially concerning the role of kinase activities – and that a temporal program of the S phase progression must be a fundamental feature to ensure a precise DNA synthesis, the mechanisms underlying the observed pattern of replication remain unknown.

TIMING IN ORIGIN ACTIVATION: TEMPORAL PROGRAM OR STOCHASTIC FIRING?

Since molecular mechanisms underlying the regulation of origin activation or origin firing are strongly conserved across evolution from budding yeast to human cells [1], unicellular model organisms have been chosen to investigate the mechanisms underlying the regulation of replication origins and the temporal program of replication. In particular, patterns of replication timing have been studied in *S. cerevisiae* [5, 52, 55, 68, 69], with the advantages that the location of origins is mapped through the whole genome and the average firing time of each origin has been calculated [5, 7]. These studies show that origins in budding yeast fire at characteristic times, with some origins firing on average earlier and others firing on average later. Budding yeast has well-defined, site-specific origins, many of which are efficient and fire in as many as 90 % of S phases [70, 71], and it is apparently more close to the mechanism of bacterial replication, in which origins are well-defined sites that fire once in each cell cycle, leading to uniform replication that is identical in every cell [72]. On the other side, in fission yeast and vertebrates, less than 50 % of potential origins fire in each S phase [8, 73-78]. This inefficient origin firing leads to an asynchronous initiation of DNA replication throughout S phase at indiscriminate sites and a different subset of origins firing in each cell, causing a random pattern of replication in each cell cycle. The excess origins may serve as backups to ensure efficient replication if origin firing is compromised or replication forks stall [79-82]. Interestingly, recent studies have revealed an intrinsic temporal disorder in the replication of yeast chromosome VI [83], suggesting that also stochastic events could influence the replication program, as it is the case for the fission yeast *Schizosaccharomyces pombe*, *Drosophila melanogaster* and *Xenopus laevis*.

Stochastic events related to the firing of specific origins can possibly determine a different pattern of origin activation. The earlier origin establishes a replication fork that replicates the later origin before it has a chance to fire [61], and direct experiments support a role for such an "origin interference" mechanism in origin efficiency [84-86]. Presumably, this inactivates the competent initiation complex assembled at the late origin [6], despite studies have shown that the assembly of this protein complex occurs – regardless of their initiation time – at all origins during mitosis [87, 88]. The reproducible replication times observed for origins in budding yeast were interpreted to demonstrate that origins fire at predetermined times in S phase [5, 7]. However, these analyses assay the average behavior of individual origins over a population of cells and it is possible that replication timing is heterogeneous on a single-cell level. Thus, a peak in a replication profile indicates a firing of an origin at a specific chromosomal location, but because the profile is generated from a population of cells, it is not possible to know whether a specific origin fires in every cell in the population [89, 90]. Czajkowsky and colleagues demonstrated heterogeneous patterns of origin firing suggesting that replication origins fire inefficiently and stochastically, instead of firing efficiently and at defined times during S phase. Moreover, replication origins show a wide efficiency, some firing in almost every cell cycle while others fire in less than 10% of the cells [5, 58, 59]. Strikingly, the replication profiles observed by Czajkowsky and colleagues were similar to the previous ones obtained from microarrays experiments [5, 7, 91]. This comparison and further computational studies demonstrate that stochastic firing is compatible with defined replication timing [90, 92-94].

Origin properties, deterministic (chromosomal position and firing time) and stochastic (firing efficiency), possibly influence DNA replication. In particular, the efficiency of an origin has a stochastic influence because it changes the shape of the replication profiles representing activation of the origins (the more efficient ones) in time. However, replication profiles discard much of the information from the experiments and cannot distinguish late, efficient origins from early, inefficient origins. Moreover, an observed firing time does not necessarily represent the time at which a replication origin fires in S phase, but reflects the fact that an origin that replicates late in S phase could do this either because it fires efficiently late in S phase or because it fires early but is inefficient and, thus, passively replicated later in S phase. As illustrated in Fig. (**2**), this results in different landscapes observed for an origin activation because an origin will not necessarily fire in every cell cycle, being characterized by a variable firing competence, *i.e.* the percentage of cells in which an origin is biochemically competent to fire [89, 90]. If it does, the origin is considered to be efficient, having a high competence to fire; if it does not fire regularly, then it is called inefficient, being characterized by a low firing competence. In the example reported, the arbitrary replication profile shows that origin number 2 has a high competence of firing whereas origin number 5 has a low competence.

The bulk of evidence presented suggests that possibly neither of the two mechanisms of replication, the temporal program of origin activation and the stochastic firing, reflects the reality of replication in budding yeast. Conversely, a combined effect of both could represent it [89, 90]. It has been proposed that an alterative model based on the stochastic firing of origins may also explain replication timing, assuming varying origin efficiency instead of a strict origin-timing program. So how can these apparently conflicting data be reconciled to explain the replication pattern reported experimentally [5-7]? As nicely discussed by Rhind in 2006, stochastic firing of origins leads to the so-called "random gap" problem: randomly distributed origin firing will occasionally lead to large gaps between replication bubbles that would take a long time to replicate [53]. To reconcile the random distribution of origin firing with efficient replication, it has been proposed that origins can fire stochastically – when they fire, the firing times are maintained – but the efficiency of origin firing increases

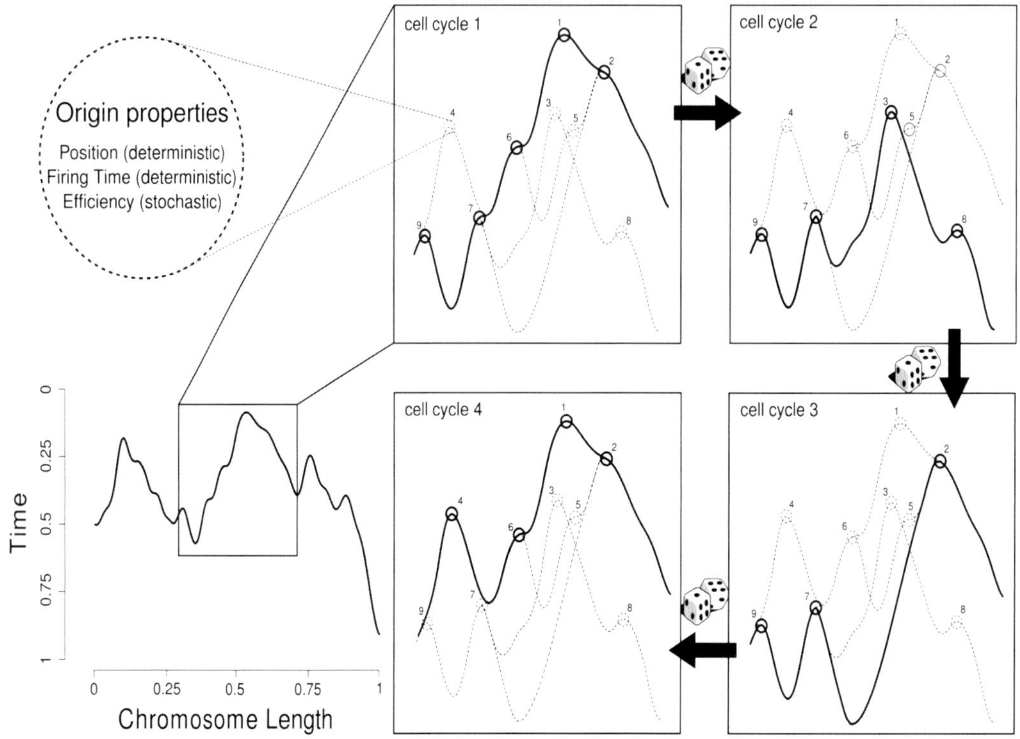

Figure 2: Stochastic events influence origin activation and replication profiles. A exemplified replication profile, *i.e.* the graph of when during S phase each locus in the genome is, on average, half replicated, is presented. In the replication profile, origins are peaks and peak height correlates with origin timing and efficiency. Focusing on the stochastic processes that can impinge the landscape of origin activation in four consecutive rounds of DNA replication, each panel represents a specific pattern of origins firing (solid lines) that can be observed out of the all possible firing landscapes (dotted lines). Replication origins are numbered from 1 to 9 according to the increasing firing times. Considering a competence of firing of an origin equal to 100 % if it is activated over four rounds of replication, the competence will be reduced in the case that the origin fires in some but not in every cell cycle: 75 % (origins 2, 7 and 9), 50 % (origins 1 and 6), 25 % (origins 3, 4 and 8) and 0 % (origin 5).

as S phase progresses [53, 90]. Therefore, the longer a large random gap persists, the more likely origins within it are to fire. This has been firstly proposed for *Xenopus* embryos [79, 95]. This approach was already faced by Takahashi in 1987, where a model for the spatiotemporal organization of DNA replication in mammalian cells was generated [96]. It could account for a partially deterministic and partially stochastic order of DNA replication in chromosomes, giving the possible solution to reconcile the apparently different characteristics of having both organized replication timing and stochastic origin firing.

DETERMINISTIC MODEL OF REPLICATION DYNAMICS IN BUDDING YEAST

The distinction between deterministic program of origin activation and stochastic firing is only an apparent conflict. Biology is inherently stochastic, and the question that has been pointed out is not whether origin firing is stochastic, but how important is the probabilistic nature of origin firing in the regulation of replication timing [90]. More importantly, how can stochastic origin firing be accommodated in realistic models that predict the patterns of replication timing observed *in vivo*? The proposed deterministic/stochastic approach to investigate the replication program considering the increasing efficiency of the replication origins during the DNA synthesis is, at the moment, not really applicable for budding yeast due to the availability of only few data about individual origin efficiencies [57, 58]. In this case, a fully stochastic model could be developed to explore the temporal program of origin activation, as it has been proposed both for fission yeast [97] and *Xenopus* embryos [98]. Modeling approaches are, at least initially, most fruitful in simple systems where more of the parameters are known, and *S. cerevisiae* represent a good example of study because quite an amount of data are available about the replication origins. For this unicellular organism, origin location of replication origins on the chromosomes [14] and initiation time of origin firing for a subset of replication origins [5, 7] are reported in the *S. cerevisiae* OriDB database [54], and replication fork rate values are available [5, 7, 16].

We used the accessible information to generate the first, fully deterministic, mathematical model of DNA replication in *S. cerevisiae* in order to investigate the temporal sequence of origin activation [13]. We used the length of the chromosomes, the position of selected replication origins, their initiation time and a fork migration rate assumed to be constant to recalculate the replication profiles of all 16 chromosomes, assuming that the efficiency of the selected origins was 100 %. Specifically, two studies report the timing for different replication origins, but they are not consistent if compared to each other. In the first study, Raghuraman and colleagues report the timing for 454 origins [5] whereas, in the second one, Yabuki and colleagues provide such information for 260 origins [7]. Considering that the first study provides information for a bigger subset of replication origins, we included these data in our study. We wanted to test whether the assumptions considered to build the deterministic model were acceptable. Essentially, we: (i) considered both chromosomal location and firing time derived from the microarray-based heavy::light (HL) analysis for 454 replication origins (~ 60 % of the total origins – 732 – included in the *S. cerevisiae* OriDB database), (ii) assumed that replication forks migrate constantly throughout S phase, as also considered in a later study [99], at an approximate rate of 3 kb/min, averaging the experimental data available (mean of 2.9 kb/min and a median of 2.3 kb/min) and (iii) assumed an efficiency equal to 100 % for the 454 origins (this corresponds to a single replication event in a cell with 732 origins that fire with an at efficiency of about 60 %).

As shown in Fig. (**3**), in the deterministic model the DNA is divided into units (u) of equal length (500 bp). Hence, in the simulation each chromosome is composed of a series of DNA units, corresponding to its original size (L_{org}) divided by 500 to yield the internal resolution size. A two-dimensional array element (A) of size L_{res} is assigned to every chromosome, and two DNA units are added to account for the left and right end of the chromosomes. The array element A contains all discrete DNA unit positions and the status of the replication for a specific position. This is represented by a Boolean variable, which is set equal to 0 when the DNA unit has not been replicated at this position yet, and set to 1 when the DNA unit has been replicated. Another two-dimensional array element stores origin information: origin name, origin position on the virtual chromosome A, origin activation time in seconds and the origin activation status, a Boolean variable set to 0 by default, indicating that the origin has not been activated yet. A variable T represents the replication time. Each of the discrete time steps needed to complete DNA replication is equal to the time (Δt) that the replication fork needs to process one DNA unit (Δu): $\Delta t = \Delta u / \Delta v$, where $\Delta u = 500$ bp and $\Delta v = 3$ kb/min and, therefore: $\Delta t = 10$ sec. The simulation proceeds in the way that at each time point T_j the algorithm reviews the array to find the origins that fire at that time. If so, the Boolean variables in the array A for these origins are set to 1, indicating that they have been activated and cannot fire anymore. Every origin issues two replication forks upon activation, each travelling in opposite directions on the DNA and therefore, at each time point T_j, the algorithm checks whether the positions left and right of a replicated region have been replicated or not. In this way, replication forks migrate in both directions until they encounter either a region that has been already replicated or the end of the chromosome. During the simulation every replication fork through the genome can be retraced and their final positions and times can be mapped.

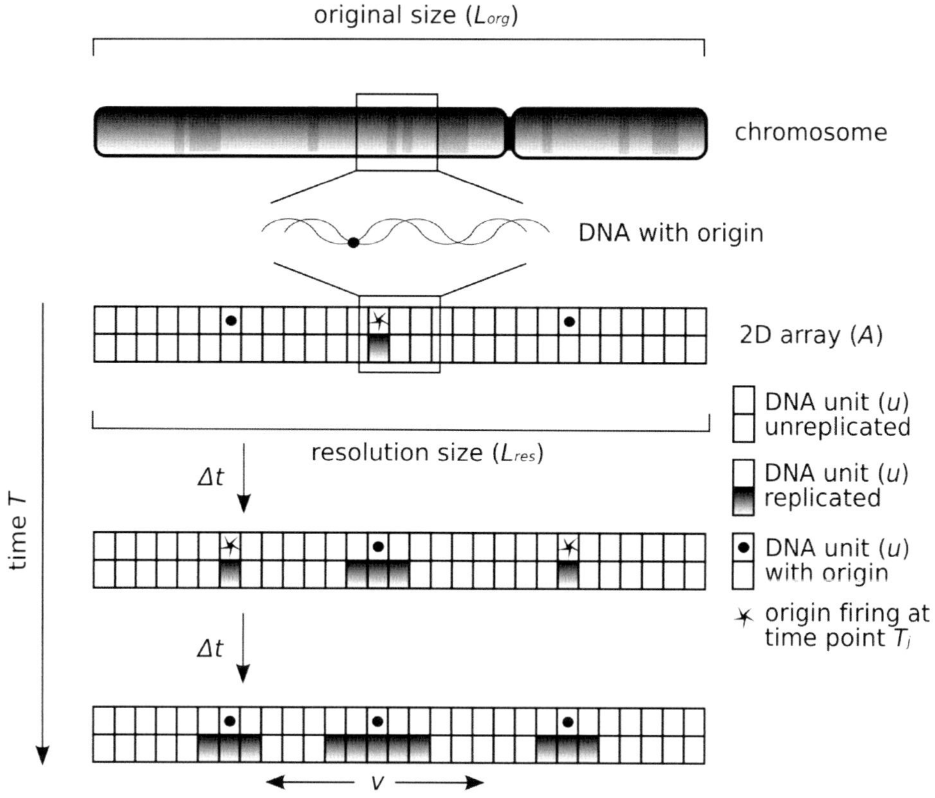

Figure 3: Scheme of the deterministic model of DNA replication and its parametrization. The features and the algorithm are explained in the text (reproduced with kind permission from Springer Science+Business Media: *Mol. Genet. Genomics*, A model for the spatiotemporal organization of DNA replication in Saccharomyces cerevisiae, *282*, 2009, 25-35, Spiesser, T.W., Klipp, E., Barberis, M., Fig. **1**, and any original (first) copyright notice displayed with material).

In spite of the approximations on which we generated the model, the recalculated replication profiles of origin activation matched the experimental profiles surprisingly well for 10 out of 16 chromosomes (see [13] for detailed comparison between computed and experimental replication profiles obtained from Raghuraman and colleagues). However, significant differences were observed for the other 6 chromosomes, with many sequences replicating later than predicted. One possible explanation is that whereas the slope of the recalculated profiles is constant due to the constant fork rate implemented, the experimental curve is smooth with a varying slope, reflecting a changing of the fork progression rate – due to the folded structures of the DNA or binding of specific protein factors to the origins which drive accessibility and motion of the polymerase – or the activation of inefficient origins that were not included, as also highlighted out from Hyrien and Goldar [100].

FROM A DETERMINISTIC MODEL TO A STOCHASTIC OUTPUT

The identified and mapped potential origins of replication on the genome of budding yeast [5-10] are redundant. Due to the intriguing fact that excess of origins may ensure efficient replication if origin firing is compromised [79-82], we investigated the relation between the number of active origins and the replication time.

By using our deterministic model, we assessed the impact of particular sets of origins on the replication time computing the replication kinetics in which the DNA is duplicated in time for all 16 yeast chromosomes. We obtained the classical "switch" for the origin activation, characteristic of the prompt availability of the Cdk1-Clb5,6 activity at the entry into S phase of the cell cycle [101]. Both Clb5 and Clb6 are essential in this process, and activate different subsets of replication origins to complete the replication program [52, 66, 102]. Then, we computed replication kinetics in perturbed conditions, *i.e.* considering different subsets of origins randomly deleted for 50 % of the total number used in the study (454), to simulate the wild type, mimicking in this way stress conditions, inefficient firing or checkpoint activation due to possible damage to the DNA and stalling of the replication fork [103]. Intriguingly, the perturbation showed only a small influence on the replication kinetics (delay in the termination of the replication), indicating that the precise timing of DNA replication was apparently not depending on specific origins potentially lost in the random deletion [13]. This means that DNA replication is robust against failure of origin firing or changes in origin efficiency, suggesting that the DNA could be duplicated through S phase independently from the temporal program of origin activation.

So, why having a temporal program if it is apparently not needed? Is this the pattern of evolution from a fix, deterministic program to a redundant, stochastic process? The result previously obtained stimulated us to further test the spatiotemporal model for understanding the relation between origin activation and replication time. We computed the time of the replication kinetics for each chromosome decreasing randomly the number of active replication origins, mimicking how the replication time could change in the case that a certain percentage of the origins would be deleted, defective or inefficient. As logically expected, systematically deactivating an increasing number of origins (from 10 % to 90 %) yielded an increasing time needed to complete DNA replication. However, strikingly, this analysis showed that for many chromosomes the replication times measured experimentally can be obtained using only subsets of activated origins [13], which are different for every chromosome and composed randomly.

The analyses performed suggest that in budding yeast a temporal program of origin activation might be not enough to describe all events occurring during replication, revealing how stochasticity in origin usage confers robustness and reliability to the DNA replication process. If the temporal activation of the origins is influenced by stochastic events that select randomly the origins, the process of DNA replication could be more efficient when perturbations affect the normal state of the cells. Considering that the density of active replication origins determines S phase dynamics [3], wheather the observed robustness of the replication is maintained also *in vivo* is an essential task to be proven. Experimentally, this feature might be tested by progressively deleting the known replication origins and evaluating the consequent replication time for each chromosome. Moreover, in the light of this evidence, the deterministic model could be now suitable to include values for origin efficiencies to test whether an increasing efficiency of firing through S phase could increase the reproducibility of the replication profile experimentally observed [5]. Since only few data are available about origin firing efficiency [57, 58], and more have still to be produced, a model which considers an *in silico* progressive increase of the efficiencies could be developed, to generate a partially deterministic and partially stochastic system in order to investigate whether this is the proper strategy to solve the "random gap" problem [53].

ROLE OF CDK1-CLB ACTIVITY IN MATHEMATICAL MODELS OF DNA REPLICATION: TIMING OF ORIGIN ACTIVATION

Entry into S phase of the cell cycle is driven by the B-type cyclin (Clb)/cyclin-dependent kinase (CDK) complexes Cdk1-Clb5 and Cdk1-Clb6 [101], the activity of which plays an important role in the control of replication timing, being a direct activator of origin firing. As previously described, Cdk1-Clb5 can activate both early and late origins, whereas Cdk1-Clb6 is only capable of activating early origins of replication [52, 66, 102].

We have recently investigated computationally the role of the Cdk1-Clb activity in the activation of the replication origins, compiling a huge amount of experimental data available in the literature and modelled the G1/S transition in *S. cerevisiae* [104]. The mathematical model was implemented by ordinary differential equations (ODEs) taking into account both nuclear and cytoplasmic compartments. Model tests revealed the main known regulatory events that impinge the functionality of this window of the cell cycle. In this study, we correlated the main output of the model, the nuclear concentration of Cdk1-Clb5,6, with the temporal activation of the replication origins and, thus, the onset of DNA replication. Considering the large number of replication origins present in a yeast nucleus [5-7] and the reported role of Cdk1-Clb5,6 in inducing firing, we associated the probability of firing for each replication origin to the nuclear concentration of Cdk1-Clb5,6. The influence of Cdk1-Clb5,6 on the licensing of replication origins has been described by a probabilistic three-step model as represented in Fig. (**1**), which does not consider molecular details of this highly regulated process [104, 105], and the distance between the considered average of replication origins (440) has been fixed. Step 1 lumps all events from free origin on the DNA to the pre-RC formation. The transition time for each of the replication origins considered was taken from a normal distribution with mean of 15 min and standard deviation of 2 min. The probability for performing Step 2 at a certain time is determined by the concentration of nuclear Cdk1-Clb5,6 at that time. The duration of this step is necessary for Cdk1-Clb5,6 to exceed a value taken from a normal distribution with a mean of 0.03 μM and standard deviation of 0.01 μM. The transition time for Step 3 to reach the "fired" state of a replication origin is again taken from a normal distribution with a mean of 1 min and standard deviation of 0.01 min. When an origin has fired, then DNA replication proceeds bidirectionally from multiple replication origins, as experimentally reported [106, 107]. If the replication reaches the neighboring origin before it fires on its own, that origin is set to the state "fired". The probabilistic model of the origin activation successfully explains the activation of DNA replication (i) in cells grown in different nutritional conditions (glucose and ethanol media) and (ii) of different yeast background, *i.e.* single deletion or overexpression mutants of central players at the G1/S network [105]. This supported the fact that the Cdk1-Clb5,6 activity sets the basis for the prompt starting of the DNA replication machinery.

In addition, we were interested in studying in more detail the role of Clb5 in activating the origins by our spatiotemporal model of DNA replication [13], in order to test whether our computational tool could be useful to reproduce the chronological program of origin activation, which has been shown to be reproducible even under altered conditions [91]. In fact, it has been demonstrated that *clb5Δ* mutant suffers a significant decrease in firing efficiency of the late origins in S phase [66]. Investigating the activation of replication origins in the *clb5Δ* mutant, McCune and colleagues pointed out that DNA replication in budding yeast follows a regular temporal program of origin activation rather than a disordered firing [52] and this evidence suggested us to look into the deterministic model of origin activation to test the reproducibility of the experiment. To this purpose, we divided the replication origins in an early (Clb5-unaffected) and in a late half (Clb5-affected) by stopping origin firing at the time corresponding to the mean value of the distribution of the experimentally determined origin firing times (27 min, about the midpoint of a normal S phase) [5, 13]. The replication profiles computed for the 16 yeast chromosomes in the *clb5Δ* mutant show the general agreement of the replication kinetics between the computed profiles and the experimental ones [13, 52]. The chromosomal portions suffering significant delays in replication correspond to the CLB5-dependent regions (CDRs) reported by McCune and colleagues, as illustrated in Fig. (**4**). In detail, we found a perfect match of the *in vivo* data for chromosomes I to VIII and XI, a good fit in the majority of the sequence length for chromosomes IX, X and

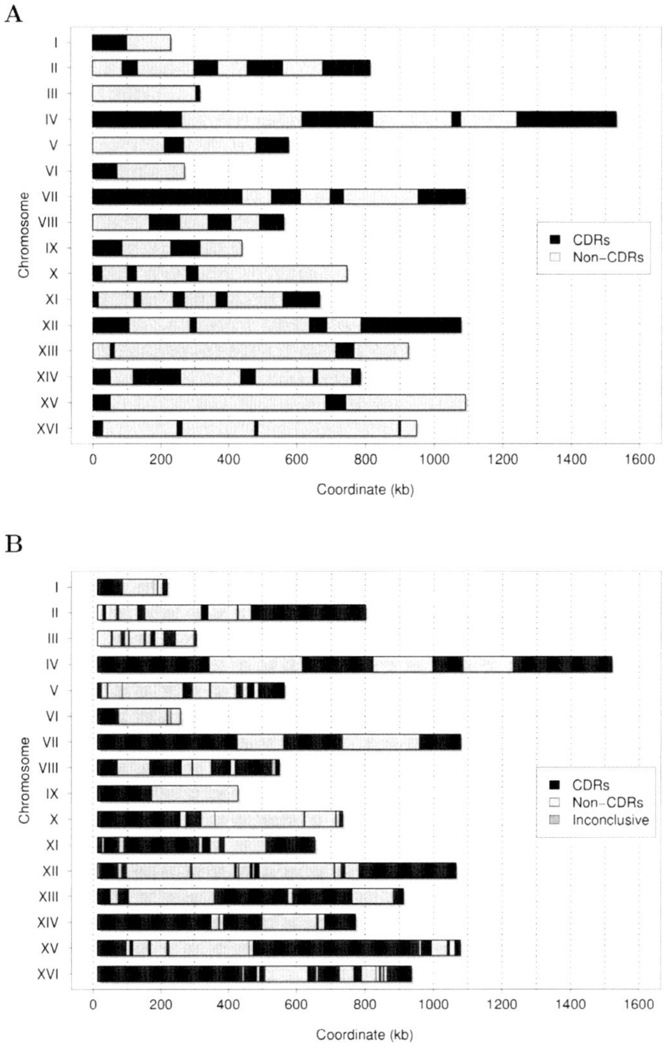

Figure 4: Computed Clb5-dependent regions in the budding yeast genome. The computed profile for the *clb5Δ* mutant is reported, showing the organization of the genome in Clb5-dependent and -independent regions (CDRs and Non-CDRs, respectively). CDRs are visualized in black and Non-CDRs in gray. The genomic regions of the CDRs for the XVI yeast chromosomes generally correspond to the ones identified by McCune and colleagues [37]. In detail, a perfect match is found for nine chromosomes (from I to VIII, and XI), a good fit for chromosomes IX, X and XIV, and a small or no match for chromosomes XII, XIII, XV and XVI [13] (compare Figs. **2** and **4** in [37]).

XIV, and a very poor match for chromosomes XII, XIII, XV and XVI. Intrinsic noise might affect the firing time of the Clb5-dependent origins, therefore the considered value of about 27 min is an approximation, which for some chromosomes could be quite accurate, but for others it could be not the case. This might result in the fact that the chromosomes containing more early origins would be less sensitive to *CLB5* deletion, whereas the chromosomes with more late origins would be more sensitive. However, this analysis is in agreement with the fact that the *clb5Δ* mutant only affects late origins, whereas the early origins fire normally, and indicates that our deterministic model is indeed able to reproduce the temporal program of origin activation [13, 52].

Altogether, the results presented here suggest that our deterministic spatiotemporal model for DNA replication in budding yeast is able both to reproduce the chronological program of activation of the replication origins and to capture possible stochastic events in their temporal firing. This feature has been never considered before and, for the first time to date, we propose that also for budding yeast the probabilistic nature of origin firing might affect the regulation of replication timing. Due to the high degree of conservation of the DNA replication process from yeast to higher eukaryotes, this study opens the possibility to study the regulation of the origin activation introducing specific components whose deregulation is often fatal and can lead to severe genetic disease and cancer in humans.

REGULATING CDK1-CLB ACTIVITY PREVENTING GENOME INSTABILITY

The evolution of multiple mechanisms to prevent DNA re-replication by the Cdk1-Clb5 activity clearly indicates the importance of avoiding this type of aberrant condition, which in humans has been associated with genomic instability and cancer [108]. The concept of genomic (or genetic) instability refers to a series of chromosomal changes occurring at an accelerated rate in cell populations [109] and indicates the increased tendency of tumor cells to acquire new mutations with each cell division. The importance of maintaining normal levels of origin efficiency during DNA replication is suggested by the fact that both increasing and decreasing origin efficiency correlates with decreased genomic stability, as it has been demonstrated both in mammalian [110] and yeast [111] cells. Increasing of origin efficiency could be caused by a diffusible, rate-limiting activator, such as CDK or DDK kinases required throughout S phase for origin firing [1, 30]. If one of them is rate limiting, this would limit the efficiency of origins during S phase. At first, the number of firing origins would be a small faction of potential origins, but as S phase progresses and the number of potential origins declines, that number would become an ever larger fraction of potential origins and the efficiency of the remaining origins would increase [90]. Other regulators of origin firing, such as the stable fork component Cdc45 or the mini-chromosome maintenance complex Mcm2-7, could also function as rate-limiting activators. For example, in mammalian cells, Cdc45 can recruit Cdk2 at the replication foci [112] and such interaction may explain the correlation between fork density and origin firing [98], although it is unknown whether it is conserved in yeast. Moreover, both in budding yeast and higher eukaryotes, the putative helicase Mcm is present in a large excess compared to the amount needed to replicate the genome [79, 113, 114]. Some of that excess Mcm is loaded at origins that will not fire, but some of the excess is loaded as multiple Mcm complexes at individual origins [115, 116]. Thus, the efficiency with which ORC loads Mcm at a given origin, or the amount of time ORC is bound to an origin and able to load Mcm, could determine an origin's firing probability [90].

Both in yeast and in higher eukaryotes, genomic instability often ensues when the G1/S transition of the cell cycle is deregulated and cells are forced to enter S phase prematurely. This acquired mutability is important since a majority of genes mutated in human cancers influence the G1/S transition [117]. In *S. cerevisiae*, a central regulatory component that controls Cdk1-Clb5,6 activation at the G1/S transition is the stoichiometric Cdk-Clb inhibitor Sic1 [118-121]. The initiation of DNA replication is regulated by an irreversible switch in which Sic1, the functional and structural homologous of the CDK inhibitor $p27^{Kip1}$ in mammalian cells [122], is multi-phosphorylated during late G1 phase by the G1 cyclin (Cln)/cyclin-dependent kinase (CDK) complexes Cdk1-Cln1,2 [123, 124] and, then, degraded at the G1/S transition by the ubiquitin/proteasome pathway before initiation of DNA replication [125-127]. Accumulation of active Cdk1-Clb5,6, as well as of the other Cdk1-Clb complexes active from S to M phases, enhances Sic1 destruction through the same mechanism and advances cells into S phase [128]. Sic1 is important for maintenance of a Cdk1-Clb-free windows of time which is critical for origin licensing, and recent studies highlighted that Sic1 plays a role in timing and oscillatory behaviour of Cdk1-Clb complexes [129-131]. *sic1Δ* mutant initiate DNA replication from fewer origins, S phase initiates early and is extended compared to wild type, but mitosis is not delayed [111]. Moreover, chromosome combing experiment showed that on average the distance between replicons is 1.5 times longer in *sic1Δ* cells compared to wild type [132]. As a consequence, chromosomes break and rearrange at a very high frequency, and cells exhibit about a 100-fold increase in minichromosome loss and gross chromosomal rearrangements (GCRs) compared to wild type [111, 133]. This clearly shows that origin density is crucial for genome integrity. The precocious Cdk1-Clb5,6 activation causes chromosome rearrangements and severe genome instability through its inhibitory effect on pre-RC formation in late G1 phase. Thus, by inhibiting any residual Cdk1-Clb activity in G1 phase, Sic1 promotes efficient pre-RC formation, probably also on sites that serve as dormant origins [134].

With the purpose to investigate genomic instability in budding yeast, we recently modelled the network of the G1/S transition analyzing the implications of the balance between Cdk1-Clb5,6 and Sic1 for cell cycle progression [104, 135] and studied the efficiency of this control in activating the replication origins at the entry into S phase [105]. According to the current scenario reported in literature, Sic1 is involved in the control of DNA replication as a negative regulator of the Cdk activity, and the mathematical models of the cell cycle of course are taking into account only this function [136]. This leaves unanswered a major phenotype of the *sic1Δ* mutant, namely sparse origin firing [111]. Thus, considering that Sic1 acts not only as a stoichiometric inhibitor of the Cdk1-Clb complexes [119, 122] but also as a promoter of Cdk1-Clb5,6 entry into the nucleus [104], as shown experimentally [137], we gave a rationale explanation for the sparse origin firing phenotype observed in the *sic1Δ* mutant [111]. As shown in Fig. (**5**), we observed an early firing of the replication origins in the mutant compared to the wild type, since no Sic1 degradation is required and Cdk1-Clb5 activity can fire the origins as soon as it becomes available, and proceeds slowly, as experimentally observed [111]. These results strongly suggest that the fine tuning of the Cdk1-Clb5,6 activity by Sic1 controls the precise temporal activation of the replication origins for the correct completion of S phase events. Whether the role of Sic1 in regulating origin firing is exploited only through inhibition of the Cdk1-Clb activity or in addition *via* direct binding to- or regulation of- the components of the DNA replication machinery has still to be investigated.

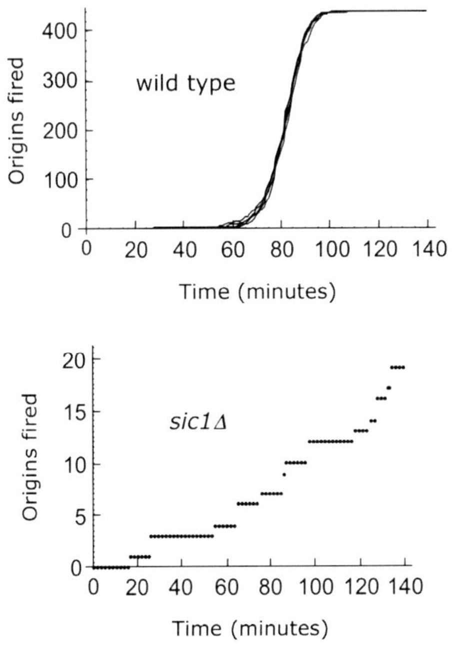

Figure 5: Simulation of origin firing for wild type and *sic1*Δ cells. The cumulative number of fired origins was calculated based on the probabilistic model for firing of origins in wild type cells (upper panel) and the *sic1*Δ mutant. To be noted the different scales on the y-axis.

The fact that cells replicating their chromosomes from a sub-optimal number of origins are karyotypically unstable is highly important for the understanding of tumorigenesis, consistent with the prevalence of G1/S regulator mutations in cancer. The mathematical approaches here presented provide useful tools to implement components which deregulation drives an abnormal replication dynamics and to predict further scenarios in the temporal activation of the replication origins which in turn can be tested experimentally.

CONCLUSIONS

Plasticity is an inherent feature of chromosomal DNA replication in eukaryotes. In budding yeast, origins of replication are made in excess and are potentially activated in a partly chronological programmed order and partly stochastically throughout the S phase of the cell cycle. In addition, origin activity is influenced by accessibility of replication origins, availability of regulatory proteins, chromatin structure and epigenetic factors, which contribute to the observed timing of replication. Since mathematical analysis explaining the temporal activation of the replication origins in budding yeast is still not available, our probabilistic/deterministic models of DNA replication dynamics are the first tool enabling to address this question and to suggest that stochastic events can influence the profile of the replication program.

CONFLICT OF INTEREST

The author(s) confirm that this chapter content has no conflict of interest.

ACKNOWLEDGMENTS

Although we have tried to represent the ever-increasing seminal contributions into the field, we deeply apologize in advance to any of our colleagues whose work was not cited owing to space limitations. MB is funded by the Network of Excellence of the European Commission (Project ENFIN, contract number LSHG-CT-2005-518254) and by UNICELLSYS (contract number HEALTH-2007-201142). TWS is funded by the German Research Foundation (DFG) through IRTG 1360 (International Research Training Group).

AUTHOR CONTRIBUTIONS

Matteo Barberis conceived the computational studies described and wrote the manuscript. Thomas W. Spiesser helped in the generation of the figures. Matteo Barberis and Edda Klipp proofread the manuscript.

DISCLOSURE

"This is an updated report of our earlier review article" Replication Origins and Timing of Temporal Replication in Budding Yeast: How to Solve the Conundrum?". Current Genomics, 2010, 11, 199-211. Matteo Barberis, Thomas W. Spiesser and Edda Klipp.

REFERENCES

[1] Bell, S.P.; Dutta, A. DNA replication in eukaryotic cells. *Annu. Rev. Biochem.*, **2002**, *71*, 333-374.

[2] Machida, Y.J.; Hamlin, J.L.; Dutta, A. Right place, right time, and only once: replication initiation in metazoans. *Cell*, **2005**, *123*, 13-24.

[3] Bielinsky, A.K. Replication origins: why do we need so many? *Cell Cycle*, **2003**, *2*, 307-309.

[4] van Brabant, A.J.; Buchanan, C.D.; Charboneau, E.; Fangman, W.L.; Brewer, B.J. An origin-deficient yeast artificial chromosome triggers a cell cycle checkpoint. *Mol. Cell*, **2001**, *7*, 705-713.

[5] Raghuraman, M.K.; Winzeler, E.A.; Collingwood, D.; Hunt, S.; Wodicka, L.; Conway, A.; Lockhart, D.J.; Davis, R.W.; Brewer, B.J.; Fangman, W.L. Replication dynamics of the yeast genome. *Science*, **2001,** *294*, 115-121.

[6] Wyrick, J.J.; Aparicio, J.G.; Chen, T.; Barnett, J.D.; Jennings, *E.G.*; Young, R.A.; Bell, S.P.; Aparicio, O.M. Genome-wide distribution of ORC and MCM proteins in S. cerevisiae: high-resolution mapping of replication origins. *Science*, **2001,** *294*, 2357-2360.

[7] Yabuki, N.; Terashima, H.; Kitada, K. Mapping of early firing origins on a replication profile of budding yeast. *Genes Cells*, **2002**, *7*, 781-789.

[8] Feng, W.; Collingwood, D.; Boeck, M.E.; Fox, L.A.; Alvino, G.M.; Fangman, W.L.; Raghuraman, M.K.; Brewer, B.J. Genomic mapping of single-stranded DNA in hydroxyurea-challenged yeasts identifies origins of replication. *Nat. Cell Biol.*, **2006**, *8*, 148-155.

[9] Nieduszynski, C.A.; Knox, Y.; Donaldson, A.D. Genome-wide identification of replication origins in yeast by comparative genomics. *Genes Dev.*, **2006**, *20*, 1874-1879.

[10] Xu, W.; Aparicio, J.G.; Aparicio, O.M.; Tavare, S. Genome-wide mapping of ORC and Mcm2p binding sites on tiling arrays and identification of essential ARS consensus sequences in S. cerevisiae. *BMC Genomics*, **2006**, *7*, 276.

[11] Dershowitz, A.; Newlon, C.S. The effect on chromosome stability of deleting replication origins. *Mol. Cell. Biol.*, **1993,** *13*, 391-398.

[12] Dershowitz, A.; Snyder, M.; Sbia, M.; Skurnick, J.; Ong, L.; Newlon, C. Linear derivatives of Saccharomyces cerevisiae chromosome III can be maintained in the absence of autonomously replicating sequence elements. *Mol. Cell. Biol.*, **2007,** *27*, 4652-4663.

[13] Spiesser, T.W.; Klipp, E.; Barberis, M. A model for the spatiotemporal organization of DNA replication in Saccharomyces cerevisiae. *Mol. Genet. Genomics*, **2009**, *282*, 25-35.

[14] Newlon, C.S.; Theis, J.F. The structure and function of yeast ARS elements. *Curr. Opin. Genet. Dev.*, **1993**, *3*, 752-758.

[15] Poloumienko, A.; Dershowitz, A.; De, J.; Newlon, C.S. Completion of replication map of Saccharomyces cerevisiae chromosome III. *Mol. Biol. Cell*, **2001,** *12*, 3317-3327.

[16] Rivin, C.J.; Fangman, W.L. Replication fork rate and origin activation during the S phase of Saccharomyces cerevisiae. *J. Cell Biol.*, **1980,** *85*, 108-115.

[17] Morales, V.; Giamarchi, C.; Chailleux, C.; Moro, F.; Marsaud, V.; Le Ricousse, S.; Richard-Foy, H. Chromatin structure and dynamics: functional implications. *Biochimie*, **2001**, *83*, 1029-1039.

[18] Goren, A.; Cedar, H. Replicating by the clock. *Nat. Rev. Mol. Cell. Biol.*, **2003,** *4*, 25-32.

[19] Donaldson, A.D. Shaping time: chromatin structure and the DNA replication programme. *Trends Genet.*, **2005**, *21*, 444-449.

[20] Vogelauer, M.; Rubbi, L.; Lucas, I.; Brewer, B.J.; Grunstein, M. Histone acetylation regulates the time of replication origin firing. *Mol. Cell*, **2002**, *10*, 1223-1233.

[21] Knott, S.R.; Viggiani, C.J.; Tavaré, S.; Aparicio, O.M. Genome-wide replication profiles indicate an expansive role for Rpd3L in regulating replication initiation timing or efficiency, and reveal genomic loci of Rpd3 function in Saccharomyces cerevisiae. *Genes Dev.*, **2009**, *23*, 1077-1090.

[22] Gilbert, D.M. Replication timing and transcriptional control: beyond cause and effect. *Curr. Opin. Cell Biol.*, **2002**, *14*, 377-383.

[23] Schwaiger, M.; Schübeler, D. A question of timing: emerging links between transcription and replication. *Curr. Opin. Genet. Dev.*, **2006**, *16*, 177-183.

[24] Hiratani, I.; Takebayashi, S.; Lu, J.; Gilbert, D.M. Replication timing and transcriptional control: beyond cause and effect--part II. *Curr. Opin. Genet. Dev.*, **2009**, *19*, 142-149.

[25] Heun, P.; Laroche, T.; Raghuraman, M.K.; Gasser, S.M. The positioning and dynamics of origins of replication in the budding yeast nucleus. *J. Cell Biol.*, **2001**, *152*, 385-400.

[26] Heun, P.; Laroche, T.; Shimada, K.; Furrer, P.; Gasser, S.M. Chromosome dynamics in the yeast interphase nucleus. *Science*, **2001**, *294*, 2181-2186.

[27] Hamlin, J.L.; Mesner, L.D.; Lar, O.; Torres, R.; Chodaparambil, S.V.; Wang, L. A revisionist replicon model for higher eukaryotic genomes. *J. Cell. Biochem.*, **2008**, *105*, 321-329.

[28] Hiratani, I.; Gilbert, D.M. Replication timing as an epigenetic mark. *Epigenetics*, **2009**, *4*, 93-97.

[29] Göndör, A.; Ohlsson, R. Replication timing and epigenetic reprogramming of gene expression: a two-way relationship? *Nat. Rev. Genet.*, **2009**, *10*, 269-276.

[30] Takeda, D.Y.; Dutta, A. DNA replication and progression through S phase. *Oncogene*, **2005**, *24*, 2827-2843.

[31] Bousset, K.; Diffley, J.F. The Cdc7 protein kinase is required for origin firing during S phase. *Genes Dev.*, **1998**, *12*, 480-490.

[32] Nougarède, R.; Della Seta, F.; Zarzov, P.; Schwob, E. Hierarchy of S-phase-promoting factors: yeast Dbf4-Cdc7 kinase requires prior S-phase cyclin-dependent kinase activation. *Mol. Cell. Biol.*, **2000**, *20*, 3795-3806.

[33] Nishitani, H.; Lygerou, Z. Control of DNA replication licensing in a cell cycle. *Genes Cells*, **2002**, *7*, 523-534.

[34] Masumoto, H.; Muramatsu, S.; Kamimura, Y.; Araki, H. S-Cdk- dependent phosphorylation of Sld2 essential for chromosomal DNA replication in budding yeast. *Nature*, **2002**, *415*, 651-655.

[35] Tanaka, S.; Umemori, T.; Hirai, K.; Muramatsu, S.; Kamimura, Y.; Araki, H. CDK-dependent phosphorylation of Sld2 and Sld3 initiates DNA replication in budding yeast. *Nature*, **2007**, *445*, 328-332.

[36] Zegerman, P.; Diffley, J.F. Phosphorylation of Sld2 and Sld3 by cyclin-dependent kinases promotes DNA replication in budding yeast. *Nature*, **2007**, *445*, 281-285.

[37] Tanaka, S.; Tak, Y.S.; Araki, H. The role of CDK in the initiation step of DNA replication in eukaryotes. *Cell Div.*, **2007**, *2*, 16.

[38] Devault, A.; Gueydon, E.; Schwob, E. Interplay between S-cyclin-dependent kinase and Dbf4-dependent kinase in controlling DNA replication through phosphorylation of yeast Mcm4 N-terminal domain. *Mol. Biol. Cell*, **2008**, *19*, 2267-2277.

[39] Francis, L.I.; Randell, J.C.; Takara, T.J.; Uchima, L.; Bell, S.P. Incorporation into the prereplicative complex activates the Mcm2-7 helicase for Cdc7-Dbf4 phosphorylation. *Genes Dev.*, **2009**, *23*, 643-654.

[40] Nguyen, V.Q.; Co, C.; Li, J.J. Cyclin-dependent kinases prevent DNA re-replication through multiple mechanisms. *Nature*, **2001**, *411*, 1068-1073.

[41] Drury, L.S.; Perkins, G.; Diffley, J.F. The Cdc4/34/53 pathway targets Cdc6p for proteolysis in budding yeast. *EMBO J.*, **1997**, *16*, 5966-5976.

[42] Elsasser, S.; Chi, Y.; Yang, P.; Campbell, J.L. Phosphorylation controls timing of Cdc6p destruction: a biochemical analysis. *Mol. Biol. Cell*, **1999**, *10*, 3263-3277.

[43] Drury, L.S.; Perkins, G.; Diffley, J.F. The cyclin dependent kinase Cdc28p regulates distinct modes of Cdc6p proteolysis during the budding yeast cell cycle. *Curr. Biol.*, **2000**, *10*, 231-240.

[44] Calzada, A.; Sanchez, M.; Sanchez, E.; Bueno, A. The stability of the Cdc6 protein is regulated by cyclin-dependent kinase/cyclin B complexes in Saccharomyces cerevisiae. *J. Biol. Chem.*, **2000**, *275*, 9734-9741.

[45] Perkins, G.; Drury, L.S.; Diffley, J.F. Separate SCF(CDC4) recognition elements target Cdc6 for proteolysis in S phase and mitosis. *EMBO J.*, **2001**, *20*, 4836-4845.

[46] Labib, K.; Diffley, J.F.; Kearsey, S.E. G1-phase and B-type cyclins exclude the DNA-replication factor Mcm4 from the nucleus. *Nat. Cell Biol.*, **1999**, *1*, 415-422.

[47] Nguyen, V.Q.; Co, C.; Irie, K.; Li, J.J. Clb/Cdc28 kinases promote nuclear export of the replication initiator proteins Mcm2-7. *Curr. Biol.*, **2000**, *10*, 195-205.

[48] Labib, K.; Kearsey, S.E.; Diffley, J.F. MCM2-7 proteins are essential components of prereplicative complexes, that accumulate co-operatively in the nucleus during G1-phase, and are required to establish, but not maintain, the S-phase checkpoint. *Mol. Biol. Cell*, **2001**, *12*, 3658-3667.

[49] Tanaka, S.; Diffley, J.F. Interdependent nuclear accumulation of budding yeast Cdt1 and Mcm2-7 during G1 phase. *Nat. Cell Biol.*, **2002**, *4*, 198-207.

[50] Dahmann, C.; Diffley, J.F.; Nasmyth, K.A. S-phase-promoting cyclin-dependent kinases prevent re-replication by inhibiting the transition of replication origins to a pre-replicative state. *Curr. Biol.*, **1995**, *5*, 1257-1269.

[51] Wilmes, G.M.; Archambault, V.; Austin, R.J.; Jacobson, M.D.; Bell, S.P.; Cross, F.R. Interaction of the S-phase cyclin Clb5 with an "RXL" docking sequence in the initiator protein Orc6 provides an origin-localized replication control switch. *Genes Dev.*, **2004**, *18*, 981-991.

[52] McCune, H.J.; Danielson, L.S.; Alvino, G.M.; Collingwood, D.; Delrow, J.J.; Fangman W.L.; Brewer, B.J.; Raghuraman, M.K. The temporal program of chromosome replication: genome-wide replication in clb5D Saccharomyces cerevisiae. *Genetics*, **2008**, *180*, 1833-1847.

[53] Rhind, N. DNA replication timing: random thoughts about origin firing. *Nat. Cell Biol.*, **2006**, *8*, 1313-1316.

[54] Nieduszynski, C.A.; Hiraga, S.; Ak, P.; Benham, C.J.; Donaldson, A.D. OriDB: a DNA replication origin database. *Nucleic Acids Res.*, **2007**, *35*, D40-46.

[55] McCarroll, R.M.; Fangman, W.L. Time of replication of yeast centromeres and telomeres. *Cell*, **1988**, *54*, 505-513.

[56] Ferguson, B.M.; Brewer, B.J.; Reynolds, A.E.; Fangman, W.L. A yeast origin of replication is activated late in S phase. *Cell*, **1991**, *65*, 507-515.

[57] Friedman, K.L.; Brewer, B.J.; Fangman, W.L. Replication profile of Saccharomyces cerevisiae chromosome VI. *Genes Cells*, **1997**, *2*, 667-678.

[58] Yamashita, M.; Hori, Y.; Shinomiya, T.; Obuse, C.; Tsurimoto, T.; Yoshikawa, H.; Shirahige, K. The efficiency and timing of initiation of replication of multiple replicons of Saccharomyces cerevisiae chromosome VI. *Genes Cells*, **1997**, *2*, 655-665.

[59] Friedman, K.L.; Diller, J.D.; Ferguson, B.M.; Nyland, S.V.; Brewer, B.J.; Fangman, W.L. Multiple determinants controlling activation of yeast replication origins late in S phase. *Genes Dev.*, **1996**, *10*, 1595-1607.

[60] Santocanale, C.; Sharma, K.; Diffley, J.F. Activation of dormant origins of DNA replication in budding yeast. *Genes Dev.*, **1999**, *13*, 2360-2364.

[61] Vujcic, M.; Miller, C.A.; Kowalski, D. Activation of silent replication origins at autonomously replicating sequence elements near the HML locus in budding yeast. *Mol. Cell. Biol.*, **1999**, *19*, 6098-6109.

[62] Weinreich, M.; Palacios DeBeer, M.A.; Fox, C.A. The activities of eukaryotic replication origins in chromatin. *Biochim. Biophys. Acta*, **2004,** *1677*, 142-157.

[63] Machida, Y.J.; Hamlin, J.L.; Dutta, A. Right place, right time, and only once: replication initiation in metazoans. *Cell*, **2005**, *123*, 13-24.

[64] Katsuno, Y.; Suzuki, A.; Sugimura, K.; Okumura, K.; Zineldeen, D.H.; Shimada, M.; Niida, H.; Mizuno, T.; Hanaoka, F.; Nakanishi, M. Cyclin A-Cdk1 regulates the origin firing program in mammalian cells. *Proc. Natl. Acad. Sci. USA*, **2009**, *106*, 3184-3189.

[65] Jackson, L.P.; Reed, S.I.; Haase, S.B. Distinct mechanisms control the stability of the related S-phase cyclins Clb5 and Clb6. *Mol. Cell. Biol.*, **2006**, *26*, 2456-2466.

[66] Donaldson, A.D.; Raghuraman, M.K.; Friedman, K.L.; Cross, F.R.; Brewer, B.J.; Fangman, W.L. CLB5-dependent activation of late replication origins in S. cerevisiae. *Mol. Cell*, **1998**, *2*, 173-182.

[67] Fangman, W.L.; Brewer, B.J. A question of time: replication origins of eukaryotic chromosomes. *Cell*, **1992**, *71*, 363-366.

[68] Fangman, W.L.; Hice, R.H.; Chlebowicz-Sledziewska, E. ARS replication during the yeast S phase. *Cell*, **1983**, *32*, 831-838.

[69] Reynolds, A.E.; McCarroll, R.M.; Newlon, C.S.; Fangman, W.L. Time of replication of ARS elements along yeast chromosome III. *Mol. Cell. Biol.*, **1989**, *9*, 4488-4494.

[70] Fangman, W.L.; Brewer, B.J. Activation of replication origins within yeast chromosomes. *Annu. Rev. Cell Biol.*, **1991**, *7*, 375-402.

[71] Newlon, C.S.; Lipchitz, L.R.; Collins, I.; Deshpande, A.; Devenish, R.J.; Green, R.P.; Klein, H.L.; Palzkill, T.G.; Ren, R.B.; Synn, S.; Woody, S.T. Analysis of a circular derivative of Saccharomyces cerevisiae chromosome III: a physical map and identification and location of ARS elements. *Genetics*, **1991,** *129*, 343-357.

[72] Jacob, F.; Brenner, S. On the regulation of DNA synthesis in bacteria: the hypothesis of the replicon. *C. R. Hebd. Seances. Acad. Sci.*, **1963**, *256*, 298-300.

[73] Dijkwel, P.A.; Wang, S.; Hamlin, J.L. Initiation sites are distributed at frequent intervals in the Chinese hamster dihydrofolate reductase origin of replication but are used with very different efficiencies. *Mol. Cell. Biol.*, **2002**, *22*, 3053-3065.

[74] Jeon, Y.; Bekiranov, S.; Karnani, N.; Kapranov, P.; Ghosh, S.; MacAlpine, D.; Lee, C.; Hwang, D.S.; Gingeras, T.R.; Dutta, A. Temporal profile of replication of human chromosomes. *Proc. Natl. Acad. Sci. USA*, **2005**, *102*, 6419-6424.

[75] Heichinger, C.; Penkett, C.J.; Bahler, J.; Nurse, P. Genome-wide characterization of fission yeast DNA replication origins. *EMBO J.*, **2006**, *25*, 5171-5179.

[76] Lebofsky, R.; Heilig, R.; Sonnleitner, M.; Weissenbach, J.; Bensimon, A. DNA Replication origin interference increases the spacing between initiation events in human cells. *Mol. Biol. Cell*, **2006**, *17*, 5337-5345.

[77] Patel, P.K.; Arcangioli, B.; Baker, S.P.; Bensimon, A.; Rhind, N. DNA replication origins fire stochastically in fission yeast. *Mol. Biol. Cell*, **2006**, *17*, 308-316.

[78] Eshaghi, M.; Karuturi, R.K.; Li, J.; Chu, Z.; Liu, E.T.; Liu, J. Global profiling of DNA replication timing and efficiency reveals that efficient replication/firing occurs late during S-phase in S. pombe. *PLoS ONE*, **2007**, *2*, e722.

[79] Hyrien, O.; Marheineke, K.; Goldar, A. Paradoxes of eukaryotic DNA replication: MCM proteins and the random completion problem. *Bioessays*, **2003**, *25*, 116-125.

[80] Legouras, I.; Xouri, G.; Dimopoulos, S.; Lygeros, J.; Lygerou, Z. DNA replication in the fission yeast: robustness in the face of uncertainty. *Yeast*, **2006**, *23*, 951-962.

[81] Woodward, A.M.; Gohler, T.; Luciani, M.G.; Oehlmann, M.; Ge, X.; Gartner, A.; Jackson, D.A.; Blow, J.J. Excess Mcm2-7 license dormant origins of replication that can be used under conditions of replicative stress. *J. Cell Biol.*, **2006**, *173*, 673-683.

[82] Ge, X.Q.; Jackson, D.A.; Blow, J.J. Dormant origins licensed by excess Mcm2–7 are required for human cells to survive replicative stress. *Genes Dev.*, **2007**, *21*, 3331-3341.

[83] Czajkowsky, D.M.; Liu, J.; Hamlin, J.L.; Shao, Z. DNA combing reveals intrinsic temporal disorder in the replication of yeast chromosome VI. *J. Mol. Biol.*, **2008**, *375*, 12-19.

[84] Brewer, B.J.; Fangman, W.L. Initiation at closely spaced replication origins in a yeast chromosome. *Science*, **1993**, *262*, 1728-1731.

[85] Brewer. B.J.; Fangman, W.L. Initiation preference at a yeast origin of replication. *Proc. Natl. Acad. Sci. USA*, **1994**, *91*, 3418-3422.

[86] Marahrens, Y.; Stillman, B. Replicator dominance in a eukaryotic chromosome. *EMBO J.*, **1994**, *13*, 3395-3400.

[87] Diffley, J.F. Replication control: choreographing replication origins. *Curr. Biol.*, **1998**, *8*, R771-773.

[88] Lei, M.; Tye, B.K. Initiating DNA synthesis: from recruiting to activating the MCM complex. *J. Cell Sci.*, **2001**, *114*, 1447-1454.

[89] Raghuraman, M.K.; Brewer, B.J. Molecular analysis of the replication program in unicellular model organisms. *Chromosome Res.*, **2010**, *18*, 19-34.

[90] Rhind, N.; Yang, S.C.; Bechhoefer, J. Reconciling stochastic origin firing with defined replication timing. *Chromosome Res.* **2010**, *18*, 35-43.

[91] Alvino, G.M.; Collingwood, D.; Murphy, J.M.; Delrow, J.; Brewer, B.J.; Raghuraman, M.K. Replication in hydroxyurea: it's a matter of time. *Mol. Cell. Biol.*, **2007**, *27*, 6396-6406.

[92] Yang, S.C.; Rhind, N.; Bechhoefer, J. Modeling genome-wide replication kinetics reveals a mechanism for regulation of replication timing. *Mol. Syst. Biol.*, **2010**, *6*, 404.

[93] de Moura, A.P.; Retkute, R.; Hawkins, M.; Nieduszynski, C.A. Mathematical modelling of whole chromosome replication. *Nucleic Acids Res.*, **2010**, *38*, 5623-5633.

[94] Retkute, R.; Nieduszynski, C.A.; de Moura, A. Dynamics of DNA replication in yeast. *Phys. Rev. Lett.*, **2011**, *107*, 068103.

[95] Lucas, I.; Chevrier-Miller, M.; Sogo, J.M.; Hyrien, O. Mechanisms ensuring rapid and complete DNA replication despite random initiation in Xenopus early embryos. *J. Mol. Biol.*, **2000**, *296*, 769-786.

[96] Takahashi, M. A model for the spatio-temporal organization of DNA replication in mammalian cells. *J. Theor. Biol.*, **1987**, *129*, 91-115.

[97] Lygeros, J.; Koutroumpas, K.; Dimopoulos, S.; Legouras, I.; Kouretas, P.; Heichinger, C.; Nurse, P.; Lygerou, Z. Stochastic hybrid modeling of DNA replication across a complete genome. *Proc. Natl. Acad. Sci. USA*, **2008**, *105*, 12295-12300.

[98] Goldar, A.; Labit, H.; Marheineke, K.; Hyrien, O. A dynamic stochastic model for DNA replication initiation in early embryos. *PLoS ONE*, **2008**, *3*, e2919.

[99] Sekedat, M.D.; Fenyö, D.; Rogers, R.S.; Tackett, A.J.; Aitchison, J.D.; Chait, B.T. GINS motion reveals replication fork progression is remarkably uniform throughout the yeast genome. *Mol. Syst. Biol.*, **2010**, *6*, 353.

[100] Hyrien, O.; Goldar, A. Mathematical modelling of eukaryotic DNA replication. *Chromosome Res.* **2010**, *18*, 147-161.

[101] Schwob, E.; Nasmyth, K. CLB5 and CLB6, a new pair of B cyclins involved in DNA replication in Saccharomyces cerevisiae. *Genes Dev.*, **1993**, *7*, 1160-1175.

[102] Stuart, D.; Wittenberg, C. CLB5 and CLB6 are required for premeiotic DNA replication and activation of the meiotic S/M checkpoint. *Genes Dev.*, **1998**, *12*, 2698-2710.

[103] Zegerman, P.; Diffley, J.F. DNA replication as a target of the DNA damage checkpoint. *DNA Repair*, **2009**, *8*, 1077-1088.

[104] Barberis, M.; Klipp, E.; Vanoni, M.; Alberghina, L. Cell size at S phase initiation: an emergent property of the G1/S network. *PLoS Comput. Biol.*, **2007**, *3*, e64.

[105] Barberis, M.; Klipp, E. Insights into the network controlling the G1/S transition in budding yeast. *Genome Inform.*, **2007**, *18*, 85-99.

[106] Newlon, C.S.; Petes, T.D.; Hereford, L.M.; Fangman, W.L. Replication of yeast chromosomal DNA. *Nature*, **1974**, *247*, 32-35.

[107] Petes, T.D.; Williamson, D.H. Fiber autoradiography of replicating yeast DNA. *Exp. Cell Res.*, **1975**, *95*, 103-110.

[108] Hook, S.S.; Lin, J.J.; Dutta, A. Mechanisms to control rereplication and implications for cancer. *Curr. Opin. Cell Biol.*, **2007**, *19*, 663-671.

[109] Bayani, J.; Selvarajah, S.; Maire, G.; Vukovic, B.; Al-Romaih, K.; Zielenska, M.; Squire, J.A. Genomic mechanisms and measurement of structural and numerical instability in cancer cells. *Semin. Cancer Biol.*, **2007**, *17*, 5-18.

[110] Spruck, C.H.; Won, K.A.; Reed, S.I. Deregulated cyclin E induces chromosome instability. *Nature*, **1999**, *401*, 297-300.

[111] Lengronne, A.; Schwob, E. The yeast CDK inhibitor Sic1 prevents genomic instability by promoting replication origin licensing in late G(1). *Mol. Cell*, **2002**, *9*, 1067-1078.

[112] Alexandrow, M.G.; Hamlin, J.L. Chromatin decondensation in S-phase involves recruitment of Cdk2 by Cdc45 and histone H1 phosphorylation. *J. Cell Biol.*, **2005**, *168*, 875-886.

[113] Lei, M.; Kawasaki, Y.; Tye, B.K. Physical interactions among Mcm proteins and effects of Mcm dosage on DNA replication in Saccharomyces cerevisiae. *Mol. Cell. Biol.*, **1996**, *16*, 5081-5090.

[114] Donovan, S.; Harwood, J.; Drury, L.S.; Diffley, J.F. Cdc6p-dependent loading of Mcm proteins onto pre-replicative chromatin in budding yeast. *Proc. Natl. Acad. Sci. USA*, **1997**, *94*, 5611-5616.

[115] Edwards, M.C.; Tutter, A.V.; Cvetic, C.; Gilbert, C.H.; Prokhorova, T.A.; Walter, J.C. MCM2–7 complexes bind chromatin in a distributed pattern surrounding the origin recognition complex in Xenopus egg extracts. *J. Biol. Chem.*, **2002**, *277*, 33049-33057.

[116] Bowers, J.L.; Randell, J.C.; Chen, S.; Bell, S.P. ATP hydrolysis by ORC catalyzes reiterative Mcm2–7 assembly at a defined origin of replication. *Mol. Cell*, **2004**, *16*, 967-978.

[117] Sidorova, J.M.; Breeden, L.L. Precocious G1/S transitions and genomic instability: the origin connection. *Mutat. Res.*, **2003**, *532*, 5-19.

[118] Mendenhall, M.D. An inhibitor of p34CDC28 protein kinase activity from Saccharomyces cerevisiae. *Science*, **1993**, *259*, 216-219.

[119] Schwob, E.; Bohm, T.; Mendenhall, M.D.; Nasmyth, K. The B-type cyclin kinase inhibitor p40SIC1 controls the G1 to S transition in S. cerevisiae. *Cell*, **1994**, *79*, 233-244.
[120] Tyers, M. The cyclin-dependent kinase inhibitor p40SIC1 imposes the requirement for Cln G1 cyclin function at Start. *Proc. Natl. Acad. Sci. USA*, **1996**, *93*, 7772-7776.
[121] Schneider, B.L.; Yang, Q.H.; Futcher, A.B. Linkage of replication to start by the Cdk inhibitor Sic1. *Science*, **1996**, *272*, 560-562.
[122] Barberis, M.; De Gioia, L.; Ruzzene, M.; Sarno, S.; Coccetti, P.; Fantucci, P.; Vanoni, M.; Alberghina, L. The yeast cyclin-dependent kinase inhibitor Sic1 and mammalian p27Kip1 are functional homologues with a structurally conserved inhibitory domain. *Biochem. J.*, **2005**, *387*, 639-647.
[123] Verma, R.; Annan, R.S.; Huddleston, M.J.; Carr, S.A.; Reynard, G.; Deshaies, R.J. Phosphorylation of Sic1p by G1 Cdk required for its degradation and entry into S phase. *Science*, **1997**, *278*, 455-460.
[124] Nash, P.; Tang, X.; Orlicky, S.; Chen, Q.; Gertler, F.B.; Mendenhall, M.D.; Sicheri, F.; Pawson, T.; Tyers, M. Multisite phosphorylation of a CDK inhibitor sets a threshold for the onset of DNA replication. *Nature*, **2001**, *414*, 514-521.
[125] Verma, R.; Feldman, R.M.; Deshaies, R.J. SIC1 is ubiquitinated *in vitro* by a pathway that requires CDC4, CDC34, and cyclin/CDK activities. *Mol. Biol. Cell*, **1997**, *8*, 1427-1437.
[126] Skowyra, D.; Craig, K.L.; Tyers, M.; Elledge, S.J.; Harper, J.W. F-box proteins are receptors that recruit phosphorylated substrates to the SCF ubiquitin-ligase complex. *Cell*, **1997**, *91*, 209-219.
[127] Feldman, R.M.; Correll, C.C.; Kaplan, K.B.; Deshaies, R.J. A complex of Cdc4p, Skp1p, and Cdc53p/cullin catalyzes ubiquitination of the phosphorylated CDK inhibitor Sic1p. *Cell*, **1997**, *91*, 221-230.
[128] Nasmyth, K. At the heart of the budding yeast cell cycle. *Trends Genet.*, **1996**, *12*, 405-412.
[129] Barberis, M.; Linke, C.; Adrover, M.À.; González-Novo, A.; Lehrach, H.; Krobitsch, S.; Posas, F.; Klipp, E. Sic1 plays a role in timing and oscillatory behaviour of B-type cyclins. *Biotechnol. Adv.*, **2012**, *30*, 108-130.
[130] Barberis, M. Molecular systems biology of Sic1 in yeast cell cycle regulation through multiscale modeling. *Adv. Exp. Med. Biol.*, **2012**, *736*, 135-167.
[131] Barberis, M. Sic1 as a timer of Clb cyclin waves in the yeast cell cycle - design principle of not just an inhibitor. *FEBS J.*, **2012**, *279*, 3386-3410.
[132] Caburet, S.; Conti, C.; Bensimon, A. Combing the genome for genomic instability. *Trends Biotechnol.*, **2002**, *20*, 344-350.
[133] Nugroho, T.T.; Mendenhall, M.D. An inhibitor of yeast cyclin-dependent protein kinase plays an important role in ensuring the genomic integrity of daughter cells. *Mol. Cell. Biol.*, **1994**, *14*, 3320-3328.
[134] Schwob, E. Flexibility and governance in eukaryotic DNA replication. *Curr. Opin. Microbiol.*, **2004**, *7*, 680-690.
[135] Barberis, M.; Beck, C.; Amoussouvi, A.; Schreiber, G.; Diener, C.; Herrmann, A.; Klipp, E. A low number of SIC1 mRNA molecules ensures a low noise level in cell cycle progression of budding yeast. *Mol. Biosyst.*, **2011**, *7*, 2804-2812.
[136] Chen, K.C.; Calzone, L.; Csikasz-Nagy, A.; Cross, F.R.; Novak, B.; Tyson, J.J. Integrative analysis of cell cycle control in budding yeast, *Mol. Biol. Cell*, **2004**, *15*, 3841-3862.
[137] Rossi, R.L.; Zinzalla, V.; Mastriani, A.; Vanoni, M.; Alberghina, L. Subcellular localization of the cyclin dependent kinase inhibitor Sic1 is modulated by the carbon source in budding yeast. *Cell Cycle*, **2005**, *4*, 1798-1807.

Send Orders of Reprints at reprints@benthamscince.net
Advances in Genome Science, Vol. 1, 2013, 45-65

CHAPTER 3

The Role of Androgen Receptor Mutations in Progression of Prostate Cancer

G.N. Brooke, J. Waxman and C. L.Bevan*

Androgen Signalling Laboratory, Department of Surgery and Cancer, Imperial College London, London W12 0NN, UK

Abstract: Prostate cancer growth is regulated by the androgen receptor pathway and so therapies to block androgen production and action are commonly used for the treatment of this disease. Although initially successful in the majority of patients, resistance to hormonal therapy generally emerges and patients' tumours progress to the 'castrate resistant' phase, for which few therapeutic options exist. Expression of the androgen receptor is maintained in castrate resistant prostate cancer and there is a significant body of evidence that demonstrates that the androgen receptor is driving growth, even in the androgen-depleted environment. Several mechanisms have been described to explain how the androgen receptor is active under such circumstances and this current review summarises how mutations of the androgen receptor provide a growth advantage through mechanisms that may involve cofactor recruitment, reduced ligand specificity or promoting constitutive activity. Indivividual mutations appear to differentially regulate gene expression, therefore tumours may behave differently dependent upon which mutation is present and which ligand is driving growth.

Keywords: Prostate cancer, Androgen receptor, Mutation, Therapy resistance, Castration resistance, Hormone therapy.

1. THE ANDROGEN RECEPTOR

The Nuclear Receptor family is the largest group of eukaryotic transcription factors, with 48 human receptors presently described [1]. The androgen receptor (AR) is a steroid receptor; a sub-family of receptors characterised as ligand dependent, sequence specific transcription factors [2] and like other nuclear receptors the AR has a modular structure (reviewed in [1]) – see Fig. (**1**). The AR gene is situated at Xq11-12 and consists of 8 exons with exon 1 encoding the N-terminal domain and the entire 5' untranslated region, exons 2 and 3 encoding

***Address correspondence to C. L. Bevan:** Androgen Signalling Laboratory, Department of Surgery and Cancer, Imperial College London, London W12 0NN, UK; E-mail: charlotte.bevan@imperial.ac.uk the DNA binding domain (DBD) and exons 4-8 encoding the "hinge" region and ligand binding domain (LBD) [3].

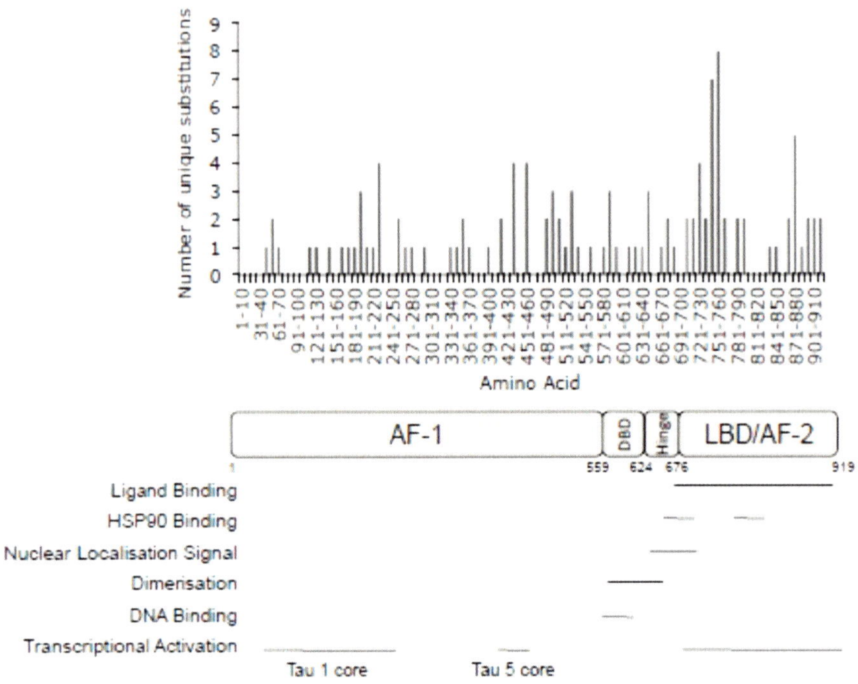

Figure 1: Frequency and location of androgen receptor mutations associated with prostate cancer. The location and frequency of different single base mutations are given, highlighting where these substitutions lie in relation to functional domains of the androgen receptor. Substitutions identified from the Androgen Receptor Mutations database [84].

1.1. Activation Function 1

The N-terminal domain contains activation function-1 (AF-1), which is composed of 2, to some extent separable, transcription activation units (TAUs) – TAU1 and TAU5 (residues 1-485 and 360-528 respectively) [3]. Although the locations of the TAUs overlap, the cores of these units (containing over 80% activity) are in separate regions and have individual characteristics. TAU1 appears to be important in ligand-dependent activation, whereas deletion of the LBD reduces the activity of TAU1; instead, most activity is *via* TAU5 [4]. Further, TAU1 acts as an autonomous activation function independent of p160 coactivators, binding of which to AF-1 occurs *via* TAU5 [5-7].

A polymorphic polyglutamine (CAG) tract is present in the N-terminus of the AR and its length ranges from 8 to 30 repeats in normal individuals, with a modal length of 20 [8]. The rare neuromuscular disorder, spinal and bulbar muscular atrophy (SBMA or Kennedy's disease), is caused by expansion of the CAG tract to more than 40 repeats. Receptors with this expansion form nuclear aggregates and cause neurodegredation through a toxic gain-of-function mechanism [9, 10]. The mechanism of this toxicity is not completely clear, but it appears that alterations in transcription, mitochondrial function and abnormal expression of caspases, in response to receptor aggregates, may be involved (reviewed in [10, 11]). There is an inverse correlation between CAG repeat length and AR transcriptional activity [12] and subsequently the link between this polymorphism and prostate cancer risk has been studied extensively. However, the results of these studies have been contradictory with some studies showing a correlation of shorter CAG length with increased risk (for example [13, 14]) and others not (for example [15, 16]).

The N-terminal region of the AR appears to be highly unstructured. A more folded and subsequently proteosome resistant conformation can be achieved by incubation with folding-inducing solvents (such as trimethylamine-N oxide, TMAO), and also by binding of the cofactor TFIIF (Transcription Factor IIF) [17, 18]. This increased folding is important in transcriptional activity since it enhances recruitment of accessory proteins such as SRC1 (Steroid Receptor Coactivator 1) [19].

1.2. DNA Binding Domain

The DNA binding domains (DBD) of nuclear receptors have high sequence homology, with differences in the amino acids that contact the DNA eliciting specificity [20]. The DBD contains 9 cysteine residues, 8 of which form 2 tetrahedral conformations each around a single zinc atom, forming 2 zinc finger-like modules through which the receptors interact with DNA [21]. The first zinc finger contains the P-box (Gly^{577}, Ser^{578} Cys^{579}, Lys^{580} and Val^{581} of the AR), which from studies using hybrid receptors has been found to confer receptor response element specificity [20, 22-24]. The second zinc finger appears to stabilise binding *via* interaction of the D-box (amino acids Ala^{596}, Ser^{597}, Lys^{598}, Asn^{599} and Asp^{600}) with the sugar-phosphate backbone of DNA [24].

1.3. Hinge

The hinge region of nuclear receptors was originally thought to be a flexible linker region important in DNA binding and dimerisation. Detailed analysis of the region, however, has demonstrated that the region plays a more diverse role than initially thought [25, 26]. Using deletion mutants and substitutions, it has been shown that the hinge region, in particular amino acids ^{629}RKLKK^{633}, is important in nuclear localisation, receptor stability, DNA binding, transcriptional activity and intra and inter-receptor interactions [25-27].

1.4. Ligand Binding Domain/ Activation Function 2

The ligand binding domain (LBD) of nuclear receptors is comprised of 11-12 α-helices and 1 β-sheet that fold to form a triple-layered anti-parallel α-helical sandwich. This folding forms a hydrophobic pocket into which the ligand fits; crystallographic analysis of various members of the family have revealed a highly conserved structure with the exact dimensions of the pocket varying according to the cognate ligand [28-38]. Ligand binding promotes the relocalisation of helix 12, which realigns to create a surface consisting of helices 3, 4, 5 and 12 that is important in cofactor binding [39-41].

The predominant activation function in the AR appears to be AF-1, since deletion of AF-2 and the LBD results in a constitutively active receptor with similar activity to the wild-type receptor in the presence of agonist [3]. However, it has been demonstrated that AF-2 plays a significant role in chromatin and hence its importance in transcriptional regulation may have been underestimated [42].

1.5. The Androgen Receptor Pathway

The unliganded AR is predominantly cytoplasmic [43, 44] and associated with a large heterocomplex, including chaperone and heat-shock proteins, that holds the receptor in a ligand binding competent state [45] (Fig. **2**). Ligand binding promotes dissociation from this complex, dimerisation, nuclear localisation and intra- and inter-receptor interactions [46]. The AR, like several other nuclear receptors (*e.g.* oestrogen receptor α and the progesterone receptor) undergoes a ligand-dependent N- and C-terminal interaction. In the AR, this interaction is predominantly mediated by an N-terminal ^{23}FQNLF^{27} motif interacting with the

AF-2 cofactor-interaction surface. It is important for transcriptional activity since it increases receptor stability, reduces ligand off-rate and creates interaction sites for accessory proteins [47, 48].

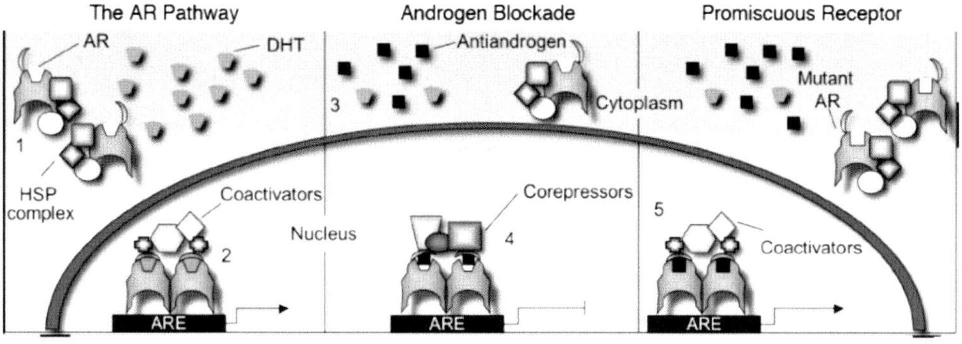

Figure 2: Hormone therapy failure caused by AR mutation. (1) The unliganded AR exists in the cytoplasm associated with a large heterocomplex that holds it in a ligand binding competent state. **(2)** Upon ligand binding the receptor dimerises, translocates to the nucleus where it binds DNA and promotes gene transcription *via* the recruitment of accessory proteins, such as coactivators. **(3)** Non organ-confined prostate cancer is usually treated by blocking the AR pathway. This is achieved by blocking the production of androgen (using LHRH analogues) and/or antiandrogens. **(4)** Antiandrogens bind to the AR but do not promote an active conformation and instead block receptor function, at least in part, by promoting the recruitment of corepressors to the regulatory regions of target genes. **(5)** This therapy selects for cells that have mechanisms by which the tumour can grow in the androgen depleted environment. One such mechanism is that of AR mutation and in some cases these mutants provide a growth advantage because they reduce the ligand specificity of the receptor. Hence other ligands, such as the antiandrogens being used in treatment, are now able to promote an active conformation, the recruitment of coactivators and subsequently gene expression.

The AR has been found to interact with a large number of proteins that enhance receptor activity, termed coactivators. Such factors often have, or recruit proteins that have, histone acetyltransferase activity and are hence believed to increase receptor activity in part by relaxing chromatin structure. The best characterised are the p160 family of coactivators, consisting of SRC-1/NCOA1 (Steroid Receptor Coactivator-1 / Nuclear Receptor Coactivator 1), SRC-2/NCOA2/TIF-2/GRIP-1 (Transcriptional Intermediary Factor 2 / Glucocorticoid Receptor Interacting Protein-1) and SRC-3/NCOA3/AIB1/pCIP/RAC-3 (Amplified In Breast Cancer-1 Protein / CBP-Interacting Protein / Receptor-Associated Coactivator 3) [49]. Many coactivators interact with nuclear receptors *via* LxxLL motifs (where L is leucine and x is any amino acid), which form an amphipathic α-helix that binds directly to the hydrophobic AF-2 coactivator groove [50, 51]. Unlike the other steroid receptors, the AR AF-2 can also interact with and has higher affinity for phenylalanine rich motifs, for example the FQNLF motif found in the N-terminus of the AR and those found in some coactivators, for example Androgen Receptor Activator 70 (ARA70) [52, 53]. Study of the crystal structure of the AR AF-2 surface has demonstrated the coactivator groove to be much deeper than that found in the other steroid receptors [53, 54]. This deeper groove appears to be able to accommodate the bulkier phenylalanine residues hence explaining the differences in receptor-interaction motif specificity.

Conversely, corepressor proteins have also been identified, for example Nuclear Co-Repressor Protein (NCoR) and Silencing Mediator for Retinoid and Thyroid Hormone Receptor (SMRT), which have been found to bind to both agonist- and antagonist-bound receptors and reduce trancriptional activity [55]. Many repressors of the AR have been described and some of these have defined roles in other pathways, for example HEY proteins [56, 57]. There are multiple mechanisms by which corepressors appear to inhibit receptor signalling - for example, *via* recruitment of histone deacetylases to condense chromatin structure, and *via* nuclear exclusion of the receptor [58].

The AR interacts with DNA *via* response elements located within the regulatory regions of target genes, usually consisting of an imperfect inverted repeat of a 5'-TGTTCT-3' half site (termed the core recognition sequence) separated by 3 base pairs [59-61]. This consensus sequence is not specific for the AR, but also acts as a response element for glucocorticoid, mineralocorticoid and progesterone receptors [60, 62]. A second class of response elements, which appear to be highly AR specific, have been described and these consist of a direct repeat of the core recognition sequence [63]. Both inverted and direct repeats have been found in the regulatory regions of many androgen responsive genes, such as *PSA* (Prostate Specific Antigen) and *SC* (Secretory Component) [64, 65]. Androgen target genes are involved in varied processes, notably prostate function such as secretion of kallikreins, and cell division resulting in prostate growth [66-69].

The temporal and spatial sequence of intra and inter-molecular events that occur following ligand binding have been deciphered using detailed FRET and FRAP analysis. Following ligand binding the N-/C-terminal interaction occurs, both intra-

and inter-receptor, whilst the receptor is mobile. Upon DNA binding this interaction is largely lost and instead AR coregulator interactions are favoured [70].

2. THE PROSTATE

The prostate is a secretory gland located at the base of the bladder with a composition that is approximately 70% glandular elements (acini that empty into multiple small ductules) and 30% fibromuscular stroma [71]. The stroma is continuous with the capsule that encases the prostate, consisting of collagen, elastin and smooth muscle. The muscle contracts upon ejaculation, forcing prostatic secretions, important in events such as semen coagulation and liquefaction, into the urethra [72]. The first link between androgens and prostate development was made by John Hunter, who in 1786 noted that the size of the gland in castrated animals was significantly reduced compared to that in intact animals [73]. The prostate has since been demonstrated to develop from the urogenital sinus in response to fetal testicular androgens (reviewed in [74]). At maturity growth of the gland ceases, but androgens continue to play an important role in prostate function. In some men, androgen dependent growth of the prostate resumes, resulting in benign prostatic hyperplasia (BPH), premalignant prostatic intraepithelial neoplasia (PIN) or prostate cancer (PCa).

2.1. Prostate Cancer

Prostate cancer is the most common cancer in men in the United Kingdom, with approximately 35,000 men diagnosed every year [75]. The biggest risk factor in prostate cancer is age, with more than 60% of cases occurring in men over 75. Incidence of prostate cancer have been consistently rising and hence it is likely that prostate cancer will overtake lung cancer as the leading cause of cancer related death in Western men.

2.2. Treatment

Approximately 25% of prostate cancer patients have organ-confined disease upon presentation and for such patients radical prostatectomy (complete removal of the prostate) offers the highest likelihood of long-term disease free survival [76]. Unfortunately the majority of patients present with disease that has spread from the prostate capsule and hence surgery is not an option. Since the growth of the prostate is almost always dependent upon the AR pathway, therapies to treat non-organ confined disease often target this signalling axis. Up to the 1950s orchiectomy was routinely used to reduce circulating levels of androgens, since the main site of production is the testis. Since then chemical castration has been the preferred method and this is achieved by using leuteinising hormone releasing hormone (LHRH) analogues, which act *via* the pituitary-hypothalamus signalling axis to block androgen production. These analogues successfully reduce circulating levels of testosterone by more than 95%, but levels of adrenally produced androgen precursors such as dehydroepiandrosterone remain unaffected, and these can be effectively converted in the prostate into the potent androgen dihydrotestosterone [77]. Hence androgen levels within the prostate may only be reduced by approximately 60% and for this reason antiandrogens are often also administered to reach "total androgen blockade".

Antiandrogens are ligands that can bind to the AR and hold it in an inactive state. The exact mechanisms of antiandrogen action are not completely understood, but they appear to function at least in part *via* the recruitment of corepressors to the regulatory regions of target genes. Shang *et al.*, for example, found that the corepressors NCoR and SMRT were recruited to the promoter region of the *PSA* gene following treatment with the antiandrogen Bicalutamide [78]. In contrast, coactivators such as SRC1 were present following treatment with an agonist.

2.3. Castrate Resistant Prostate Cancer (CRPC)

Hormone therapy is successful in the majority of patients, resulting in both symptomatic and pathological improvement. Unfortunately this therapy invariably fails after a median of 2 years and the tumour progresses to a more aggressive "castrate resistant" stage. Other terms for this stage of the disease include androgen- or hormone-independent prostate cancer, but these are perhaps misleading since in most cases the AR is expressed and much data suggests that the receptor is still functional. Several mechanisms have been proposed to explain how the AR may still be driving growth even in the androgen-depleted environment and these include AR amplification, alterations in cofactor levels and AR mutation [79].

2.4. AR Mutations

In early stages of prostate cancer, mutations of the AR are rare but their frequency is significantly increased in advanced, androgen-independent tumours suggesting that AR mutations play a role in tumour progression [80, 81]. Marcelli *et al.*, for example, found that out of 99 patients in the early stages of prostate cancer, none had mutation(s) in the AR coding sequence [81]. In advanced stages of the disease, however, 8 out of the 38 patients studied (21%) were found to have mutation(s) of the AR. There is a great variation in the frequency of AR mutations between studies [82], which appears to be due to, for example, sequencing of only part of the receptor or differences in the sensitivity to detect variants in sub-populations of cells.

More than 150 different AR mutations have been described in patients with prostate cancer, most of which are somatic single base substitutions [83, 84]. Cells carrying such mutations are likely to provide a growth advantage in the androgen-depleted environment and hence be selected for during therapy [85]. Not all mutations of the AR associated with prostate cancer result in a gain of function in the presence of androgen [86, 87] and instead appear to offer a growth advantage as a result of, for example, reduced ligand specificity and alterations in cofactor binding.

2.4.1. Alterations in Cofactor Binding

Although most substitutions associated with PCa cluster in the LBD, more than 40 are also found in the N-terminal domain (Fig. **1**). Buchanan *et al.*, for example, identified dual somatic missense mutations within the N-terminal polyglutamine tract [88], which resulted in interruption of the tract by two leucine residues. These substitutions were found to reduce the ligand induced N- and C-terminal interaction, but paradoxically led to a receptor with greater activity than the wild-type AR. Coactivators have been described that interact with the CAG repeat, for example ARA24 (Androgen Receptor-Associated Protein 24, also known as Ran) [89], and it is believed that the dual somatic missense mutations confer greater transcriptional activity due to increased stability and folding of the tract, enhancing such interactions. This was demonstrated by the finding that ARA24 was found to enhance activity of the mutant to a greater extent than the wild-type receptor [88]. It is possible that these enhanced interactions with coactivators may potentiate AR signalling in low levels of androgen or in the presence of weaker agonists and thus contribute to therapy failure.

Mutations have also been identified which disrupt an inhibitory domain located within the AR. The activity of AF-2 has been demonstrated to be inhibited by the hinge region [90]. Haelens *et al.* have studied the effect of two mutations, R629Q and K630T, located in the hinge region of the AR [26]. These mutations, which lie in the bipartite nuclear localisation signal, reduced nuclear localisation and DNA binding of the receptor. Surprisingly, however, the mutants had greater transcriptional activity than the wild-type receptor. Similarly Buchanan *et al.* found that mutations Gln668Arg and Ile670Thr, also within the hinge region, also had increased activity and reduced ligand specificity compared to the wild-type receptor without changes in receptor levels, ligand binding or DNA binding [91]. The data therefore suggests that mutations of the hinge region provide a growth advantage by increasing AR activity through disruption of an inhibitory domain. The exact mechanisms by which the hinge region inhibits the activity of AF-2 is yet to be elucidated, however, it has been postulated that the region may be inhibiting coactivator interactions [26]. Hence, mutations of the hinge region may increase receptor activity by attenuating coactivator recruitment.

2.4.2. Mutations Resulting in Constitutively Active Receptor

As described above, deletion of the AR LBD results in a receptor with similar activity to the full-length AR in the presence of androgen [92, 93]. Interestingly splice variants of the *AR* have been described which give rise to C-terminal truncated AR protein, whereby a novel exon (2b) is spliced downstream of either exon 2 or 3 giving rise to ARs containing the N-terminal domain, one or both zinc fingers of the DBD, followed by 11 amino acids encoded by exon 2b [94]. Such variants are constitutively active and have been shown to be up-regulated in androgen independent prostate cancer [94, 95]. Point mutations of the AR, which similarly result in a constitutively active receptor have also been reported in prostate cancer. Ceraline *et al.*, for example, identified a substitution that generated a stop codon at Q640 resulting in a receptor truncated in the hinge region [96]. Such truncations, as a result of mutation, however, appear to be rare.

Mutations that lead to constitutive activation of the AR have also been demonstrated in the full-length receptor. In a study of 45 different AR mutations,

Hay *et al.* identified 7 mutations that lead to constitutive gain of function [87]. For example, mutations M523V and G524D were found to have constitutive activity in the absence of ligand and a slight gain of function in the presence of androgen. Although the mechanism of constitutive activity has not been defined, the authors suggest that these mutations may affect sumoylation of nearby amino acid 520, a modification which has been demonstrated to repress AR activity [97]. Hence it is possible that these mutations may reduce/inhibit sumoylation and subsequently lead to an active receptor [87].

2.4.3. Alterations in Ligand Specificity

The first AR variant with loss of ligand specificity to be described was a threonine to alanine substitution at amino acid 877 [98]. This mutant has since been frequently found in advanced prostatic carcinomas - Taplin *et al.* for example, found the mutation in 30% of bone marrow metastases [80]. The receptor not only responds to androgens but is also activated by oestrogens, progestins and the antiandrogens cyproterone acetate and hydroxyflutamide (the active form of flutamide) [99]. Crystal structure analysis of the AR LBD has revealed that threonine 877 forms hydrogen bonds with the 17β-hydroxyl group of androgen [38]. Further modelling has demonstrated that substitutions to the smaller alanine affects the size and shape of the receptor such that other ligands can fit into the pocket and induce an active conformation [100].

Not all prostate cancer-associated substitutions in the LBD reduce ligand specificity by altering the dimensions of the pocket. The H874Y mutant AR, for example, is also activated by hydroxyflutamide, oestradiol, progesterone, and cyproterone acetate, but the side chain of this residue points away from the pocket and is buried in a cavity between helices 11 and 12, which is formed following ligand induced activation [101]. This cavity is large enough to accommodate the tyrosine aromatic ring and hence it appears unlikely that the altered ligand specificity is as a result of steric alterations at this site. Instead, it appears that the more hydrophobic tyrosine side chain strengthens the interaction of helix 12 with this groove. It has been proposed that this stronger interaction could promote the relocation of helix 12 to the active position even in the presence of a ligand that does not optimally fit into the pocket, thus reducing ligand specificity [101]. Interestingly, the mutant has been found to have enhanced binding to the p160 coactivators, suggesting that, in addition to broadened ligand specificity, the mutant receptor also has enhanced coactivator recruitment [102, 103].

Differences in coactivator binding have also been found dependent upon which 'agonist' is activating the mutant receptor. As described previously, the AR can bind both LxxLL and phenylalanine-rich motifs (such as FxxLY), but has higher affinity for the latter. We have studied the preference of several of the most commonly identified mutant receptors (H874Y, T877A and T877S) for these motifs and found striking differences in motif utilisation dependent upon which ligand is activating the receptor [104]. In the presence of cyproterone acetate, for example, the mutants specifically interact with the LxxLL motif whereas in the presence of hydroxyflutamide the receptor interacts with the FxxLY motif. Using chromatin immunoprecipitation, siRNA and target gene expression analysis, we were able to show that this selectivity extended to coactivator recruitment to endogenous genes demonstrating that dependent upon ligand and interaction motif, the mutant receptors may utilise different subsets of coactivators to potentiate gene expression.

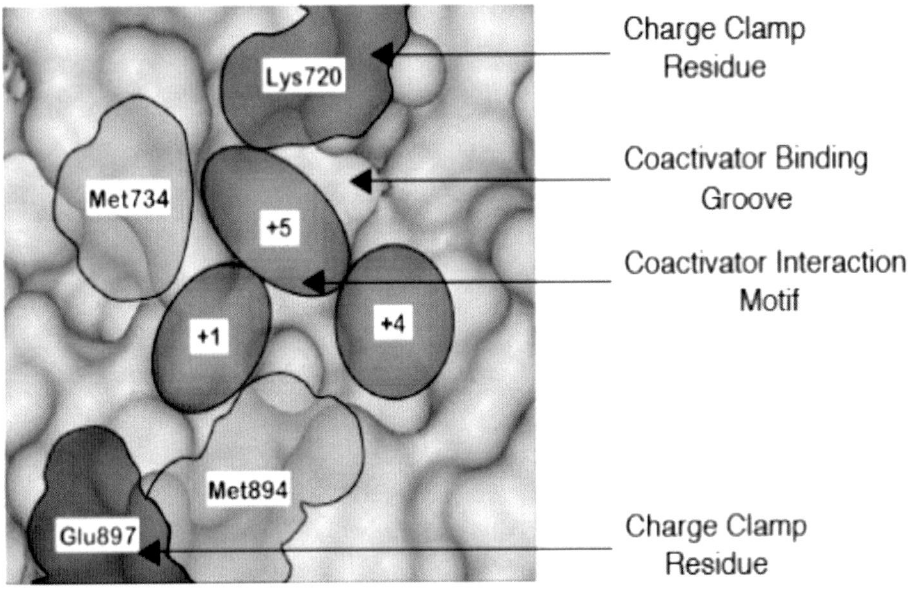

Figure 3: Surface representation of the AR coactivator groove. Representation of the AR AF-2 surface highlighting the L-shaped cleft and key residues important in coactivator binding. +1, +4 and +5 refer to the regions in which the 1st, 4th and 5th amino acids of LxxLL and FxxLF-like motifs lie following binding. Image created using RasMol V2.6 using co-ordinates from[38].

The coactivator interaction groove of the AR is comprised of an L-shaped cleft comprised of three distinct subsites (formed from helices 3, 4, 5 and 12 of the LBD) that bind hydrophobic groups at the +1, +4 and +5 positions in cognate peptides [53]. The conserved charge residues at either end of the cleft, Lys702 and Glu897, form what is referred to as the "charge clamp" (Fig. **3**). The charge clamp residues form electrostatic interactions with the main chain atoms at either end of phenylalanine-rich motifs whereas LxxLL motifs only form hydrogen bonds with Lys720. Charge clamp residue Glu897 is located in helix 12, the positioning of which is likely to be affected by the agonist bound. Therefore if an agonist is bound that does not induce the correct positioning of the charge clamp residues for interaction with phenylalanine rich motifs, then the site appears to be available for LxxLL motif binding.

Several studies have demonstrated that inter- and intra-receptor interactions made by the AR have differential effects dependent upon promoter context [105-107]. We hypothesised that a mutant receptor activated by different agonists could therefore regulate different subsets of genes, hence we studied the expression levels of genes involved in prostate differentiation (*Kallikrein 2, KLK2,* and *Differentiation Regulated Gene-1, DRG-1*) and cell cycle progression (*Cyclin Dependent Kinases 2 and 4, CDK2 and CDK4*) in the LNCaP prostate cancer cell line, which endogenously expresses the T877A mutant AR [104]. The expression of *KLK2* and *CDK2* in response to different ligands was similar – induced most strongly by androgen, then the antiandrogen hydroxyflutamide then cyproterone acetate. Evidence of *CDK4* being an androgen-regulated gene is contradictory, with upregulation in response to androgen reported in some studies [108] but not found in others [109]. In agreement with the latter we found no induction of *CDK4* in response to androgen, but interestingly the two antiandrogens did induce expression. Even more striking was the regulation of *DRG-1,* which we found to be highly upregulated by androgen (more than 12-fold) and only weakly by hydroxyflutamide (approximately 1.7-fold) [104]. Hence the mutant AR induced different "patterns" of regulation of a subset of androgen-regulated genes according to the ligand. Since this includes genes that are involved in tumour growth, there may be implications for tumour progression and treatment.

3. EMERGING THERAPIES TO TARGET ANDROGEN SIGNALLING IN CASTRATE RESISTANT PROSTATE CANCER

The AR remains a valid target in men with CRPC, and a recent advance in the treatment of this stage of the disease has been the development of the 17 α-hydroxylase/C17,20 lyase (CYP17) inhibitor Abiraterone, which blocks two rate-limiting steps in the testosterone synthesis pathway (the conversion of progesterone and pregnenolone to their 17-hydroxylated derivatives, and the subsequent conversion of these to deyhdroepiandrosterone and androsterone, precursors of testosterone). Although chemical castration successfully ablates testicular androgens, adrenal production of androgens remains. Abiraterone, by targeting the synthesis pathway, reduces adrenal-derived androgens and CYP17-dependent androgen synthesis within the tumour [110]. Since CYP17 inhibition also blocks cortisol synthesis, glucocorticoids are co-administered to reduce side effects; some patients also benefit from the use of mineralocorticoid receptor antagonists such as eplerenone to reduce the effects of mineralcorticoid excess associated with this treatment. Phase III trials have demonstrated Abiraterone to have an average of 4 months survival benefit for men with advanced prostate cancer [111]. However resistance, both acquired and intrinsic, to this therapy has been described. Abiraterone resistance, as with other therapies, is often associated with a rise in PSA, suggesting that the AR pathway is again becoming active. One possible mechanism is that the co-therapies being used, and the changes in hormone levels within the patients, exert selective pressure for mutant ARs since, for example, glucocorticoid and eplerenone are able to activate certain AR variants [112].

Another emerging therapy is the antiandrogen MDV3100, which has shown promising results in clinical trials (reviewed in [110]). MDV3100 acts differently from other antiandrogens in that it reduces AR nuclear localisation and blocks DNA binding [113]. Importantly, it has also been demonstrated to successfully inhibit mutant ARs with reduced ligand specificity that are activated by, for example, the antiandrogens flutamide and bicalutamide (T877A and W741C) [113]. However, not all patients respond to MDV3100 and of those that do, many become resistant. This resistance is associated with a rise in PSA [110], again suggesting that the AR pathway is re-activated. Whether drug-resistant mutations of the AR can explain this relapse remains to be seen.

In recent years inhibitors of the chaperone proteins involved in AR folding, notably heat shock protein 90 and its co-chaperones, have been developed as therapies for solid tumours including prostate cancer [114-116]. Trials are ongoing with variable results, and it is too early to know whether the same issues of resistance will arise. However such therapies would be expected to exert their effects on AR signalling prior to ligand binding, hence it would be predicted that these therapeutics would remain active against many mutant forms of the AR.

4. IMPLICATIONS FOR THE TREATMENT OF PROSTATE CANCER

The findings summarised here have important implications for the treatment of prostate cancer. The data suggest that tumours may behave differently dependent upon (i) which mutation, if any, is present and (ii) which ligand is driving growth, since different subsets of genes may be regulated. We suggest that it would therefore be useful to screen patients for AR mutations following hormone therapy failure, so that subsequent treatment could be adjusted accordingly. Understanding of how AR mutations alter androgen signalling at the molecular level will also be useful in the development of novel therapies, in particular Selective Androgen Receptor Modulators (SARMs). Knowledge of how such molecules affect the structure of the AR, interactions that the receptor makes and subsequently gene expression could aid in designing drugs that regulate certain subsets of genes. For example, it would be desirable to design a SARM that blocks expression of androgen target genes that promote tumour growth, whilst up-regulating transcription of beneficial genes (for example those important for maintenance of bone density).

CONFLICT OF INTEREST

The author(s) confirm that this chapter content has no conflict of interest.

ACKNOWLEDGEMENTS

The authors wish to thank all members of the Androgen Signalling Laboratory, and to acknowledge the support of The Martin Harris Research Fellowship, The Medical Research Council, Prostate Action and The Prostate Cancer Charity.

DISCLOSURE

"This is an updated report of the earlier review article By G.N. Brooke and C. L. Bevan, "The role of androgen receptor mutations in prostate cancer progression". Current Genomics, 2009, 10: 18-25".

REFERENCES

1] Robinson-Rechavi, M. H.; Garcia, E.; Laudet, V. The nuclear receptor superfamily. *J. Cell Sci.*, **2003**, *116*(Pt 4): 585-586.

[2] Mangelsdorf, D.J., *et al.*, The nuclear receptor superfamily: the second decade. *Cell*, **1995**. *83*(6): 835-839.

[3] Jenster, G., *et al.*, Functional domains of the human androgen receptor. *J. Steroid Biochem. Mol. Biol.*, **1992**, *41*(3-8): 671-675.

[4] Jenster, G., *et al.*, Identification of two transcription activation units in the N-terminal domain of the human androgen receptor. *J. Biol. Chem.*, **1995**. *270*(13): 7341-7346.

[5] Callewaert, L.; Van Tilborgh, N.; Claessens, F. Interplay between two hormone-independent activation domains in the androgen receptor. *Cancer Res.*, **2006**, *66*(1): 543-553.

[6] Bevan, C.L., *et al.*, The AF1 and AF2 domains of the androgen receptor interact with distinct regions of SRC1. *Mol. Cell Biol.*, **1999**, *19*(12): 8383-8392.

[7] Christiaens, V., *et al.*, Characterization of the two coactivator-interacting surfaces of the androgen receptor and their relative role in transcriptional control. *J. Biol. Chem.*, **2002**, *277*(51): 49230-49237.

[8] Hardy, D.O., *et al.*, Androgen receptor CAG repeat lengths in prostate cancer: correlation with age of onset. *J. Clin. Endocrinol. Metab.*, **1996**, *81*(12): 4400-4405.

[9] La Spada, A.R., *et al.*, Androgen receptor gene mutations in X-linked spinal and bulbar muscular atrophy. *Nature*, **1991**, *352*(6330): 77-79.

[10] Lieberman, A.P.; Fischbeck, K.H. Triplet repeat expansion in neuromuscular disease. *Muscle Nerve*, **2000**, *23*(6): 843-850.

[11] Parodi, S.; Pennuto, M. Neurotoxic effects of androgens in spinal and bulbar muscular atrophy. *Front Neuroendocrinol.*, **2011**, *32*(4): 416-425.

[12] Beilin, J., *et al.*, Effect of the androgen receptor CAG repeat polymorphism on transcriptional activity: specificity in prostate and non-prostate cell lines. *J. Mol. Endocrinol.*, **2000**, *25*(1): 85-96.

[13] Irvine, R.A., *et al.*, The CAG and GGC microsatellites of the androgen receptor gene are in linkage disequilibrium in men with prostate cancer. *Cancer Res.*, **1995**, *55*(9): 1937-1940.

[14] Hsing, A.W., *et al.*, Polymorphic CAG and GGN repeat lengths in the androgen receptor gene and prostate cancer risk: a population-based case-control study in China. *Cancer Res.*, **2000**, ***60***(18): 5111-5116.

[15] Price, D.K., *et al.*, Androgen receptor CAG repeat length and association with prostate cancer risk: results from the prostate cancer prevention trial. *J. Urol.*, **2010**, *184*(6): 2297-2302.

[16] Correa-Cerro, L., *et al.*, (CAG)nCAA and GGN repeats in the human androgen receptor gene are not associated with prostate cancer in a French-German population. *Eur. J. Hum. Genet.*, **1999**, *7*(3): 357-362.

[17] Lavery, D.N.; McEwan, I.J. The human androgen receptor AF1 transactivation domain: interactions with transcription factor IIF and molten-globule-like structural characteristics. *Biochem. Soc. Trans.*, **2006**, *34*(Pt 6): 1054-1057.

[18] McEwan, I.J., *et al.*, Natural disordered sequences in the amino terminal domain of nuclear receptors: lessons from the androgen and glucocorticoid receptors. *Nucl. Recept. Signal*, **2007**, 5, e001.

[19] Kumar, R., *et al.*, Induced alpha-helix structure in AF1 of the androgen receptor upon binding transcription factor TFIIF. *Biochemistry*, **2004**, *43*(11): 3008-3013.

[20] Freedman, L.P. Luisi, B.F. On the mechanism of DNA binding by nuclear hormone receptors: a structural and functional perspective. *J. Cell Biochem.*, **1993**, *51*(2): 140-150.

[21] Shaffer, P.L., *et al.*, Structural basis of androgen receptor binding to selective androgen response elements. Proc. Natl. Acad. Sci. USA, **2004**, *101*(14), 4758-4763.

[22] Umesono, K.; Evans, R.M. Determinants of target gene specificity for steroid/thyroid hormone receptors. *Cell*, **1989**, *57*(7), 1139-1146.

[23] Mader, S., *et al.*, Three amino acids of the oestrogen receptor are essential to its ability to distinguish an oestrogen from a glucocorticoid-responsive element. *Nature*, **1989**, *338*(6212): 271-274.

[24] Dahlman-Wright, K., *et al.*, Interaction of the glucocorticoid receptor DNA-binding domain with DNA as a dimer is mediated by a short segment of five amino acids. *J. Biol. Chem.*, **1991**, 266(5): 3107-3112.

[25] Tanner, T.; Claessens, F.; Haelens, A. The hinge region of the androgen receptor plays a role in proteasome-mediated transcriptional activation. *Ann. N. Y. Acad. Sci.*, **2004**, *1030*, 587-592.

[26] Haelens, A., *et al.*, The hinge region regulates DNA binding, nuclear translocation, and transactivation of the androgen receptor. *Cancer Res.*, **2007**. *67*(9), 4514-4523.

[27] Clinckemalie, L., *et al.*, The hinge region in androgen receptor control. *Mol. Cell Endocrinol.*, **2012**.

[28] Williams, S.P. Sigler, P.B. Atomic structure of progesterone complexed with its receptor. *Nature*, **1998**, *393*(6683), 392-396.

[29] Uppenberg, J., *et al.*, Crystal structure of the ligand binding domain of the human nuclear receptor PPARgamma. *J. Biol. Chem.*, **1998**, *273*(47), 31108-31112.

[30] Rochel, N., *et al.*, The crystal structure of the nuclear receptor for vitamin D bound to its natural ligand. *Mol. Cell*, **2000**, *5*(1): 173-179.

[31] Renaud, J.P., *et al.*, Crystal structure of the RAR-gamma ligand-binding domain bound to all-trans retinoic acid. *Nature*, **1995**, *378*(6558): 681-689.

[32] Pike, A.C., *et al.*, Structure of the ligand-binding domain of oestrogen receptor beta in the presence of a partial agonist and a full antagonist. *EMBO J.*, **1999**, *18*(17), 4608-4618.

[33] Kumar, R.; Thompson, E.B. The structure of the nuclear hormone receptors. *Steroids*, **1999**, *64*(5), 310-319.

[34] Egea, P.F., *et al.*, Crystal structure of the human RXRalpha ligand-binding domain bound to its natural ligand: 9-cis retinoic acid. *EMBO J.*, **2000**, *19*(11): 2592-2601.

[35] Brzozowski, A.M., *et al.*, Molecular basis of agonism and antagonism in the oestrogen receptor. *Nature*, **1997**, *389*(6652):753-758.

[36] Bourguet, W., *et al.*, Purification, functional characterization, and crystallization of the ligand binding domain of the retinoid X receptor. *Protein Expr Purif*, **1995**, *6*(5): 604-608.

[37] Sack, J.S., *et al.*, Crystallographic structures of the ligand-binding domains of the androgen receptor and its T877A mutant complexed with the natural agonist dihydrotestosterone. *Proc. Natl. Acad. Sci. USA*, **2001**, *98*(9): 4904-4909.

[38] Matias, P.M., *et al.*, Structural evidence for ligand specificity in the binding domain of the human androgen receptor. Implications for pathogenic gene mutations. *J. Biol. Chem.*, **2000**, *275*(34): 26164-26171.

[39] Mak, H.Y., *et al.*, Molecular determinants of the estrogen receptor-coactivator interface. *Mol. Cell Biol.*, **1999**, *19*(5): 3895-3903.

[40] Parker, M.G.; White, R. Nuclear receptors spring into action. *Nat. Struct. Biol.*, **1996**, *3*(2): 113-115.

[41] Wurtz, J.M., *et al.*, A canonical structure for the ligand-binding domain of nuclear receptors. *Nat. Struct. Biol.*, **1996**, *3*(1): 87-94.

[42] Li, J., *et al.*, Structural and Functional Analysis of Androgen Receptor in Chromatin. *Mol. Endocrinol.*, **2007**.

[43] Whitaker, H.C., *et al.*, Androgen receptor is targeted to distinct subcellular compartments in response to different therapeutic antiandrogens. *Clin. Cancer Res.*, **2004**, *10*(21): 7392-7401.

[44] Tyagi, R.K., *et al.*, Dynamics of intracellular movement and nucleocytoplasmic recycling of the ligand-activated androgen receptor in living cells. *Mol. Endocrinol.*, **2000**, **14**(8): 1162-1174.

[45] Pratt, W.B.; Toft, D.O. Steroid receptor interactions with heat shock protein and immunophilin chaperones. *Endocr. Rev.*, **1997**, *18*(3): 306-360.

[46] Dehm, S.M.; Tindall, D.J. Androgen receptor structural and functional elements: role and regulation in prostate cancer. *Mol. Endocrinol.*, **2007**, *21*(12): 2855-2863.

[47] Langley, E., Kemppainen, J.A.; Wilson, E.M. Intermolecular NH2-/carboxyl-terminal interactions in androgen receptor dimerization revealed by mutations that cause androgen insensitivity. *J. Biol. Chem.*, **1998**, *273*(1): 92-101.

[48] He, B., Kemppainen, J.A.; Wilson, E.M. FXXLF and WXXLF sequences mediate the NH2-terminal interaction with the ligand binding domain of the androgen receptor. *J. Biol. Chem.*, **2000**, *275*(30): 22986-22994.

[49] Chen, J.D. Steroid/nuclear receptor coactivators. *Vitam. Horm.*, **2000**, *58*: 391-448.

[50] Darimont, B.D., *et al.*, Structure and specificity of nuclear receptor-coactivator interactions. *Genes Dev.*, **1998**, *12*(21): 3343-3356.

[51] Heery, D.M., *et al.*, A signature motif in transcriptional co-activators mediates binding to nuclear receptors. *Nature*, **1997**, *387*(6634): 733-736.

[52] Ozers, M.S., *et al.*, The androgen receptor T877A mutant recruits LXXLL and FXXLF peptides differently than wild-type androgen receptor in a time-resolved fluorescence resonance energy transfer assay. *Biochemistry*, **2007**, *46*(3): 683-695.

[53] Hur, E., *et al.*, Recognition and accommodation at the androgen receptor coactivator binding interface. *PLoS Biol.*, **2004**, *2*(9): E274.

[54] Dubbink, H.J., *et al.*, Distinct recognition modes of FXXLF and LXXLL motifs by the androgen receptor. *Mol. Endocrinol.*, **2004**, *18*(9): 2132-2150.

[55] Yoon, H.G.; Wong, J. The corepressors silencing mediator of retinoid and thyroid hormone receptor and nuclear receptor corepressor are involved in agonist- and antagonist-regulated transcription by androgen receptor. *Mol. Endocrinol.*, **2006**, *20*(5): 1048-1060.

[56] Lavery, D.N., *et al.*, Repression of androgen receptor activity by HEYL, a third member of the Hairy/Enhancer-of-split-related family of Notch effectors. *J. Biol. Chem.*, **2011**, *286*(20): 17796-17808.

[57] Belandia, B., *et al.*, Hey1, a mediator of notch signaling, is an androgen receptor corepressor. *Mol. Cell Biol.*, **2005**, *25*(4): 1425-1436.

[58] Wang, L.; Hsu, C.L.; Chang, C. Androgen receptor corepressors: an overview. *Prostate*, **2005**, *63*(2): 117-130.
[59] Cato, A.C.; Henderson, D.; Ponta, H. The hormone response element of the mouse mammary tumour virus DNA mediates the progestin and androgen induction of transcription in the proviral long terminal repeat region. *EMBO J.*, **1987**, *6*(2): 363-368.
[60] Ham, J., *et al.*, Characterization of response elements for androgens, glucocorticoids and progestins in mouse mammary tumour virus. *Nucleic Acids Res.*, **1988**, *16*(12): 5263-5276.
[61] Beato, M. Gene regulation by steroid hormones. *Cell*, **1989**, *56*(3): 335-344.
[62] Tsai, M.J.; O'Malley, B.W. Molecular mechanisms of action of steroid/thyroid receptor superfamily members. *Annu. Rev. Biochem.*, **1994**, *63*: 451-486.
[63] Claessens, F., *et al.*, Selective DNA binding by the androgen receptor as a mechanism for hormone-specific gene regulation. *J. Steroid Biochem. Mol. Biol.*, **2001**, *76*(1-5): 23-30.
[64] Riegman, P.H., *et al.*, The promoter of the prostate-specific antigen gene contains a functional androgen responsive element. *Mol. Endocrinol.*, **1991**, *5*(12): 1921-1930.
[65] Verrijdt, G., *et al.*, Androgen specificity of a response unit upstream of the human secretory component gene is mediated by differential receptor binding to an essential androgen response element. *Mol. Endocrinol.*, **1999**, *13*(9): 1558-1570.
[66] Wolf, D.A., Schulz, P.; Fittler, F. Transcriptional regulation of prostate kallikrein-like genes by androgen. *Mol. Endocrinol.*, **1992**, *6*(5): 753-762.
[67] Knudsen, K.E.; Arden, K.C.; Cavenee, W.K. Multiple G1 regulatory elements control the androgen-dependent proliferation of prostatic carcinoma cells. *J. Biol. Chem.*, **1998**, *273*(32): 20213-20222.
[68] Brooke, G.N., *et al.*, FUS/TLS is a novel mediator of androgen-dependent cell-cycle progression and prostate cancer growth. *Cancer Res.*, **2011**, *71*(3): 914-924.
[69] Gamble, S.C., *et al.*, Androgens target prohibitin to regulate proliferation of prostate cancer cells. *Oncogene*, **2004**, *23*(17): 2996-3004.
[70] van Royen, M.E., *et al.*, Stepwise androgen receptor dimerization. *J. Cell Sci.*, **2012**.
[71] Brooks, J. *Anatomy of the lower urinary tract and male genitalia*, in *Campbell's Urology*, A.R. P. Walsh, E. Darracott Vaughan, A. Wein, Editor 2002, Saunders: London. pp. 41-80.
[72] Aumuller, G.; Seitz, J. Protein secretion and secretory processes in male accessory sex glands. *Int. Rev. Cytol.*, **1990**, *121*: 127-231.
[73] Hunter, J., *Observations on Certain Parts of the Animal Oeconomy*1786, Sold at: 13 Castle-Street, Leister-Square, London, England.
[74] Cunha, G.R., *et al.*, Hormonal, cellular, and molecular regulation of normal and neoplastic prostatic development. *J. Steroid Biochem. Mol. Biol.*, **2004**, *92*(4): 221-236.
[75] CRUK, *Cancer Stats*.
[76] Shepherd, D.L.; Harris, M.; Thompson, I. *Radical Prostatectomy in the management of localised prostate cancer*, in *Prostate Cancer: Biology, Diagnosis and Management*, K. Syrigos, Editor 2001, Oxford University Press: Oxford. pp. 229-242.
[77] Labrie, F., *et al.*, Science behind total androgen blockade: from gene to combination therapy. *Clin. Invest. Med.*, **1993**, *16*(6): 475-492.
[78] Shang, Y.; Myers, M.; Brown, M. Formation of the androgen receptor transcription complex. *Mol. Cell*, **2002**, *9*(3): 601-610.
[79] Feldman, B.J.; Feldman, D. The development of androgen-independent prostate cancer. *Nat. Rev. Cancer*, **2001**, *1*(1): 34-45.
[80] Taplin, M.E., *et al.*, Mutation of the androgen-receptor gene in metastatic androgen-independent prostate cancer. *N. Engl. J. Med.*, **1995**, *332*(21): 1393-1398.

[81] Marcelli, M., *et al.*, Androgen receptor mutations in prostate cancer. *Cancer Res.*, **2000**, *60*(4): 944-949.

[82] Koochekpour, S. Androgen receptor signaling and mutations in prostate cancer. *Asian J. Androl.*, **2010**, *12*(5): 639-657.

[83] Gottlieb, B., *et al.*, The androgen receptor gene mutations database (ARDB): 2004 update. *Hum. Mutat.*, **2004**, *23*(6): 527-533.

[84] Gottlieb, B., *et al.*, The androgen receptor gene mutations database: 2012 update. *Hum. Mutat.*, **2012**, *33*(5): 887-894.

[85] Steinkamp, M.P., *et al.*, Treatment-dependent androgen receptor mutations in prostate cancer exploit multiple mechanisms to evade therapy. *Cancer Res.*, **2009**, *69*(10): 4434-4442.

[86] Shi, X.B., *et al.*, Functional analysis of 44 mutant androgen receptors from human prostate cancer. *Cancer Res.*, **2002**, *62*(5): 1496-1502.

[87] Hay, C.W.; McEwan, I.J. The impact of point mutations in the human androgen receptor: classification of mutations on the basis of transcriptional activity. *PLoS One*, **2012**, *7*(3): p. e32514.

[88] Buchanan, G., *et al.*, Structural and functional consequences of glutamine tract variation in the androgen receptor. *Hum. Mol. Genet.*, **2004**, *13*(16): 1677-1692.

[89] Hsiao, P.W., *et al.*, The linkage of Kennedy's neuron disease to ARA24, the first identified androgen receptor polyglutamine region-associated coactivator. *J. Biol. Chem.*, **1999**, *274*(29): 20229-20234.

[90] Moilanen, A., *et al.*, The presence of a transcription activation function in the hormone-binding domain of androgen receptor is revealed by studies in yeast cells. *FEBS Lett.*, **1997**, *412*(2): 355-358.

[91] Buchanan, G., *et al.*, Mutations at the boundary of the hinge and ligand binding domain of the androgen receptor confer increased transactivation function. *Mol. Endocrinol.*, **2001**, *15*(1): 46-56.

[92] Zhou, Z.X., *et al.*, A ligand-dependent bipartite nuclear targeting signal in the human androgen receptor. Requirement for the DNA-binding domain and modulation by NH2-terminal and carboxyl-terminal sequences. *J. Biol. Chem.*, **1994**, *269*(18): 13115-13123.

[93] Doesburg, P., *et al.*, Functional *in vivo* interaction between the amino-terminal, transactivation domain and the ligand binding domain of the androgen receptor. *Biochemistry*, **1997**, *36*(5): 1052-1064.

[94] Dehm, S.M., *et al.*, Splicing of a novel androgen receptor exon generates a constitutively active androgen receptor that mediates prostate cancer therapy resistance. *Cancer Res.*, **2008**, *68*(13): 5469-5477.

[95] Hu, R., *et al.*, Ligand-independent androgen receptor variants derived from splicing of cryptic exons signify hormone-refractory prostate cancer. *Cancer Res.*, **2009**, *69*(1): 16-22.

[96] Ceraline, J., *et al.*, Constitutive activation of the androgen receptor by a point mutation in the hinge region: a new mechanism for androgen-independent growth in prostate cancer. *Int. J. Cancer*, **2004**, *108*(1): 152-157.

[97] Callewaert, L., *et al.*, Differential effect of small ubiquitin-like modifier (SUMO)-ylation of the androgen receptor in the control of cooperativity on selective *versus* canonical response elements. *Mol. Endocrinol.*, **2004**, *18*(6): 1438-1449.

[98] Veldscholte, J., *et al.*, The androgen receptor in LNCaP cells contains a mutation in the ligand binding domain which affects steroid binding characteristics and response to antiandrogens. *J. Steroid Biochem. Mol. Biol.*, **1992**, *41*(3-8): 665-669.

[99] Berrevoets, C.A.; Veldscholte, J.; Mulder, E. Effects of antiandrogens on transformation and transcription activation of wild-type and mutated (LNCaP) androgen receptors. *J. Steroid Biochem. Mol. Biol.*, **1993**, *46*(6): 731-736.

[100] Han, G., *et al.*, Hormone status selects for spontaneous somatic androgen receptor variants that demonstrate specific ligand and cofactor dependent activities in autochthonous prostate cancer. *J. Biol. Chem.*, **2001**, *276*(14): 11204-11213.

[101] McDonald, S., *et al.*, Ligand responsiveness in human prostate cancer: structural analysis of mutant androgen receptors from LNCaP and CWR22 tumors. *Cancer Res.*, **2000**, *60*(9): 2317-2322.

[102] Duff, J.; McEwan, I.J. Mutation of histidine 874 in the androgen receptor ligand-binding domain leads to promiscuous ligand activation and altered p160 coactivator interactions. *Mol. Endocrinol.*, **2005**, *19*(12): 2943-2954.

[103] Askew, E.B., *et al.*, Modulation of androgen receptor activation function 2 by testosterone and dihydrotestosterone. *J. Biol. Chem.*, **2007**, *282*(35): 25801-25816.

[104] Brooke, G.N.; Parker, M.G.; Bevan, C.L. Mechanisms of androgen receptor activation in advanced prostate cancer: differential co-activator recruitment and gene expression. *Oncogene*, **2008**, *27*(21): 2941-2950.

[105] Powell, S.M., *et al.*, Mechanisms of androgen receptor signalling *via* steroid receptor coactivator-1 in prostate. *Endocr. Relat. Cancer*, **2004**, *11*(1): 117-130.

[106] He, B., *et al.*, Dependence of selective gene activation on the androgen receptor NH2- and COOH-terminal interaction. *J. Biol. Chem.*, **2002**, *277*(28): 25631-25639.

[107] Callewaert, L., *et al.*, Dual function of an amino-terminal amphipatic helix in androgen receptor-mediated transactivation through specific and nonspecific response elements. *J. Biol. Chem.*, **2003**, *278*(10): 8212-8218.

[108] Lu, S.; Tsai, S.Y.; Tsai, M.J. Regulation of androgen-dependent prostatic cancer cell growth: androgen regulation of CDK2, CDK4, and CKI p16 genes. *Cancer Res.*, **1997**, *57*(20): 4511-4516.

[109] Gregory, C.W., *et al.*, Androgen receptor regulation of G1 cyclin and cyclin-dependent kinase function in the CWR22 human prostate cancer xenograft. *J. Androl.*, **2001**, *22*(4): 537-548.

[110] Chen, Y.; Clegg, N.J.; Scher, H.I. Anti-androgens and androgen-depleting therapies in prostate cancer: new agents for an established target. *Lancet Oncol.*, **2009**, *10*(10): 981-991.

[111] de Bono, J.S., *et al.*, Abiraterone and increased survival in metastatic prostate cancer. *N. Engl. J. Med.*, **2011**, *364*(21): 1995-2005.

[112] Richards, J., *et al.*, Interactions of abiraterone, eplerenone and prednisolone with wild-type and mutant androgen receptor: a rationale for increasing abiraterone exposure or combining with MDV3100. *Cancer Res.*, **2012**.

[113] Tran, C., *et al.*, Development of a second-generation antiandrogen for treatment of advanced prostate cancer. *Science*, **2009**, *324*(5928): 787-790.

[114] Neckers, L.;Workman, P. Hsp90 molecular chaperone inhibitors: are we there yet? *Clin. Cancer Res.*, **2012**, *18*(1): 64-76.

[115] Lamoureux, F., *et al.*, A novel HSP90 inhibitor delays castrate-resistant prostate cancer without altering serum PSA levels and inhibits osteoclastogenesis. *Clin. Cancer Res.*, **2011**, *17*(8): 2301-2313.

[116] Zoubeidi, A.; Chi, K.; Gleave, M. Targeting the cytoprotective chaperone, clusterin, for treatment of advanced cancer. *Clin. Cancer Res.*, **2010**, *16*(4): 1088-1093.

Send Orders of Reprints at reprints@benthamscince.net

Advances in Genome Science, Vol. 1, 2013, 66-100

CHAPTER 4

MUTYH Associated Polyposis (MAP)

M.L.M. Binderup* and M.L. Bisgaard

Department of Cellular and Molecular Medicine, University of Copenhagen, Denmark

Abstract: MUTYH Associated Polyposis (MAP) is an autosomal recessive predisposition to polyposis and colorectal cancer (CRC) caused by biallelic mutations in the base excision repair (BER) gene *MUTYH*. Phenotypically, MAP is difficult to distinguish from other hereditary CRC syndromes like Familial Adenomatous Polyposis (FAP) and to a lesser extent Hereditary Non-Polyposis Colorectal Cancer (HNPCC), known to be caused by germline mutations in the *APC* and *Mismatch Repair* (*MMR*) genes, respectively. Here we review the main genetic and clinical aspects of MAP and aim to provide a broad, up-to-date overview of existing findings regarding MAP.

The exact role of *MUTYH* in CRC tumorgenesis is not fully determined, but a defective BER system due to germline *MUTYH* mutations leads to somatic mutations in the *APC* gene. Furthermore, cooperation between the BER and MMR systems exists, and possibly monoallelic defects in both pathways are of significance to CRC development. The two most common *MUTYH* variants, *Y179C* and *G396D*, both generate dysfunctional gene products, but *Y179C* has the most severe functional consequences and possibly causes a more severe phenotype.

Median clinical onset of MAP is 47 years of age, and due to the recessive mode of inheritance, most have already developed CRC. Typically MAP patients develop between 10-100 colorectal adenomas, although polyposis may not be obligatory for MAP. Typical FAP and HNPCC associated extracolonic manifestations are not common in MAP, except for upper gastro-duodenal polyposis which can be part of the phenotype, but more seldom than in FAP. Recommended MAP surveillance is colonoscopy with polypectomy from 20-25 years of age. In the future, genetic testing as well as surveillance may be targeted according to specific ethnic *MUTYH* mutation variants and genotype-phenotype correlations.

Keywords: Colorectal Cancer, MUTYH Associated Polyposis, *The MUTYH gene*, Base excision repair, (Attenuated) Familial Adenomatous Polyposis, Hereditary Non-Polyposis Colorectal Cancer.

INTRODUCTION

Worldwide, colorectal cancer (CRC) is the second and third most prevalent cancer

***Address correspondence to Marie Louise Mølgaard Binderup:** Department of Cellular and Molecular Medicine, University of Copenhagen, Panum Institute 24.4, Blegdamsvej 3, 2200 Copenhagen N, Denmark; Tel: +4526363991; Fax: +4535327845; E-mail: mlmp@sund.ku.dk

Christian Neri (Ed)

among women and men, respectively [1]. Hereditary factors are found in about 30% of CRC patients [2, 3], and for some the specific genetic background is well characterized, major CRC syndromes being Familial Adenomatous Polyposis (FAP), Hereditary Non-Polyposis Colorectal Cancer (HNPCC), and *MUTYH* Associated Polyposis (MAP) the latter being the focus of this review. Fig. (**1**) shows a delimitation of the patient groups referred to.

In this text the patient groups are defined as follows:

On a clinical basis:

Polyposis patients: Patients presenting with adenomas in the colon and/or rectum.

CRC patients: Patients presenting with carcinomas in the colon and/or rectum, with or without previous/synchronous occurrence of polyposis.

On a mixture of clinical and genetic basis:

MAP patients: (OMIM 608456)
Polyposis and/or CRC patients with detected biallelic germline mutations in the *MUTYH* gene.

FAP patients: (OMIM 175100)
Polyposis patients presenting with > 100 adenomas in the colon and/or rectum and – in this paper - with a detected germline mutation in the *APC* gene.

AFAP patients: (OMIM 175100)
Polyposis patients presenting with < 100 adenomas in the colon and/or rectum.

HNPCC (Hereditary Non-Polyposis patients Colorectal Cancer)
(OMIM: HNPCC1: 120435, HNPCC2: 609310)
(Also known as Lynch syndrome):
Several guidelines for the clinical diagnosis exist (e.g. The Amsterdam criteria I and II and the Bethesda criteria). Familial disposition for CRC and possibly a detected germline mutation in a *Mismatch Repair (MMR)* gene.

Figure 1: Delimitation of patient groups

FAP is an autosomal dominant disease which entails a near 100% risk of CRC and development of more than 100 and up to thousands of colorectal adenomas at young ages, as well as extracolonic manifestations such as gastric and duodenal adenomas, desmoid tumors, and congenital hypertrophy of the retinal pigment epithelium (reviewed in [4-8]). FAP is caused by germline mutations in the *APC* gene, and genotype-phenotype correlations, in which specific *APC* gene mutations are associated with particular manifestations, occur [9-11]. The milder phenotypical FAP variant, Attenuated Familial Adenomatous Polyposis (AFAP), found in about 8% of registered FAP families [12], is mainly associated with *APC* mutations in the extreme ends of the gene, and in an alternatively spliced region of exon 9 [10, 13]. AFAP is distinguished from FAP by the development of less than 100 colorectal adenomas, fewer extracolonic manifestations and later CRC development [5-8, 10, 12]. In as many as 30% of both FAP and AFAP patients no germline mutations can be found in the *APC* gene [7, 12, 13]. HNPCC is an autosomal dominant disease clinically characterized by particularly CRC and endometrial cancer development at a young age [14-17]. HNPCC can be caused by germline mutations in one of the *Mismatch Repair (MMR)* genes, but in more than 50% of HNPCC patients no causative mutation can be found [18]. Clinical and genetic features of HNPCC have previously been thoroughly reviewed in [4, 15-17, 19].

In 2002, the significance of mutations in the *MUTYH* gene and the CRC and Polyposis predisposition syndrome *MUTYH* Associated Polyposis (MAP) was discovered [20]. Since then, many aspects of MAP have been investigated, and the question of how, this new knowledge can be clinically implemented in genetic counseling and screening of individuals at risk of developing MAP, now stands to be answered. Here we review the main genetic aspects of MAP, including analysis of mutational functional consequences, current nomenclature for *MUTYH* germline mutations, and outlining of specific ethnic allelic frequencies of *MUTYH* variants. In addition, we give an overview of clinical aspects of the syndrome, the differential diagnoses as well as clinical surveillance recommendations. We believe that this review provides a broad, up-to-date overview of existing findings regarding MAP.

GENETIC BACKGROUND AND PATHOGENESIS OF MAP

The MUTYH Gene

The MUTYH glycosylase is encoded by the *MUTYH* gene, located on the short arm of chromosome 1 (1p32.1-p34.3). The gene consists of 16 exons and encodes a protein of 535 amino acids, the MUTYH glycosylase. Characterization of the *MUTYH* gene and functional variants has been reviewed in [21-26].

Exon 3 in *MUTYH* is alternatively spliced in various ways, generating different MUTYH transcripts [21, 22, 27]. In accordance with the Human Genome Variation

Society nomenclature rules, the longest existing *MUTYH* transcript (NM_001128425.1) will be used as coding DNA reference sequence, as outlined by Leiden Open Variation Database (LOVD) at http://www.LOVD.nl/MUTYH [27, 28]. The use of the longest existing *MUTYH* transcript as coding DNA reference sequence has replaced the previous commonly used *MUTYH* transcript (NM_001048171). Consequently, the amino acid numbering in this paper differs from the one used in many previous and even some current papers. An overview of the new and old terms regarding some of the most common *MUTYH* mutations mentioned in this paper can be found in Table **1**.

Table 1: Overview of nomenclature for some of the most common *MUTYH* germline mutations

Nomenclature from reference sequence NM_ 001048171	Nomenclature from reference sequence NM_001128425.1*
p.V22M	*p.V22M*
p.Y90X	*p.Y104X*
p.Y114H	*p.Y128H*
p. Y165C	*p.Y179C*
p.R168H	*p.R182H*
p.R168C	*p.R182C*
p.R171Q	*p.R185Q*
p.G175E	*p.G189E*
p.R227W	*p.R241W*
p.R231C	*p.R245C*
p.V232F	*p.V246F*
p.Q324H	*p.Q338H*
p.E369fsX437	*p.E383fsX451*
p. G382D	*p. G396D*
p.P391L	*p.P405L*
p.A459T	*p.A473T*
p.E466X	*p.E480X*
p.S501F	*p.S515F*
c.934-2A>G	*c.976-2A>G*
c.1391delAGG	*c.1433delAGG*
c.1103delGGA	*c.1145delGGA*
c.1395delGGA	*c.1437delGGA*
c.1103delC	*c.1145delC*
c.1186_1187insGG	*c.1228_1229dupGG*
c.IVS10-2A>G	*c.IVS10-2A>G*

*The nomenclature for *MUTYH* mutations is based on referral to the longest existing *MUTYH* transcript and were found at http://www.LOVD.nl/MUTYH

Base Excision Repair

The *MUTYH* gene product, the MUTYH glycosylase, is part of the Base Excision Repair (BER) system, an important part of cells′ defense against oxidative DNA damage. The BER system and the functional role of MUTYH have previously been reviewed in detail by [21-26, 29-32]. Oxidative DNA damage, caused by reactive oxygen species (ROS) produced during aerobic metabolism and exposure to certain chemicals or radiation, constantly threatens the integrity of cellular DNA. The oxidized base 7,8-dihydroxy-8-oxoguanine (8-oxo-G) is one of the most stable and mutagenic products of oxidative DNA damage. 8-oxo-G is often mistakenly paired with adenine (A), resulting in the appearance of Guanine:Cytosine > Thymidine:Adenine (G:C > T:A) transversions at the next round of DNA replication, as the detection of stable 8-oxo-G:A base-pairs is missed by the replicative DNA polymerases [21-26, 29-32]. The human DNA-damage-specific glycosylases OGG1, MUTYH and MTH1, which are central enzymes in BER pathway, function by specifically recognizing and facilitating the removal of 8-oxo-G [21-24, 26, 29-32]. The MUTYH glycosylase acts as the third level of BER, as it postreplicativly excises the misincorporated A opposite 8-oxo-G [21, 22, 24, 26, 31, 32]. For this reason, defective MUTYH function is associated with an increased frequency of G:C > T:A transversions [21]. A number of studies of the human MUTYH glycosylase, both *in vitro* and *in vivo*, have demonstrated that it directly interacts with various proteins involved in other DNA repair pathways (reviewed by [21, 32].) Inactivation of MUTYH has been associated with various cancer forms including lung cancer, bladder cancer, duodenal cancer, gastric cancer, CRC, and endometrial cancer [21, 22, 33-38].

Several studies have also screened the *OGG1* and *MHT1* genes in Polyposis patients without any significant findings of association with Polyposis or CRC phenotypes. [20, 39-43]. However, *OGG1* variants have been demonstrated to be significantly associated with a multiple adenoma phenotype [44] and the development of sporadic CRC [45], although the latter association was of borderline significance and should be further investigated. Also, one study identified a Polyposis and CRC patient as being compound heterozygous for a *OGG1* and a *MUTYH* mutation, indicating the possibility of digenic inheritance for these predispositions [46].

Development of Colorectal Adenomas and CRC Tumorgenesis in MAP

Mutations in the *APC* tumor suppressor gene play a large role in CRC tumorgenesis due to the gene's gatekeeper role and its involvement in many cellular processes [9, 10]. Development of colorectal adenomas is likely to be initiated by an *APC* gene left dysfunctional as a result of germline or somatic mutations [10].

In several studies of CRC patients (N= 2,268), a significant association between biallelic germline mutations in the *MUTYH* gene and the development of CRC has been determined [47, 48]. At present it has not been fully determined why *MUTYH* germline mutations predispose to colorectal adenomas in particular. However, the number of spontaneous somatic C:G > T:A transversions in the *APC* gene is significantly greater in tumor cells with biallelic *MUTYH* germline mutations compared to tumor cells without *MUTYH* mutations [20, 39, 41, 49, 50]. The DNA sequence adjacent to the sites of G:C > T:A transversions in the 3' end of the *APC* gene appears to be of significance to the specificity of the MUTYH glycosylase. A significantly higher occurrence of GAA sites has repeatedly been demonstrated 3' to the G:C > T:A transversions, thus creating stop codons, and resultantly a truncated APC protein. These results are seen even though the G:C > T:A transversions theoretically could occur at any other G:C site in the *APC* gene [20, 39, 51]. However, the significance of these findings and the question as to why GAA sites seem to be prone to G:C > T:A transversions, remain unclear and call for further investigations.

Both the high number of GAA sites in the *APC* gene compared to other key tumorgenesis genes, as well as the considerable exposure to ROS in the gastrointestinal tract, could be part of the explanation of why germline *MUTYH* mutations, and subsequently somatic APC mutations, are associated with development of particularly CRC [26].

Molecular features in MAP tumors are characteristic compared to carcinomas from sporadic CRC, FAP or HNPCC tumors [49], which could potentially be used in CRC classification [52]. Characteristic molecular profiles of colorectal tumors, both adenomas and carcinomas taken from MAP patients include: C:G > T:A transversions in the *APC* gene and the proto-oncogene *K-Ras* [49, 52]. Comparable with somatic *APC* mutations, significantly larger numbers of somatic mutations in the *K-Ras* gene have been found in MAP tumors compared to tumors without

MUTYH mutations [50, 52]. The described *K-Ras* mutations have all been identical G > C transversions in codon 12, *G12C* [49, 50].

The issue of a possible significance of the *MUTYH* gene in sporadic CRC has been only sparsely addressed, and with contradictory outcomes [40, 53, 54].

Interaction Between the *MUTYH* Gene and the Mismatch Repair Genes

The Mismatch Repair (MMR) system functions postreplicativly correcting DNA errors, which occur during DNA replication; key proteins include MutL homologs (MLH1 and PMS2) and MutS homologs (MSH2, MSH3, MSH6), the latter group forming two heterodimeric complexes [17, 21, 55].

Presumably cooperation between the BER and the MMR systems exists, as *in vitro* experiments have shown that MUTYH physically interacts with the MSH2/MSH6 heterodimeric complex [56, 57]. The MSH2/MSH6 complex stimulates the activity of the MUTYH glycosylase by enhancing the affinity of MUTYH for 8-oxo-G:A mismatched base pairs in the DNA [56, 57]. Several *MUTYH* germline mutations are shown to influence the interaction between MUTYH and MSH6, followed by a massive decrease in activity of the MUTYH glycosylase [57, 58]. Mutations in one or more of the genes involved in the two systems possibly affect the repair of DNA damage caused by 8-oxo-G and increases risk of CRC development. Two studies have found significantly higher frequencies of monoallelic *MUTYH* mutation carriers among CRC patients who also had a *MSH6* mutation (overall 7/46 = 15%) in comparison to CRC patients without MMR mutations (1/184 =0.1%) [59, 60]. In contrast, several studies have found no *MUTYH* mutations in groups of CRC patients with *MMR* mutations (overall N= 354) or have found *MUTYH* mutations in the same frequencies as in healthy controls [61-64]. However, due to small numbers the mentioned studies do not provide substantial data for any conclusions, for which reason further investigations need to be carried out.

Allelic Frequencies of MUTYH Germline Mutations in Differing Populations

An overview of the most commonly identified germline mutations in the *MUTYH* gene to date can be found at the LOVD (http://www.LOVD.nl/MUTYH).

Specific mutations in the *MUTYH* gene are found in different populations at various allelic frequencies, see Fig. **2**. In European populations the two missense mutations

Y179C and *G396D* are most frequent, and the allelic frequencies of these variants found among MAP patients are much higher compared to background populations, see Fig. **3**.

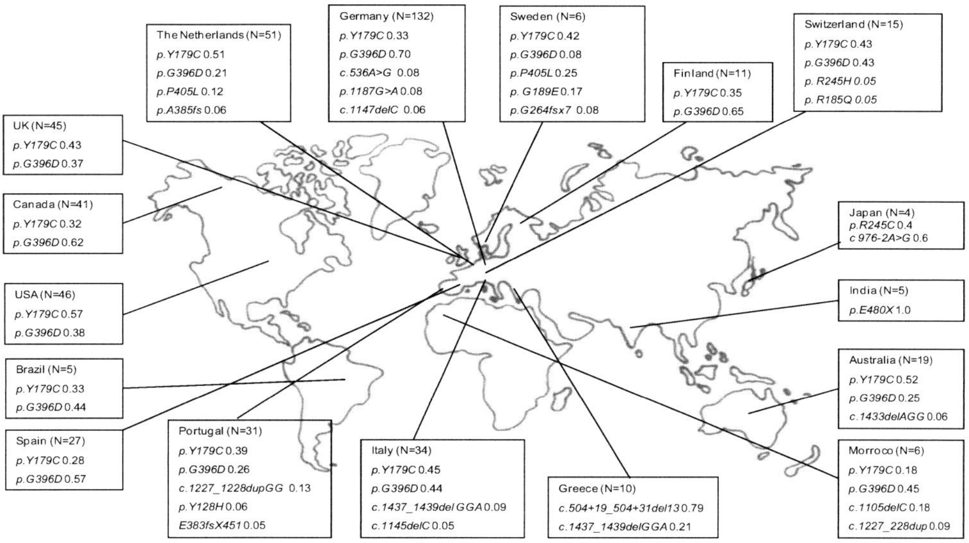

N= Number of biallelic and monoallelic *MUTYH* mutation carriers pooled from studies of *MUTYH* mutation carriers of the respective nationalities. Included only data with pooled number of persons > 3 and only mutations believed to be pathogenetic, and which are found with an allelic frequency of > 0,05 in mutation carriers from the respective countries.

Based on data colleced from the following studies: Al-Tassan et al., 2002; Jones et al., 2002; Enholm et al., 2003; Sieber et al., 2003; Sampson et al., 2003; Halford et al., 2004; Fleischmann et al., 2004; Gismondi et al., 2004; Isidro et al., 2004; Venesio et al., 2004; Wang et al., 2004; Aceto et al., 2005; Jo et al., 2005; Kairupan et al., 2005; Miyaki et al., 2005; Nielsen et al., 2005; Leite et al., 2005; Aretz et al., 2006; Kanter-Smoler et al., 2006; Niessen et al., 2006; Russel et al., 2006, Ajith Kumar et al., 2007; Balaguer et al., 2007; Fostira et al., 2010; Morak et al., 2010; Nascimbeni et al., 2010; Casper et al., 2011; Pitroski et al., 2011; Laarabi et al., 2012

Figure 2: Mean allelic frequencies of various *MUTYH* germline mutations in carriers

In Asian populations, *Y179C* and *G396D* do not seem to be of significance to Polyposis, as neither has been found in Asian Polyposis patients or in corresponding background populations [44, 45, 58, 65-67]. In studies of Korean and Japanese Polyposis patients (N=97), 7.2% were biallelic carriers of other germline mutations in the *MUTYH* gene [58, 67]. Other studies of Korean and Singaporean Polyposis patients (N= 63) failed to find any *MUTYH* mutations after screening of coding regions in the entire gene, although these results may be biased due to small sample size (Overall N=63) [45, 65, 66]. To date, five unrelated Indian MAP patients have been identified, all were homozygote for the missense mutation *E480X* [39, 68, 69]. However, *MUTYH* mutations do not seem to be of significant importance to Polyposis development among Indian individuals, as a case-control study of 120 Polyposis patients and 100 controls, merely one case and one control were heterozygote for *E480X*, while no other *MUTYH* mutations were found [70].

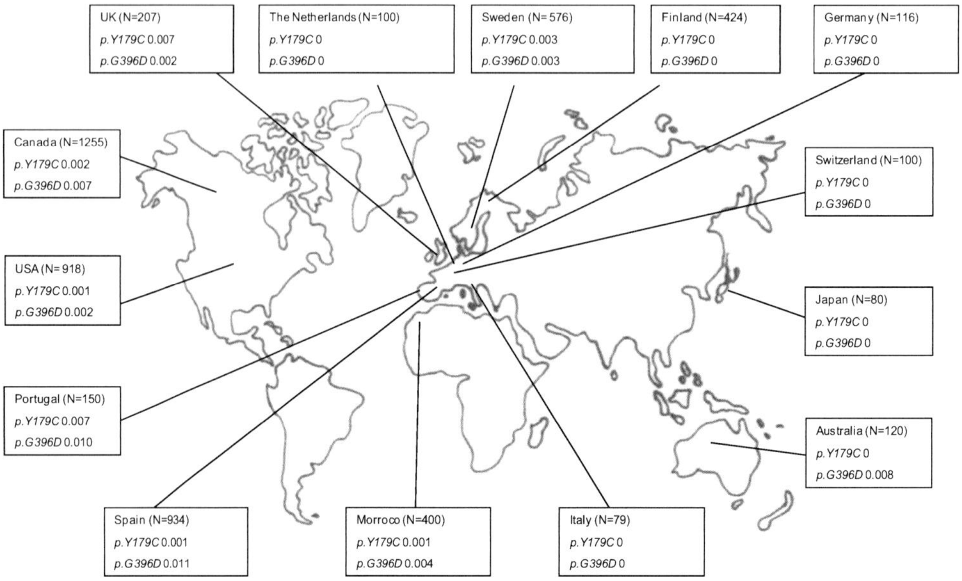

N= Number of tested individuals in the population. The individuals tested belong to control groups without Polyposis or CRC.

Based on data colleced from the following studies : Al-Tassan et al., 2002; Enholm et al., 2003; Sieber et al., 2003; Aceto et al., 2004; Croitoru et al., 2004; Isidro et al., 2004; Kairupan et al., 2005; Leite et al. 2005; Miyaki et al., 2005; Peterlongo et al., 2005; Zhou et al., 2005; Aretz et al., 2006; Kanter-Smoler et al., 2006; Niessen et al., 2006; Russel et al., 2006; Balaguer et al., 2007; Laarabi et al., 2011

Figure 3: Mean Allelic Frequencies of *p.Y179C* and *p.G396D* in Background Populations

In addition, other *MUTYH* polymorphisms without apparent pathogenic importance have been found, the most frequent *V22M*, *Q388H* and *S515F,* which have been found in equivalent allelic frequencies in healthy control groups and MAP patients [20, 40-42, 47, 51, 59, 71-80]. Knowledge of the characteristic population specific mutations variants makes it possible to target the genetic MAP analysis in accordance with ethnic background [62, 70].

Functional Consequences of Specific Germline MUTYH Mutations

The two most common *MUTYH* mutation variants, the *Y179C* and *G396D*, are both situated in regions of importance to the MUTYH glycosylase's A:8-oxo-G mispair recognition and binding [32, 81-84]. Accordingly, functional studies of corresponding *E.coli* and murin variants have indicated compromised substrate recognition and glycosylase activity as mutational consequences [20, 83, 85].

Several studies have found marked differences in functional consequences of the *Y179C* and *G396D* variants: the defective MUTYH function in human cell lines with biallelic *Y179C* mutations results from both significantly reduced levels of

MUTYH protein (protein levels of 5-10% compared to wild-type levels) as well as from reduced binding and cleavage ability towards the mispaired substrates [86, 87]. In contrast, protein instability does not seem to be a consequence of *G396D* mutations, as MUTYH protein levels in human cell lines homozygous for this variant have been found to be equivalent to the levels of wild type protein [86]. The consequences of biallelic *G396D* mutations are both reduced binding activity of mispaired substrates (about 50% of the wild-type MUTYH activity), as well as lower rates of repair, compared to the wild type [86, 87]. Several studies describe large differences in the repair activities of *Y179C* and *G396D* variant glycosylases [20, 83, 87, 88]. Goto *et al.* demonstrated a difference in repair activity of the *Y179C* and *G396D* variants of 5% and 15%, respectively [88]. Likewise, the variants corresponding to *G396D* and *Y179C* show a 6-fold and 80-fold slower rate of adenine removal from a G: A substrate, respectively [83]. The different functional effects of these mutations are likely to account for some of the observed genotype-phenotype correlations among MAP patients (see later).

Functional studies of additional *MUTYH* mutations have also been conducted, although none as comprehensive as for *Y179C* or *G396D*. The variants *R182C*, *R182H*, *R185Q* and *G189E*, all are located in the MUTYH catalytic domain, are considered to induce functional MUTYH changes comparable to those observed in cell lines with *Y179C* variants [47, 62, 73, 77]. Functional analyses of *MUTYH* mutation variants located close to or within the MSH6-binding domain, *R241W*, *V246F*, and *R245C*, have been performed. The variants are all thought to compromise the interaction between the MUTYH and MSH6 and consequently reduce MUTYH glycosylase activity [57, 58].

The Significance of the *MUTYH* Gene in Relation to Development of Polyposis

Results from several studies indicate, that mutations in the *MUTYH* gene are often the disease causing factor in Polyposis. In AFAP, the contributions of germline mutations in the *APC* gene and the *MUTYH* gene seem to be fairly equal or even with an overweight of *MUTYH* mutations as the causative genetic background [12, 41].

In numerous studies of Polyposis patients (overall N=1,210), it has been found that 5-22% of patients with 3-100 adenomas and about 8-17% of those with over 100 adenomas had biallelic mutations in the *MUTYH* gene [12, 41, 89, 90].

Clinical Features of MAP

In recent years it has become clear that MAP has a variable phenotypic expression, and mapping of its clinical range and natural history is still the object of profound investigation worldwide. Table **2** provides an overview of phenotypic characteristics of MAP, using information gathered from all reported studies of MAP patients in which more than five MAP patients were included and clinically thoroughly described.

Table 2: Clinical features of identified MAP Patients

	Sieber *et al.*, 2003, [41]	**Sampson *et al.*, 2003, [68]**	**Kanter-Smoler *et al.*, 2003, [77]**	**Isidro *et al.* 2004, [73]**	**Wang *et al.* 2004, [62]**	**Nielsen *et al.*, 2005, [94]**	**Russell *et al.*, 2006, [47]**	**Balaguer *et al.*, 2007, [105]**	**Vogt *et al.*, 2009, [33]**	**Nascimbeni *et al.*, 2010, [102]**	**Morak *et al.*, 2010, [90]**	**Nieuwenhuis *et al.*, 2011, [103]**
Number of biallelic *MUTYH* germline mutation carriers	N=14	N=25	N= 6	N=21	N=18	N=40	N=7	N = 8	N = 276	N = 11	N = 33	N = 254
Mean age in years at polyposis diagnosis (Range)	56 (45-59)*[1] 48 (30-70)*[2]	46 (13-65)	39.5 (23 – 59)	50 (36-68)	46.2 (37-63)	45 (21-67)	48 (33-60)	57.7 *[3]	45 (0 -70) *[4]	50 (40 – 69)	43 (28-64)	44.8 (12-77)
Number of colorectal adenomas	43%: 18-100 57%: 100-1000	44%: 10-100 36%: >100 20%: Unspecified	33%: 20-50 33%: 100-1000 17%: >1000 17%: "Multiple"	71%:10-100 19%:100-1000 5%: >1000	35%: Unspecified 25%: 100-500 19%: 20-49 19%: 50-99	42%: "Multiple" 29%: 10-99 29% 100-1000	<100* (Only patients with 5-99 adenomas were tested)	50%: < 15 25%: > 15	21%: 50-100 20%: 10-50 20%:"Multiple" or "polyposis" 17%: 100-500 8%: "Unknown number" 6%: 1-10 3%: 500-1000	45%: 10- 50 36%: 100-200 9%: 50-100 9%: 500-1000	Not reported	63%: < 100 20%: > 100 15%: Exact number unknown 2%: no adenomas
Number of patients with CRC (%)	7 (50%)	12 (48%)	4 (67%)	13 (62%)	6 (38%)	26 (65%)	5 (71%)	- *[5]	158 (43%)	8 (73%)	11 (33%)	147 (58%)

In this table only studies in which the numbers of tested MAP patients are > 5 are included.
The tested individuals are all *APC* germline mutation negative. The individuals tested are either probands or call-up patients with colorectal adenomas.
*[1] Subjects with 10-100 adenomas, *[2] Subjects with 100-1000 adenomas, *[3] Age at CRC diagnosis *[4] Age at molecular diagnosis (confirmation of biallelic *MUTYH* mutations), *[5] Only subjects with CRC included

Mode of Inheritance

Due to the autosomal recessive mode of inheritance, MAP patients often present as sporadic cases, and in cases with family histories typically with affected siblings but unaffected parents [12, 20, 39-41, 47, 68, 73, 77, 89, 91-95]. Some studies have however reported family histories appearing to follow a dominant mode of inheritance in about 15-30% of the studied MAP patients [64, 73, 77, 80, 89, 93]. It is possible that a recessive trait as MAP could display a pseudo-dominant mode of inheritance, mimicking dominant inheritance, due to a relatively high frequency of heterozygote mutation carriers in some populations, especially in cases of consanguinity in several generations. On the other hand, some have proposed a co-dominant model for mode of inheritance, suggesting an increased CRC risk for monoallelic *MUTYH* germline mutation carriers compared to non-carriers, as will be discussed further [48, 51, 62, 71, 72, 96-100]. Furthermore, existence of phenocopies in some of these families (*i.e.* family members presenting with a similar clinical appearances as a mutation carrier, but without having inherited the family mutation) may influence the apparent mode of inheritance.

Table 3: Frequencies of *MUTYH* germline *Y179C* and *G396D* mutation carriers among CRC patients and back-ground populations

	Enholm *et al.*, 2003, [100]	**Croitoru *et al.*, 2004, [71]**	**Fleischmann *et al.*, 2004, [72]**	**Wang *et al.*, 2004, [62]**	**Peterlongo *et al.*, 2005, [99]**	**Webb *et al.*, 2006, [123]**	**Küry *et al.*, 2007, [42]**	**Balaguer *et al.*, 2007, [105]**
Population	Finland	Canada	United Kingdom	The USA	The USA	United Kingdom	France	Spain
Carriers among CRC patients								
Number of tested CRC patients	N =1,042	N=1,238	N =358	N=444	N=238	N=2,561	N=1,024	N= 1,116
Monoallelic carriers	0.5 %	2.3 %	2.2 %	2.3 %	1.7 %	2.1%	2.3%	1.7%
Biallelic carriers	0.4 %	1.9 %	0.6 %	0.5 %	0.8 %	0.2%	0.1%	0.4%
Total percentage of *MUTYH* germline mutations carriers	0.9%	4.2%	2.8%	2.8%	2.5%	2.2%	2.4%	2.1%
Number of tested individuals	N= 424	N= 1,255	N=207*	Not studied	N=918	N=2,695	N=1,121	N=934
Monoallelic carriers	0	1.7%	1.9%	Not studied	0.8%	2.1%	1.8%	2.3%
Biallelic carriers	0	0	0	Not studied	0	0	0	0

*Based on data from studies by Al-Tassan *et al.*, 2002 and Sieber *et al.*, 2003

Determination of the true penetrance of biallelic germline *MUTYH* mutations would require large prospective population-based studies. However, so far biallelic

MUTYH mutations seem to be highly penetrant, as no unaffected individuals who were homozygote for germline *MUTYH* mutations have been found in studies of healthy controls [62, 71, 72, 98-100] and Table **3**.

Age at Clinical Diagnosis of MAP

The average age at clinical diagnosis among MAP patients is typically around 47 years (range: 13-72 years, N=106) [12, 62, 64, 68, 78, 79, 89, 90, 94, 95], Table **2**.

Characteristically, the age at clinical diagnosis is higher among patients with biallelic *MUTYH* germline mutations compared to Polyposis patients without *MUTYH* mutations [41, 62, 77, 92, 101]. This applies both to studied Polyposis patients with <100 adenomas [62, 77, 92] as well as to Polyposis patients with >100 adenomas, who are also negative for *APC* germline mutations [41, 101].

Development of Colorectal Adenomas

Studies of Polyposis patients have found the highest incidence of biallelic *MUTYH* mutations in groups of patients with between 15-100 adenomas [33, 41, 47, 67, 68, 73, 89, 102, 103], Table **2**. In studies of Polyposis patients, the phenotypes with regard to the number of adenomas are generally more severe in MAP patients compared to AFAP patients, but milder compared to those of classic FAP patients [41, 67, 77, 89, 93, 104]. One study of AFAP patients (N=140), comparing the clinical features of patients with *APC* (N=93) and biallelic *MUTYH* germline mutations (N=26), found no significant differences between the two groups [12]. However, this result might be biased due to the smaller sample size of the MAP patients compared to the *APC* mutation carriers. The morphology of the adenomas appears to be similar regardless of whether their occurrence is caused by germline mutations in the *MUTYH* or in the *APC* gene [41, 92]. Likewise, microadenomas have been found in patients with both genotypes [41, 49, 68].

It has been suggested that Polyposis may not be an obligatory phenotype of MAP, as some studies have found that almost half of the studied MAP patients (overall 15 of 35 patients) develop CRC without preceding Polyposis [48, 71, 100, 105]. Furthermore, some studies have reported biallelic *MUTYH* mutations in CRC patients with family histories suggestive of HNPCC (overall N=15) [62, 64, 90]. However larger systematic studies are needed to fully clarify the phenotypic variability of MAP and its clinical consequences.

Development of CRC

In several studies of CRC patients biallelic germline *MUTYH* mutations were found in 0.1-1.9% of all cases (Table **3**). Based on these results, the contribution of biallelic *MUTYH* mutations to CRC corresponds to or is even greater than that of FAP, FAP accounting for less than 0.1% of all CRC cases [106].

Among MAP patients, the average age of CRC onset is found to be about 48 years (range: 21-72 years) [12, 33, 47, 64, 68, 89, 93, 94, 103]. The frequency of patients with a synchronous CRC at time of diagnosis is greater among MAP patients compared to FAP patients [73, 77, 101]. Furthermore, FAP patients generally are diagnosed earlier than MAP patients facilitated by the dominant mode of inheritance of FAP and the use of Polyposis Registers in many countries. Therefore, the prophylactic treatment of FAP prevents CRC development in a higher number of patients.

There have been inconsistent results regarding the typical location of carcinoma in MAP patients. Over-representation of both right and left sided CRC has been demonstrated [12, 40, 47, 49, 72, 77, 94, 96, 105, 107].

Associated Cancers and Extracolonic Manifestations

Knowledge of natural history and association of MAP to extracolonic manifestations is important to further target MAP screening. Several extracolonic manifestations are associated with FAP and HNPCC, and have given rise to specific extracolonic screening measures [8, 19].

Extracolonic manifestations are reported to be less frequent among MAP patients compared to among FAP patients [77], and several studies have even failed to report any extracolonic features among examined MAP patients [39, 41, 62, 67, 72, 73, 91, 92, 101]. However, the methods of clinical examination were not specified in the applied studies, and bias could be suspected, as patient information is often gathered from several different databases without assurance of systematic clinical evaluation.

Conversely, in recent years, several studies have begun to focus on occurrence of extracolonic features in MAP patients. [38, 41, 47, 64, 68, 69, 77, 80, 89, 90, 94, 95, 100, 102, 104, 108-111], Table **4**. Unfortunately the numbers of examined MAP patients in the majority of these studies were small, and bias could result from difference in examination methods used in the individual studies, as these are only sparsely described. Also some studies were case reports [38, 69, 108-111], giving rise to possible report bias overestimating the frequency of some manifestations.

Vogt *et al.* recently performed a systematic examination of extracolonic manifestations in the largest cohort of MAP patients (N= 276) to date [33]. They found that MAP patients have an overall lifetime risk of 38% of developing an extraintestinal malignancy [33]. Even though data was retrospectively collected from three different countries with slightly different methods of examination, the study was very thorough and systematically performed, and all patients in this extensive cohort were molecularly confirmed biallelic *MUTYH* carriers [33].

Results of studies of extracolonic MAP manifestations suggest that upper gastro-duodenal polyposis can be a common part of the MAP phenotype, but in much lower frequencies than seen in FAP patients [33]. The most frequent extracolonic manifestation found in MAP patients was duodenal polyposis found in about a sixth of patients and gastric polyps found in about 12% (Table **4**). The observed incidence of gastric cancer in MAP patients was not significantly increased compared to the general population [33].

Table 4: Extracolonic manifestations reported in MAP Patients

Extracolonic manifestation	Overall number of MAP patients with manifestation/ number of MAP patients examined	Overall percentage of MAP patients with extracolonic manifestation	Reported by
Duodenal polyposis	52/ 301	17%	[33, 41,47, 77, 64, 90, 95, 96, 102, 104, 108]
Gastric polyps	23/ 187	12%	[33, 64, 95, 104]
Gallbladder polyps	1/33	3%	[90]
Duodenal cancer	2/ 150	1%	[33]
Gastric cancer	5/ 210	2%	[33, 96, 102]
Oesophagus cancer	2/ 150	1%	[33]
Breast cancer	10/ 334	3%	[33, 64, 96]
Benign breast tumors*[1]	2/ 48	4%	[33]

Table 4: contd...

Endometrial cancer	6/ 170	4%	[33, 38, 96, 111]
Benign endometrial tumors	3/ 48	6%	[33]
Ovary cancer	3/ 118	3%	[33]
Malignant skin lesions*2	16/ 329	5%	[33, 96, 100]
Benign skin lesions*3	18/ 170	11%	[33, 38, 69, 101, 109, 110]
Bladder cancer	4/ 276	1%	[33]
Pancreas cancer	2/ 17	12%	[79, 102]
Carcinoid tumors in the bowel	4 / 150	3%	[33]
Osteomas	2/ 14	14%	[101]
Dental cysts	1/ 14	7%	[101]
CNS tumor	2/ 55	4%	[79, 95]
Leukemia	1/ 6	17%	[101]
Lung cancer	2/ 276	1%	[33]
Lipomas	7/ 98	7%	[33]
CHPE (Congenital hypertrophy of the retinal pigment epithelium)	5/ 31	16%	[41, 64, 101]
Thyroid cancer	1/1*4	-	[109]

Studies of MAP patients where patients have later been reported in subsequent studies are not included.
*1 Only female MAP patients included, *2 Includes melanoma and basal cell carcinoma *3 Includes sebaceous gland tumors, pilomatricomas, and dermoid cysts,
*4 Case report

Increased risks of breast cancer and gynecological cancers among MAP patients have been suggested [94, 111, 112]. Breast cancer risk was not confirmed in 118 female MAP patients, where a trend towards an increased breast cancer risk was found, but the overall incidence not significantly increased compared to the general population [33]. Furthermore, no biallelic *MUTYH* carriers have been found in two studies of 691 and 303 breast cancer patients [113, 114]. Vogt *et al.* found a significantly increased risk of ovarian cancer, but no increased risk of endometrial cancer among female MAP patients as is seen for HNPCC patients [33]. Also, only one biallelic *MUTYH Y179C* or *G392D* mutation carrier has been found among overall 416 endometrial cancer patients, suggesting that no such association exists or is too small to be clinically relevant [38, 115]. To our knowledge no other MAP studies have reported ovarian cancers, and a possible association should be further investigated.

Many of the other reported extracolonic manifestations were reported in very low frequencies, and are therefore more likely to be present by chance, rather than being associated with *MUTYH* mutations. However, further prospective studies with more systematic and thorough investigation of MAP patients are needed to address this issue, especially determining risks of urological cancers and skin manifestations consistent with those seen in some HNPCC patients [116]. Also, further knowledge of especially upper gastro-duodenal MAP manifestations and its consequences is needed for adequate surveillance measures to be taken.

GENOTYPE-PHENOTYPE CORRELATIONS

Different genotypes may account for phenotypic variability among MAP patients, and of the two most common mutation variants, *G396D* seems to be associated with a milder phenotype compared to *Y179C* [105, 107, 117]. Two large meta-analyses of CRC case-control studies found that the *Y179C* variant was associated with a higher risk of CRC compared to the *G396D* variant [105, 117] (the smallest study was later included in the other study of 20,565 CRC cases and 15,524). When compared to *G396D* homozygotes and *Y179C/G396D* compound heterozygous, *Y179C* homozygotes both have earlier clinical debut and significantly higher risk of CRC [105, 107]. In an analysis including both homozygotes and heterozygotes (N=1116 CRC patients and 934 controls), having a *Y179C* mutation was significantly associated with an increased risk of developing CRC (OR 6.70, CI95% 1.08 – 41.30), this correlation was not found for *G396D* (OR 0.79 CI95% 0.42 – 1.48) [105]. With regard to development of extracolonic manifestations, no genotype-phenotype correlations with *Y179C* and *G396D* have been found [33].

In a study of a variety of different variants of *MUTYH* mutations other than *Y179C* and *G396D* (N=42 biallelic *MUTYH* carriers), no genotype-phenotype correlations were found [118].

Heterozygote and CRC Risk

There has been much debate of whether monoallelic *MUTYH* mutations entail an increased CRC or Polyposis risk compared to the general population. However after numerous large studies have addressed this issue, it seems that a monoallelic effect –in the worst of cases- can only be a matter of a minimally increased CRC risk compared to the risk of the general population, which is not clinically relevant in a diagnostic or prophylactic setting.

Frequencies of monoallelic *MUTYH* mutation carriers are generally higher among CRC patients compared to background populations (Table **3**) and several large studies of Polyposis patients have found frequencies of monoallelic *MUTYH* mutation carriers that correspond fairly well to those of biallelic carriers (Table **5**). Also several case-control studies have shown a tendency for a slightly elevated CRC risk [42, 62, 71, 72, 89, 96, 98, 119-121]. However, reservations towards these studies should be taken, as results have failed to be convincing despite large populations investigated, and some studies have even been criticized for the statistical methods used [122, 123]. In one study with a significant result, bias may be present, as the increased risk was found in first-degree relatives assumed to be *MUTYH* heterozygote in whom genetic status was not confirmed [119], and could therefore have been due to misclassified biallelic *MUTYH* mutation carriers. Several population-based studies and meta-analyses of CRC case-control studies have found inconsistent results with regard to monoallelic CRC effect [97, 117, 123-125], finding either a slightly significant monoallelic risk (relative risk = 1.27 (95%CI: 1.01-1.61) [124], or no increased risk [99, 105, 125].

Some suggest an age-specific monoallelic effect with excess CRC risk in especially heterozygotes over 55 years of age [42, 48, 97, 123]. Others have proposed differing monoallelic effects of the *Y179C* and *G396D* variants, as greater CRC risks for monoallelic *Y179C* carriers compared to *G396D* carriers have been found [71, 105, 117]. Recently the largest multicenter meta-analysis to date of CRC cases (N= 20.565) and controls (N= 15.524), found a slight monoallelic effect of combined identified *MUTYH* mutation variants (OR=1.16, 95% CI: 1.00 – 1.34) and for the*Y179C* variant alone (OR: 1.34, 95% CI: 1.00 – 1.80) [117]. However the exact effect and possible clinical relevance of monoallelic *Y179C* variants remain to be fully determined.

Table 5: *MUTYH* germline mutations among Polyposis patients who are negative for *APC* germline mutations

	Sieber *et al.*, 2003, [41]	**Sampson *et al.*, 2003, [68]**	**Isidro *et al.* 2004, [73]**	**Wang *et al.* 2004, [62]**	**Nielsen *et al.*, 2005, [94]**	**Russell *et al.*, 2006, [47]**	**Slová *et al.*, 2007, [78]**	**Kim *et al.*, 2007, [67]**	**Morak *et al.*, 2010, [90]**
Population	United Kingdom	United Kingdom	Portugal	USA	The Netherlands	Switzerland	Czech Republic	Korea	Germany*[3]
Number of tested Polyposis patients (Number of adenomas)*[1]	N=259 (3 to >100)	N =111 (At least 10)	N =53 (10 to >1000)	N=140 (4 to >500)	N=170	N=61 (5-99)	N= 82 (3 to >100)	N=62 (10 to >100)	N=215 (>100: 30%, 10-100: 40%, <10:30%)
Method of screening of the *MUTYH* gene*[2]	All exons and exon-intron boundaries	Caucasians: Exon 7 +13, Hetero-zygote/Non-Caucasians: All exons	All exons	Exon 7 + 13	All exons	Exon 7 + 13, Hetero-zygote: all exons	All exons and exon-intron boundaries	All exons	All exons and flanking regions
Carriers of monoallelic *MUTYH* germline mutations	4 %	3%	0%	3%	24%	10%	1%	4.8%	15%
Carriers of biallelic *MUTYH* germline mutations	5 %	23%	40%	11%	24%	10%	2%	3.2%	3.7%

In this table only studies in which the numbers of tested individuals are > 50 are included. The tested individuals are all *APC* germline mutation- negative and diagnosed with colorectal adenomas.
*[1] The groups of tested Polyposis patients included both patients with < and > 100 colorectal adenomas. *[2] The common mutations *Y179C* and *G396D* are found in exon 7 and 13, respectively
*[3] 195 were Caucasian and 20 of Turkish descent

CLINICAL MAP MANAGEMENT

Genetic Counseling and Testing

Frequently, MAP patients have already developed CRC at the time of clinical diagnosis before prophylactic treatment can be initiated. Genetic testing and counseling of family members at risk of developing MAP is essential for initiation of prophylactic screening.

Based on the recessive mode of inheritance, siblings of an affected individual have a 25% a priori risk of disease. Consequently, determination of the genotype is especially important in these individuals [47, 48, 68]. However, many cases will appear sporadic, and in some MAP families, the mode of inheritance may even appear to be dominant [12, 77, 93, 94]. Some therefore suggest *MUTYH* genetic testing in all patients with early-onset CRC or two events of CRC with or without polyposis [90, 126]. As biallelic *MUTYH* carriers have been found among patients fulfilling diagnostic criteria for HNPCC [62, 64, 90], genetic testing for *MUTYH* mutations could also be considered in CRC families which fulfill the clinical criteria for HNPCC were no *MMR* mutations are found [64, 90]. Balaguer *et al.* have suggested a set of clinical criteria to identify which CRC patients should be screened for *MUTYH* mutations: 1) presence of more than 15 synchronous colorectal adenomas or 2) CRC diagnosis before 50 years of age [105]. In their case-control study of CRC patients, Balaguer *et al.* found the suggested characteristics to be independent predictors of biallelic *MUYTH* mutations, but did not find presence of a family history of CRC to be an independent predictor [105].

In our judgment, there is no indication for prophylactic surveillance in heterozygote carriers of *MUTYH* germline mutations at the present time. However, we believe that siblings and other relatives at risk of developing MAP should be searched for, if *MUTYH* germline mutations are discovered, in order for appropriate prophylactic measures to be made.

Determination of the Genotype

In practice it is important to search for germline mutations in both the *APC* and the *MUTYH* gene in Polyposis patients. In cases with a positive family history, the most probable mode of inheritance can guide the assessment of which gene to start with [12].

An obvious strategy for genetic *MUTYH* analysis would be to specifically screen for the most frequent variants, *Y179C* and *G396D*, in Caucasian individuals [92]. A disadvantage of such a selective screening would however be, that MAP patients, who are compound heterozygote for just one of these variants, or homozygote for other *MUTYH* variants, would be missed [62, 79, 95, 127]. Eliason *et al.* demonstrated an increased clinical sensitivity for the detection of *MUYTH* mutations when all exons and intron-exon boundaries of the *MUTYH* gene were screened compared to sole testing for *Y179C* and *G396D* [127]. Accordingly, all coding regions of the *MUTYH* gene should be screened in individuals found to be heterozygote for *Y179C* or *G396D* [89, 95, 127]. Piccioli *et al.* have designed specific assays for detecting the 6 most frequent *MUTYH* mutations using a multiplex T-ARMS-PCR method [128] and new advances in for example next generation sequencing are expected to contribute to development of new approaches in the future.

In patients presenting with <10 colorectal adenomas or familial mismatch repair proficient CRCs, van Puijenbroek *et al.* have proposed a cost-effective prescreening method consisting of the screening of tumors for *KRAS2 c.34G > T*, a somatic mutation in the *KRAS2* gene shown to be more common in MAP patients compared to sporadic CRC cases [129]. This should be followed by screening for population specific *MUTYH* mutations in cases positive for the aforementioned *KRAS2* mutation [129].

The frequencies of *Y179C* and *G396D* in the general population are low compared to the occurrence among MAP patients (Figs. **2** and **3**), and for this reason, there is at present no indications for MAP screening of the general population [47, 89, 100]. However, genetic testing of spouses of *MUTYH* mutation carriers to assess the genotype and corresponding disease risk of offspring has been recommended [79].

Surveillance and Treatment

Prophylactic screening of HNPCC patients, FAP, AFAP patients and their families is shown to reduce the development of CRC and CRC-associated mortality markedly [5, 8, 15, 19, 106, 130]. For HNPCC patients colonoscopy is recommended from about 20-25 years of age in intervals of 1-3 years [15, 17, 19, 130]. The benefit of screening for endometrial cancer and other HNPCC associated cancers is still debated, and extent of extracolonic screening based on individual assessment [17, 19, 130]. In FAP families, sigmoidoscopy is advised from early teens every 1-2 years, adjusted according to clinical manifestations [5, 8]. Additionally, FAP patients are offered upper gastrointestinal endoscopy every 1-5 years from about 25-30 years of age due to risk of duodenal polyposis [8]. Especially in patients with an early first appearance of the disease, prophylactic colectomy is usually advisable due to the uncontrollable extent of Polyposis [5, 8]. For AFAP patients, colonoscopy every 2 years is recommended from 18-20 years of age, due to later CRC development and a typically more proximal location of adenomas compared to FAP patients [8, 12].

For biallelic *MUTYH* mutation carriers a surveillance protocol in accordance with the recommendations for AFAP patients is recomended [8]. However, some recommend MAP surveillance initiation later than for AFAP patients, at the age of 20-25 years [12]. Furthermore, upper gastro-intestinal endoscopy starting from the age of 25-30 years is advised in MAP patients [8], even though the question of upper gastrointestinal endoscopy ought to be further investigated. The outlined recommendations have previously been advised to be adjusted according to the number, size and degree of dysplasia of the adenomas in the individual patient [62]. However, recent observations argue that screening should not be guided according to number of adenomas found, but instead be done at regular intervals [103]. A large study of MAP patients found that the risk of CRC in MAP patients does not seem to be associated with number of colorectal adenomas, as seen in FAP patients [103]. Further, several studies have reported biallelic *MUTYH* mutation carriers who developed CRC without previous or synchronous adenomas [48, 62, 64, 71, 100, 103, 105].

Since MAP patients typically develop less adenomas than FAP patients, the prophylactic treatment of MAP patients should as a starting point be aimed at colonoscopy with polypectomy. It could however be appropriate to apply colectomy in MAP patients; for instance for those developing a particularly large number or advanced adenomas [8, 12, 93, 102]. Nieuwenhuis *et al.* suggest that if surgery is needed, colectomy with ileo-rectal anastomosis should be performed. This approach could be a better choice in the long run, as reoperation frequency after hemi-colectomy is often high, and the cumulative risk of developing a metachronous tumor 15 years after the primary CRC is as high as 20% for MAP patients [103].

It seems that other factors than prophylactic surveillance influence prognosis among MAP patients. A recent European study (147 MAP patients and 272 CRC patients) found that five-year survival is significantly better for MAP patients when compared to CRC patients in the general population, even after exclusion of MAP patients in which CRC was diagnosed due to surveillance [131]. The reason for this difference is unknown and needs to be further investigated.

It is important that knowledge about clinical presentation of the disease and surveillance approaches is continuously searched for and utilized for the organization of guidelines, which we believe will assure the best treatment for these patients.

CONCLUSION AND FUTURE PERSPECTIVES

At times, it can be difficult to distinguish between the phenotypes of FAP, AFAP, HNPCC and MAP. Further characterization of clinical aspects of MAP, especially of genotype-phenotype correlations for specific *MUTYH* mutations, is needed to secure optimal differential diagnosis, surveillance and treatment for affected families.

Germline mutations found in the *MUTYH* gene have shown a great ethnic variability, and further knowledge in this area could target the genetic screening of Polyposis patients towards specific population groups.

Until further studies provide evidence to the contrary, we recommend:

- Mutation screening of the *MUTYH* gene based on Balaguer et al's proposed clinical criteria, in patients with 1) more than 15 synchronous colorectal adenomas or 2) CRC diagnosis before 50 years of age [105].
- MAP should also be considered as a differential diagnosis in HNPCC families without a detectable *MMR* mutation.
- *MUTYH* genetic testing strategy starting with *Y179C* and *G382D* variants in Caucasian populations, followed by screening of all coding regions in heterozygote or patients negative for these common variants.
- Prophylactic surveillance of MAP patients should be performed with colonoscopy with polypectomy in mind from 20-25 years of age as well as upper gastro-intestinal endoscopy every second year.
- In the future, more MAP patients could be identified before developing CRC by establishing MAP registers and finding call-up patients based on family history, as it is currently done for FAP patients in many countries.

SEARCH CRITERIA

The applied papers were all found in the PubMed database using the following terms: MYH Associated Polyposis /MUTYH Associated Polyposis /MutYH

Associated Polyposis. The search was only for material published in English and was made without time-limitation. All papers used as references were published between 01/01 2002 and 01/03 2012.

ABBREVIATIONS

A	=	Adenine
AFAP	=	Attenuated Familial Adenomatous Polyposis
APC	=	Adenomatous Polyposis Coli
BER	=	Base Excision Repair
C	=	Cytosine
CRC	=	Colorectal Cancer
FAP	=	Familial Adenomatous Polyposis
G	=	Guanine
HhH	=	Helix-Hairpin-Helix
HNPCC	=	Hereditary Non-Polyposis Colorectal Cancer
LOH	=	Loss of Heterozygosity
LOVD	=	Leiden Open Variation Database
MAP	=	*MUTYH* Associated Polyposis
MMR	=	Mismatch Repair
MSI	=	Microsatellite Instability
MUTYH gene	=	*MYH* gene, *MutY Human Homolog* gene
OR	=	Odds ratio

ROS	=	Reactive Oxygen Species
T	=	Thymine
8-oxoG	=	8-oxo-7,8-dihydroxy-2′-deoxyguanosine
95% CI	=	95% Confidence interval

CONFLICT OF INTEREST

The authors confirm that this chapter content has no conflict of interest.

ACKNOWLEDGEMENT

Declared none.

DISCLOSURE

This is an updated report of our earlier review article, "*MUTYH* Associated Polyposis (MAP)". Current Genomics, 2008, 9, 420-35. Poulsen ML and Bisgaard ML.

REFERENCES

[1] Jemal, A.; Bray, F.; Center, M. M.; Ferlay, J.; Ward, E.; Forman, D. Global cancer statistics. *CA: a cancer journal for clinicians,* **2011,** *61* (2), 69-90.

[2] Lichtenstein, P.; Holm, N. V.; Verkasalo, P. K.; Iliadou, A.; Kaprio, J.; Koskenvuo, M.; Pukkala, E.; Skytthe, A.; Hemminki, K. Environmental and heritable factors in the causation of cancer--analyses of cohorts of twins from Sweden, Denmark, and Finland. *The New England journal of medicine,* **2000,** *343* (2), 78-85.

[3] Johns, L. E.; Houlston, R. S. A systematic review and meta-analysis of familial colorectal cancer risk. *The American journal of gastroenterology,* **2001,** *96* (10), 2992-3003.

[4] Abdel-Rahman, W. M.; Peltomaki, P. Molecular basis and diagnostics of hereditary colorectal cancers. *Annals of medicine,* **2004,** *36* (5), 379-88.

[5] Galiatsatos, P.; Foulkes, W. D. Familial adenomatous polyposis. *The American journal of gastroenterology,* **2006,** *101* (2), 385-98.

[6] Lipton, L.; Tomlinson, I. The genetics of FAP and FAP-like syndroms. *Familial Cancer,* **2006,** *4*, 221-226.

[7] Varesco, L. Familial adenomatous polyposis: genetics and epidemiology. *Techniques in coloproctology,* **2004,** *8 Suppl 2*, s305-8.

[8] Vasen, H. F.; Moeslein, G.; Alonso, A.; Aretz, S.; Bernstein, I.; Bertario, L.; Blanco, I.; Bulow, S.; Burn, J.; Capella, G.; Colas, C.; Engel, C.; Frayling, I.; Friedl, W.; Hes, F.; Hodgson, S.; Jarvinen, H.; Mecklin, J. P.; Moller, P.; Myrhoj, T.; Nagengast, F. M.; Parc, Y.; Phillips, R.; Clark, S.; Ponz de Leon, M.; Renkonen-Sinisalo, L.; Sampson, J.; Stormorken, A.; Tejpar, S.; Thomas, H.; Wijnen, J. Guidelines for the clinical management of familial adenomatous polyposis (FAP). *Gut,* **2008**.

[9] Näthke APC at a glance. *Journal of Cell Science,* **2004,** (117), 3.

[10] Fearnhead, N. S.; Britton, M. P.; Bodmer, W. F. The ABC of APC. *Human molecular genetics,* **2001,** *10* (7), 721-33.

[11] Nieuwenhuis, M. H.; Vasen, H. F. Correlations between mutation site in APC and phenotype of familial adenomatous polyposis (FAP): a review of the literature. *Critical reviews in oncology/hematology,* **2007,** *61* (2), 153-61.

[12] Nielsen, M.; Hes, F. J.; Nagengast, F. M.; Weiss, M. M.; Mathus-Vliegen, E. M.; Morreau, H.; Breuning, M. H.; Wijnen, J. T.; Tops, C. M.; Vasen, H. F. Germline mutations in APC and MUTYH are responsible for the majority of families with attenuated familial adenomatous polyposis. *Clinical genetics,* **2007,** *71* (5), 427-33.

[13] Lamlum, H.; Al Tassan, N.; Jaeger, E.; Frayling, I.; Sieber, O.; Reza, F. B.; Eckert, M.; Rowan, A.; Barclay, E.; Atkin, W.; Williams, C.; Gilbert, J.; Cheadle, J.; Bell, J.; Houlston, R.; Bodmer, W.; Sampson, J.; Tomlinson, I. Germline APC variants in patients with multiple colorectal adenomas, with evidence for the particular importance of E1317Q. *Human molecular genetics,* **2000,** *9* (15), 2215-21.

[14] Vasen, H. F.; Watson, P.; Mecklin, J. P.; Lynch, H. T. New clinical criteria for hereditary nonpolyposis colorectal cancer (HNPCC, Lynch syndrome) proposed by the International Collaborative group on HNPCC. *Gastroenterology,* **1999,** *116* (6), 1453-6.

[15] Lynch, H. T.; Lynch, J. Lynch syndrome: genetics, natural history, genetic counseling, and prevention. *J Clin Oncol,* **2000,** *18* (21 Suppl), 19S-31S.

[16] Vasen, H. F. Clinical description of the Lynch syndrome [hereditary nonpolyposis colorectal cancer (HNPCC)]. *Fam Cancer,* **2005,** *4* (3), 219-25.

[17] Abdel-Rahman, W. M.; Mecklin, J. P.; Peltomaki, P. The genetics of HNPCC: application to diagnosis and screening. *Critical reviews in oncology/hematology,* **2006,** *58* (3), 208-20.

[18] Wijnen, J. T.; Vasen, H. F.; Khan, P. M.; Zwinderman, A. H.; van der Klift, H.; Mulder, A.; Tops, C.; Moller, P.; Fodde, R. Clinical findings with implications for genetic testing in families with clustering of colorectal cancer. *The New England journal of medicine,* **1998,** *339* (8), 511-8.

[19] Vasen, H. F.; Moslein, G.; Alonso, A.; Bernstein, I.; Bertario, L.; Blanco, I.; Burn, J.; Capella, G.; Engel, C.; Frayling, I.; Friedl, W.; Hes, F. J.; Hodgson, S.; Mecklin, J. P.; Moller, P.; Nagengast, F.; Parc, Y.; Renkonen-Sinisalo, L.; Sampson, J. R.; Stormorken, A.; Wijnen, J. Guidelines for the clinical management of Lynch syndrome (hereditary non-polyposis cancer). *Journal of medical genetics,* **2007,** *44* (6), 353-62.

[20] Al-Tassan, N.; Chmiel, N. H.; Maynard, J.; Fleming, N.; Livingston, A. L.; Williams, G. T.; Hodges, A. K.; Davies, D. R.; David, S. S.; Sampson, J. R.; Cheadle, J. P. Inherited variants of MYH associated with somatic G:C-->T:A mutations in colorectal tumors. *Nature genetics,* **2002,** *30* (2), 227-32.

[21] Parker, A. R.; Eshleman, J. R. Human MutY: gene structure, protein functions and interactions, and role in carcinogenesis. *Cell Mol Life Sci,* **2003,** *60* (10), 2064-83.

[22] Nohmi, T.; Kim, S. R.; Yamada, M. Modulation of oxidative mutagenesis and carcinogenesis by polymorphic forms of human DNA repair enzymes. *Mutation research,* **2005,** *591* (1-2), 60-73.

[23] David, S. S.; O'Shea, V. L.; Kundu, S. Base-excision repair of oxidative DNA damage. *Nature,* **2007,** *447* (7147), 941-50.

[24] Sampson, J. R.; Jones, S.; Dolwani, S.; Cheadle, J. P. MutYH (MYH) and colorectal cancer. *Biochemical Society transactions,* **2005,** *33* (Pt 4), 679-83.

[25] Cheadle, J. P.; Sampson, J. R. MUTYH-associated polyposis--from defect in base excision repair to clinical genetic testing. *DNA repair,* **2007,** *6* (3), 274-9.

[26] Cheadle, J. P.; Sampson, J. R. Exposing the MYtH about base excision repair and human inherited disease. *Human molecular genetics,* **2003,** *12 Spec No 2,* R159-65.

[27] Fokkema, I. F.; den Dunnen, J. T.; Taschner, P. E. LOVD: easy creation of a locus-specific sequence variation database using an "LSDB-in-a-box" approach. *Human mutation,* **2005,** *26* (2), 63-8.

[28] Out, A. A.; Tops, C. M.; Nielsen, M.; Weiss, M. M.; van Minderhout, I. J.; Fokkema, I. F.; Buisine, M. P.; Claes, K.; Colas, C.; Fodde, R.; Fostira, F.; Franken, P. F.; Gaustadnes, M.; Heinimann, K.; Hodgson, S. V.; Hogervorst, F. B.; Holinski-Feder, E.; Lagerstedt-Robinson, K.; Olschwang, S.; van den Ouweland, A. M.; Redeker, E. J.; Scott, R. J.; Vankeirsbilck, B.; Gronlund, R. V.; Wijnen, J. T.; Wikman, F. P.; Aretz, S.; Sampson, J. R.; Devilee, P.; den Dunnen, J. T.; Hes, F. J. Leiden Open Variation Database of the MUTYH gene. *Human mutation,* **2010,** *31* (11), 1205-15.

[29] Chow, E.; Thirlwell, C.; Macrae, F.; Lipton, L. Colorectal cancer and inherited mutations in base-excision repair. *The lancet oncology,* **2004,** *5* (10), 600-6.

[30] Lipton, L.; Tomlinson, I. The multiple colorectal adenoma phenotype and MYH, a base excision repair gene. *Clin Gastroenterol Hepatol,* **2004,** *2* (8), 633-8.

[31] Guan, Y.; Manuel, R. C.; Arvai, A. S.; Parikh, S. S.; Mol, C. D.; Miller, J. H.; Lloyd, S.; Tainer, J. A. MutY catalytic core, mutant and bound adenine structures define specificity for DNA repair enzyme superfamily. *Nature structural biology,* **1998,** *5* (12), 1058-64.

[32] Lu, A. L.; Bai, H.; Shi, G.; Chang, D. Y. MutY and MutY homologs (MYH) in genome maintenance. *Front Biosci,* **2006,** *11*, 3062-80.

[33] Vogt, S.; Jones, N.; Christian, D.; Engel, C.; Nielsen, M.; Kaufmann, A.; Steinke, V.; Vasen, H. F.; Propping, P.; Sampson, J. R.; Hes, F. J.; Aretz, S. Expanded extracolonic tumor spectrum in MUTYH-associated polyposis. *Gastroenterology,* **2009,** *137* (6), 1976-85 e1-10.

[34] Shinmura, K.; Yamaguchi, S.; Saitoh, T.; Kohno, T.; Yokota, J. Somatic mutations and single nucleotide polymorphisms of base excision repair genes involved in the repair of 8-hydroxyguanine in damaged DNA. *Cancer letters,* **2001,** *166* (1), 65-9.

[35] Tao, H.; Shinmura, K.; Hanaoka, T.; Natsukawa, S.; Shaura, K.; Koizumi, Y.; Kasuga, Y.; Ozawa, T.; Tsujinaka, T.; Li, Z.; Yamaguchi, S.; Yokota, J.; Sugimura, H.; Tsugane, S. A novel splice-site variant of the base excision repair gene MYH is associated with production of an aberrant mRNA transcript encoding a truncated MYH protein not localized in the nucleus. *Carcinogenesis,* **2004,** *25* (10), 1859-66.

[36] Kim, C. J.; Cho, Y. G.; Park, C. H.; Kim, S. Y.; Nam, S. W.; Lee, S. H.; Yoo, N. J.; Lee, J. Y.; Park, W. S. Genetic alterations of the MYH gene in gastric cancer. *Oncogene,* **2004,** *23* (40), 6820-2.

[37] Yamaguchi, S.; Shinmura, K.; Saitoh, T.; Takenoshita, S.; Kuwano, H.; Yokota, J. A single nucleotide polymorphism at the splice donor site of the human MYH base excision repair genes results in reduced translation efficiency of its transcripts. *Genes Cells,* **2002,** *7* (5), 461-74.

[38] Barnetson, R. A.; Devlin, L.; Miller, J.; Farrington, S. M.; Slater, S.; Drake, A. C.; Campbell, H.; Dunlop, M. G.; Porteous, M. E. Germline mutation prevalence in the base excision repair gene, MYH, in patients with endometrial cancer. *Clinical genetics,* **2007,** *72* (6), 551-5.

[39] Jones, S.; Emmerson, P.; Maynard, J.; Best, J. M.; Jordan, S.; Williams, G. T.; Sampson, J. R.; Cheadle, J. P. Biallelic germline mutations in MYH predispose to multiple colorectal adenoma and somatic G:C-->T:A mutations. *Human molecular genetics,* **2002,** *11* (23), 2961-7.

[40] Halford, S. E.; Rowan, A. J.; Lipton, L.; Sieber, O. M.; Pack, K.; Thomas, H. J.; Hodgson, S. V.; Bodmer, W. F.; Tomlinson, I. P. Germline mutations but not somatic changes at the MYH locus contribute to the pathogenesis of unselected colorectal cancers. *The American journal of pathology,* **2003,** *162* (5), 1545-8.

[41] Sieber, O. M.; Lipton, L.; Crabtree, M.; Heinimann, K.; Fidalgo, P.; Phillips, R. K.; Bisgaard, M. L.; Orntoft, T. F.; Aaltonen, L. A.; Hodgson, S. V.; Thomas, H. J.; Tomlinson, I. P. Multiple colorectal adenomas, classic adenomatous polyposis, and germ-line mutations in MYH. *The New England journal of medicine,* **2003,** *348* (9), 791-9.

[42] Kury, S.; Buecher, B.; Robiou-du-Pont, S.; Scoul, C.; Colman, H.; Lelievre, B.; Olschwang, S.; Le Houerou, C.; Le Neel, T.; Faroux, R.; Ollivry, J.; Lafraise, B.; Chupin, L. D.; Bezieau, S. The Thorough Screening of the MUTYH Gene in a Large French Cohort of Sporadic Colorectal Cancers. *Genetic testing,* **2007**.

[43] Dallosso, A. R.; Dolwani, S.; Jones, N.; Jones, S.; Colley, J.; Maynard, J.; Idziaszczyk, S.; Humphreys, V.; Arnold, J.; Donaldson, A.; Eccles, D.; Ellis, A.; Evans, D. G.; Frayling, I. M.; Hes, F. J.; Houlston, R. S.; Maher, E. R.; Nielsen, M.; Parry, S.; Tyler, E.; Moskvina, V.; Cheadle, J. P.; Sampson, J. R. Inherited predisposition to colorectal adenomas caused by multiple rare alleles of MUTYH but not OGG1, NUDT1, NTH1 or NEIL 1, 2 or 3. *Gut,* **2008,** *57* (9), 1252-5.

[44] Kim, J. C.; Ka, I. H.; Lee, Y. M.; Koo, K. H.; Kim, H. C.; Yu, C. S.; Jang, S. J.; Kim, Y. S., Lee, H. I.; Lee, K. H. MYH, OGG1, MTH1, and APC alterations involved in the colorectal tumorigenesis of Korean patients with multiple adenomas. *Virchows Arch,* **2007,** *450* (3), 311-9.

[45] Kim, I. J.; Ku, J. L.; Kang, H. C.; Park, J. H.; Yoon, K. A.; Shin, Y.; Park, H. W.; Jang, S. G.; Lim, S. K.; Han, S. Y.; Shin, Y. K.; Lee, M. R.; Jeong, S. Y.; Shin, H. R.; Lee, J. S.; Kim, W. H.; Park, J. G. Mutational analysis of OGG1, MYH, MTH1 in FAP, HNPCC and sporadic colorectal cancer patients: R154H OGG1 polymorphism is associated with sporadic colorectal cancer patients. *Human genetics,* **2004,** *115* (6), 498-503.

[46] Morak, M.; Massdorf, T.; Sykora, H.; Kerscher, M.; Holinski-Feder, E. First evidence for digenic inheritance in hereditary colorectal cancer by mutations in the base excision repair genes. *Eur J Cancer,* **2011,** *47* (7), 1046-55.

[47] Russell, A. M.; Zhang, J.; Luz, J.; Hutter, P.; Chappuis, P. O.; Berthod, C. R.; Maillet, P.; Mueller, H.; Heinimann, K. Prevalence of MYH germline mutations in Swiss APC mutation-negative polyposis patients. *International journal of cancer,* **2006,** *118* (8), 1937-40.

[48] Farrington, S. M.; Tenesa, A.; Barnetson, R.; Wiltshire, A.; Prendergast, J.; Porteous, M.; Campbell, H.; Dunlop, M. G. Germline susceptibility to colorectal cancer due to base-excision repair gene defects. *American journal of human genetics,* **2005,** *77* (1), 112-9.

[49] Lipton, L.; Halford, S. E.; Johnson, V.; Novelli, M. R.; Jones, A.; Cummings, C.; Barclay, E.; Sieber, O.; Sadat, A.; Bisgaard, M. L.; Hodgson, S. V.; Aaltonen, L. A.; Thomas, H. J.; Tomlinson, I. P. Carcinogenesis in MYH-associated polyposis follows a distinct genetic pathway. *Cancer research,* **2003,** *63* (22), 7595-9.

[50] Jones, S.; Lambert, S.; Williams, G. T.; Best, J. M.; Sampson, J. R.; Cheadle, J. P. Increased frequency of the k-ras G12C mutation in MYH polyposis colorectal adenomas. *British journal of cancer,* **2004,** *90* (8), 1591-3.

[51] Kambara, T.; Whitehall, V. L.; Spring, K. J.; Barker, M. A.; Arnold, S.; Wynter, C. V.; Matsubara, N.; Tanaka, N.; Young, J. P.; Leggett, B. A.; Jass, J. R. Role of inherited defects of MYH in the development of sporadic colorectal cancer. *Genes, chromosomes & cancer,* **2004,** *40* (1), 1-9.

[52] Johnson, V.; Lipton, L. R.; Cummings, C.; Eftekhar Sadat, A. T.; Izatt, L.; Hodgson, S. V.; Talbot, I. C.; Thomas, H. J.; Silver, A. J.; Tomlinson, I. P. Analysis of somatic molecular changes, clinicopathological features, family history, and germline mutations in colorectal cancer families: evidence for efficient diagnosis of HNPCC and for the existence of distinct groups of non-HNPCC families. *Journal of medical genetics,* **2005,** *42* (10), 756-62.

[53] Vasovcak, P.; Pavlikova, K.; Sedlacek, Z.; Skapa, P.; Kouda, M.; Hoch, J.; Krepelova, A. Molecular genetic analysis of 103 sporadic colorectal tumours in Czech patients. *PloS one,* **2011,** *6* (8), e24114.

[54] Bougatef, K.; Marrakchi, R.; Kourda, N.; Ben Lahely, Y. B.; Jileni, S. B.; El Khil, H. K.; Soubrier, F.; Ben Ammar Elgaaied, A. Somatic mutation of MutYH in Tunisian patients with sporadic colorectal cancer. *Journal of clinical laboratory analysis,* **2007,** *21* (6), 372-4.

[55] Edelmann, L.; Edelmann, W. Loss of DNA mismatch repair function and cancer predisposition in the mouse: animal models for human hereditary nonpolyposis colorectal cancer. *American journal of medical genetics,* **2004,** *129* (1), 91-9.

[56] Gu, Y.; Parker, A.; Wilson, T. M.; Bai, H.; Chang, D. Y.; Lu, A. L. Human MutY homolog, a DNA glycosylase involved in base excision repair, physically and functionally interacts with mismatch repair proteins human MutS homolog 2/human MutS homolog 6. *The Journal of biological chemistry,* **2002,** *277* (13), 11135-42.

[57] Bai, H.; Jones, S.; Guan, X.; Wilson, T. M.; Sampson, J. R.; Cheadle, J. P.; Lu, A. L. Functional characterization of two human MutY homolog (hMYH) missense mutations (R227W and V232F) that lie within the putative hMSH6 binding domain and are associated with hMYH polyposis. *Nucleic acids research,* **2005,** *33* (2), 597-604.

[58] Miyaki, M.; Iijima, T.; Yamaguchi, T.; Hishima, T.; Tamura, K.; Utsunomiya, J.; Mori, T. Germline mutations of the MYH gene in Japanese patients with multiple colorectal adenomas. *Mutation research,* **2005,** *578* (1-2), 430-3.

[59] Niessen, R. C.; Sijmons, R. H.; Ou, J.; Olthof, S. G.; Osinga, J.; Ligtenberg, M. J.; Hogervorst, F. B.; Weiss, M. M.; Tops, C. M.; Hes, F. J.; de Bock, G. H.; Buys, C. H.; Kleibeuker, J. H.; Hofstra, R. M. MUTYH and the mismatch repair system: partners in crime? *Human genetics,* **2006,** *119* (1-2), 206-11.

[60] Giraldez, M. D.; Balaguer, F.; Caldes, T.; Sanchez-de-Abajo, A.; Gomez-Fernandez, N.; Ruiz-Ponte, C.; Munoz, J.; Garre, P.; Gonzalo, V.; Moreira, L.; Ocana, T.; Clofent, J.; Carracedo, A.; Andreu, M.; Jover, R.; Llor, X.; Castells, A.; Castellvi-Bel, S.; Gastrointestinal Oncology Group of the Spanish Gastroenterological, A. Association of MUTYH and MSH6 germline mutations in colorectal cancer patients. *Fam Cancer,* **2009,** *8* (4), 525-31.

[61] Ashton, K. A.; Meldrum, C. J.; McPhillips, M. L.; Kairupan, C. F.; Scott, R. J. Frequency of the Common MYH Mutations (G382D and Y165C) in MMR Mutation Positive and Negative HNPCC Patients. *Hereditary cancer in clinical practice,* **2005,** *3* (2), 65-70.

[62] Wang, L.; Baudhuin, L. M.; Boardman, L. A.; Steenblock, K. J.; Petersen, G. M.; Halling, K. C.; French, A. J.; Johnson, R. A.; Burgart, L. J.; Rabe, K.; Lindor, N. M.; Thibodeau, S. N. MYH mutations in patients with attenuated and classic polyposis and with young-onset colorectal cancer without polyps. *Gastroenterology,* **2004,** *127* (1), 9-16.

[63] Steinke, V.; Rahner, N.; Morak, M.; Keller, G.; Schackert, H. K.; Gorgens, H.; Schmiegel, W.; Royer-Pokora, B.; Dietmaier, W.; Kloor, M.; Engel, C.; Propping, P.; Aretz, S.; German, H. C. No association between MUTYH and MSH6 germline mutations in 64 HNPCC patients. *Eur J Hum Genet,* **2008,** *16* (5), 587-92.

[64] Jo, W. S.; Bandipalliam, P.; Shannon, K. M.; Niendorf, K. B.; Chan-Smutko, G.; Hur, C.; Syngal, S.; Chung, D. C. Correlation of polyp number and family history of colon cancer with germline MYH mutations. *Clin Gastroenterol Hepatol,* **2005,** *3* (10), 1022-8.

[65] Cao, X.; Hong, Y.; Eu, K. W.; Loi, C.; Cheah, P. Y. Singapore familial adenomatous polyposis (FAP) patients with classical adenomatous polyposis but undetectable APC mutations have accelerated cancer progression. *The American journal of gastroenterology,* **2006,** *101* (12), 2810-7.

[66] Kim, H.; Kim, H. J.; Chi, S. G.; Lee, S. K.; Joo, G. R.; Dong, S. H.; Kim, B. H.; Chang, Y. W.; Lee, J. I.; Chang, R. Absence of MutY homologue mutation in patients with multiple sporadic adenomatous polyps in Korea. *World J Gastroenterol,* **2006,** *12* (6), 951-5.

[67] Kim, D. W.; Kim, I. J.; Kang, H. C.; Jang, S. G.; Kim, K.; Yoon, H. J.; Ahn, S. A.; Han, S. Y.; Hong, S. H.; Hwang, J. A.; Sohn, D. K.; Jeong, S. Y.; Choi, H. S.; Hong, C. W.; Lim, S. B.; Park, J. G. Germline mutations of the MYH gene in Korean patients with multiple colorectal adenomas. *International journal of colorectal disease,* **2007,** *22* (10), 1173-8.

[68] Sampson, J. R.; Dolwani, S.; Jones, S.; Eccles, D.; Ellis, A.; Evans, D. G.; Frayling, I.; Jordan, S.; Maher, E. R.; Mak, T.; Maynard, J.; Pigatto, F.; Shaw, J.; Cheadle, J. P. Autosomal recessive colorectal adenomatous polyposis due to inherited mutations of MYH. *Lancet,* **2003,** *362* (9377), 39-41.

[69] Ajith Kumar, V. K.; Gold, J. A.; Mallon, E.; Thomas, S.; Hodgson, S. V. Sebaceous adenomas in an MYH associated polyposis patient of Indian (Gujarati) origin. *Fam Cancer,* **2007**.

[70] Dolwani, S.; Williams, G. T.; West, K. P.; Newman, J.; Stock, D.; Griffiths, A. P.; Best, J.; Cheadle, J. P.; Sampson, J. R. Analysis of inherited MYH/(MutYH) mutations in British Asian patients with colorectal cancer. *Gut,* **2007,** *56* (4), 593.

[71] Croitoru, M. E.; Cleary, S. P.; Di Nicola, N.; Manno, M.; Selander, T.; Aronson, M.; Redston, M.; Cotterchio, M.; Knight, J.; Gryfe, R.; Gallinger, S. Association between biallelic and monoallelic germline MYH gene mutations and colorectal cancer risk. *Journal of the National Cancer Institute,* **2004,** *96* (21), 1631-4.

[72] Fleischmann, C.; Peto, J.; Cheadle, J.; Shah, B.; Sampson, J.; Houlston, R. S. Comprehensive analysis of the contribution of germline MYH variation to early-onset colorectal cancer. *International journal of cancer,* **2004,** *109* (4), 554-8.

[73] Isidro, G.; Laranjeira, F.; Pires, A.; Leite, J.; Regateiro, F.; Castro e Sousa, F.; Soares, J.; Castro, C.; Giria, J.; Brito, M. J.; Medeira, A.; Teixeira, R.; Morna, H.; Gaspar, I.; Marinho, C.; Jorge, R.; Brehm, A.; Ramos, J. S.; Boavida, M. G. Germline MUTYH (MYH) mutations in Portuguese individuals with multiple colorectal adenomas. *Human mutation,* **2004,** *24* (4), 353-4.

[74] Alhopuro, P.; Parker, A. R.; Lehtonen, R.; Enholm, S.; Jarvinen, H. J.; Mecklin, J. P.; Karhu, A.; Eshleman, J. R.; Aaltonen, L. A. A novel functionally deficient MYH variant in individuals with colorectal adenomatous polyposis. *Human mutation,* **2005,** *26* (4), 393.

[75] Kairupan, C. F.; Meldrum, C. J.; Crooks, R.; Milward, E. A.; Spigelman, A. D.; Burgess, B.; Groombridge, C.; Kirk, J.; Tucker, K.; Ward, R.; Williams, R.; Scott, R. J. Mutation analysis of the MYH gene in an Australian series of colorectal polyposis patients with or without germline APC mutations. *International journal of cancer,* **2005,** *116* (1), 73-7.

[76] Zhou, X. L.; Djureinovic, T.; Werelius, B.; Lindmark, G.; Sun, X. F.; Lindblom, A. Germline mutations in the MYH gene in Swedish familial and sporadic colorectal cancer. *Genetic testing,* **2005,** *9* (2), 147-51.

[77] Kanter-Smoler, G.; Bjork, J.; Fritzell, K.; Engwall, Y.; Hallberg, B.; Karlsson, G.; Gronberg, H.; Karlsson, P.; Wallgren, A.; Wahlstrom, J.; Hultcrantz, R.; Nordling, M. Novel findings in Swedish patients with MYH-associated polyposis: mutation detection and clinical characterization. *Clin Gastroenterol Hepatol,* **2006,** *4* (4), 499-506.

[78] Sulova, M.; Zidkova, K.; Kleibl, Z.; Stekrova, J.; Kebrdlova, V.; Bortlik, M.; Lukas, M.; Kohoutova, M. Mutation analysis of the MYH gene in unrelated Czech APC mutation-negative polyposis patients. *Eur J Cancer,* **2007,** *43* (10), 1617-21.

[79] Croitoru, M. E.; Cleary, S. P.; Berk, T.; Di Nicola, N.; Kopolovic, I.; Bapat, B.; Gallinger, S. Germline MYH mutations in a clinic-based series of Canadian multiple colorectal adenoma patients. *Journal of surgical oncology,* **2007,** *95* (6), 499-506.

[80] Lejeune, S.; Guillemot, F.; Triboulet, J. P.; Cattan, S.; Mouton, C.; Porchet, N.; Manouvrier, S.; Buisine, M. P. Low frequency of AXIN2 mutations and high frequency of MUTYH mutations in patients with multiple polyposis. *Human mutation,* **2006,** *27* (10), 1064.

[81] Fromme, J. C.; Banerjee, A.; Huang, S. J.; Verdine, G. L. Structural basis for removal of adenine mispaired with 8-oxoguanine by MutY adenine DNA glycosylase. *Nature,* **2004,** *427* (6975), 652-6.

[82] Chmiel, N. H.; Golinelli, M. P.; Francis, A. W.; David, S. S. Efficient recognition of substrates and substrate analogs by the adenine glycosylase MutY requires the C-terminal domain. *Nucleic acids research,* **2001,** *29* (2), 553-64.

[83] Chmiel, N. H.; Livingston, A. L.; David, S. S. Insight into the functional consequences of inherited variants of the hMYH adenine glycosylase associated with colorectal cancer: complementation assays with hMYH variants and pre-steady-state kinetics of the corresponding mutated E.coli enzymes. *Journal of molecular biology,* **2003,** *327* (2), 431-43.

[84] Volk, D. E.; House, P. G.; Thiviyanathan, V.; Luxon, B. A.; Zhang, S.; Lloyd, R. S.; Gorenstein, D. G. Structural similarities between MutT and the C-terminal domain of MutY. *Biochemistry,* **2000,** *39* (25), 7331-6.

[85] Pope, M. A.; Chmiel, N. H.; David, S. S. Insight into the functional consequences of hMYH variants associated with colorectal cancer: distinct differences in the adenine glycosylase activity and the response to AP endonucleases of Y150C and G365D murine MYH. *DNA repair,* **2005,** *4* (3), 315-25.

[86] Parker, A. R.; Sieber, O. M.; Shi, C.; Hua, L.; Takao, M.; Tomlinson, I. P.; Eshleman, J. R. Cells with pathogenic biallelic mutations in the human MUTYH gene are defective in DNA damage binding and repair. *Carcinogenesis,* **2005,** *26* (11), 2010-8.

[87] Ali, M.; Kim, H.; Cleary, S.; Cupples, C.; Gallinger, S.; Bristow, R. Characterization of mutant MUTYH proteins associated with familial colorectal cancer. *Gastroenterology,* **2008,** *135* (2), 499-507.

[88] Goto, M.; Shinmura, K.; Nakabeppu, Y.; Tao, H.; Yamada, H.; Tsuneyoshi, T.; Sugimura, H. Adenine DNA glycosylase activity of 14 human MutY homolog (MUTYH) variant proteins found in patients with colorectal polyposis and cancer. *Human mutation,* **2010,** *31* (11), E1861-74.

[89] Aretz, S.; Uhlhaas, S.; Goergens, H.; Siberg, K.; Vogel, M.; Pagenstecher, C.; Mangold, E.; Caspari, R.; Propping, P.; Friedl, W. MUTYH-associated polyposis: 70 of 71 patients with biallelic mutations present with an attenuated or atypical phenotype. *International journal of cancer,* **2006,** *119* (4), 807-14.

[90] Morak, M.; Laner, A.; Bacher, U.; Keiling, C.; Holinski-Feder, E. MUTYH-associated polyposis - variability of the clinical phenotype in patients with biallelic and monoallelic MUTYH mutations and report on novel mutations. *Clinical genetics,* **2010,** *78* (4), 353-63.

[91] Aceto, G.; Cristina Curia, M.; Veschi, S.; De Lellis, L.; Mammarella, S.; Catalano, T.; Stuppia, L.; Palka, G.; Valanzano, R.; Tonelli, F.; Casale, V.; Stigliano, V.; Cetta, F.; Battista, P.; Mariani-Costantini, R.; Cama, A. Mutations of APC and MYH in unrelated Italian patients with adenomatous polyposis coli. *Human mutation,* **2005,** *26* (4), 394.

[92] Venesio, T.; Molatore, S.; Cattaneo, F.; Arrigoni, A.; Risio, M.; Ranzani, G. N. High frequency of MYH gene mutations in a subset of patients with familial adenomatous polyposis. *Gastroenterology,* **2004,** *126* (7), 1681-5.

[93] Leite, J. S.; Isidro, G.; Martins, M.; Regateiro, F.; Albuquerque, O.; Amaro, P.; Romaozinho, J. M.; Boavida, G.; Castro-Sousa, F. Is prophylactic colectomy indicated in patients with MYH-associated polyposis? *Colorectal Dis,* **2005,** *7* (4), 327-31.

[94] Nielsen, M.; Franken, P. F.; Reinards, T. H.; Weiss, M. M.; Wagner, A.; van der Klift, H.; Kloosterman, S.; Houwing-Duistermaat, J. J.; Aalfs, C. M.; Ausems, M. G.; Brocker-Vriends, A. H.; Gomez Garcia, E. B.; Hoogerbrugge, N.; Menko, F. H.; Sijmons, R. H.; Verhoef, S.; Kuipers, E. J.; Morreau, H.; Breuning, M. H.; Tops, C. M.; Wijnen, J. T.; Vasen, H. F.; Fodde, R.; Hes, F. J. Multiplicity in polyp count and extracolonic manifestations in 40 Dutch patients with MYH associated polyposis coli (MAP). *Journal of medical genetics,* **2005,** *42* (9), e54.

[95] Bouguen, G.; Manfredi, S.; Blayau, M.; Dugast, C.; Buecher, B.; Bonneau, D.; Siproudhis, L.; David, V.; Bretagne, J. F. Colorectal adenomatous polyposis Associated with MYH mutations: genotype and phenotype characteristics. *Diseases of the colon and rectum,* **2007,** *50* (10), 1612-7.

[96] Olschwang, S.; Blanche, H.; de Moncuit, C.; Thomas, G. Similar colorectal cancer risk in patients with monoallelic and biallelic mutations in the MYH gene identified in a population with adenomatous polyposis. *Genetic testing,* **2007,** *11* (3), 315-20.

[97] Jenkins, M. A.; Croitoru, M. E.; Monga, N.; Cleary, S. P.; Cotterchio, M.; Hopper, J. L.; Gallinger, S. Risk of colorectal cancer in monoallelic and biallelic carriers of MYH mutations: a population-based case-family study. *Cancer Epidemiol Biomarkers Prev,* **2006,** *15* (2), 312-4.

[98] Peterlongo, P.; Mitra, N.; Sanchez de Abajo, A.; de la Hoya, M.; Bassi, C.; Bertario, L.; Radice, P.; Glogowski, E.; Nafa, K.; Caldes, T.; Offit, K.; Ellis, N. A. Increased frequency of disease-causing MYH mutations in colon cancer families. *Carcinogenesis,* **2006,** *27* (11), 2243-9.

[99] Peterlongo, P.; Mitra, N.; Chuai, S.; Kirchhoff, T.; Palmer, C.; Huang, H.; Nafa, K.; Offit, K.; Ellis, N. A. Colorectal cancer risk in individuals with biallelic or monoallelic mutations of MYH. *International journal of cancer,* **2005,** *114* (3), 505-7.

[100] Enholm, S.; Hienonen, T.; Suomalainen, A.; Lipton, L.; Tomlinson, I.; Karja, V.; Eskelinen, M.; Mecklin, J. P.; Karhu, A.; Jarvinen, H. J.; Aaltonen, L. A. Proportion and phenotype of MYH-associated colorectal neoplasia in a population-based series of Finnish colorectal cancer patients. *The American journal of pathology,* **2003,** *163* (3), 827-32.

[101] Gismondi, V.; Meta, M.; Bonelli, L.; Radice, P.; Sala, P.; Bertario, L.; Viel, A.; Fornasarig, M.; Arrigoni, A.; Gentile, M.; Ponz de Leon, M.; Anselmi, L.; Mareni, C.; Bruzzi, P.; Varesco, L. Prevalence of the Y165C, G382D and 1395delGGA germline mutations of the MYH gene in Italian patients with adenomatous polyposis coli and colorectal adenomas. *International journal of cancer,* **2004,** *109* (5), 680-4.

[102] Nascimbeni, R.; Pucciarelli, S.; Di Lorenzo, D.; Urso, E.; Casella, C.; Agostini, M.; Nitti, D.; Salerni, B. Rectum-sparing surgery may be appropriate for biallelic MutYH-associated polyposis. *Diseases of the colon and rectum,* **2010,** *53* (12), 1670-5.

[103] Nieuwenhuis, M. H.; Vogt, S.; Jones, N.; Nielsen, M.; Hes, F. J.; Sampson, J. R.; Aretz, S.; Vasen, H. F. Evidence for accelerated colorectal adenoma-carcinoma progression in MUTYH-associated polyposis? *Gut,* **2011**.

[104] Lefevre, J. H.; Parc, Y.; Svrcek, M.; Kerneis, S.; Colas, C.; Shields, C.; Flejou, J. F.; Parc, R.; Tiret, E. APC, MYH, and the correlation genotype-phenotype in colorectal polyposis. *Annals of surgical oncology,* **2009,** *16* (4), 871-7.

[105] Balaguer, F.; Castellvi-Bel, S.; Castells, A.; Andreu, M.; Munoz, J.; Gisbert, J. P.; Llor, X.; Jover, R.; de Cid, R.; Gonzalo, V.; Bessa, X.; Xicola, R. M.; Pons, E.; Alenda, C.; Paya, A.; Pique, J. M.; Gastrointestinal Oncology Group of the Spanish Gastroenterological, A. Identification of MYH mutation carriers in colorectal cancer: a multicenter, case-control, population-based study. *Clin Gastroenterol Hepatol,* **2007,** *5* (3), 379-87.

[106] Bulow, S. Results of national registration of familial adenomatous polyposis. *Gut,* **2003,** *52* (5), 742-6.

[107] Nielsen, M.; Joerink-van de Beld, M. C.; Jones, N.; Vogt, S.; Tops, C. M.; Vasen, H. F.; Sampson, J. R.; Aretz, S.; Hes, F. J. Analysis of MUTYH genotypes and colorectal phenotypes in patients With MUTYH-associated polyposis. *Gastroenterology,* **2009,** *136* (2), 471-6.

[108] Nielsen, M.; Poley, J. W.; Verhoef, S.; van Puijenbroek, M.; Weiss, M. M.; Burger, G. T.; Dommering, C. J.; Vasen, H. F.; Kuipers, E. J.; Wagner, A.; Morreau, H.; Hes, F. J. Duodenal carcinoma in MUTYH-associated polyposis. *Journal of clinical pathology,* **2006,** *59* (11), 1212-5.

[109] Ponti, G.; Ponz de Leon, M.; Maffei, S.; Pedroni, M.; Losi, L.; Di Gregorio, C.; Gismondi, V.; Scarselli, A.; Benatti, P.; Roncari, B.; Seidenari, S.; Pellacani, G.; Varotti, C.; Prete, E.; Varesco, L.; Roncucci, L. Attenuated familial adenomatous polyposis and Muir-Torre syndrome linked to compound biallelic constitutional MYH gene mutations. *Clinical genetics,* **2005,** *68* (5), 442-7.

[110] Baglioni, S.; Melean, G.; Gensini, F.; Santucci, M.; Scatizzi, M.; Papi, L.; Genuardi, M. A kindred with MYH-associated polyposis and pilomatricomas. *Am J Med Genet A,* **2005,** *134* (2), 212-4.

[111] Tricarico, R.; Bet, P.; Ciambotti, B.; Di Gregorio, C.; Gatteschi, B.; Gismondi, V.; Toschi, B.; Tonelli, F.; Varesco, L.; Genuardi, M. Endometrial cancer and somatic G>T KRAS transversion in patients with constitutional MUTYH biallelic mutations. *Cancer letters,* **2009,** *274* (2), 266-70.

[112] Wasielewski, M.; Out, A. A.; Vermeulen, J.; Nielsen, M.; van den Ouweland, A.; Tops, C. M.; Wijnen, J. T.; Vasen, H. F.; Weiss, M. M.; Klijn, J. G.; Devilee, P.; Hes, F. J.; Schutte, M.

Increased MUTYH mutation frequency among Dutch families with breast cancer and colorectal cancer. *Breast cancer research and treatment,* **2010,** *124* (3), 635-41.

[113] Beiner, M. E.; Zhang, W. W.; Zhang, S.; Gallinger, S.; Sun, P.; Narod, S. A. Mutations of the MYH gene do not substantially contribute to the risk of breast cancer. *Breast cancer research and treatment,* **2009,** *114* (3), 575-8.

[114] Out, A. A.; Wasielewski, M.; Huijts, P. E.; van Minderhout, I. J.; Houwing-Duistermaat, J. J.; Tops, C. M.; Nielsen, M.; Seynaeve, C.; Wijnen, J. T.; Breuning, M. H.; van Asperen, C. J.; Schutte, M.; Hes, F. J.; Devilee, P. MUTYH gene variants and breast cancer in a Dutch case-control study. *Breast cancer research and treatment,* **2012.**

[115] Ashton, K. A.; Proietto, A.; Otton, G.; Symonds, I.; Scott, R. J. Genetic variants in MUTYH are not associated with endometrial cancer risk. *Hereditary cancer in clinical practice,* **2009,** *7* (1), 3.

[116] Sampson, J. R.; Jones, N. MUTYH-associated polyposis. *Best practice & research. Clinical gastroenterology,* **2009,** *23* (2), 209-18.

[117] Theodoratou, E.; Campbell, H.; Tenesa, A.; Houlston, R.; Webb, E.; Lubbe, S.; Broderick, P.; Gallinger, S.; Croitoru, E. M.; Jenkins, M. A.; Win, A. K.; Cleary, S. P.; Koessler, T.; Pharoah, P. D.; Kury, S.; Bezieau, S.; Buecher, B.; Ellis, N. A.; Peterlongo, P.; Offit, K.; Aaltonen, L. A.; Enholm, S.; Lindblom, A.; Zhou, X. L.; Tomlinson, I. P.; Moreno, V.; Blanco, I.; Capella, G.; Barnetson, R.; Porteous, M. E.; Dunlop, M. G.; Farrington, S. M. A large-scale meta-analysis to refine colorectal cancer risk estimates associated with MUTYH variants. *British journal of cancer,* **2010,** *103* (12), 1875-84.

[118] Lefevre, J. H.; Colas, C.; Coulet, F.; Baert-Desurmont, S.; Mongin, C.; Tiret, E.; Frebourg, T.; Soubrier, F.; Parc, Y. Frequent mutation in North African patients with MUTYH-associated polyposis. *Clinical genetics,* **2010.**

[119] Jones, N.; Vogt, S.; Nielsen, M.; Christian, D.; Wark, P. A.; Eccles, D.; Edwards, E.; Evans, D. G.; Maher, E. R.; Vasen, H. F.; Hes, F. J.; Aretz, S.; Sampson, J. R. Increased colorectal cancer incidence in obligate carriers of heterozygous mutations in MUTYH. *Gastroenterology,* **2009,** *137* (2), 489-94, 494 e1; quiz 725-6.

[120] Cleary, S. P.; Cotterchio, M.; Jenkins, M. A.; Kim, H.; Bristow, R.; Green, R.; Haile, R.; Hopper, J. L.; LeMarchand, L.; Lindor, N.; Parfrey, P.; Potter, J.; Younghusband, B.; Gallinger, S. Germline MutY human homologue mutations and colorectal cancer: a multisite case-control study. *Gastroenterology,* **2009,** *136* (4), 1251-60.

[121] Win, A. K.; Cleary, S. P.; Dowty, J. G.; Baron, J. A.; Young, J. P.; Buchanan, D. D.; Southey, M. C.; Burnett, T.; Parfrey, P. S.; Green, R. C.; Le Marchand, L.; Newcomb, P. A.; Haile, R. W.; Lindor, N. M.; Hopper, J. L.; Gallinger, S.; Jenkins, M. A. Cancer risks for monoallelic MUTYH mutation carriers with a family history of colorectal cancer. *International journal of cancer,* **2010.**

[122] Tenesa, A.; Farrington, S. M.; Dunlop, M. G. Re: Association between biallelic and monoallelic germline MYH gene mutations and colorectal cancer risk. *Journal of the National Cancer Institute,* **2005,** *97* (4), 320-1; author reply 321-2.

[123] Webb, E. L.; Rudd, M. F.; Houlston, R. S. Colorectal cancer risk in monoallelic carriers of MYH variants. *American journal of human genetics,* **2006,** *79* (4), 768-71; author reply 771-2.

[124] Tenesa, A.; Campbell, H.; Barnetson, R.; Porteous, M.; Dunlop, M.; Farrington, S. M. Association of MUTYH and colorectal cancer. *British journal of cancer,* **2006,** *95* (2), 239-42.

[125] Lubbe, S. J.; Di Bernardo, M. C.; Chandler, I. P.; Houlston, R. S. Clinical implications of the colorectal cancer risk associated with MUTYH mutation. *J Clin Oncol,* **2009,** *27* (24), 3975-80.

[126] Casper, M.; Plotz, G.; Juengling, B.; Zeuzem, S.; Lammert, F.; Raedle, J. MUTYH hotspot mutations in unselected colonoscopy patients. *Colorectal Dis,* **2011**.

[127] Eliason, K.; Hendrickson, B. C.; Judkins, T.; Norton, M.; Leclair, B.; Lyon, E.; Ward, B.; Noll, W.; Scholl, T. The potential for increased clinical sensitivity in genetic testing for polyposis colorectal cancer through the analysis of MYH mutations in North American patients. *Journal of medical genetics,* **2005,** *42* (1), 95-6.

[128] Piccioli, P.; Serra, M.; Gismondi, V.; Pedemonte, S.; Loiacono, F.; Lastraioli, S.; Bertario, L.; De Angioletti, M.; Varesco, L.; Notaro, R. Multiplex tetra-primer amplification refractory mutation system PCR to detect 6 common germline mutations of the MUTYH gene associated with polyposis and colorectal cancer. *Clinical chemistry,* **2006,** *52* (4), 739-43.

[129] van Puijenbroek, M.; Nielsen, M.; Tops, C. M.; Halfwerk, H.; Vasen, H. F.; Weiss, M. M.; van Wezel, T.; Hes, F. J.; Morreau, H. Identification of Patients with (Atypical) MUTYH-Associated Polyposis by KRAS2 c.34G > T Prescreening Followed by MUTYH Hotspot Analysis in Formalin-Fixed Paraffin-Embedded Tissue. *Clin Cancer Res,* **2008,** *14* (1), 139-42.

[130] Mecklin, J. P.; Jarvinen, H. J. Surveillance in Lynch Syndrom. *Familial Cancer,* **2005,** *4*, 5.

[131] Nielsen, M.; van Steenbergen, L. N.; Jones, N.; Vogt, S.; Vasen, H. F.; Morreau, H.; Aretz, S.; Sampson, J. R.; Dekkers, O. M.; Janssen-Heijnen, M. L.; Hes, F. J. Survival of MUTYH-associated polyposis patients with colorectal cancer and matched control colorectal cancer patients. *Journal of the National Cancer Institute,* **2010,** *102* (22), 1724-30.

Send Orders of Reprints at reprints@benthamscince.net
Advances in Genome Science, Vol. 1, 2013, 101-145

CHAPTER 5

Gene Clusters, Molecular Evolution and Disease: A Speculation

Alireza Baradaran-Heravi[1], Leah I. Elizondo[2,3,4] and Cornelius F. Boerkoel[1,*]

[1]Department of Medical Genetics, Child and Family Research Institute, University of British Columbia, Vancouver, British Columbia, Canada; [2]Interdepartmental Program in Cellular and Molecular Biology, [3]Department of Molecular and Human Genetics, Baylor College of Medicine, Houston, Texas, USA and [4]Centre for Molecular Medicine and Therapeutics, Child and Family Research Institute, Department of Medical Genetics, University of British Columbia, Vancouver, Canada

Abstract: Traditionally eukaryotic genes are considered independently expressed under the control of their promoters and *cis*-regulatory domains. However, recent studies in worms, flies, mice and humans have shown that genes co-habiting a chromatin domain or "genomic neighborhood" are frequently co-expressed. Often these co-expressed genes neither constitute part of an operon nor function within the same biological pathway. The mechanisms underlying the partitioning of the genome into transcriptional genomic neighborhoods are poorly defined. However, cross-species analyses find that the linkage among the co-expressed genes of these clusters is significantly conserved and that the expression patterns of genes within clusters have co-evolved with the clusters. Such selection could be mediated by chromatin interactions with the nuclear matrix and long-range remodeling of chromatin structure. In the context of human disease, we propose that dysregulation of gene expression across genomic neighborhoods will cause highly pleiotropic diseases. Candidate genomic neighborhood diseases include the nuclear laminopathies, chromosomal translocations and genomic instability disorders, imprinting disorders of errant insulator function, syndromes from impaired cohesin complex assembly, as well as diseases of global covalent histone modifications and DNA methylation. The alteration of transcriptional genomic neighborhoods provides a model for studying epigenetic alterations as quantitative traits in complex common human diseases.

Keywords: Gene clusters, genomic neighborhoods, nuclear organization, gene expression.

INTRODUCTION

Gene expression is epigenetically regulated by DNA methylation, histone modifications, and chromatin remodeling [1]. Along with DNA helicity, the covalent

***Address correspondence to Cornelius F. Boerkoel:** Provincial Medical Genetics Program, Department of Medical Genetics, Children's and Women's Health Centre of BC, 4500 Oak St., Rm. C234, Vancouver, B.C. V6H 3N1 Canada; Tel: 604-875-2157; Fax: 604-875-2376; E-mail: boerkoel@interchange.ubc.ca modification of DNA and histones alters chromatin structure changing promoter accessibility to transcription factors [2-6]. These DNA and histone modifications impose a level of transcriptional regulation in addition to the nuclear concentration and covalent modification of transcription factors. Therefore, in the context of a gene promoter, these DNA and histone modifications influence the binding of transcription factors and the regulation of gene expression [1, 7].

Except for a few clusters of functionally related genes such as the rRNA, histone, Hox, and globin genes that are frequently co-regulated, the general perception has been that functionally unrelated genes are not co-regulated but are independently expressed under the control of their promoters and associated regulatory domains. However, in contrast to this perception, several studies in yeast, worms, flies, mice and humans have shown that genes co-habiting a chromatin domain or "genomic neighborhood" are frequently co-expressed [8-16]. Often these co-expressed genes do not function within the same biological pathway [16-18].

The mechanisms for partitioning the genome into clusters of co-expressed genes are poorly understood. However, the conservation of clustering across species suggests that this organization did not arise randomly but by selection and therefore that perturbation of this organization could be deleterious. We review the evidence for co-expressed gene clusters, or genomic neighborhoods, discuss possible regulators of these clusters, and review a group of human disorders potentially disrupting genomic neighborhoods.

Genomic Neighborhoods

Microarray, Serial Analysis of Gene Expression (SAGE), Expressed-Sequence Tag (EST) and transcriptome analyses of gene expression have shown that many genes are organized into co-expressed clusters. Simplistically, these co-expressed gene clusters can be organized into three groups: 1) paralogous genes arising from gene duplications, 2) functionally related, non-duplicated genes, and 3) unrelated, non-duplicated genes that are co-expressed. Paralogous gene clusters contain genes that arose through tandem duplications resulting in genes that are often co-expressed as is observed for duplicate gene clusters in *C. elegans* [11]. However, as illustrated by the clustered Hox genes, which are temporospatially co-regulated, not all clustered paralogous genes are necessarily co-expressed [19].

Clusters of functionally related genes include genes within the same pathway, genes encoding interacting proteins, or genes that affect the same trait. Not all functionally related genes are clustered within or among species. For example, genes within the same metabolic pathway, as defined by the Kyoto Encyclopedia of Genes and Genomes (KEGG) cluster in humans, whereas genes within the same Gene Ontology do not cluster in humans [20, 21]. Clusters of co-expressed unrelated, non-duplicated genes can be subdivided into those that are co-expressed at a fixed developmental time and/or location or into those that are co-expressed at all times and in all locations. For example, within the human genome, co-expression of housekeeping gene clusters has been found in addition to tissue-specific gene clustering of cartilage-expressed transcripts [22, 23].

S. Cerevisiae

Genomic clustering of co-expressed genes was first identified in studies of *S. cerevisiae* genes with cell-cycle dependent expression patterns [10]. About 25% of the genes expressed during the same phase of the cell cycle are arranged as pairs [24].

C. Elegans

Unlike most other eukaryotes, *C. elegans* has approximately 15 percent of its genes organized into operons, which are transcribed as polycistronic messages. These messages are subsequently processed into monocistronic mRNAs by trans-splicing [25]. Thus the genes within these operons are transcriptionally co-regulated. Additionally, tandem duplication of genes results in co-expression of many paralogues. Interestingly however, transcriptional co-expression is not limited to operons and tandem duplications because when the transcriptional data from such genes are excluded, the remaining genes are still transcriptionally clustered [11, 13, 15]. Some of this clustering reflects co-regulation of tissue-specific genes such as those expressed during spermatogenesis and in larval muscle [13, 15]. These observations support the idea that genomic neighborhoods are structural features that are, at least in part, independent of their occupants.

Drosophila

In *Drosophila melanogaster*, about 20% of genes are arranged into clusters spanning 20 to 200 kb and containing 10 to 30 genes each [16]. Spellman and

Rubin found that although paralogues and genes encoding functionally related products cluster, the gene composition of many other clusters is not defined by these identifiers [16]. In the transcriptional co-expression of adjacent genes observed by Boutanaev *et al.*, a third of testes-specific genes were expressed in clusters of two to six genes yet only about 16% of these clusters contained two or more paralogues [8]. Therefore, gene duplications account for many two-gene clusters but are not characteristic of clusters containing three or more co-expressed genes [8]. Moreover, Meadows *et al.* showed that disruption of the linear organization of genes in a neighborhood in *Drosophila* testis does not alter gene expression within the neighborhood [26].

Mouse

Profiling of murine testes-specific ESTs showed that the mouse genome contains clusters of genes that are specifically expressed in the testis [12]. As observed in *C. elegans* and *Drosophila*, non-paralogous genes co-expressed in the testes were organized into large clusters on all mouse chromosomes [12]. Similarly, the profiling of murine oocyte-specific ESTs identified oocyte-specific clusters of co-expressed genes adjacent to the telomeres of chromosomes 9, 12, 14 and 19 [14]. Because oocyte-specific genes outside of these clusters were not biased toward a telomeric location, the authors suggested that the oocyte-specific telomeric clusters were nonrandom [14]. Additionally they suggested that this organization facilitates silencing of oocyte-specific genes in non-ovarian tissues because murine telomeres were highly heterochromatic [14]. Therefore these oocyte-specific clusters may constitute genomic neighborhoods that are regulated and controlled in a tissue-specific manner by higher order genomic organization.

Human

In humans, highly expressed genes cluster in large domains called RIDGEs (Regions of IncreaseD Gene Expression) and, interestingly, 40-50% of these RIDGEs map to telomeres [9]. Caron *et al.* observed that chromosomes with high gene density contain an increased number of RIDGEs when compared to chromosomes with low gene density [9]. More specifically, RIDGEs themselves have high gene, short interspersed nuclear repetitive DNA element (SINE), and GC content [9, 27].

Examining human Serial Analysis of Gene Expression (SAGE) data from 14 different types of tissues, Lercher *et al.* identified co-expression clusters of highly expressed housekeeping genes but not of tissue-specific genes [22]. Similar to RIDGEs, these clusters were highly correlated with regions of high GC content [27, 28]. Based on these observations, they hypothesized that housekeeping genes would be clustered in open chromatin to facilitate transcription. Consistent with this, they found that highly expressed gene clusters associated cytogenetically with the lightest staining Giemsa bands, *i.e.*, relaxed chromatin [28].

While mapping cartilage-expressed transcripts in the human genome, Yager *et al.* was able to identify clustering of co-expressed non-housekeeping genes, and in the liver and colon, respectively, Yamashita *et al.* identified six and five domains of genes highly expressed [29]. Moreover, it has been shown that human embryonic stem cells contain co-expression domains that are distinguishable from co-expression domains identified in the embryoid body [30]. These data suggest that not all clusters of co-expressed genes consist of housekeeping genes, but rather some clusters also constitute tissue-specific genes, or even differentiation-specific genes.

Further refining the co-expression clusters within the human genome, Sémon and Duret found that ~65% of the human genes belong to co-expressed gene clusters and that the distribution of cluster size, as judged by the number of genes, is biased toward small clusters [31]. The clusters that they defined also were not limited to functionally related or housekeeping genes.

In summary, clustering of co-expressed genes occurs with each species and across species [20, 31-33]. Interestingly, this is not only applicable to paralogues, functionally related genes, tissue-specific genes, or housekeeping genes but also to apparently unrelated genes. Studies in several species show that clusters of co-expressed genes fall within genomic regions containing low recombination rates, demonstrate co-evolution of gene expression with gene clustering, and conservation of clusters between species. These observations suggest that these co-expressed clusters are maintained by natural selection [34-37]. Therefore, if the assembly of co-expressed genes into clusters is nonrandom, then this organization must be highly regulated within the three-dimensional structure of the nucleus, and one could hypothesize that the loss of this organization would be deleterious to the organism.

Three-Dimensional Nuclear Organization and Gene Regulation

Although the mechanisms responsible for partitioning the genome into regions of co-expressed clusters are incompletely defined, our current understanding of the three-dimensional organization of the nucleus suggests possible candidate mechanisms. Factors contributing to the formation of genomic neighborhoods likely include chromosomal spatial orientation, chromatin interaction with the nuclear lamina, and the association of chromatin with subnuclear structures.

Oliver *et al.* proposed three models for co-expression of non-paralogous genes: 1) the incidental expression model, 2) the chromatin domain model and 3) the three-dimensional space model [38]. The incidental expression model predicts that when a transcription factor binds a target gene, it also incidentally activates neighboring genes and that this incidental co-expression of neighboring genes may have little biological relevance. The structural domain model proposes that the opening of an entire genomic neighborhood facilitates the binding of the transcriptional machinery to any gene within the neighborhood and that these neighborhoods would be demarcated and regulated by boundaries such as insulator complexes. Finally, according to the three-dimensional space model, a target gene is recruited to a subnuclear location such as a transcription factory and adjacent genes would also be exposed to the transcription machinery and therefore also transcribed [38]. Each of these non-exclusive models emphasizes a different aspect of nuclear architecture and thus provides a slightly different mechanistic view of the role for nuclear architecture in regulation of genomic neighborhoods.

Nuclear Envelope

The lamins and associated proteins are the major components of the nuclear lamina, which underlies the inner membrane of the nuclear envelope. These proteins maintain the integrity of the nuclear envelope [39], provide a structural attachment point for chromatin [40], help define DNA replication sites [39, 41-43], localize nuclear bodies [44], regulate epigenetic modifications of chromatin [45], and facilitate DNA repair and transcription [46-52]. Spann *et al.*

showed that interfering with lamin organization inhibited RNA polymerase II activity and therefore suggested that the nuclear lamina is required for the assembly of basal transcription factors and RNA polymerase II [48]. A role for the nuclear lamina in gene expression is further substantiated by interaction of lamin-associated proteins with the transcriptional apparatus. For example, emerin, an inner nuclear envelope protein, not only binds lamins A and B but also interacts with the transcriptional factors GCL, Btf, Lmo7, BAF, and YT521-B [46, 49, 53-56]. Also, silencing of lamin B1 expression leads to lamin A/C-rich nuclear envelope blebs, hyperphosphorylated RNA polymerase II and histone marks of active transcription although the transcription is likely ineffective and associated with polymerase stalling at promoter-proximal sites [51]. Lastly, profiling of gene expression changes in cells with a dominant *LMNA* mutation showed alterations in gene expression and intranuclear chromosomal positions [52]. Therefore, although, the roles of the nuclear lamina in defining regions of transcriptional co-expression have yet to be fully discovered, the nuclear lamina does modulate global and tissue-specific gene expression.

Nuclear Genomic Organization

Rabl first proposed a territorial organization of interphase chromosomes in 1885 [57] and in 1909 Boveri introduced the term "chromosome territory" [58]. Interphase nuclei have regions of high density (heterochromatin), regions of lower density (euchromatin), and regions of lowest density (nucleoli) [59]. However, the DNA from the various chromosomes is not randomly intertwined, rather chromosomes occupy non-overlapping territories of irregular shape [60-64]. There are two existing models for the distribution of chromosome territories [65]. One model suggests that gene-dense chromosomes are located more internally than gene-poor ones [66, 67] and that this localization also correlates with transcriptional activity, replication timing, and GC content [68-71]. Although not consistently observed [72], this arrangement has been frequently reported for mammalian and chicken cells [62, 73-77] as well as for homologous chromosomes in higher primates [78] and syntenic chromosomes in humans and mice [79-81]. Another model proposes that chromosomal territories are organized by gene function. For example, chromosomal organization arises from interactions between chromosome territories or domains required for DNA repair and homologous recombination [65, 82].

Development-specific and cell cycle-specific spatial alignment of homologous chromosome segments occurs in *Drosophila* [83-85] and for mammalian chromosomes in a variety of cells [86, 87]. Interestingly, the positioning of interphase chromosomes is largely inherited from mother to daughter nuclei in mammals suggesting that this may be a high order epigenetic regulatory mechanism [88-92].

Subnuclear Structures and the Interchromosomal Domains

Chromatin from separate territories minimally intermingles [62, 93-95]. Electron microscopy and polymerization of probes such as vimentin clearly define an interchromosomal domain or space between chromosomal territories [96-100]. Additionally, supporting separation of the chromosomal territories, higher resolution analysis of nuclear genomic organization using chromosome conformation capture techniques has identified a predominance of intra-chromosomal physical interactions compared to inter-chromosomal interactions [101, 102]. Nuclear bodies, such as Cajal and PML bodies, speckles, and specific nascent RNA accumulations lie in the interchromosomal domains along the surface of chromosome territories and are excluded from the chromosome territories [97, 100, 103-105]. Genes are generally distributed along the periphery of chromosome territories and invaginating channels where they can loop out into interchromosomal domains upon the induction of expression [79, 80, 106-114]. Thus, transcription may predominate along the periphery of a chromosome territory to allow access for transcription factors and to facilitate processing and transport of mRNA [62, 99, 100, 107]. Moreover, these findings suggest that the topology of genes relative to the interchromosomal domain compartment might provide a framework for the clustering of co-expressed genes and contribute to the overall regulation of gene expression [62, 93, 99, 100, 107, 115, 116].

Nuclear Matrix Interactions and Boundary Elements

Interactions between the nuclear lamina and chromatin have been mapped by DamID in *Drosophila*, mouse and human cell lines [117-119]. The lamina-associated domains are generally gene-poor and transcriptionally inactive, and when those domains dissociate from the lamina, the genes within them generally become transcriptionally active or poised for transcription [117-119].

Observations from yeast, *Drosophila*, chicken, mouse and human studies show that insulators compartmentalize the genome into separate regions of gene expression through interactions with DNA, the nuclear matrix, and other protein components [120-122]. Through these interactions, insulator complexes are proposed to facilitate the formation of higher order genomic structures that can be independently regulated and are defined by two activities: 1) they can inhibit the spread of heterochromatin, and/or 2) they can block a transcriptional enhancer from activating a promoter when located between the two [122].

One of the best-studied insulators is the *gypsy* element found in the *Drosophila gypsy* retrotransposon. Proteins that directly or indirectly bind the *gypsy* insulator include suppressor of hairy wing, modifier of mdg4, Centrosomal protein 190 kD, Trithorax-like, and dTopors (Drosophila Topoisomerase I-interacting RS proteins); the last binds the *gypsy* insulator to the nuclear lamina [123-127]. Modulation of the interaction between these proteins and the gypsy insulator sequences defines higher order chromatin structures, such as chromatin loops, and the functionality of each gypsy sequence [123, 124, 126].

In vertebrates, our understanding of insulator function primarily comes from studies of imprinted genomic domains such as 11p15.5 and from the beta-globin gene cluster [128, 129]. CCCTC-binding factor (CTCF), which binds the insulator sequences in these regions as well as many other sites, is the best-characterized vertebrate insulator-binding protein. The 5′ chicken HS4 insulator sequence, which is located upstream of the beta-globin cluster, has two separable functions: 1) it inhibits the adjacent heterochromatin from spreading into the beta-globin locus, and 2) it blocks enhancer-promoter interactions [130]. In addition to binding sequences such as HS4, CTCF binds nucleophosmin, a nucleolar protein and a component of the nuclear matrix [131-133]. Binding of CTCF to the HS4 and other insulators is proposed to facilitate the formation of chromatin loops and to tether these loops to subnuclear structures, such as the nucleolus and the nuclear matrix [133]. In cultured human lung fibroblasts the lamin B1-associated domains are flanked by CTCF [118], and association of CTCF with lamin A is required for localization of telomeres to nuclear periphery [134]. Lastly, CTCF is not only involved in chromosome-matrix interactions but also in intra- and inter-chromosomal interactions [135-138]. Therefore, by maintaining three-dimensional nuclear organization, insulators modulate various basic cellular functions including transcription (for a review see [139, 140]).

Another chromatin structural motif that is potentially involved in the establishment of genomic neighborhoods is the cohesin complex. Cohesin complexes were originally identified for their role in maintaining sister chromatid cohesion prior to their separation during anaphase [141]. The *Drosophila* protein Nipped-B loads the cohesin complex onto the chromosomes and is also required for facilitating enhancer-promoter communication [142-144] and regulating gene expression [142, 145]. These findings as well as the association of cohesin complexes with CTCF target sequences in mammalian cells suggest that cohesin complexes are structural elements defining and regulating gene co-expression domains. Indeed, impaired or errant localization of cohesin complexes affect RNA transcription and putatively do so across large genomic regions [146-155].

In summary, the three-dimensional structure of the nucleus provides a context for understanding and elucidating transcriptional genomic neighborhoods and an experimental basis for understanding the incidental expression, chromatin domain and three-dimensional space models for clustering of co-expressed genes. The intricate architecture of the nucleus and the nonrandomness of genomic neighborhoods suggest that perturbations at many different levels could cause dysregulation of genomic neighborhoods resulting in altered spatial localization of chromatin within the nucleus, and ultimately predisposing to diseases that affect multiple organ systems.

Deregulation of Genomic Organization and Disease

To date, most epigenetic research in human disease has focused on the histone and DNA modifications occurring at discrete gene loci. Increasingly, however, genome-wide alterations are being identified in human cancers and in rare Mendelian disorders. These disorders affect gene expression through disruption of higher order genomic organization, and we review a set of diseases illustrating this. We have grouped the disorders in a hierarchy extending from three-dimensional intramolecular positioning of chromosomes to those resulting in molecular alterations of DNA across genomic neighborhoods.

Chromosomal Rearrangements: Disorders of Chromatin Localization?

As reviewed above, several studies have suggested that there are nuclear territories specific for each chromosome within a given cell type [65, 73, 156-159]. Additionally, studies have shown that there are specific transcriptional regions or neighborhoods within the nucleus [62, 160]. These two observations suggest that reciprocal translocation of chromosomal segments would alter gene expression by placing the gene in a different transcriptional environment within the nucleus. Thus, the pathology arising from the translocations is not limited to disrupting genes, separating genes from *cis* regulatory elements, or placing genes within a new *cis* chromatin environment [161-164] but could also arise by changing the position of a gene or group of genes relative to the inter-chromosomal domain compartment and associated transcription factors. To date, this model has not been extensively investigated in somatic or germline human diseases associated with chromosomal rearrangements and genome instability.

Nuclear Envelopathies: Disorders of Nuclear Envelop Integrity and Maintenance of Chromosomal and Inter-Chromosomal Domains?

The lamins and associated proteins constitute major components of the nuclear lamina and are involved in maintaining the integrity of the nuclear envelope [39], providing a structural attachment point for chromatin [40], defining DNA replication sites [39, 41-43], localizing nuclear bodies [44], regulating epigenetic modifications of chromatin [45], and mediating transcription [46-52]. Based on these identified functions, disruption of the nuclear matrix would impair the ability of chromatin to attach to the nuclear lamina and thereby impair the ability of a cell to define and to maintain the domains necessary for appropriate gene expression and genomic stability. Mutations of nuclear lamina proteins have been identified in many different diseases including Emery-Dreifuss muscular dystrophy (OMIM 310300, 181350, 604929), dilated cardiomyopathy (OMIM 115200), Charcot-Marie-Tooth disease type 2B1 (OMIM 605588), mandibuloacral dysplasia (OMIM 248370, 608612), Dunnigan-type familial partial lipodystrophy (OMIM 151660), Hutchinson-Gilford progeria (OMIM 176670), atypical Werner syndrome (OMIM 277700), Pelger-Huët anomaly (OMIM 169400), Greenberg skeletal dysplasia (OMIM 215140), restrictive dermopathy (OMIM 275210), melorheostosis (OMIM 155950), Buschke-Ollendorff syndrome (OMIM 166700), and lipoatrophy with diabetes, hepatic steatosis, hypertrophic cardiomyopathy, and leukomelanodermic papules (OMIM 608056) [165].

Given the role the nuclear lamina plays in regulating transcription [46-49, 51, 52], we propose that disruption of the nuclear lamina might affect transcription by altering the nuclear organization defining genomic neighborhoods. Consistent with this hypothesis, studies of gene expression in mandibuloacral dysplasia and Emery-Dreifuss muscular dystrophy, dilated cardiomyopathy, and Hutchinson-Gilford progeria showed deregulation of gene expression at multiple loci [166-170]. Therefore, superimposed on the impaired DNA repair and nuclear envelope stability, alterations in transcription help explain the pleiotropy, variable penetrance, and variable expressivity of diseases associated with mutant lamina proteins.

Schimke Immuno-Osseous Dysplasia: A Disorder of Genomic Neighborhood Establishment and Maintenance?

Mutations in *SMARCAL1* (SWI/SNF-related, matrix-associated, actin-dependent regulator of chromatin, subfamily A-like 1) cause Schimke immuno-osseous dysplasia (SIOD: OMIM 242900). SIOD is an autosomal recessive disorder of T-cell immunodeficiency, spondyloepiphyseal dysplasia, renal failure, hypothyroidism, episodic cerebral ischemia, and bone marrow failure. Biallelic loss-of-function mutations of *SMARCAL1* are associated with the disease [171]. This gene encodes a DNA-dependent, RNA-independent ATPase that binds DNA at single-strand to double-strand transitions and has annealing helicase activity [172-174]. Both clinical and cell culture studies suggest that functional SMARCAL1 protein is involved in DNA replication, repair, recombination and transcription [7, 175-186]. The SMARCAL1 protein is predominantly localized to the open chromatin and loss of functional SMARCAL1 alters transcription identifying one mechanism for cell autonomous disease [7, 171, 174]. In SMARCAL1 mutant tissues, transcription is altered across large regions of the genome suggesting that SMARCAL1 is necessary either for the establishment or maintenance of transcriptional genomic neighborhoods [7, Boerkoel CF and co-workers unpublished results].

Cornelia de Lange Syndrome: A Disorder of Cohesin-Regulated Transcription?

Mutations in *NIPBL* (Nipped-B-Like), which encodes a cohesin regulator, and *SMC1A* and *SMC3*, which encode cohesin structural components, result in Cornelia de Lange syndrome (CDLS: OMIM 122470) [187-190]. CDLS is an autosomal dominant (*NIPBL* and *SMC3*) or X-linked (*SMC1A*) multisystem disease characterized by distinctive facial features, growth retardation, hirsutism, upper limb reduction defects, cognitive impairment, and behavioral abnormalities as well as other congenital malformations. In *Saccharomyces cerevisiae, S. pombe* and *Xenopus*, the *NIPBL* homologues (*Scc2*, *Mis4* and *Xscc2*), known collectively as adherins, load the cohesin protein complex onto chromosomes [191-195] and are required for sister chromatid cohesion as well as for DNA repair and genome stability [189, 196-198]. As described above, the *Drosophila NIPBL* homologue, *Nipped-B*, is not only required for sister chromatid cohesion but also for facilitating enhancer-promoter communication [142-144] and is associated with transcriptionally active regions [145]. Similarly, in murine stem cells, Nipbl, in combination with the cohesin and mediator complexes contributes to DNA loop formation between enhancers and core promoters [149]. The sum of these observations suggests that the pathology of CDLS is not only due to precocious sister chromatid separation [199] but also due to aberrant transcription. Since the cohesin complex binds approximately every 10kb along the *S. cerevisiae and S. pombe* interphase chromosome [200, 201], impaired or errant localization of cohesin complexes have the potential to profoundly affect RNA transcription across large genomic regions, and indeed, recent studies in human, mouse, *Drosophila,* and zebrafish tissues find that reduction in the dosage of *NIPBL* homologues causes modest global alterations in gene expression [150, 202-204].

Roberts syndrome: A Disorder of Cohesin-Regulated Transcription?

Mutations in *ESCO2* (establishment of cohesion 1 homolog 2 (S. cerevisiae)) [205, 206], which encodes an acetyltransferase, has been shown to cause Roberts syndrome (RBS: OMIM 268300) and SC phocomelia (OMIM 269000). RBS is an autosomal recessive disease characterized by hypomelia, growth deficiency, craniofacial anomalies, microcephaly, and mental deficiency. Severely affected infants may be stillborn or die shortly after birth. SC phocomelia, a milder disorder, has less limb reduction and includes flexion contractures, midfacial hemangioma, hypoplastic cartilage of ears and nose, scant silvery-blond hair, and cloudy corneae. In *S. cerevisiae, S. pombe, Drosophila* and *H. sapiens*, the *ESCO2* homologues are required for sister chromatid cohesion [207-211] and are bound to chromosomes throughout the cell cycle [207, 208]. In addition to affecting mitotic progression and ploidy [212], RBS cells are hypersensitive to DNA damaging agents [213-215] and Esco2 deficient zebrafish have abnormal expression of many genes, particularly those regulating the cell cycle [205]. We hypothesize therefore that, like Cornelia de Lange syndrome, mutations of ESCO2 could give rise to aberrant transcription across large genomic regions and that this partially accounts for the pathology of RBS and SC phocomelia.

Rubenstein Taybi Syndrome: A Local or Global Disorder of Histone Acetylation?

Mutations in *CREBBP* (cAMP-responsive element binding protein (CREB) binding protein) and in *EP300* (E1A-binding protein p300) cause Rubinstein Taybi syndrome (RTS: OMIM 180849) [216, 217]. RTS is an autosomal dominant disorder of growth retardation, facial abnormalities, broad thumbs and toes, and mental retardation. As histone acetylases, CREBBP and EP300 promote the decondensation of chromatin and interact with many transcription factors to facilitate RNA transcription [218, 219]. More specifically, CREBBP-mediated histone acetylation plays a role in synaptic plasticity, long term memory and in ameliorating neuronal degeneration and thus can partially explain the mental retardation seen in RTS [220-223]. Suggesting an explanation for the malformations of RTS, mouse embryonic fibroblasts deficient for both CREBBP and EP300 exhibit altered transcriptional responses to developmental morphogens and other stimuli [224, 225]. If CREBBP and EP300 are generally involved in facilitating the transcription of many or most genes, then decreased dosage of CREBBP and EP300 could globally impair the RNA transcription including that of genomic neighborhoods.

ICF Syndrome: A Global Disorder of DNA Methylation?

Mutations in *DNMT3B* (DNA methyltransferase 3B) and *ZBTB24* (zinc-finger- and BTB (bric-a-bric, tramtrack, broad complex)-domain-containing 24) cause immunodeficiency-centromeric instability-facial anomalies syndrome-1 and 2, respectively (ICF1: OMIM 242860, ICF2: OMIM 614069) [226, 227]. Patients with ICF1 and ICF2 are clinically indistinguishable [228]. ICF syndrome is an autosomal recessive disorder of genomic methylation that mainly affects satellites 2 and 3 of constitutive heterochromatin, although centromeric alpha satellites [229], Alu sequences [230, 231], D4Z4 and NBL2 repeats [232], and certain imprinted genes [227, 233] are also involved. In addition, the inactive X chromosome (Xi), which consists of facultative heterochromatin, is globally undermethylated although all sequences are not equally affected in all patients [231, 234]. Interestingly, Xi CpG island demethylation is not accompanied by significant biallelic transcriptional reactivation [230, 235], and consistent with this observation, histone modifications along the Xi are normal [236].

Gene expression array analysis of ICF1 cell lines has identified global alterations in gene expression that are not consistently correlated with altered methylation at the regulatory regions of the dysregulated genes [237, 238]. The transcriptional pathology of ICF may arise from 1) translocation of chromosomal fragments to inappropriate chromosome territories [239-241]; 2) loss of repetitive DNA methylation impairing heterochromatin formation which is a key long-range transcriptional regulator; and 3) alteration of methylation of repetitive elements that regulate chromatin boundary elements which modulate transcription.

Rett Syndrome: A Disorder of Methylated DNA Recognition or of RNA Processing?

Mutations in *MECP2* (methyl-CpG-binding protein 2) cause a spectrum of X-linked disorders including Rett syndrome (OMIM 312750), neonatal-onset encephalopathy, mental retardation, autism, and an Angelman-like syndrome [242]. MECP2 binds methylated promoter sites and recruits the mSinA/Histone Deacetylase (HDAC) 1/2 corepressor complex [243] and thereby mediates dynamic repression of gene expression [244, 245]. These data and transcriptome and biochemical analyses suggest that the pathology associated with MECP2 mutations results at least partially from inappropriate gene expression as well as from dysregulation of alternative splicing [246-250]. Consistent with this hypothesis, tissues of some patients with mutations in *MECP2* exhibit histone H4 hyperacetylation [251] and findings in mouse neurons suggest that MeCP2 acts as a linker histone [252]. Thus loss of functional MECP2 may cause widespread alterations in gene expression by altering of RNA transcription and mRNA processing.

CHARGE Syndrome: A Disorder of Chromatin Remodeling and Transcription?

Mutations in *CHD7* (chromodomain helicase DNA-binding protein-7) are associated with coloboma, heart defect, atresia choanae, retarded growth and development, genital abnormality and ear abnormality (CHARGE) syndrome (OMIM 214800) [253, 254]. Most cases are sporadic but occasionally it occurs with autosomal dominant inheritance [255]. CHD7 is a member of CHD family of enzymes that belongs to SNF2 superfamily of ATP-dependent chromatin remodelers [256]. In addition to chromatin remodeling CHD7 modulates other basic cellular functions including transcription. Indeed, Kismet, the *Drosophila* orthologue for CHD7, has been shown to function downstream of the positive transcription elongation factor b (P-TEFb) and facilitate an early step in transcriptional elongation by RNA polymerase II [257]. Using chromatin immunoprecipitation and tiling microarrays (ChIP-chip), Schnetz *et al.* showed that CHD7 localization along chromatin is cell type specific and correlates with the location of methylated lysine 4 of histone H3 (H3K4) [258]. Furthermore, studies in mouse ES cells showed that CHD7 targets active gene enhancer elements with high levels of H3K4 mono-methylation and co-localizes with enhancer binding protein P300 and transcription factors OCT4, SOX2, and NANOG [259]. Analysis of lymphoblastoid cell lines derived from patients with CHARGE syndrome has revealed alterations in gene expression across various neighborhoods [260].

Coffin-Siris Syndrome: A Disorder of Chromatin Remodeling and Transcription?

Coffin-Siris syndrome (CSS: OMIM 135900) is a rare autosomal dominant congenital disorder characterized by intellectual disability, developmental delay, growth deficiency, coarse facial features and hypoplastic or absent fifth fingernails and/or toenails [261]. It has recently been shown that CSS is caused by *de novo* mutations in genes encoding several components of the chromatin remodeling SWI/SNF (SWItch/Sucrose NonFermenting) complex including

SMARCB1, *SMARCA4*, *SMARCA2*, *SMARCE1*, *ARID1A* and *ARID1B* [262, 263]. Components of SWI/SNF complex remodel chromatin and participate in a variety of cellular processes including DNA replication, DNA repair and recombination, gene expression, chromosome segregation, and nuclear organization [264]. Subunits of SWI/SNF complex associate with nuclear matrix proteins and transcription factors [265, 266]. Also, genome-wide mapping of several SWI/SNF components showed that they frequently occupy transcription start sites and enhancers, CTCF-binding domains and regions occupied by RNA polymerase II [265]. The sum of these data suggests that the pathophysiology of CSS arises at least partially from alterations in nuclear organization and gene expression. Further studies are required to delineate the transcriptional alterations in CSS.

ATRX Syndrome: A Disorder of Chromatin Remodeling and Transcription?

Mutations in the α-thalassemia mental retardation X-linked protein (ATRX) are associated with ATRX syndrome (OMIM 301040), which is characterized by mental retardation with facial features, mild α-thalassemia, and genital abnormalities [267]. ATRX is a member of the SNF2 family of ATP-dependent chromatin remodeling proteins and is involved in DNA replication and maintenance of genomic integrity during mitosis and meiosis [268-272], gene expression [273-276], telomere maintenance [277], and X chromosome inactivation [278]. It modulates DNA methylation and associates with heterochromatic or heterochromatic associated regions at pericentromeric repeats, ribosomal DNA arrays, telomeric repeats, and PML bodies [279-281]. These chromatin interactions of ATRX could be in part due to its direct interaction with trimethylated H3K9 and unmethylated H3K4 as well as interactions with heterochromatin protein (HP1α), cohesion complex MECP2, and death-associated protein (Daxx) [273, 276, 280-283]. The ATRX-Daxx complex participates in replication-independent deposition of H3.3, a nucleosome modification contributing to active chromatin [284-286]. ATRX can also repress transcription [273]. Genome-wide analysis of ATRX binding sites using ChIP-sequencing in human erythroid and mouse ES cells revealed that ATRX mostly targets GC-rich variable number tandem repeats (VNTRs) and that the degree of perturbation in the gene expression of ATRX target genes was related to the size of these repeats [287]. In the peripheral blood mononuclear cells of ATRX patients and mice tissues deficient for ATRX, expression of many genes is transcriptionally dysregulated [274, 275].

Cancer: A Panoply of Disorders Affecting all Levels of Genomic Organization?

Cancerous tissue and tumor-derived cell lines exhibit errors at nearly all levels of nuclear organization including chromosomal aberrations [288, 289], altered nucleosome spacing [164], covalent histone modifications [290], and DNA methylation [291]. Besides the established roles in loss of heterozygosity for tumor suppressor genes and oncogene activation, we propose chromosomal rearrangements might also play a role in cancer by altering the expression of genes across large domains by relocating them to a different chromosomal territory.

The cancer-associated ATP-dependent chromatin remodeling enzymes *SMARCB1* (SWI/SNF-related, matrix-associated, actin-dependent regulator of chromatin, subfamily B, member 1), *BRG1* (brahma-related gene 1), and *BRM* (brahma) regulate gene expression by altering nucleosome spacing [292-296]. Consistently, transcriptome analyses of mammary tumors from *Brg1* heterozygous mice show altered expression of hundreds of genes [297].

Abnormalities of global and targeted histone acetylation occur in several human neoplasms. In acute promylelocytic leukemia, acute lymphocytic leukemia, and non-Hodgkins lymphoma, mutations of proteins that recruit HDACs to DNA errantly activate and target interacting HDACs [292]. In several human cancers, this results in a reduction of overall histone acetylation [298]. In addition, histone acetylases are deregulated in some human neoplasias; this dysregulation plays a crucial role in tumor development and progression [299-302]. Underscoring the role of histone acetylation in cancer, HDAC inhibitors slow tumor cell proliferation and induce differentiation. These changes are associated with global affects on transcription [303, 304] and suggest that the tumor phenotype results from the altered expression of many genes.

Human tumor DNA is globally hypomethylated [305, 306]. Malignant cells can have 20%–60% less genomic 5-methyl-cytosine than normal counterpart cells [306]. However, genomic hypomethylation does not associate with overexpression of oncogenes. Rather three mechanisms have been invoked for the contribution of global DNA hypomethylation to carcinogenesis: chromosomal instability, reactivation of transposable elements, and loss of imprinting [307-310]. Each of these processes is involved in regulation of chromatin structure and likely in defining transcriptional genomic neighborhoods. Additionally, in contrast to the global hypomethylation, CpG islands in the promoter regions of many tumor suppressor genes undergo hypermethylation in cancer cells leading to gene silencing [311, 312], and the number of hypermethylated tumor suppressor genes increases with the malignant potential [312-315]. Thus at the level of DNA methylation, the cancer phenotype appears to be a combination of global chromatin structural changes through overall DNA hypomethylation and of targeted gene inactivation through promoter hypermethylation.

Defects in the nuclear matrix proteins also are associated with cancer and probably contribute to its pathogenesis by disrupting chromosome architecture and transcription. High and low expression of SATB1 (special AT-rich binding protein 1), which is a matrix attachment region (MAR)-binding protein that modulates chromatin architecture and regulates gene expression in various cellular processes [316-319], correlate with poor prognosis for breast cancer and lung cancer, respectively [320, 321]. Although the oncogenic consequences of the following remain to be defined, loss of CTCF binding upstream of the several tumor suppressor genes including *BRCA1*, *RB*, $p16^{INK4a}$, *RASSF1A* and *CDH1* correlates with aberrant DNA methylation of the respective promoters [322-324], and altered expression of CTCF has been observed in lung and breast cancer cell lines [325-327].

Genomic Organization and Natural Selection

The correlation of perturbed genomic neighborhoods with disease suggests that they are functionally significant and not purely coincidental. This hypothesis is also supported by several lines of evidence that these clusters arose and are preserved by natural selection. First, co-expression clusters contain fewer chromosomal breakpoints between human and mouse and *Drosophila* than expected by chance [36, 328]. Second, highly co-expressed clusters of genes are phylogenetically conserved between human and mouse and consist mainly of non-paralogous genes [329, 330]. Third, the genomic neighborhoods in *Drosophila melanogaster* are conserved across 12 studied *Drosophila* species [37, 331]. Fourth, comparison of the human and chicken genomes shows that there is more linkage retention among genes within clusters than outside of clusters [31]. Fifth, analysis of co-expression of neighboring genes within the mouse and human genomes suggests that there is co-evolution of the pattern of expression in neighboring genes [31]. Sixth, comparison of the human and chimpanzee genome shows that genes within conserved neighborhoods undergo less expression divergence than genes located in a structurally divergent neighborhood [332]. Lastly, co-expression of gene neighbors in *Caenorhabditis elegans* and *C. briggsae* are divergent when the genomic neighborhood is not conserved [333].

CONCLUSIONS

Advances in our understanding of gene expression and distribution in the genome make the supposition of random gene order in eukaryotes increasingly untenable. Recent advances in nuclear biology have not only initiated novel views of the role of genomic organization in gene expression regulation but also new insights into the cause of disease.

The nonrandom organization of the genome into a series of chromosomal blocks that are transcriptionally regulated provides possible insight into some recent unexplained observations during development. For example, during differentiation of hematopoietic stem cells, particular chromosomal regions are inactivated [334], and during differentiation of non-neuronal tissues, specific chromosomal intervals are transcriptionally down-regulated [335].

According to this model of genomic organization, the pathophysiology underlying many diseases could arise as a result of alterations in genomic organization causing quantitative transcriptional disturbances. The cumulative effect of these transcriptional disturbances or the confluence of aberrant transcription of several genes would then sufficiently perturb homeostasis to predispose to or cause disease. Thus, for the rare Mendelian disorders considered above, the chromosomal aberration or mutant protein could affect the disease phenotype by causing quantitative changes in gene transcription through altering the nuclear localization or structure of large chromatin domains. The variable expressivity of many of these disorders reflects the quantitative differences in transcription, and the phenotypic features can be considered quantitative traits.

Considering the features of these rare Mendelian disorders as quantitative traits provides a model for incomplete penetrance, variable expressivity and pleiotropism; characteristics which cannot be explained simply by dysregulation of a single pathway or gene interaction network. It also explains why segregation of gene mutations for these diseases frequently predispose to rather than cause disease and why many features of these diseases have genocopies. Lastly, since genomic neighborhoods are likely regulated by both genetic and epigenetic mechanisms, this model integrates both in the causation of human diseases.

Viewing co-expression of clustered genes as a quantitative trait also provides an explanation of why genomic neighborhoods would be a substrate for and not only a product of natural selection. In contrast to strong single gene mutations that result in dramatic changes in fitness, alterations in expression of genes, and likely gene clusters, allow gradual and persistent selection of offspring that are more fit than their parents [336]. Moreover, since mRNA levels for many genes are heritable [337], gene expression and co-expressed gene clusters could be the subject of selection over many generations. Consequently, alterations in the genomic neighborhoods would not only influence the genetic and phenotypic diversity in the populations but also likely contribute to speciation during evolution [332, 338].

In this context, the features of rare Mendelian disorders provide a novel method for modeling and studying quantitative traits and multifactorial disorders. Such disorders include mental retardation, diabetes, hypertension, atherosclerosis, vascular cognitive impairment, and Metabolic Syndrome (type 2 diabetes, cardiovascular disease, and hypertension). For example, mutations of *SMARCAL1* alter transcription across chromatin domains creating a chromatin environment permissive for other genetic and environmental insults to push the expression of several genes such as *ELN* beyond the threshold for disease [7, 185].

FUTURE DIRECTIONS

Defining better the selective pressure for and on genomic neighborhoods and determining whether transcriptional alterations of these are a common cause of complex disorders requires further testing. These future studies will provide an exciting opportunity to define the heritable epigenetic variation contributing to complex diseases and address many of the questions regarding the contribution of epigenetic variation to quantitative traits. Additionally, these studies will allow investigators to clarify whether the increasing incidences of diseases such as asthma, Metabolic Syndrome, and some neoplasias are due in part from environmental influences on epigenetic traits. Answering these questions is essential for defining the contribution of genomic organization in human disease.

ACKNOWLEDGEMENTS

This work was supported in part by a Ruth L. Kirschstein National Research Service Award (L.I.E.) and grants from the March of Dimes (C.F.B.), the Gillson Longenbaugh Foundation (C.F.B.), the Dana Foundation (C.F.B) and the New Development Award, Microscopy, and Administrative Cores of the Mental Retardation and Developmental Disabilities Research Center at Baylor College of Medicine (C.F.B.), the Burroughs Wellcome Foundation (C.F.B.), the National Institute of Diabetes, Digestive, and Kidney Diseases, NIH (C.F.B.), the Association Autour D'Emeric et D'Anthony (C.F.B.), the Michael Smith Foundation for Health Research (C.F.B.), Little Giants Foundation (C.F.B.), and the Child & Family Research Institute of British Columbia Children's Hospital (C.F.B.). C.F.B. is a scholar of the Michael Smith Foundation for Health Research.

CONFLICT OF INTEREST

The author(s) confirm that this chapter content has no conflict of interest.

DISCOLUSURE

"This is an updated report of the earlier review article By L.I. Elizondo, P. Jafar-Nejad, J.M. Clewing and C.F. Boerkoel" Gene clusters, molecular evolution and disease: a speculation ". Current Genomics, 2009 10(1):64-75".

REFERENCES

[1] Jiang, Y. H.; Bressler, J.; Beaudet, A. L., Epigenetics and human disease. *Annu Rev Genomics Hum Genet* **2004,** *5*, 479-510.

[2] Felsenfeld, G.; Groudine, M., Controlling the double helix. *Nature* **2003,** *421* (6921), 448-53.

[3] Hatfield, G. W.; Benham, C. J., DNA topology-mediated control of global gene expression in Escherichia coli. *Annu Rev Genet* **2002,** *36*, 175-203.

[4] Kouzine, F.; Sanford, S.; Elisha-Feil, Z.; Levens, D., The functional response of upstream DNA to dynamic supercoiling *in vivo*. *Nat Struct Mol Biol* **2008,** *15* (2), 146-54.

[5] Lim, H. M.; Lewis, D. E.; Lee, H. J.; Liu, M.; Adhya, S., Effect of varying the supercoiling of DNA on transcription and its regulation. *Biochemistry* **2003,** *42* (36), 10718-25.

[6] Pruss, G. J.; Drlica, K., DNA supercoiling and prokaryotic transcription. *Cell* **1989,** *56* (4), 521-3.

[7] Baradaran-Heravi, A.; Cho, K. S.; Tolhuis, B.; Sanyal, M.; Morozova, O.; Morimoto, M.; Elizondo, L. I.; Bridgewater, D.; Lubieniecka, J.; Beirnes, K.; Myung, C.; Leung, D.; Fam, H. K.; Choi, K.; Huang, Y.; Dionis, K. Y.; Zonana, J.; Keller, K.; Stenzel, P.; Mayfield, C.; Lucke, T.; Bokenkamp, A.; Marra, M. A.; van Lohuizen, M.; Lewis, D. B.; Shaw, C.; Boerkoel, C. F., Penetrance of biallelic SMARCAL1 mutations is associated with environmental and genetic disturbances of gene expression. *Human molecular genetics* **2012,** *21* (11), 2572-87.

[8] Boutanaev, A. M.; Kalmykova, A. I.; Shevelyov, Y. Y.; Nurminsky, D. I., Large clusters of co-expressed genes in the Drosophila genome. *Nature* **2002,** *420* (6916), 666-9.

[9] Caron, H.; van Schaik, B.; van der Mee, M.; Baas, F.; Riggins, G.; van Sluis, P.; Hermus, M. C.; van Asperen, R.; Boon, K.; Voute, P. A.; Heisterkamp, S.; van Kampen, A.; Versteeg, R., The human transcriptome map: clustering of highly expressed genes in chromosomal domains. *Science* **2001,** *291* (5507), 1289-92.

[10] Cho, R. J.; Campbell, M. J.; Winzeler, E. A.; Steinmetz, L.; Conway, A.; Wodicka, L.; Wolfsberg, T. G.; Gabrielian, A. E.; Landsman, D.; Lockhart, D. J.; Davis, R. W., A genome-wide transcriptional analysis of the mitotic cell cycle. *Molecular cell* **1998,** *2* (1), 65-73.

[11] Lercher, M. J.; Blumenthal, T.; Hurst, L. D., Coexpression of neighboring genes in Caenorhabditis elegans is mostly due to operons and duplicate genes. *Genome research* **2003,** *13* (2), 238-43.

[12] Li, Q.; Lee, B. T.; Zhang, L., Genome-scale analysis of positional clustering of mouse testis-specific genes. *BMC genomics* **2005,** *6* (1), 7.

[13] Miller, M. A.; Cutter, A. D.; Yamamoto, I.; Ward, S.; Greenstein, D., Clustered organization of reproductive genes in the C. elegans genome. *Curr Biol* **2004,** *14* (14), 1284-90.

[14] Paillisson, A.; Dade, S.; Callebaut, I.; Bontoux, M.; Dalbies-Tran, R.; Vaiman, D.; Monget, P., Identification, characterization and metagenome analysis of oocyte-specific genes organized in clusters in the mouse genome. *BMC genomics* **2005,** *6* (1), 76.

[15] Roy, P. J.; Stuart, J. M.; Lund, J.; Kim, S. K., Chromosomal clustering of muscle-expressed genes in Caenorhabditis elegans. *Nature* **2002,** *418* (6901), 975-9.

[16] Spellman, P. T.; Rubin, G. M., Evidence for large domains of similarly expressed genes in the Drosophila genome. *J Biol* **2002,** *1* (1), 5.

[17] Kalmykova, A. I.; Nurminsky, D. I.; Ryzhov, D. V.; Shevelyov, Y. Y., Regulated chromatin domain comprising cluster of co-expressed genes in Drosophila melanogaster. *Nucleic acids research* **2005,** *33* (5), 1435-44.

[18] Chen, N.; Stein, L. D., Conservation and functional significance of gene topology in the genome of Caenorhabditis elegans. *Genome research* **2006,** *16* (5), 606-17.

[19] Deschamps, J.; van den Akker, E.; Forlani, S.; De Graaff, W.; Oosterveen, T.; Roelen, B.; Roelfsema, J., Initiation, establishment and maintenance of Hox gene expression patterns in the mouse. *Int J Dev Biol* **1999,** *43* (7), 635-50.

[20] Fukuoka, Y.; Inaoka, H.; Kohane, I. S., Inter-species differences of co-expression of neighboring genes in eukaryotic genomes. *BMC genomics* **2004,** *5* (1), 4.

[21] Lee, J. M.; Sonnhammer, E. L., Genomic gene clustering analysis of pathways in eukaryotes. *Genome research* **2003,** *13* (5), 875-82.

[22] Lercher, M. J.; Urrutia, A. O.; Hurst, L. D., Clustering of housekeeping genes provides a unified model of gene order in the human genome. *Nature genetics* **2002,** *31* (2), 180-3.

[23] Yager, T. D.; Dempsey, A. A.; Tang, H.; Stamatiou, D.; Chao, S.; Marshall, K. W.; Liew, C. C., First comprehensive mapping of cartilage transcripts to the human genome. *Genomics* **2004,** *84* (3), 524-35.

[24] Cohen, B. A.; Mitra, R. D.; Hughes, J. D.; Church, G. M., A computational analysis of whole-genome expression data reveals chromosomal domains of gene expression. *Nature genetics* **2000,** *26* (2), 183-6.

[25] Zorio, D. A.; Cheng, N. N.; Blumenthal, T.; Spieth, J., Operons as a common form of chromosomal organization in C. elegans. *Nature* **1994,** *372* (6503), 270-2.

[26] Meadows, L. A.; Chan, Y. S.; Roote, J.; Russell, S., Neighbourhood continuity is not required for correct testis gene expression in Drosophila. *PLoS biology* **2010,** *8* (11), e1000552.

[27] Versteeg, R.; van Schaik, B. D.; van Batenburg, M. F.; Roos, M.; Monajemi, R.; Caron, H.; Bussemaker, H. J.; van Kampen, A. H., The human transcriptome map reveals extremes in gene density, intron length, GC content, and repeat pattern for domains of highly and weakly expressed genes. *Genome research* **2003,** *13* (9), 1998-2004.

[28] Lercher, M. J.; Urrutia, A. O.; Pavlicek, A.; Hurst, L. D., A unification of mosaic structures in the human genome. *Human molecular genetics* **2003,** *12* (19), 2411-5.

[29] Yamashita, T.; Honda, M.; Takatori, H.; Nishino, R.; Hoshino, N.; Kaneko, S., Genome-wide transcriptome mapping analysis identifies organ-specific gene expression patterns along human chromosomes. *Genomics* **2004,** *84* (5), 867-75.

[30] Li, H.; Liu, Y.; Shin, S.; Sun, Y.; Loring, J. F.; Mattson, M. P.; Rao, M. S.; Zhan, M., Transcriptome coexpression map of human embryonic stem cells. *BMC genomics* **2006,** *7*, 103.

[31] Semon, M.; Duret, L., Evolutionary origin and maintenance of coexpressed gene clusters in mammals. *Molecular biology and evolution* **2006,** *23* (9), 1715-23.

[32] Michalak, P., Coexpression, coregulation, and cofunctionality of neighboring genes in eukaryotic genomes. *Genomics* **2008,** *91* (3), 243-8.

[33] Hurst, L. D.; Pal, C.; Lercher, M. J., The evolutionary dynamics of eukaryotic gene order. *Nature reviews. Genetics* **2004,** *5* (4), 299-310.

[34] Hurst, L. D.; Williams, E. J.; Pal, C., Natural selection promotes the conservation of linkage of co-expressed genes. *Trends Genet* **2002,** *18* (12), 604-6.

[35] Pal, C.; Hurst, L. D., Evidence for co-evolution of gene order and recombination rate. *Nature genetics* **2003,** *33* (3), 392-5.

[36] Singer, G. A.; Lloyd, A. T.; Huminiecki, L. B.; Wolfe, K. H., Clusters of co-expressed genes in mammalian genomes are conserved by natural selection. *Molecular biology and evolution* **2005,** *22* (3), 767-75.

[37] Mezey, J. G.; Nuzhdin, S. V.; Ye, F.; Jones, C. D., Coordinated evolution of co-expressed gene clusters in the Drosophila transcriptome. *BMC evolutionary biology* **2008,** *8*, 2.

[38] Oliver, B.; Parisi, M.; Clark, D., Gene expression neighborhoods. *J Biol* **2002,** *1* (1), 4.

[39] Goldman, R. D.; Gruenbaum, Y.; Moir, R. D.; Shumaker, D. K.; Spann, T. P., Nuclear lamins: building blocks of nuclear architecture. *Genes & development* **2002,** *16* (5), 533-47.

[40] Gant, T. M.; Harris, C. A.; Wilson, K. L., Roles of LAP2 proteins in nuclear assembly and DNA replication: truncated LAP2beta proteins alter lamina assembly, envelope formation, nuclear size, and DNA replication efficiency in Xenopus laevis extracts. *The Journal of cell biology* **1999,** *144* (6), 1083-96.

[41] Gruenbaum, Y.; Goldman, R. D.; Meyuhas, R.; Mills, E.; Margalit, A.; Fridkin, A.; Dayani, Y.; Prokocimer, M.; Enosh, A., The nuclear lamina and its functions in the nucleus. *Int Rev Cytol* **2003,** *226*, 1-62.

[42] Jenkins, H.; Whitfield, W. G.; Goldberg, M. W.; Allen, T. D.; Hutchison, C. J., Evidence for the direct involvement of lamins in the assembly of a replication competent nucleus. *Acta Biochim Pol* **1995,** *42* (2), 133-43.

[43] Martins, S.; Eikvar, S.; Furukawa, K.; Collas, P., HA95 and LAP2 beta mediate a novel chromatin-nuclear envelope interaction implicated in initiation of DNA replication. *The Journal of cell biology* **2003,** *160* (2), 177-88.

[44] Jagatheesan, G.; Thanumalayan, S.; Muralikrishna, B.; Rangaraj, N.; Karande, A. A.; Parnaik, V. K., Colocalization of intranuclear lamin foci with RNA splicing factors. *J Cell Sci* **1999,** *112 (Pt 24)*, 4651-61.

[45] Shumaker, D. K.; Dechat, T.; Kohlmaier, A.; Adam, S. A.; Bozovsky, M. R.; Erdos, M. R.; Eriksson, M.; Goldman, A. E.; Khuon, S.; Collins, F. S.; Jenuwein, T.; Goldman, R. D., Mutant nuclear lamin A leads to progressive alterations of epigenetic control in premature aging. *Proceedings of the National Academy of Sciences of the United States of America* **2006,** *103* (23), 8703-8.

[46] Haraguchi, T.; Holaska, J. M.; Yamane, M.; Koujin, T.; Hashiguchi, N.; Mori, C.; Wilson, K. L.; Hiraoka, Y., Emerin binding to Btf, a death-promoting transcriptional repressor, is disrupted by a missense mutation that causes Emery-Dreifuss muscular dystrophy. *Eur J Biochem* **2004,** *271* (5), 1035-45.

[47] Nili, E.; Cojocaru, G. S.; Kalma, Y.; Ginsberg, D.; Copeland, N. G.; Gilbert, D. J.; Jenkins, N. A.; Berger, R.; Shaklai, S.; Amariglio, N.; Brok-Simoni, F.; Simon, A. J.; Rechavi, G., Nuclear membrane protein LAP2beta mediates transcriptional repression alone and together with its binding partner GCL (germ-cell-less). *J Cell Sci* **2001,** *114* (Pt 18), 3297-307.

[48] Spann, T. P.; Goldman, A. E.; Wang, C.; Huang, S.; Goldman, R. D., Alteration of nuclear lamin organization inhibits RNA polymerase II-dependent transcription. *The Journal of cell biology* **2002,** *156* (4), 603-8.

[49] Wilkinson, F. L.; Holaska, J. M.; Zhang, Z.; Sharma, A.; Manilal, S.; Holt, I.; Stamm, S.; Wilson, K. L.; Morris, G. E., Emerin interacts *in vitro* with the splicing-associated factor, YT521-B. *Eur J Biochem* **2003,** *270* (11), 2459-66.

[50] Warren, D. T.; Shanahan, C. M., Defective DNA-damage repair induced by nuclear lamina dysfunction is a key mediator of smooth muscle cell aging. *Biochem Soc Trans* **2011,** *39* (6), 1780-5.

[51] Shimi, T.; Pfleghaar, K.; Kojima, S.; Pack, C. G.; Solovei, I.; Goldman, A. E.; Adam, S. A.; Shumaker, D. K.; Kinjo, M.; Cremer, T.; Goldman, R. D., The A- and B-type nuclear lamin networks: microdomains involved in chromatin organization and transcription. *Genes & development* **2008,** *22* (24), 3409-21.

[52] Puckelwartz, M. J.; Depreux, F. F.; McNally, E. M., Gene expression, chromosome position and lamin A/C mutations. *Nucleus* **2011,** *2* (3), 162-7.

[53] Clements, L.; Manilal, S.; Love, D. R.; Morris, G. E., Direct interaction between emerin and lamin A. *Biochem Biophys Res Commun* **2000,** *267* (3), 709-14.

[54] Holaska, J. M.; Kowalski, A. K.; Wilson, K. L., Emerin caps the pointed end of actin filaments: evidence for an actin cortical network at the nuclear inner membrane. *PLoS biology* **2004,** *2* (9), E231.

[55] Holaska, J. M.; Lee, K. K.; Kowalski, A. K.; Wilson, K. L., Transcriptional repressor germ cell-less (GCL) and barrier to autointegration factor (BAF) compete for binding to emerin *in vitro*. *The Journal of biological chemistry* **2003,** *278* (9), 6969-75.

[56] Lee, K. K.; Haraguchi, T.; Lee, R. S.; Koujin, T.; Hiraoka, Y.; Wilson, K. L., Distinct functional domains in emerin bind lamin A and DNA-bridging protein BAF. *J Cell Sci* **2001,** *114* (Pt 24), 4567-73.

[57] Rabl, C., Über Zelltheilung. *Morphol Jarbuch* **1885,** *10*, 214-330.

[58] Boveri, T., Die Blastomerenkerne von Ascaris megalocephala und die Theorie der Chromosomenindividualität. *Arch Zellforsch* **1909,** *3*, 181–268.

[59] Heitz, E., Heterochromatin der Moose I. *Jahrb Wiss Bot* **1928,** *69*, 762-818.

[60] Lichter, P.; Cremer, T.; Borden, J.; Manuelidis, L.; Ward, D. C., Delineation of individual human chromosomes in metaphase and interphase cells by *in situ* suppression hybridization using recombinant DNA libraries. *Hum Genet* **1988,** *80* (3), 224-34.

[61] Pinkel, D.; Landegent, J.; Collins, C.; Fuscoe, J.; Segraves, R.; Lucas, J.; Gray, J., Fluorescence *in situ* hybridization with human chromosome-specific libraries: detection of trisomy 21 and translocations of chromosome 4. *Proceedings of the National Academy of Sciences of the United States of America* **1988,** *85* (23), 9138-42.

[62] Cremer, T.; Cremer, C., Chromosome territories, nuclear architecture and gene regulation in mammalian cells. *Nature reviews. Genetics* **2001,** *2* (4), 292-301.

[63] Bischoff, A.; Albers, J.; Kharboush, I.; Stelzer, E.; Cremer, T.; Cremer, C., Differences of size and shape of active and inactive X-chromosome domains in human amniotic fluid cell nuclei. *Microsc Res Tech* **1993,** *25* (1), 68-77.

[64] Cremer, T.; Lichter, P.; Borden, J.; Ward, D. C.; Manuelidis, L., Detection of chromosome aberrations in metaphase and interphase tumor cells by *in situ* hybridization using chromosome-specific library probes. *Hum Genet* **1988,** *80* (3), 235-46.

[65] Parada, L.; Misteli, T., Chromosome positioning in the interphase nucleus. *Trends Cell Biol* **2002,** *12* (9), 425-32.

[66] Cremer, M.; Kupper, K.; Wagler, B.; Wizelman, L.; von Hase, J.; Weiland, Y.; Kreja, L.; Diebold, J.; Speicher, M. R.; Cremer, T., Inheritance of gene density-related higher order chromatin arrangements in normal and tumor cell nuclei. *The Journal of cell biology* **2003,** *162* (5), 809-20.

[67] Murmann, A. E.; Gao, J.; Encinosa, M.; Gautier, M.; Peter, M. E.; Eils, R.; Lichter, P.; Rowley, J. D., Local gene density predicts the spatial position of genetic loci in the interphase nucleus. *Exp Cell Res* **2005,** *311* (1), 14-26.

[68] Federico, C.; Scavo, C.; Cantarella, C. D.; Motta, S.; Saccone, S.; Bernardi, G., Gene-rich and gene-poor chromosomal regions have different locations in the interphase nuclei of cold-blooded vertebrates. *Chromosoma* **2006,** *115* (2), 123-8.

[69] Goetze, S.; Mateos-Langerak, J.; Gierman, H. J.; de Leeuw, W.; Giromus, O.; Indemans, M. H.; Koster, J.; Ondrej, V.; Versteeg, R.; van Driel, R., The three-dimensional structure of

human interphase chromosomes is related to the transcriptome map. *Mol Cell Biol* **2007,** *27* (12), 4475-87.

[70] Grasser, F.; Neusser, M.; Fiegler, H.; Thormeyer, T.; Cremer, M.; Carter, N. P.; Cremer, T.; Muller, S., Replication-timing-correlated spatial chromatin arrangements in cancer and in primate interphase nuclei. *J Cell Sci* **2008,** *121* (Pt 11), 1876-86.

[71] Hepperger, C.; Mannes, A.; Merz, J.; Peters, J.; Dietzel, S., Three-dimensional positioning of genes in mouse cell nuclei. *Chromosoma* **2008,** *117* (6), 535-51.

[72] Bridger, J. M.; Boyle, S.; Kill, I. R.; Bickmore, W. A., Re-modelling of nuclear architecture in quiescent and senescent human fibroblasts. *Curr Biol* **2000,** *10* (3), 149-52.

[73] Croft, J. A.; Bridger, J. M.; Boyle, S.; Perry, P.; Teague, P.; Bickmore, W. A., Differences in the localization and morphology of chromosomes in the human nucleus. *The Journal of cell biology* **1999,** *145* (6), 1119-31.

[74] Neusser, M.; Schubel, V.; Koch, A.; Cremer, T.; Muller, S., Evolutionarily conserved, cell type and species-specific higher order chromatin arrangements in interphase nuclei of primates. *Chromosoma* **2007,** *116* (3), 307-20.

[75] Koehler, D.; Zakhartchenko, V.; Froenicke, L.; Stone, G.; Stanyon, R.; Wolf, E.; Cremer, T.; Brero, A., Changes of higher order chromatin arrangements during major genome activation in bovine preimplantation embryos. *Exp Cell Res* **2009,** *315* (12), 2053-63.

[76] Habermann, F. A.; Cremer, M.; Walter, J.; Kreth, G.; von Hase, J.; Bauer, K.; Wienberg, J.; Cremer, C.; Cremer, T.; Solovei, I., Arrangements of macro- and microchromosomes in chicken cells. *Chromosome Res* **2001,** *9* (7), 569-84.

[77] Kozubek, S.; Lukasova, E.; Jirsova, P.; Koutna, I.; Kozubek, M.; Ganova, A.; Bartova, E.; Falk, M.; Pasekova, R., 3D Structure of the human genome: order in randomness. *Chromosoma* **2002,** *111* (5), 321-31.

[78] Tanabe, H.; Muller, S.; Neusser, M.; von Hase, J.; Calcagno, E.; Cremer, M.; Solovei, I.; Cremer, C.; Cremer, T., Evolutionary conservation of chromosome territory arrangements in cell nuclei from higher primates. *Proceedings of the National Academy of Sciences of the United States of America* **2002,** *99* (7), 4424-9.

[79] Mahy, N. L.; Perry, P. E.; Bickmore, W. A., Gene density and transcription influence the localization of chromatin outside of chromosome territories detectable by FISH. *The Journal of cell biology* **2002,** *159* (5), 753-63.

[80] Mahy, N. L.; Perry, P. E.; Gilchrist, S.; Baldock, R. A.; Bickmore, W. A., Spatial organization of active and inactive genes and noncoding DNA within chromosome territories. *The Journal of cell biology* **2002,** *157* (4), 579-89.

[81] Clemson, C. M.; Hall, L. L.; Byron, M.; McNeil, J.; Lawrence, J. B., The X chromosome is organized into a gene-rich outer rim and an internal core containing silenced nongenic sequences. *Proceedings of the National Academy of Sciences of the United States of America* **2006,** *103* (20), 7688-93.

[82] Rieger, R.; Michaelis, A.; Schubert, I.; Meister, A., Somatic interphase pairing of Vicia chromosomes as inferred from the hom/het ratio of induced chromatid interchanges. *Mutation research* **1973,** *20,* 295-298.

[83] Csink, A. K.; Henikoff, S., Large-scale chromosomal movements during interphase progression in Drosophila. *The Journal of cell biology* **1998,** *143* (1), 13-22.

[84] Fung, J. C.; Marshall, W. F.; Dernburg, A.; Agard, D. A.; Sedat, J. W., Homologous chromosome pairing in Drosophila melanogaster proceeds through multiple independent initiations. *The Journal of cell biology* **1998,** *141* (1), 5-20.

[85] Hiraoka, Y.; Dernburg, A. F.; Parmelee, S. J.; Rykowski, M. C.; Agard, D. A.; Sedat, J. W., The onset of homologous chromosome pairing during Drosophila melanogaster embryogenesis. *The Journal of cell biology* **1993,** *120* (3), 591-600.

[86] Chandley, A. C.; Speed, R. M.; Leitch, A. R., Different distributions of homologous chromosomes in adult human Sertoli cells and in lymphocytes signify nuclear differentiation. *J Cell Sci* **1996,** *109 (Pt 4)*, 773-6.

[87] Nagele, R. G.; Freeman, T.; McMorrow, L.; Thomson, Z.; Kitson-Wind, K.; Lee, H., Chromosomes exhibit preferential positioning in nuclei of quiescent human cells. *J Cell Sci* **1999,** *112 (Pt 4)*, 525-35.

[88] Bickmore, W. A.; Chubb, J. R., Dispatch. Chromosome position: now, where was I? *Curr Biol* **2003,** *13* (9), R357-9.

[89] Gerlich, D.; Beaudouin, J.; Kalbfuss, B.; Daigle, N.; Eils, R.; Ellenberg, J., Global chromosome positions are transmitted through mitosis in mammalian cells. *Cell* **2003,** *112* (6), 751-64.

[90] Parada, L. A.; Roix, J. J.; Misteli, T., An uncertainty principle in chromosome positioning. *Trends Cell Biol* **2003,** *13* (8), 393-6.

[91] Walter, J.; Schermelleh, L.; Cremer, M.; Tashiro, S.; Cremer, T., Chromosome order in HeLa cells changes during mitosis and early G1, but is stably maintained during subsequent interphase stages. *The Journal of cell biology* **2003,** *160* (5), 685-97.

[92] Williams, R. R.; Fisher, A. G., Chromosomes, positions please! *Nature cell biology* **2003,** *5* (5), 388-90.

[93] Visser, A. E.; Jaunin, F.; Fakan, S.; Aten, J. A., High resolution analysis of interphase chromosome domains. *J Cell Sci* **2000,** *113 (Pt 14)*, 2585-93.

[94] Olivares-Chauvet, P.; Fennessy, D.; Jackson, D. A.; Maya-Mendoza, A., Innate structure of DNA foci restricts the mixing of DNA from different chromosome territories. *PloS one* **2011,** *6* (12), e27527.

[95] Markaki, Y.; Smeets, D.; Fiedler, S.; Schmid, V. J.; Schermelleh, L.; Cremer, T.; Cremer, M., The potential of 3D-FISH and super-resolution structured illumination microscopy for studies of 3D nuclear architecture: 3D structured illumination microscopy of defined chromosomal structures visualized by 3D (immuno)-FISH opens new perspectives for studies of nuclear architecture. *BioEssays : news and reviews in molecular, cellular and developmental biology* **2012,** *34* (5), 412-26.

[96] Dehghani, H.; Dellaire, G.; Bazett-Jones, D. P., Organization of chromatin in the interphase mammalian cell. *Micron* **2005,** *36* (2), 95-108.

[97] Bridger, J. M.; Herrmann, H.; Munkel, C.; Lichter, P., Identification of an interchromosomal compartment by polymerization of nuclear-targeted vimentin. *J Cell Sci* **1998,** *111 (Pt 9)*, 1241-53.

[98] Reichenzeller, M.; Burzlaff, A.; Lichter, P.; Herrmann, H., *In vivo* observation of a nuclear channel-like system: evidence for a distinct interchromosomal domain compartment in interphase cells. *J Struct Biol* **2000,** *129* (2-3), 175-85.

[99] Cremer, T.; Kurz, A.; Zirbel, R.; Dietzel, S.; Rinke, B.; Schrock, E.; Speicher, M. R.; Mathieu, U.; Jauch, A.; Emmerich, P.; Scherthan, H.; Ried, T.; Cremer, C.; Lichter, P., Role of chromosome territories in the functional compartmentalization of the cell nucleus. *Cold Spring Harb Symp Quant Biol* **1993,** *58*, 777-92.

[100] Zirbel, R. M.; Mathieu, U. R.; Kurz, A.; Cremer, T.; Lichter, P., Evidence for a nuclear compartment of transcription and splicing located at chromosome domain boundaries. *Chromosome Res* **1993,** *1* (2), 93-106.

[101] Lieberman-Aiden, E.; van Berkum, N. L.; Williams, L.; Imakaev, M.; Ragoczy, T.; Telling, A.; Amit, I.; Lajoie, B. R.; Sabo, P. J.; Dorschner, M. O.; Sandstrom, R.; Bernstein, B.;

Bender, M. A.; Groudine, M.; Gnirke, A.; Stamatoyannopoulos, J.; Mirny, L. A.; Lander, E. S.; Dekker, J., Comprehensive mapping of long-range interactions reveals folding principles of the human genome. *Science* **2009,** *326* (5950), 289-93.

[102] Duan, Z.; Andronescu, M.; Schutz, K.; McIlwain, S.; Kim, Y. J.; Lee, C.; Shendure, J.; Fields, S.; Blau, C. A.; Noble, W. S., A three-dimensional model of the yeast genome. *Nature* **2010,** *465* (7296), 363-7.

[103] Clemson, C. M.; Lawrence, J. B., Multifunctional compartments in the nucleus: insights from DNA and RNA localization. *J Cell Biochem* **1996,** *62* (2), 181-90.

[104] Lampel, S.; Bridger, J. M.; Zirbel, R. M.; Mathieu, U. R.; Lichter, P., Nuclear RNA accumulations contain released transcripts and exhibit specific distributions with respect to Sm antigen foci. *DNA Cell Biol* **1997,** *16* (10), 1133-42.

[105] Xing, Y.; Johnson, C. V.; Moen, P. T., Jr.; McNeil, J. A.; Lawrence, J., Nonrandom gene organization: structural arrangements of specific pre-mRNA transcription and splicing with SC-35 domains. *The Journal of cell biology* **1995,** *131* (6 Pt 2), 1635-47.

[106] Dietzel, S.; Schiebel, K.; Little, G.; Edelmann, P.; Rappold, G. A.; Eils, R.; Cremer, C.; Cremer, T., The 3D positioning of ANT2 and ANT3 genes within female X chromosome territories correlates with gene activity. *Exp Cell Res* **1999,** *252* (2), 363-75.

[107] Kurz, A.; Lampel, S.; Nickolenko, J. E.; Bradl, J.; Benner, A.; Zirbel, R. M.; Cremer, T.; Lichter, P., Active and inactive genes localize preferentially in the periphery of chromosome territories. *The Journal of cell biology* **1996,** *135* (5), 1195-205.

[108] Scheuermann, M. O.; Tajbakhsh, J.; Kurz, A.; Saracoglu, K.; Eils, R.; Lichter, P., Topology of genes and nontranscribed sequences in human interphase nuclei. *Exp Cell Res* **2004,** *301* (2), 266-79.

[109] Volpi, E. V.; Chevret, E.; Jones, T.; Vatcheva, R.; Williamson, J.; Beck, S.; Campbell, R. D.; Goldsworthy, M.; Powis, S. H.; Ragoussis, J.; Trowsdale, J.; Sheer, D., Large-scale chromatin organization of the major histocompatibility complex and other regions of human chromosome 6 and its response to interferon in interphase nuclei. *J Cell Sci* **2000,** *113 (Pt 9)*, 1565-76.

[110] Williams, R. R.; Broad, S.; Sheer, D.; Ragoussis, J., Subchromosomal positioning of the epidermal differentiation complex (EDC) in keratinocyte and lymphoblast interphase nuclei. *Exp Cell Res* **2002,** *272* (2), 163-75.

[111] Ferrai, C.; Xie, S. Q.; Luraghi, P.; Munari, D.; Ramirez, F.; Branco, M. R.; Pombo, A.; Crippa, M. P., Poised transcription factories prime silent uPA gene prior to activation. *PLoS biology* **2010,** *8* (1), e1000270.

[112] Chambeyron, S.; Bickmore, W. A., Chromatin decondensation and nuclear reorganization of the HoxB locus upon induction of transcription. *Genes & development* **2004,** *18* (10), 1119-30.

[113] Morey, C.; Da Silva, N. R.; Perry, P.; Bickmore, W. A., Nuclear reorganisation and chromatin decondensation are conserved, but distinct, mechanisms linked to Hox gene activation. *Development* **2007,** *134* (5), 909-19.

[114] Chambeyron, S.; Da Silva, N. R.; Lawson, K. A.; Bickmore, W. A., Nuclear re-organisation of the Hoxb complex during mouse embryonic development. *Development* **2005,** *132* (9), 2215-23.

[115] Osborne, C. S.; Chakalova, L.; Brown, K. E.; Carter, D.; Horton, A.; Debrand, E.; Goyenechea, B.; Mitchell, J. A.; Lopes, S.; Reik, W.; Fraser, P., Active genes dynamically colocalize to shared sites of ongoing transcription. *Nature genetics* **2004,** *36* (10), 1065-71.

[116] Shopland, L. S.; Johnson, C. V.; Byron, M.; McNeil, J.; Lawrence, J. B., Clustering of multiple specific genes and gene-rich R-bands around SC-35 domains: evidence for local euchromatic neighborhoods. *The Journal of cell biology* **2003,** *162* (6), 981-90.

[117] Pickersgill, H.; Kalverda, B.; de Wit, E.; Talhout, W.; Fornerod, M.; van Steensel, B., Characterization of the Drosophila melanogaster genome at the nuclear lamina. *Nature genetics* **2006,** *38* (9), 1005-14.

[118] Guelen, L.; Pagie, L.; Brasset, E.; Meuleman, W.; Faza, M. B.; Talhout, W.; Eussen, B. H.; de Klein, A.; Wessels, L.; de Laat, W.; van Steensel, B., Domain organization of human chromosomes revealed by mapping of nuclear lamina interactions. *Nature* **2008,** *453* (7197), 948-51.

[119] Peric-Hupkes, D.; Meuleman, W.; Pagie, L.; Bruggeman, S. W.; Solovei, I.; Brugman, W.; Graf, S.; Flicek, P.; Kerkhoven, R. M.; van Lohuizen, M.; Reinders, M.; Wessels, L.; van Steensel, B., Molecular maps of the reorganization of genome-nuclear lamina interactions during differentiation. *Molecular cell* **2010,** *38* (4), 603-13.

[120] Capelson, M.; Corces, V. G., Boundary elements and nuclear organization. *Biol Cell* **2004,** *96* (8), 617-29.

[121] Felsenfeld, G.; Burgess-Beusse, B.; Farrell, C.; Gaszner, M.; Ghirlando, R.; Huang, S.; Jin, C.; Litt, M.; Magdinier, F.; Mutskov, V.; Nakatani, Y.; Tagami, H.; West, A.; Yusufzai, T., Chromatin boundaries and chromatin domains. *Cold Spring Harb Symp Quant Biol* **2004,** *69*, 245-50.

[122] Wei, G. H.; Liu de, P.; Liang, C. C., Chromatin domain boundaries: insulators and beyond. *Cell Res* **2005,** *15* (4), 292-300.

[123] Byrd, K.; Corces, V. G., Visualization of chromatin domains created by the gypsy insulator of Drosophila. *The Journal of cell biology* **2003,** *162* (4), 565-74.

[124] Melnikova, L.; Juge, F.; Gruzdeva, N.; Mazur, A.; Cavalli, G.; Georgiev, P., Interaction between the GAGA factor and Mod(mdg4) proteins promotes insulator bypass in Drosophila. *Proceedings of the National Academy of Sciences of the United States of America* **2004,** *101* (41), 14806-11.

[125] Nabirochkin, S.; Ossokina, M.; Heidmann, T., A nuclear matrix/scaffold attachment region co-localizes with the gypsy retrotransposon insulator sequence. *The Journal of biological chemistry* **1998,** *273* (4), 2473-9.

[126] Pai, C. Y.; Lei, E. P.; Ghosh, D.; Corces, V. G., The centrosomal protein CP190 is a component of the gypsy chromatin insulator. *Molecular cell* **2004,** *16* (5), 737-48.

[127] Capelson, M.; Corces, V. G., The ubiquitin ligase dTopors directs the nuclear organization of a chromatin insulator. *Molecular cell* **2005,** *20* (1), 105-16.

[128] Gaszner, M.; Felsenfeld, G., Insulators: exploiting transcriptional and epigenetic mechanisms. *Nature reviews. Genetics* **2006,** *7* (9), 703-13.

[129] West, A. G.; Fraser, P., Remote control of gene transcription. *Human molecular genetics* **2005,** *14 Spec No 1*, R101-11.

[130] Recillas-Targa, F.; Pikaart, M. J.; Burgess-Beusse, B.; Bell, A. C.; Litt, M. D.; West, A. G.; Gaszner, M.; Felsenfeld, G., Position-effect protection and enhancer blocking by the chicken beta-globin insulator are separable activities. *Proceedings of the National Academy of Sciences of the United States of America* **2002,** *99* (10), 6883-8.

[131] Dunn, K. L.; Zhao, H.; Davie, J. R., The insulator binding protein CTCF associates with the nuclear matrix. *Exp Cell Res* **2003,** *288* (1), 218-23.

[132] Yusufzai, T. M.; Felsenfeld, G., The 5'-HS4 chicken beta-globin insulator is a CTCF-dependent nuclear matrix-associated element. *Proceedings of the National Academy of Sciences of the United States of America* **2004,** *101* (23), 8620-4.

[133] Yusufzai, T. M.; Tagami, H.; Nakatani, Y.; Felsenfeld, G., CTCF tethers an insulator to subnuclear sites, suggesting shared insulator mechanisms across species. *Molecular cell* **2004,** *13* (2), 291-8.

[134] Ottaviani, A.; Schluth-Bolard, C.; Rival-Gervier, S.; Boussouar, A.; Rondier, D.; Foerster, A. M.; Morere, J.; Bauwens, S.; Gazzo, S.; Callet-Bauchu, E.; Gilson, E.; Magdinier, F., Identification of a perinuclear positioning element in human subtelomeres that requires A-type lamins and CTCF. *The EMBO journal* **2009,** *28* (16), 2428-36.

[135] Ling, J. Q.; Li, T.; Hu, J. F.; Vu, T. H.; Chen, H. L.; Qiu, X. W.; Cherry, A. M.; Hoffman, A. R., CTCF mediates interchromosomal colocalization between Igf2/H19 and Wsb1/Nf1. *Science* **2006,** *312* (5771), 269-72.

[136] Zhao, Z.; Tavoosidana, G.; Sjolinder, M.; Gondor, A.; Mariano, P.; Wang, S.; Kanduri, C.; Lezcano, M.; Sandhu, K. S.; Singh, U.; Pant, V.; Tiwari, V.; Kurukuti, S.; Ohlsson, R., Circular chromosome conformation capture (4C) uncovers extensive networks of epigenetically regulated intra- and interchromosomal interactions. *Nature genetics* **2006,** *38* (11), 1341-7.

[137] Xu, N.; Donohoe, M. E.; Silva, S. S.; Lee, J. T., Evidence that homologous X-chromosome pairing requires transcription and Ctcf protein. *Nature genetics* **2007,** *39* (11), 1390-6.

[138] Donohoe, M. E.; Silva, S. S.; Pinter, S. F.; Xu, N.; Lee, J. T., The pluripotency factor Oct4 interacts with Ctcf and also controls X-chromosome pairing and counting. *Nature* **2009,** *460* (7251), 128-32.

[139] Yang, J.; Corces, V. G., Chromatin insulators: a role in nuclear organization and gene expression. *Advances in cancer research* **2011,** *110*, 43-76.

[140] Fiorentino, F. P.; Giordano, A., The tumor suppressor role of CTCF. *Journal of cellular physiology* **2012,** *227* (2), 479-92.

[141] Michaelis, C.; Ciosk, R.; Nasmyth, K., Cohesins: chromosomal proteins that prevent premature separation of sister chromatids. *Cell* **1997,** *91* (1), 35-45.

[142] Dorsett, D.; Eissenberg, J. C.; Misulovin, Z.; Martens, A.; Redding, B.; McKim, K., Effects of sister chromatid cohesion proteins on cut gene expression during wing development in Drosophila. *Development* **2005,** *132* (21), 4743-53.

[143] Rollins, R. A.; Korom, M.; Aulner, N.; Martens, A.; Dorsett, D., Drosophila nipped-B protein supports sister chromatid cohesion and opposes the stromalin/Scc3 cohesion factor to facilitate long-range activation of the cut gene. *Mol Cell Biol* **2004,** *24* (8), 3100-11.

[144] Rollins, R. A.; Morcillo, P.; Dorsett, D., Nipped-B, a Drosophila homologue of chromosomal adherins, participates in activation by remote enhancers in the cut and Ultrabithorax genes. *Genetics* **1999,** *152* (2), 577-93.

[145] Misulovin, Z.; Schwartz, Y. B.; Li, X. Y.; Kahn, T. G.; Gause, M.; MacArthur, S.; Fay, J. C.; Eisen, M. B.; Pirrotta, V.; Biggin, M. D.; Dorsett, D., Association of cohesin and Nipped-B with transcriptionally active regions of the Drosophila melanogaster genome. *Chromosoma* **2008,** *117* (1), 89-102.

[146] Parelho, V.; Hadjur, S.; Spivakov, M.; Leleu, M.; Sauer, S.; Gregson, H. C.; Jarmuz, A.; Canzonetta, C.; Webster, Z.; Nesterova, T.; Cobb, B. S.; Yokomori, K.; Dillon, N.; Aragon, L.; Fisher, A. G.; Merkenschlager, M., Cohesins functionally associate with CTCF on mammalian chromosome arms. *Cell* **2008,** *132* (3), 422-33.

[147] Rubio, E. D.; Reiss, D. J.; Welcsh, P. L.; Disteche, C. M.; Filippova, G. N.; Baliga, N. S.; Aebersold, R.; Ranish, J. A.; Krumm, A., CTCF physically links cohesin to chromatin. *Proceedings of the National Academy of Sciences of the United States of America* **2008,** *105* (24), 8309-14.

[148] Wendt, K. S.; Yoshida, K.; Itoh, T.; Bando, M.; Koch, B.; Schirghuber, E.; Tsutsumi, S.; Nagae, G.; Ishihara, K.; Mishiro, T.; Yahata, K.; Imamoto, F.; Aburatani, H.; Nakao, M.; Imamoto, N.; Maeshima, K.; Shirahige, K.; Peters, J. M., Cohesin mediates transcriptional insulation by CCCTC-binding factor. *Nature* **2008,** *451* (7180), 796-801.

[149] Kagey, M. H.; Newman, J. J.; Bilodeau, S.; Zhan, Y.; Orlando, D. A.; van Berkum, N. L.; Ebmeier, C. C.; Goossens, J.; Rahl, P. B.; Levine, S. S.; Taatjes, D. J.; Dekker, J.; Young, R. A., Mediator and cohesin connect gene expression and chromatin architecture. *Nature* **2010,** *467* (7314), 430-5.

[150] Liu, J.; Zhang, Z.; Bando, M.; Itoh, T.; Deardorff, M. A.; Clark, D.; Kaur, M.; Tandy, S.; Kondoh, T.; Rappaport, E.; Spinner, N. B.; Vega, H.; Jackson, L. G.; Shirahige, K.; Krantz, I. D., Transcriptional dysregulation in NIPBL and cohesin mutant human cells. *PLoS biology* **2009,** *7* (5), e1000119.

[151] Nitzsche, A.; Paszkowski-Rogacz, M.; Matarese, F.; Janssen-Megens, E. M.; Hubner, N. C.; Schulz, H.; de Vries, I.; Ding, L.; Huebner, N.; Mann, M.; Stunnenberg, H. G.; Buchholz, F., RAD21 cooperates with pluripotency transcription factors in the maintenance of embryonic stem cell identity. *PloS one* **2011,** *6* (5), e19470.

[152] Stedman, W.; Kang, H.; Lin, S.; Kissil, J. L.; Bartolomei, M. S.; Lieberman, P. M., Cohesins localize with CTCF at the KSHV latency control region and at cellular c-myc and H19/Igf2 insulators. *The EMBO journal* **2008,** *27* (4), 654-66.

[153] Nativio, R.; Wendt, K. S.; Ito, Y.; Huddleston, J. E.; Uribe-Lewis, S.; Woodfine, K.; Krueger, C.; Reik, W.; Peters, J. M.; Murrell, A., Cohesin is required for higher-order chromatin conformation at the imprinted IGF2-H19 locus. *PLoS genetics* **2009,** *5* (11), e1000739.

[154] Hou, C.; Dale, R.; Dean, A., Cell type specificity of chromatin organization mediated by CTCF and cohesin. *Proceedings of the National Academy of Sciences of the United States of America* **2010,** *107* (8), 3651-6.

[155] Hadjur, S.; Williams, L. M.; Ryan, N. K.; Cobb, B. S.; Sexton, T.; Fraser, P.; Fisher, A. G.; Merkenschlager, M., Cohesins form chromosomal cis-interactions at the developmentally regulated IFNG locus. *Nature* **2009,** *460* (7253), 410-3.

[156] Cremer, M.; von Hase, J.; Volm, T.; Brero, A.; Kreth, G.; Walter, J.; Fischer, C.; Solovei, I.; Cremer, C.; Cremer, T., Non-random radial higher-order chromatin arrangements in nuclei of diploid human cells. *Chromosome Res* **2001,** *9* (7), 541-67.

[157] LaSalle, J. M.; Lalande, M., Homologous association of oppositely imprinted chromosomal domains. *Science* **1996,** *272* (5262), 725-8.

[158] Parada, L. A.; McQueen, P. G.; Misteli, T., Tissue-specific spatial organization of genomes. *Genome biology* **2004,** *5* (7), R44.

[159] Parada, L. A.; McQueen, P. G.; Munson, P. J.; Misteli, T., Conservation of relative chromosome positioning in normal and cancer cells. *Curr Biol* **2002,** *12* (19), 1692-7.

[160] Misteli, T., Concepts in nuclear architecture. *BioEssays : news and reviews in molecular, cellular and developmental biology* **2005,** *27* (5), 477-87.

[161] Kleinjan, D. A.; van Heyningen, V., Long-range control of gene expression: emerging mechanisms and disruption in disease. *American journal of human genetics* **2005,** *76* (1), 8-32.

[162] Tufarelli, C.; Stanley, J. A.; Garrick, D.; Sharpe, J. A.; Ayyub, H.; Wood, W. G.; Higgs, D. R., Transcription of antisense RNA leading to gene silencing and methylation as a novel cause of human genetic disease. *Nature genetics* **2003,** *34* (2), 157-65.

[163] Bickmore, W. A.; van der Maarel, S. M., Perturbations of chromatin structure in human genetic disease: recent advances. *Human molecular genetics* **2003,** *12 Spec No 2*, R207-13.

[164] Cho, K. S.; Elizondo, L. I.; Boerkoel, C. F., Advances in chromatin remodeling and human disease. *Current opinion in genetics & development* **2004,** *14* (3), 308-15.

[165] Somech, R.; Shaklai, S.; Amariglio, N.; Rechavi, G.; Simon, A. J., Nuclear envelopathies--raising the nuclear veil. *Pediatr Res* **2005,** *57* (5 Pt 2), 8R-15R.

[166] Amati, F.; Biancolella, M.; D'Apice, M. R.; Gambardella, S.; Mango, R.; Sbraccia, P.; D'Adamo, M.; Margiotti, K.; Nardone, A.; Lewis, M.; Novelli, G., Gene expression profiling of fibroblasts from a human progeroid disease (mandibuloacral dysplasia, MAD #248370) through cDNA microarrays. *Gene Expr* **2004,** *12* (1), 39-47.

[167] Tsukahara, T.; Arahata, K., A comparative gene expression analysis of Emery-Dreifuss muscular dystrophy using a cDNA microarray. *Methods Mol Biol* **2003,** *217*, 253-62.

[168] Tsukahara, T.; Tsujino, S.; Arahata, K., CDNA microarray analysis of gene expression in fibroblasts of patients with X-linked Emery-Dreifuss muscular dystrophy. *Muscle Nerve* **2002,** *25* (6), 898-901.

[169] Mewborn, S. K.; Puckelwartz, M. J.; Abuisneineh, F.; Fahrenbach, J. P.; Zhang, Y.; MacLeod, H.; Dellefave, L.; Pytel, P.; Selig, S.; Labno, C. M.; Reddy, K.; Singh, H.; McNally, E., Altered chromosomal positioning, compaction, and gene expression with a lamin A/C gene mutation. *PloS one* **2010,** *5* (12), e14342.

[170] Csoka, A. B.; English, S. B.; Simkevich, C. P.; Ginzinger, D. G.; Butte, A. J.; Schatten, G. P.; Rothman, F. G.; Sedivy, J. M., Genome-scale expression profiling of Hutchinson-Gilford progeria syndrome reveals widespread transcriptional misregulation leading to mesodermal/mesenchymal defects and accelerated atherosclerosis. *Aging cell* **2004,** *3* (4), 235-43.

[171] Elizondo, L. I.; Cho, K. S.; Zhang, W.; Yan, J.; Huang, C.; Huang, Y.; Choi, K.; Sloan, E. A.; Deguchi, K.; Lou, S.; Baradaran-Heravi, A.; Takashima, H.; Lucke, T.; Quiocho, F. A.; Boerkoel, C. F., Schimke immuno-osseous dysplasia: SMARCAL1 loss-of-function and phenotypic correlation. *Journal of medical genetics* **2009,** *46* (1), 49-59.

[172] Coleman, M. A.; Eisen, J. A.; Mohrenweiser, H. W., Cloning and characterization of HARP/SMARCAL1: a prokaryotic HepA-related SNF2 helicase protein from human and mouse. *Genomics* **2000,** *65* (3), 274-82.

[173] Coleman, M. A.; Miller, K. A.; Beernink, P. T.; Yoshikawa, D. M.; Albala, J. S., Identification of chromatin-related protein interactions using protein microarrays. *Proteomics* **2003,** *3* (11), 2101-7.

[174] Yusufzai, T.; Kadonaga, J. T., HARP is an ATP-driven annealing helicase. *Science* **2008,** *322* (5902), 748-50.

[175] Bansbach, C. E.; Betous, R.; Lovejoy, C. A.; Glick, G. G.; Cortez, D., The annealing helicase SMARCAL1 maintains genome integrity at stalled replication forks. *Genes & development* **2009,** *23* (20), 2405-14.

[176] Bansbach, C. E.; Boerkoel, C. F.; Cortez, D., SMARCAL1 and replication stress: An explanation for SIOD? *Nucleus* **2010,** *1* (3), 245 - 248.

[177] Boerkoel, C. F.; Takashima, H.; John, J.; Yan, J.; Stankiewicz, P.; Rosenbarker, L.; Andre, J. L.; Bogdanovic, R.; Burguet, A.; Cockfield, S.; Cordeiro, I.; Frund, S.; Illies, F.; Joseph, M.; Kaitila, I.; Lama, G.; Loirat, C.; McLeod, D. R.; Milford, D. V.; Petty, E. M.; Rodrigo, F.; Saraiva, J. M.; Schmidt, B.; Smith, G. C.; Spranger, J.; Stein, A.; Thiele, H.; Tizard, J.; Weksberg, R.; Lupski, J. R.; Stockton, D. W., Mutant chromatin remodeling protein SMARCAL1 causes Schimke immuno-osseous dysplasia. *Nature genetics* **2002,** *30* (2), 215-20.

[178] Ciccia, A.; Bredemeyer, A. L.; Sowa, M. E.; Terret, M. E.; Jallepalli, P. V.; Harper, J. W.; Elledge, S. J., The SIOD disorder protein SMARCAL1 is an RPA-interacting protein involved in replication fork restart. *Genes & development* **2009,** *23* (20), 2415-25.
[179] Clewing, J. M.; Antalfy, B. C.; Lücke, T.; Najafian, B.; Marwedel, K. M.; Hori, A.; Powel, R. M.; Safo Do, A. F.; Najera, L.; SantaCruz, K.; Hicks, M. J.; Armstrong, D. L.; Boerkoel, C. F., Schimke immuno-osseous dysplasia: a clinicopathological correlation. *Journal of medical genetics* **2007,** *44* (2), 122-30.
[180] Deguchi, K.; Clewing, J. M.; Elizondo, L. I.; Hirano, R.; Huang, C.; Choi, K.; Sloan, E. A.; Lucke, T.; Marwedel, K. M.; Powell, R. D., Jr.; Santa Cruz, K.; Willaime-Morawek, S.; Inoue, K.; Lou, S.; Northrop, J. L.; Kanemura, Y.; van der Kooy, D.; Okano, H.; Armstrong, D. L.; Boerkoel, C. F., Neurologic phenotype of Schimke immuno-osseous dysplasia and neurodevelopmental expression of SMARCAL1. *J Neuropathol Exp Neurol* **2008,** *67* (6), 565-77.
[181] Postow, L.; Woo, E. M.; Chait, B. T.; Funabiki, H., Identification of SMARCAL1 as a component of the DNA damage response. *The Journal of biological chemistry* **2009,** *284* (51), 35951-61.
[182] Yuan, J.; Ghosal, G.; Chen, J., The annealing helicase HARP protects stalled replication forks. *Genes & development* **2009,** *23* (20), 2394-9.
[183] Yusufzai, T.; Kong, X.; Yokomori, K.; Kadonaga, J. T., The annealing helicase HARP is recruited to DNA repair sites *via* an interaction with RPA. *Genes & development* **2009,** *23* (20), 2400-4.
[184] Baradaran-Heravi, A.; Raams, A.; Lubieniecka, J.; Cho, K. S.; DeHaai, K. A.; Basiratnia, M.; Mari, P.; Xue, Y.; Rauth, M.; Olney, A.; Shago, M.; Choi, K.; Weksberg, R.; Nowaczyk, M. J. M.; Wang, W.; Jaspers, N. G. J.; Boerkoel, C. F., SMARCAL1 Deficiency Predisposes to Non-Hodgkin Lymphoma and Hypersensitivity to Genotoxic Agents *In vivo*. *American journal of medical genetics. Part A* **2012,** 158A(9), 2204-13.
[185] Morimoto, M.; Yu, Z.; Stenzel, P.; Clewing, J. M.; Najafian, B.; Mayfield, C.; Hendson, G.; Weinkauf, J. G.; Gormley, A. K.; Parham, D. M.; Ponniah, U.;André, J.; Asakura, Y.; Basiratnia, M.; Bogdanović, R.; Bokenkamp, A.; Bonneau, D.; Buck, A.; Charrow, J.; Cochat, P.; Cordeiro, I.; Deschenes, G.; Fenkçi, M. S.; Frange, P.; Fründ, S.; Fryssira, H.; Guillen-Navarro, E.; Keller, K.; Kirmani, S.; Kobelka, K.; Lamfers, P.; Levtchenko, E.; Lewis, D. B.; Massella, L.; McLeod, R.; Milford, D.; Nobili, F.; Saraiva, J. M.; Semerci, C. N.; Shoemaker, L.; Stajić, N.; Stein, A.; Taha, D.; Wand, D.; Zonana, J.; Lücke, T.; Boerkoel, C. F., Reduced elastogenesis: A clue to the arteriosclerosis and emphysematous changes in schimke immuno-osseous dysplasia? *Orphanet journal of rare diseases* **2012**, 7 (1), 70.
[186] Betous, R.; Mason, A. C.; Rambo, R. P.; Bansbach, C. E.; Badu-Nkansah, A.; Sirbu, B. M.; Eichman, B. F.; Cortez, D., SMARCAL1 catalyzes fork regression and Holliday junction migration to maintain genome stability during DNA replication. *Genes & development* **2012,** *26* (2), 151-62.
[187] Krantz, I. D.; McCallum, J.; DeScipio, C.; Kaur, M.; Gillis, L. A.; Yaeger, D.; Jukofsky, L.; Wasserman, N.; Bottani, A.; Morris, C. A.; Nowaczyk, M. J.; Toriello, H.; Bamshad, M. J.; Carey, J. C.; Rappaport, E.; Kawauchi, S.; Lander, A. D.; Calof, A. L.; Li, H. H.; Devoto, M.; Jackson, L. G., Cornelia de Lange syndrome is caused by mutations in NIPBL, the human homolog of Drosophila melanogaster Nipped-B. *Nature genetics* **2004,** *36* (6), 631-5.
[188] Musio, A.; Selicorni, A.; Focarelli, M. L.; Gervasini, C.; Milani, D.; Russo, S.; Vezzoni, P.; Larizza, L., X-linked Cornelia de Lange syndrome owing to SMC1L1 mutations. *Nature genetics* **2006,** *38* (5), 528-30.

[189] Schar, P.; Fasi, M.; Jessberger, R., SMC1 coordinates DNA double-strand break repair pathways. *Nucleic acids research* **2004,** *32* (13), 3921-9.
[190] Tonkin, E. T.; Wang, T. J.; Lisgo, S.; Bamshad, M. J.; Strachan, T., NIPBL, encoding a homolog of fungal Scc2-type sister chromatid cohesion proteins and fly Nipped-B, is mutated in Cornelia de Lange syndrome. *Nature genetics* **2004,** *36* (6), 636-41.
[191] Ciosk, R.; Shirayama, M.; Shevchenko, A.; Tanaka, T.; Toth, A.; Shevchenko, A.; Nasmyth, K., Cohesin's binding to chromosomes depends on a separate complex consisting of Scc2 and Scc4 proteins. *Molecular cell* **2000,** *5* (2), 243-54.
[192] Dorsett, D., Adherin: key to the cohesin ring and cornelia de Lange syndrome. *Curr Biol* **2004,** *14* (19), R834-6.
[193] Gillespie, P. J.; Hirano, T., Scc2 couples replication licensing to sister chromatid cohesion in Xenopus egg extracts. *Curr Biol* **2004,** *14* (17), 1598-603.
[194] Takahashi, T. S.; Yiu, P.; Chou, M. F.; Gygi, S.; Walter, J. C., Recruitment of Xenopus Scc2 and cohesin to chromatin requires the pre-replication complex. *Nature cell biology* **2004,** *6* (10), 991-6.
[195] Tomonaga, T.; Nagao, K.; Kawasaki, Y.; Furuya, K.; Murakami, A.; Morishita, J.; Yuasa, T.; Sutani, T.; Kearsey, S. E.; Uhlmann, F.; Nasmyth, K.; Yanagida, M., Characterization of fission yeast cohesin: essential anaphase proteolysis of Rad21 phosphorylated in the S phase. *Genes & development* **2000,** *14* (21), 2757-70.
[196] Mannini, L.; Menga, S.; Tonelli, A.; Zanotti, S.; Bassi, M. T.; Magnani, C.; Musio, A., SMC1A codon 496 mutations affect the cellular response to genotoxic treatments. *American journal of medical genetics. Part A* **2011**.
[197] Vrouwe, M. G.; Elghalbzouri-Maghrani, E.; Meijers, M.; Schouten, P.; Godthelp, B. C.; Bhuiyan, Z. A.; Redeker, E. J.; Mannens, M. M.; Mullenders, L. H.; Pastink, A.; Darroudi, F., Increased DNA damage sensitivity of Cornelia de Lange syndrome cells: evidence for impaired recombinational repair. *Human molecular genetics* **2007,** *16* (12), 1478-87.
[198] Stursberg, S.; Riwar, B.; Jessberger, R., Cloning and characterization of mammalian SMC1 and SMC3 genes and proteins, components of the DNA recombination complexes RC-1. *Gene* **1999,** *228* (1-2), 1-12.
[199] Kaur, M.; DeScipio, C.; McCallum, J.; Yaeger, D.; Devoto, M.; Jackson, L. G.; Spinner, N. B.; Krantz, I. D., Precocious sister chromatid separation (PSCS) in Cornelia de Lange syndrome. *American journal of medical genetics. Part A* **2005,** *138* (1), 27-31.
[200] Blat, Y.; Kleckner, N., Cohesins bind to preferential sites along yeast chromosome III, with differential regulation along arms *versus* the centric region. *Cell* **1999,** *98* (2), 249-59.
[201] Tanaka, T.; Cosma, M. P.; Wirth, K.; Nasmyth, K., Identification of cohesin association sites at centromeres and along chromosome arms. *Cell* **1999,** *98* (6), 847-58.
[202] Kawauchi, S.; Calof, A. L.; Santos, R.; Lopez-Burks, M. E.; Young, C. M.; Hoang, M. P.; Chua, A.; Lao, T.; Lechner, M. S.; Daniel, J. A.; Nussenzweig, A.; Kitzes, L.; Yokomori, K.; Hallgrimsson, B.; Lander, A. D., Multiple organ system defects and transcriptional dysregulation in the Nipbl(+/-) mouse, a model of Cornelia de Lange Syndrome. *PLoS genetics* **2009,** *5* (9), e1000650.
[203] Schaaf, C. A.; Misulovin, Z.; Sahota, G.; Siddiqui, A. M.; Schwartz, Y. B.; Kahn, T. G.; Pirrotta, V.; Gause, M.; Dorsett, D., Regulation of the Drosophila Enhancer of split and invected-engrailed gene complexes by sister chromatid cohesion proteins. *PloS one* **2009,** *4* (7), e6202.

[204] Muto, A.; Calof, A. L.; Lander, A. D.; Schilling, T. F., Multifactorial origins of heart and gut defects in nipbl-deficient zebrafish, a model of Cornelia de Lange Syndrome. *PLoS biology* **2011,** *9* (10), e1001181.

[205] Monnich, M.; Kuriger, Z.; Print, C. G.; Horsfield, J. A., A zebrafish model of Roberts syndrome reveals that Esco2 depletion interferes with development by disrupting the cell cycle. *PloS one* **2011,** *6* (5), e20051.

[206] Vega, H.; Waisfisz, Q.; Gordillo, M.; Sakai, N.; Yanagihara, I.; Yamada, M.; van Gosliga, D.; Kayserili, H.; Xu, C.; Ozono, K.; Jabs, E. W.; Inui, K.; Joenje, H., Roberts syndrome is caused by mutations in ESCO2, a human homolog of yeast ECO1 that is essential for the establishment of sister chromatid cohesion. *Nature genetics* **2005,** *37* (5), 468-70.

[207] Skibbens, R. V.; Corson, L. B.; Koshland, D.; Hieter, P., Ctf7p is essential for sister chromatid cohesion and links mitotic chromosome structure to the DNA replication machinery. *Genes & development* **1999,** *13* (3), 307-19.

[208] Toth, A.; Ciosk, R.; Uhlmann, F.; Galova, M.; Schleiffer, A.; Nasmyth, K., Yeast cohesin complex requires a conserved protein, Eco1p(Ctf7), to establish cohesion between sister chromatids during DNA replication. *Genes & development* **1999,** *13* (3), 320-33.

[209] Tanaka, K.; Yonekawa, T.; Kawasaki, Y.; Kai, M.; Furuya, K.; Iwasaki, M.; Murakami, H.; Yanagida, M.; Okayama, H., Fission yeast Eso1p is required for establishing sister chromatid cohesion during S phase. *Mol Cell Biol* **2000,** *20* (10), 3459-69.

[210] Williams, B. C.; Garrett-Engele, C. M.; Li, Z.; Williams, E. V.; Rosenman, E. D.; Goldberg, M. L., Two putative acetyltransferases, san and deco, are required for establishing sister chromatid cohesion in Drosophila. *Curr Biol* **2003,** *13* (23), 2025-36.

[211] Hou, F.; Zou, H., Two human orthologues of Eco1/Ctf7 acetyltransferases are both required for proper sister-chromatid cohesion. *Mol Biol Cell* **2005,** *16* (8), 3908-18.

[212] Tomkins, D. J.; Sisken, J. E., Abnormalities in the cell-division cycle in Roberts syndrome fibroblasts: a cellular basis for the phenotypic characteristics? *American journal of human genetics* **1984,** *36* (6), 1332-40.

[213] Burns, M. A.; Tomkins, D. J., Hypersensitivity to mitomycin C cell-killing in Roberts syndrome fibroblasts with, but not without, the heterochromatin abnormality. *Mutation research* **1989,** *216* (5), 243-9.

[214] Van den Berg, D. J.; Francke, U., Sensitivity of Roberts syndrome cells to gamma radiation, mitomycin C, and protein synthesis inhibitors. *Somatic cell and molecular genetics* **1993,** *19* (4), 377-92.

[215] Gordillo, M.; Vega, H.; Trainer, A. H.; Hou, F.; Sakai, N.; Luque, R.; Kayserili, H.; Basaran, S.; Skovby, F.; Hennekam, R. C.; Uzielli, M. L.; Schnur, R. E.; Manouvrier, S.; Chang, S.; Blair, E.; Hurst, J. A.; Forzano, F.; Meins, M.; Simola, K. O.; Raas-Rothschild, A.; Schultz, R. A.; McDaniel, L. D.; Ozono, K.; Inui, K.; Zou, H.; Jabs, E. W., The molecular mechanism underlying Roberts syndrome involves loss of ESCO2 acetyltransferase activity. *Human molecular genetics* **2008,** *17* (14), 2172-80.

[216] Petrij, F.; Giles, R. H.; Dauwerse, H. G.; Saris, J. J.; Hennekam, R. C.; Masuno, M.; Tommerup, N.; van Ommen, G. J.; Goodman, R. H.; Peters, D. J.; *et al.*, Rubinstein-Taybi syndrome caused by mutations in the transcriptional co-activator CBP. *Nature* **1995,** *376* (6538), 348-51.

[217] Roelfsema, J. H.; White, S. J.; Ariyurek, Y.; Bartholdi, D.; Niedrist, D.; Papadia, F.; Bacino, C. A.; den Dunnen, J. T.; van Ommen, G. J.; Breuning, M. H.; Hennekam, R. C.; Peters, D.

J., Genetic heterogeneity in Rubinstein-Taybi syndrome: mutations in both the CBP and EP300 genes cause disease. *American journal of human genetics* **2005,** *76* (4), 572-80.

[218] Janknecht, R., The versatile functions of the transcriptional coactivators p300 and CBP and their roles in disease. *Histol Histopathol* **2002,** *17* (2), 657-68.

[219] Ausio, J.; Levin, D. B.; De Amorim, G. V.; Bakker, S.; Macleod, P. M., Syndromes of disordered chromatin remodeling. *Clinical genetics* **2003,** *64* (2), 83-95.

[220] Guan, Z.; Giustetto, M.; Lomvardas, S.; Kim, J. H.; Miniaci, M. C.; Schwartz, J. H.; Thanos, D.; Kandel, E. R., Integration of long-term-memory-related synaptic plasticity involves bidirectional regulation of gene expression and chromatin structure. *Cell* **2002,** *111* (4), 483-93.

[221] McCampbell, A.; Taylor, J. P.; Taye, A. A.; Robitschek, J.; Li, M.; Walcott, J.; Merry, D.; Chai, Y.; Paulson, H.; Sobue, G.; Fischbeck, K. H., CREB-binding protein sequestration by expanded polyglutamine. *Human molecular genetics* **2000,** *9* (14), 2197-202.

[222] Nucifora, F. C., Jr.; Sasaki, M.; Peters, M. F.; Huang, H.; Cooper, J. K.; Yamada, M.; Takahashi, H.; Tsuji, S.; Troncoso, J.; Dawson, V. L.; Dawson, T. M.; Ross, C. A., Interference by huntingtin and atrophin-1 with cbp-mediated transcription leading to cellular toxicity. *Science* **2001,** *291* (5512), 2423-8.

[223] Taylor, J. P.; Taye, A. A.; Campbell, C.; Kazemi-Esfarjani, P.; Fischbeck, K. H.; Min, K. T., Aberrant histone acetylation, altered transcription, and retinal degeneration in a Drosophila model of polyglutamine disease are rescued by CREB-binding protein. *Genes & development* **2003,** *17* (12), 1463-8.

[224] Kasper, L. H.; Lerach, S.; Wang, J.; Wu, S.; Jeevan, T.; Brindle, P. K., CBP/p300 double null cells reveal effect of coactivator level and diversity on CREB transactivation. *The EMBO journal* **2010,** *29* (21), 3660-72.

[225] Kasper, L. H.; Thomas, M. C.; Zambetti, G. P.; Brindle, P. K., Double null cells reveal that CBP and p300 are dispensable for p53 targets p21 and Mdm2 but variably required for target genes of other signaling pathways. *Cell Cycle* **2011,** *10* (2), 212-21.

[226] de Greef, J. C.; Wang, J.; Balog, J.; den Dunnen, J. T.; Frants, R. R.; Straasheijm, K. R.; Aytekin, C.; van der Burg, M.; Duprez, L.; Ferster, A.; Gennery, A. R.; Gimelli, G.; Reisli, I.; Schuetz, C.; Schulz, A.; Smeets, D. F.; Sznajer, Y.; Wijmenga, C.; van Eggermond, M. C.; van Ostaijen-Ten Dam, M. M.; Lankester, A. C.; van Tol, M. J.; van den Elsen, P. J.; Weemaes, C. M.; van der Maarel, S. M., Mutations in ZBTB24 are associated with immunodeficiency, centromeric instability, and facial anomalies syndrome type 2. *American journal of human genetics* **2011,** *88* (6), 796-804.

[227] Xu, G. L.; Bestor, T. H.; Bourc'his, D.; Hsieh, C. L.; Tommerup, N.; Bugge, M.; Hulten, M.; Qu, X.; Russo, J. J.; Viegas-Pequignot, E., Chromosome instability and immunodeficiency syndrome caused by mutations in a DNA methyltransferase gene. *Nature* **1999,** *402* (6758), 187-91.

[228] Hagleitner, M. M.; Lankester, A.; Maraschio, P.; Hulten, M.; Fryns, J. P.; Schuetz, C.; Gimelli, G.; Davies, E. G.; Gennery, A.; Belohradsky, B. H.; de Groot, R.; Gerritsen, E. J.; Mattina, T.; Howard, P. J.; Fasth, A.; Reisli, I.; Furthner, D.; Slatter, M. A.; Cant, A. J.; Cazzola, G.; van Dijken, P. J.; van Deuren, M.; de Greef, J. C.; van der Maarel, S. M.; Weemaes, C. M., Clinical spectrum of immunodeficiency, centromeric instability and facial dysmorphism (ICF syndrome). *Journal of medical genetics* **2008,** *45* (2), 93-9.

[229] Miniou, P.; Jeanpierre, M.; Bourc'his, D.; Coutinho Barbosa, A. C.; Blanquet, V.; Viegas-Pequignot, E., alpha-satellite DNA methylation in normal individuals and in ICF

patients: heterogeneous methylation of constitutive heterochromatin in adult and fetal tissues. *Hum Genet* **1997,** *99* (6), 738-45.

[230] Bourc'his, D.; Miniou, P.; Jeanpierre, M.; Molina Gomes, D.; Dupont, J.; De Saint-Basile, G.; Maraschio, P.; Tiepolo, L.; Viegas-Pequignot, E., Abnormal methylation does not prevent X inactivation in ICF patients. *Cytogenet Cell Genet* **1999,** *84* (3-4), 245-52.

[231] Miniou, P.; Bourc'his, D.; Molina Gomes, D.; Jeanpierre, M.; Viegas-Pequignot, E., Undermethylation of Alu sequences in ICF syndrome: molecular and *in situ* analysis. *Cytogenet Cell Genet* **1997,** *77* (3-4), 308-13.

[232] Kondo, T.; Bobek, M. P.; Kuick, R.; Lamb, B.; Zhu, X.; Narayan, A.; Bourc'his, D.; Viegas-Pequignot, E.; Ehrlich, M.; Hanash, S. M., Whole-genome methylation scan in ICF syndrome: hypomethylation of non-satellite DNA repeats D4Z4 and NBL2. *Human molecular genetics* **2000,** *9* (4), 597-604.

[233] Schuffenhauer, S.; Bartsch, O.; Stumm, M.; Buchholz, T.; Petropoulou, T.; Kraft, S.; Belohradsky, B.; Hinkel, G. K.; Meitinger, T.; Wegner, R. D., DNA, FISH and complementation studies in ICF syndrome: DNA hypomethylation of repetitive and single copy loci and evidence for a trans acting factor. *Hum Genet* **1995,** *96* (5), 562-71.

[234] Miniou, P.; Jeanpierre, M.; Blanquet, V.; Sibella, V.; Bonneau, D.; Herbelin, C.; Fischer, A.; Niveleau, A.; Viegas-Pequignot, E., Abnormal methylation pattern in constitutive and facultative (X inactive chromosome) heterochromatin of ICF patients. *Human molecular genetics* **1994,** *3* (12), 2093-102.

[235] Hansen, R. S.; Stoger, R.; Wijmenga, C.; Stanek, A. M.; Canfield, T. K.; Luo, P.; Matarazzo, M. R.; D'Esposito, M.; Feil, R.; Gimelli, G.; Weemaes, C. M.; Laird, C. D.; Gartler, S. M., Escape from gene silencing in ICF syndrome: evidence for advanced replication time as a major determinant. *Human molecular genetics* **2000,** *9* (18), 2575-87.

[236] Gartler, S. M.; Varadarajan, K. R.; Luo, P.; Canfield, T. K.; Traynor, J.; Francke, U.; Hansen, R. S., Normal histone modifications on the inactive X chromosome in ICF and Rett syndrome cells: implications for methyl-CpG binding proteins. *BMC Biol* **2004,** *2* (1), 21.

[237] Ehrlich, M.; Sanchez, C.; Shao, C.; Nishiyama, R.; Kehrl, J.; Kuick, R.; Kubota, T.; Hanash, S. M., ICF, an immunodeficiency syndrome: DNA methyltransferase 3B involvement, chromosome anomalies, and gene dysregulation. *Autoimmunity* **2008,** *41* (4), 253-71.

[238] Jin, B.; Tao, Q.; Peng, J.; Soo, H. M.; Wu, W.; Ying, J.; Fields, C. R.; Delmas, A. L.; Liu, X.; Qiu, J.; Robertson, K. D., DNA methyltransferase 3B (DNMT3B) mutations in ICF syndrome lead to altered epigenetic modifications and aberrant expression of genes regulating development, neurogenesis and immune function. *Human molecular genetics* **2008,** *17* (5), 690-709.

[239] Matarazzo, M. R.; Boyle, S.; D'Esposito, M.; Bickmore, W. A., Chromosome territory reorganization in a human disease with altered DNA methylation. *Proceedings of the National Academy of Sciences of the United States of America* **2007,** *104* (42), 16546-51.

[240] Jefferson, A.; Colella, S.; Moralli, D.; Wilson, N.; Yusuf, M.; Gimelli, G.; Ragoussis, J.; Volpi, E. V., Altered intra-nuclear organisation of heterochromatin and genes in ICF syndrome. *PloS one* **2010,** *5* (6), e11364.

[241] Dupont, C.; Guimiot, F.; Perrin, L.; Marey, I.; Smiljkovski, D.; Le Tessier, D.; Lebugle, C.; Baumann, C.; Bourdoncle, P.; Tabet, A. C.; Aboura, A.; Benzacken, B.; Dupont, J. M., 3D position of pericentromeric heterochromatin within the nucleus of a patient with ICF syndrome. *Clinical genetics* **2011**.

[242] Zoghbi, H. Y., MeCP2 dysfunction in humans and mice. *J Child Neurol* **2005,** *20* (9), 736-40.

[243] Kriaucionis, S.; Bird, A., DNA methylation and Rett syndrome. *Human molecular genetics* **2003,** *12 Spec No 2*, R221-7.

[244] Martinowich, K.; Hattori, D.; Wu, H.; Fouse, S.; He, F.; Hu, Y.; Fan, G.; Sun, Y. E., DNA methylation-related chromatin remodeling in activity-dependent BDNF gene regulation. *Science* **2003,** *302* (5646), 890-3.

[245] Stancheva, I.; Collins, A. L.; Van den Veyver, I. B.; Zoghbi, H.; Meehan, R. R., A mutant form of MeCP2 protein associated with human Rett syndrome cannot be displaced from methylated DNA by notch in Xenopus embryos. *Molecular cell* **2003,** *12* (2), 425-35.

[246] Colantuoni, C.; Jeon, O. H.; Hyder, K.; Chenchik, A.; Khimani, A. H.; Narayanan, V.; Hoffman, E. P.; Kaufmann, W. E.; Naidu, S.; Pevsner, J., Gene expression profiling in postmortem Rett Syndrome brain: differential gene expression and patient classification. *Neurobiol Dis* **2001,** *8* (5), 847-65.

[247] Tudor, M.; Akbarian, S.; Chen, R. Z.; Jaenisch, R., Transcriptional profiling of a mouse model for Rett syndrome reveals subtle transcriptional changes in the brain. *Proceedings of the National Academy of Sciences of the United States of America* **2002,** *99* (24), 15536-41.

[248] Chahrour, M.; Jung, S. Y.; Shaw, C.; Zhou, X.; Wong, S. T.; Qin, J.; Zoghbi, H. Y., MeCP2, a key contributor to neurological disease, activates and represses transcription. *Science* **2008,** *320* (5880), 1224-9.

[249] Long, S. W.; Ooi, J. Y.; Yau, P. M.; Jones, P. L., A brain-derived MeCP2 complex supports a role for MeCP2 in RNA processing. *Bioscience reports* **2011,** *31* (5), 333-43.

[250] Young, J. I.; Hong, E. P.; Castle, J. C.; Crespo-Barreto, J.; Bowman, A. B.; Rose, M. F.; Kang, D.; Richman, R.; Johnson, J. M.; Berget, S.; Zoghbi, H. Y., Regulation of RNA splicing by the methylation-dependent transcriptional repressor methyl-CpG binding protein 2. *Proceedings of the National Academy of Sciences of the United States of America* **2005,** *102* (49), 17551-8.

[251] Wan, M.; Zhao, K.; Lee, S. S.; Francke, U., MECP2 truncating mutations cause histone H4 hyperacetylation in Rett syndrome. *Human molecular genetics* **2001,** *10* (10), 1085-92.

[252] Skene, P. J.; Illingworth, R. S.; Webb, S.; Kerr, A. R.; James, K. D.; Turner, D. J.; Andrews, R.; Bird, A. P., Neuronal MeCP2 is expressed at near histone-octamer levels and globally alters the chromatin state. *Molecular cell* **2010,** *37* (4), 457-68.

[253] Blake, K. D.; Davenport, S. L.; Hall, B. D.; Hefner, M. A.; Pagon, R. A.; Williams, M. S.; Lin, A. E.; Graham, J. M., Jr., CHARGE association: an update and review for the primary pediatrician. *Clinical pediatrics* **1998,** *37* (3), 159-73.

[254] Vissers, L. E.; van Ravenswaaij, C. M.; Admiraal, R.; Hurst, J. A.; de Vries, B. B.; Janssen, I. M.; van der Vliet, W. A.; Huys, E. H.; de Jong, P. J.; Hamel, B. C.; Schoenmakers, E. F.; Brunner, H. G.; Veltman, J. A.; van Kessel, A. G., Mutations in a new member of the chromodomain gene family cause CHARGE syndrome. *Nature genetics* **2004,** *36* (9), 955-7.

[255] Zentner, G. E.; Layman, W. S.; Martin, D. M.; Scacheri, P. C., Molecular and phenotypic aspects of CHD7 mutation in CHARGE syndrome. *American journal of medical genetics. Part A* **2010,** *152A* (3), 674-86.

[256] Marfella, C. G.; Imbalzano, A. N., The Chd family of chromatin remodelers. *Mutation research* **2007,** *618* (1-2), 30-40.

[257] Srinivasan, S.; Dorighi, K. M.; Tamkun, J. W., Drosophila Kismet regulates histone H3 lysine 27 methylation and early elongation by RNA polymerase II. *PLoS genetics* **2008,** *4* (10), e1000217.

[258] Schnetz, M. P.; Bartels, C. F.; Shastri, K.; Balasubramanian, D.; Zentner, G. E.; Balaji, R.; Zhang, X.; Song, L.; Wang, Z.; Laframboise, T.; Crawford, G. E.; Scacheri, P. C., Genomic distribution of CHD7 on chromatin tracks H3K4 methylation patterns. *Genome research* **2009,** *19* (4), 590-601.

[259] Schnetz, M. P.; Handoko, L.; Akhtar-Zaidi, B.; Bartels, C. F.; Pereira, C. F.; Fisher, A. G.; Adams, D. J.; Flicek, P.; Crawford, G. E.; Laframboise, T.; Tesar, P.; Wei, C. L.; Scacheri, P. C., CHD7 targets active gene enhancer elements to modulate ES cell-specific gene expression. *PLoS genetics* **2010,** *6* (7), e1001023.

[260] Lalani, S. R.; Safiullah, A. M.; Fernbach, S. D.; Harutyunyan, K. G.; Thaller, C.; Peterson, L. E.; McPherson, J. D.; Gibbs, R. A.; White, L. D.; Hefner, M.; Davenport, S. L.; Graham, J. M.; Bacino, C. A.; Glass, N. L.; Towbin, J. A.; Craigen, W. J.; Neish, S. R.; Lin, A. E.; Belmont, J. W., Spectrum of CHD7 mutations in 110 individuals with CHARGE syndrome and genotype-phenotype correlation. *American journal of human genetics* **2006,** *78* (2), 303-14.

[261] Coffin, G. S.; Siris, E., Mental Retardation with Absent Fifth Fingernail and Terminal Phalanx. *Am J Dis Child* **1970,** *119* (5), 433-&.

[262] Santen, G. W.; Aten, E.; Sun, Y.; Almomani, R.; Gilissen, C.; Nielsen, M.; Kant, S. G.; Snoeck, I. N.; Peeters, E. A.; Hilhorst-Hofstee, Y.; Wessels, M. W.; den Hollander, N. S.; Ruivenkamp, C. A.; van Ommen, G. J.; Breuning, M. H.; den Dunnen, J. T.; van Haeringen, A.; Kriek, M., Mutations in SWI/SNF chromatin remodeling complex gene ARID1B cause Coffin-Siris syndrome. *Nature genetics* **2012,** *44* (4), 379-80.

[263] Tsurusaki, Y.; Okamoto, N.; Ohashi, H.; Kosho, T.; Imai, Y.; Hibi-Ko, Y.; Kaname, T.; Naritomi, K.; Kawame, H.; Wakui, K.; Fukushima, Y.; Homma, T.; Kato, M.; Hiraki, Y.; Yamagata, T.; Yano, S.; Mizuno, S.; Sakazume, S.; Ishii, T.; Nagai, T.; Shiina, M.; Ogata, K.; Ohta, T.; Niikawa, N.; Miyatake, S.; Okada, I.; Mizuguchi, T.; Doi, H.; Saitsu, H.; Miyake, N.; Matsumoto, N., Mutations affecting components of the SWI/SNF complex cause Coffin-Siris syndrome. *Nature genetics* **2012,** *44* (4), 376-8.

[264] Clapier, C. R.; Cairns, B. R., The biology of chromatin remodeling complexes. *Annual review of biochemistry* **2009,** *78*, 273-304.

[265] Euskirchen, G. M.; Auerbach, R. K.; Davidov, E.; Gianoulis, T. A.; Zhong, G.; Rozowsky, J.; Bhardwaj, N.; Gerstein, M. B.; Snyder, M., Diverse roles and interactions of the SWI/SNF chromatin remodeling complex revealed using global approaches. *PLoS genetics* **2011,** *7* (3), e1002008.

[266] Reyes, J. C.; Muchardt, C.; Yaniv, M., Components of the human SWI/SNF complex are enriched in active chromatin and are associated with the nuclear matrix. *The Journal of cell biology* **1997,** *137* (2), 263-74.

[267] Gibbons, R. J.; Picketts, D. J.; Villard, L.; Higgs, D. R., Mutations in a putative global transcriptional regulator cause X-linked mental retardation with alpha-thalassemia (ATR-X syndrome). *Cell* **1995,** *80* (6), 837-45.

[268] Picketts, D. J.; Higgs, D. R.; Bachoo, S.; Blake, D. J.; Quarrell, O. W.; Gibbons, R. J., ATRX encodes a novel member of the SNF2 family of proteins: mutations point to a common mechanism underlying the ATR-X syndrome. *Human molecular genetics* **1996,** *5* (12), 1899-907.

[269] De La Fuente, R.; Viveiros, M. M.; Wigglesworth, K.; Eppig, J. J., ATRX, a member of the SNF2 family of helicase/ATPases, is required for chromosome alignment and meiotic

spindle organization in metaphase II stage mouse oocytes. *Developmental biology* **2004,** *272* (1), 1-14.

[270] Baumann, C.; Viveiros, M. M.; De La Fuente, R., Loss of maternal ATRX results in centromere instability and aneuploidy in the mammalian oocyte and pre-implantation embryo. *PLoS genetics* **2010,** *6* (9).

[271] Ritchie, K.; Seah, C.; Moulin, J.; Isaac, C.; Dick, F.; Berube, N. G., Loss of ATRX leads to chromosome cohesion and congression defects. *The Journal of cell biology* **2008,** *180* (2), 315-24.

[272] Bagheri-Fam, S.; Argentaro, A.; Svingen, T.; Combes, A. N.; Sinclair, A. H.; Koopman, P.; Harley, V. R., Defective survival of proliferating Sertoli cells and androgen receptor function in a mouse model of the ATR-X syndrome. *Human molecular genetics* **2011,** *20* (11), 2213-24.

[273] Tang, J.; Wu, S.; Liu, H.; Stratt, R.; Barak, O. G.; Shiekhattar, R.; Picketts, D. J.; Yang, X., A novel transcription regulatory complex containing death domain-associated protein and the ATR-X syndrome protein. *The Journal of biological chemistry* **2004,** *279* (19), 20369-77.

[274] Barresi, V.; Ragusa, A.; Fichera, M.; Musso, N.; Castiglia, L.; Rappazzo, G.; Travali, S.; Mattina, T.; Romano, C.; Cocchi, G.; Condorelli, D. F., Decreased expression of GRAF1/OPHN-1-L in the X-linked alpha thalassemia mental retardation syndrome. *BMC medical genomics* **2010,** *3*, 28.

[275] Levy, M. A.; Fernandes, A. D.; Tremblay, D. C.; Seah, C.; Berube, N. G., The SWI/SNF protein ATRX co-regulates pseudoautosomal genes that have translocated to autosomes in the mouse genome. *BMC genomics* **2008,** *9*, 468.

[276] Kernohan, K. D.; Jiang, Y.; Tremblay, D. C.; Bonvissuto, A. C.; Eubanks, J. H.; Mann, M. R.; Berube, N. G., ATRX partners with cohesin and MeCP2 and contributes to developmental silencing of imprinted genes in the brain. *Developmental cell* **2010,** *18* (2), 191-202.

[277] Wong, L. H.; McGhie, J. D.; Sim, M.; Anderson, M. A.; Ahn, S.; Hannan, R. D.; George, A. J.; Morgan, K. A.; Mann, J. R.; Choo, K. H., ATRX interacts with H3.3 in maintaining telomere structural integrity in pluripotent embryonic stem cells. *Genome research* **2010,** *20* (3), 351-60.

[278] De La Fuente, R.; Baumann, C.; Viveiros, M. M., Role of ATRX in chromatin structure and function: implications for chromosome instability and human disease. *Reproduction* **2011,** *142* (2), 221-34.

[279] Gibbons, R. J.; McDowell, T. L.; Raman, S.; O'Rourke, D. M.; Garrick, D.; Ayyub, H.; Higgs, D. R., Mutations in ATRX, encoding a SWI/SNF-like protein, cause diverse changes in the pattern of DNA methylation. *Nature genetics* **2000,** *24* (4), 368-71.

[280] McDowell, T. L.; Gibbons, R. J.; Sutherland, H.; O'Rourke, D. M.; Bickmore, W. A.; Pombo, A.; Turley, H.; Gatter, K.; Picketts, D. J.; Buckle, V. J.; Chapman, L.; Rhodes, D.; Higgs, D. R., Localization of a putative transcriptional regulator (ATRX) at pericentromeric heterochromatin and the short arms of acrocentric chromosomes. *Proceedings of the National Academy of Sciences of the United States of America* **1999,** *96* (24), 13983-8.

[281] Xue, Y.; Gibbons, R.; Yan, Z.; Yang, D.; McDowell, T. L.; Sechi, S.; Qin, J.; Zhou, S.; Higgs, D.; Wang, W., The ATRX syndrome protein forms a chromatin-remodeling complex with Daxx and localizes in promyelocytic leukemia nuclear bodies. *Proceedings of the National Academy of Sciences of the United States of America* **2003,** *100* (19), 10635-40.

[282] Dhayalan, A.; Tamas, R.; Bock, I.; Tattermusch, A.; Dimitrova, E.; Kudithipudi, S.; Ragozin, S.; Jeltsch, A., The ATRX-ADD domain binds to H3 tail peptides and reads the combined methylation state of K4 and K9. *Human molecular genetics* **2011,** *20* (11), 2195-203.

[283] Nan, X.; Hou, J.; Maclean, A.; Nasir, J.; Lafuente, M. J.; Shu, X.; Kriaucionis, S.; Bird, A., Interaction between chromatin proteins MECP2 and ATRX is disrupted by mutations that cause inherited mental retardation. *Proceedings of the National Academy of Sciences of the United States of America* **2007,** *104* (8), 2709-14.

[284] Drane, P.; Ouararhni, K.; Depaux, A.; Shuaib, M.; Hamiche, A., The death-associated protein DAXX is a novel histone chaperone involved in the replication-independent deposition of H3.3. *Genes & development* **2010,** *24* (12), 1253-65.

[285] Goldberg, A. D.; Banaszynski, L. A.; Noh, K. M.; Lewis, P. W.; Elsaesser, S. J.; Stadler, S.; Dewell, S.; Law, M.; Guo, X.; Li, X.; Wen, D.; Chapgier, A.; DeKelver, R. C.; Miller, J. C.; Lee, Y. L.; Boydston, E. A.; Holmes, M. C.; Gregory, P. D.; Greally, J. M.; Rafii, S.; Yang, C.; Scambler, P. J.; Garrick, D.; Gibbons, R. J.; Higgs, D. R.; Cristea, I. M.; Urnov, F. D.; Zheng, D.; Allis, C. D., Distinct factors control histone variant H3.3 localization at specific genomic regions. *Cell* **2010,** *140* (5), 678-91.

[286] Lewis, P. W.; Elsaesser, S. J.; Noh, K. M.; Stadler, S. C.; Allis, C. D., Daxx is an H3.3-specific histone chaperone and cooperates with ATRX in replication-independent chromatin assembly at telomeres. *Proceedings of the National Academy of Sciences of the United States of America* **2010,** *107* (32), 14075-80.

[287] Law, M. J.; Lower, K. M.; Voon, H. P.; Hughes, J. R.; Garrick, D.; Viprakasit, V.; Mitson, M.; De Gobbi, M.; Marra, M.; Morris, A.; Abbott, A.; Wilder, S. P.; Taylor, S.; Santos, G. M.; Cross, J.; Ayyub, H.; Jones, S.; Ragoussis, J.; Rhodes, D.; Dunham, I.; Higgs, D. R.; Gibbons, R. J., ATR-X syndrome protein targets tandem repeats and influences allele-specific expression in a size-dependent manner. *Cell* **2010,** *143* (3), 367-78.

[288] Kops, G. J.; Weaver, B. A.; Cleveland, D. W., On the road to cancer: aneuploidy and the mitotic checkpoint. *Nat Rev Cancer* **2005,** *5* (10), 773-85.

[289] Mitelman, F., *Catalog of chromosome aberrations in cancer*. 4 ed.; Wiley-Liss: New York, 1991.

[290] Gibbons, R. J., Histone modifying and chromatin remodelling enzymes in cancer and dysplastic syndromes. *Human molecular genetics* **2005,** *14 Spec No 1*, R85-92.

[291] Esteller, M., Aberrant DNA methylation as a cancer-inducing mechanism. *Annu Rev Pharmacol Toxicol* **2005,** *45*, 629-56.

[292] Davis, P. K.; Brackmann, R. K., Chromatin remodeling and cancer. *Cancer Biol Ther* **2003,** *2* (1), 22-9.

[293] Betz, B. L.; Strobeck, M. W.; Reisman, D. N.; Knudsen, E. S.; Weissman, B. E., Re-expression of hSNF5/INI1/BAF47 in pediatric tumor cells leads to G1 arrest associated with induction of p16ink4a and activation of RB. *Oncogene* **2002,** *21* (34), 5193-203.

[294] Hendricks, K. B.; Shanahan, F.; Lees, E., Role for BRG1 in cell cycle control and tumor suppression. *Mol Cell Biol* **2004,** *24* (1), 362-76.

[295] Oruetxebarria, I.; Venturini, F.; Kekarainen, T.; Houweling, A.; Zuijderduijn, L. M.; Mohd-Sarip, A.; Vries, R. G.; Hoeben, R. C.; Verrijzer, C. P., P16INK4a is required for hSNF5 chromatin remodeler-induced cellular senescence in malignant rhabdoid tumor cells. *The Journal of biological chemistry* **2004,** *279* (5), 3807-16.

[296] Bochar, D. A.; Wang, L.; Beniya, H.; Kinev, A.; Xue, Y.; Lane, W. S.; Wang, W.; Kashanchi, F.; Shiekhattar, R., BRCA1 is associated with a human SWI/SNF-related complex: linking chromatin remodeling to breast cancer. *Cell* **2000,** *102* (2), 257-65.

[297] Bultman, S. J.; Herschkowitz, J. I.; Godfrey, V.; Gebuhr, T. C.; Yaniv, M.; Perou, C. M.; Magnuson, T., Characterization of mammary tumors from Brg1 heterozygous mice. *Oncogene* **2008,** *27* (4), 460-8.
[298] Yasui, W.; Oue, N.; Ono, S.; Mitani, Y.; Ito, R.; Nakayama, H., Histone acetylation and gastrointestinal carcinogenesis. *Ann N Y Acad Sci* **2003,** *983*, 220-31.
[299] Bandyopadhyay, D.; Okan, N. A.; Bales, E.; Nascimento, L.; Cole, P. A.; Medrano, E. E., Down-regulation of p300/CBP histone acetyltransferase activates a senescence checkpoint in human melanocytes. *Cancer research* **2002,** *62* (21), 6231-9.
[300] Debes, J. D.; Sebo, T. J.; Heemers, H. V.; Kipp, B. R.; Haugen de, A. L.; Lohse, C. M.; Tindall, D. J., p300 modulates nuclear morphology in prostate cancer. *Cancer research* **2005,** *65* (3), 708-12.
[301] Debes, J. D.; Sebo, T. J.; Lohse, C. M.; Murphy, L. M.; Haugen de, A. L.; Tindall, D. J., p300 in prostate cancer proliferation and progression. *Cancer research* **2003,** *63* (22), 7638-40.
[302] Deguchi, K.; Ayton, P. M.; Carapeti, M.; Kutok, J. L.; Snyder, C. S.; Williams, I. R.; Cross, N. C.; Glass, C. K.; Cleary, M. L.; Gilliland, D. G., MOZ-TIF2-induced acute myeloid leukemia requires the MOZ nucleosome binding motif and TIF2-mediated recruitment of CBP. *Cancer Cell* **2003,** *3* (3), 259-71.
[303] de Ruijter, A. J.; van Gennip, A. H.; Caron, H. N.; Kemp, S.; van Kuilenburg, A. B., Histone deacetylases (HDACs): characterization of the classical HDAC family. *Biochem J* **2003,** *370* (Pt 3), 737-49.
[304] Mitsiades, C. S.; Mitsiades, N. S.; McMullan, C. J.; Poulaki, V.; Shringarpure, R.; Hideshima, T.; Akiyama, M.; Chauhan, D.; Munshi, N.; Gu, X.; Bailey, C.; Joseph, M.; Libermann, T. A.; Richon, V. M.; Marks, P. A.; Anderson, K. C., Transcriptional signature of histone deacetylase inhibition in multiple myeloma: biological and clinical implications. *Proceedings of the National Academy of Sciences of the United States of America* **2004,** *101* (2), 540-5.
[305] Feinberg, A. P.; Vogelstein, B., Hypomethylation distinguishes genes of some human cancers from their normal counterparts. *Nature* **1983,** *301* (5895), 89-92.
[306] Esteller, M.; Fraga, M. F.; Guo, M.; Garcia-Foncillas, J.; Hedenfalk, I.; Godwin, A. K.; Trojan, J.; Vaurs-Barriere, C.; Bignon, Y. J.; Ramus, S.; Benitez, J.; Caldes, T.; Akiyama, Y.; Yuasa, Y.; Launonen, V.; Canal, M. J.; Rodriguez, R.; Capella, G.; Peinado, M. A.; Borg, A.; Aaltonen, L. A.; Ponder, B. A.; Baylin, S. B.; Herman, J. G., DNA methylation patterns in hereditary human cancers mimic sporadic tumorigenesis. *Human molecular genetics* **2001,** *10* (26), 3001-7.
[307] Walsh, C. P.; Chaillet, J. R.; Bestor, T. H., Transcription of IAP endogenous retroviruses is constrained by cytosine methylation. *Nature genetics* **1998,** *20* (2), 116-7.
[308] Yoder, J. A.; Walsh, C. P.; Bestor, T. H., Cytosine methylation and the ecology of intragenomic parasites. *Trends Genet* **1997,** *13* (8), 335-40.
[309] Feinberg, A. P., Imprinting of a genomic domain of 11p15 and loss of imprinting in cancer: an introduction. *Cancer research* **1999,** *59* (7 Suppl), 1743s-1746s.
[310] Plass, C.; Soloway, P. D., DNA methylation, imprinting and cancer. *Eur J Hum Genet* **2002,** *10* (1), 6-16.
[311] Costello, J. F.; Fruhwald, M. C.; Smiraglia, D. J.; Rush, L. J.; Robertson, G. P.; Gao, X.; Wright, F. A.; Feramisco, J. D.; Peltomaki, P.; Lang, J. C.; Schuller, D. E.; Yu, L.; Bloomfield, C. D.; Caligiuri, M. A.; Yates, A.; Nishikawa, R.; Su Huang, H.; Petrelli, N. J.; Zhang, X.; O'Dorisio, M. S.; Held, W. A.; Cavenee, W. K.; Plass, C., Aberrant CpG-island

methylation has non-random and tumour-type-specific patterns. *Nature genetics* **2000,** *24* (2), 132-8.

[312] Esteller, M.; Corn, P. G.; Baylin, S. B.; Herman, J. G., A gene hypermethylation profile of human cancer. *Cancer research* **2001,** *61* (8), 3225-9.

[313] Fraga, M. F.; Herranz, M.; Espada, J.; Ballestar, E.; Paz, M. F.; Ropero, S.; Erkek, E.; Bozdogan, O.; Peinado, H.; Niveleau, A.; Mao, J. H.; Balmain, A.; Cano, A.; Esteller, M., A mouse skin multistage carcinogenesis model reflects the aberrant DNA methylation patterns of human tumors. *Cancer research* **2004,** *64* (16), 5527-34.

[314] Esteller, M., CpG island hypermethylation and tumor suppressor genes: a booming present, a brighter future. *Oncogene* **2002,** *21* (35), 5427-40.

[315] Lyko, F.; Brown, R., DNA methyltransferase inhibitors and the development of epigenetic cancer therapies. *J Natl Cancer Inst* **2005,** *97* (20), 1498-506.

[316] Cai, S.; Han, H. J.; Kohwi-Shigematsu, T., Tissue-specific nuclear architecture and gene expression regulated by SATB1. *Nature genetics* **2003,** *34* (1), 42-51.

[317] Cai, S.; Lee, C. C.; Kohwi-Shigematsu, T., SATB1 packages densely looped, transcriptionally active chromatin for coordinated expression of cytokine genes. *Nature genetics* **2006,** *38* (11), 1278-88.

[318] Dickinson, L. A.; Joh, T.; Kohwi, Y.; Kohwi-Shigematsu, T., A tissue-specific MAR/SAR DNA-binding protein with unusual binding site recognition. *Cell* **1992,** *70* (4), 631-45.

[319] Kumar, P. P.; Bischof, O.; Purbey, P. K.; Notani, D.; Urlaub, H.; Dejean, A.; Galande, S., Functional interaction between PML and SATB1 regulates chromatin-loop architecture and transcription of the MHC class I locus. *Nature cell biology* **2007,** *9* (1), 45-56.

[320] Han, H. J.; Russo, J.; Kohwi, Y.; Kohwi-Shigematsu, T., SATB1 reprogrammes gene expression to promote breast tumour growth and metastasis. *Nature* **2008,** *452* (7184), 187-93.

[321] Selinger, C. I.; Cooper, W. A.; Al-Sohaily, S.; Mladenova, D. N.; Pangon, L.; Kennedy, C. W.; McCaughan, B. C.; Stirzaker, C.; Kohonen-Corish, M. R., Loss of special AT-rich binding protein 1 expression is a marker of poor survival in lung cancer. *Journal of thoracic oncology : official publication of the International Association for the Study of Lung Cancer* **2011,** *6* (7), 1179-89.

[322] Davalos-Salas, M.; Furlan-Magaril, M.; Gonzalez-Buendia, E.; Valdes-Quezada, C.; Ayala-Ortega, E.; Recillas-Targa, F., Gain of DNA methylation is enhanced in the absence of CTCF at the human retinoblastoma gene promoter. *BMC cancer* **2011,** *11*, 232.

[323] Witcher, M.; Emerson, B. M., Epigenetic silencing of the p16(INK4a) tumor suppressor is associated with loss of CTCF binding and a chromatin boundary. *Molecular cell* **2009,** *34* (3), 271-84.

[324] De La Rosa-Velazquez, I. A.; Rincon-Arano, H.; Benitez-Bribiesca, L.; Recillas-Targa, F., Epigenetic regulation of the human retinoblastoma tumor suppressor gene promoter by CTCF. *Cancer research* **2007,** *67* (6), 2577-85.

[325] Docquier, F.; Farrar, D.; D'Arcy, V.; Chernukhin, I.; Robinson, A. F.; Loukinov, D.; Vatolin, S.; Pack, S.; Mackay, A.; Harris, R. A.; Dorricott, H.; O'Hare, M. J.; Lobanenkov, V.; Klenova, E., Heightened expression of CTCF in breast cancer cells is associated with resistance to apoptosis. *Cancer research* **2005,** *65* (12), 5112-22.

[326] Fiorentino, F. P.; Macaluso, M.; Miranda, F.; Montanari, M.; Russo, A.; Bagella, L.; Giordano, A., CTCF and BORIS regulate Rb2/p130 gene transcription: a novel mechanism and a new paradigm for understanding the biology of lung cancer. *Molecular cancer research : MCR* **2011,** *9* (2), 225-33.

[327] Green, A. R.; Krivinskas, S.; Young, P.; Rakha, E. A.; Paish, E. C.; Powe, D. G.; Ellis, I. O., Loss of expression of chromosome 16q genes DPEP1 and CTCF in lobular carcinoma *in situ* of the breast. *Breast cancer research and treatment* **2009,** *113* (1), 59-66.
[328] Ranz, J. M.; Maurin, D.; Chan, Y. S.; von Grotthuss, M.; Hillier, L. W.; Roote, J.; Ashburner, M.; Bergman, C. M., Principles of genome evolution in the Drosophila melanogaster species group. *PLoS biology* **2007,** *5* (6), e152.
[329] Purmann, A.; Toedling, J.; Schueler, M.; Carninci, P.; Lehrach, H.; Hayashizaki, Y.; Huber, W.; Sperling, S., Genomic organization of transcriptomes in mammals: Coregulation and cofunctionality. *Genomics* **2007,** *89* (5), 580-7.
[330] Tsaparas, P.; Marino-Ramirez, L.; Bodenreider, O.; Koonin, E. V.; Jordan, I. K., Global similarity and local divergence in human and mouse gene co-expression networks. *BMC evolutionary biology* **2006,** *6*, 70.
[331] Bhutkar, A.; Schaeffer, S. W.; Russo, S. M.; Xu, M.; Smith, T. F.; Gelbart, W. M., Chromosomal rearrangement inferred from comparisons of 12 Drosophila genomes. *Genetics* **2008,** *179* (3), 1657-80.
[332] De, S.; Teichmann, S. A.; Babu, M. M., The impact of genomic neighborhood on the evolution of human and chimpanzee transcriptome. *Genome research* **2009,** *19* (5), 785-94.
[333] Yanai, I.; Hunter, C. P., Comparison of diverse developmental transcriptomes reveals that coexpression of gene neighbors is not evolutionarily conserved. *Genome research* **2009,** *19* (12), 2214-20.
[334] Akashi, K.; He, X.; Chen, J.; Iwasaki, H.; Niu, C.; Steenhard, B.; Zhang, J.; Haug, J.; Li, L., Transcriptional accessibility for genes of multiple tissues and hematopoietic lineages is hierarchically controlled during early hematopoiesis. *Blood* **2003,** *101* (2), 383-9.
[335] Lunyak, V. V.; Burgess, R.; Prefontaine, G. G.; Nelson, C.; Sze, S. H.; Chenoweth, J.; Schwartz, P.; Pevzner, P. A.; Glass, C.; Mandel, G.; Rosenfeld, M. G., Corepressor-dependent silencing of chromosomal regions encoding neuronal genes. *Science* **2002,** *298* (5599), 1747-52.
[336] Jordan, K. W.; Carbone, M. A.; Yamamoto, A.; Morgan, T. J.; Mackay, T. F., Quantitative genomics of locomotor behavior in Drosophila melanogaster. *Genome biology* **2007,** *8* (8), R172.
[337] Li, J.; Burmeister, M., Genetical genomics: combining genetics with gene expression analysis. *Human molecular genetics* **2005,** *14 Spec No. 2*, R163-9.
[338] De, S.; Babu, M. M., Genomic neighbourhood and the regulation of gene expression. *Current opinion in cell biology* **2010,** *22* (3), 326-33.

Send Orders of Reprints at reprints@benthamscince.net

Advances in Genome Science, Vol. 1, 2013, 146-168

CHAPTER 6

Expression and Function of Kruppel Like-Factors (KLF) in Carcinogenesis

Diab Thoria[1,2], Bureau Christophe[1,2,3], Hanoun Naima[1,2], Torrisani Jérôme[1,2], Vinel Jean-Pierre[1,2,3], Buscail Louis [1,2,3] and Cordelier Pierre[1,2,*]

[1]Institut National de la Santé et de la Recherche Médicale (INSERM), Centre de Recherches en Cancérologie de Toulouse UMR1037; [2]Université Paul Sabatier, Toulouse, France and [3]Service d'Hépato-gastroentérologie, CHU Toulouse, France

Abstract: Krüppel-like factor (KLF) family members share a three C2H2 zinc finger DNA binding domain, and are involved in cell proliferation and differentiation control in normal as in pathological situations. Studies over the past several years support a significant role for this family of transcription factors in carcinogenesis. KLFs can both activate and repress genes that participate in cell-cycle regulation. Among them, many up-regulated genes are inhibitors of proliferation, whereas genes that promote cell proliferation are repressed. However, several studies do present KLFs as positive regulator of cell proliferation. KLFs can be deregulated in multiple cancers either by loss of heterozygosity (LOH), somatic mutation or transcriptional silencing by promoter hypermethylation. Accordingly, KLF mediates growth inhibition when ectopically expressed in multiple cancer-derived cell lines through the inhibition of a number of key oncogenic signalling pathways, and to reverse the tumorogenic phenotype *in vivo*. Taken together, these observations suggest that KLFs act as tumor suppressor. However, in some occasion, KLFs could act as tumor promoters, depending on "cellular context". Thus, this review will discuss the roles and the functions of KLF family members in carcinogenesis, with a special focus on cancers from epithelial origin.

Keywords: Krüppel-like factors, cell cycle, transcription factor, cancer.

1. EXPRESSION AND FUNCTION OF KLF IN GASTROINTESTINAL TRACT TUMORS

1.1. Expression and Function of KLF in Intestinal and Colon Tumors

KLF4 or gut-enriched KLF (GKLF) is expressed extensively in the gastrointestinal (GI) tract from where it was first identified to maintain enterocyte differentiation and restrain cell proliferation. In mice, KLF4 level peaks by embryonic day 17 [1].

***Address correspondence to Pierre Cordelier:** INSERM U1037, CRCT Team 10, 1, avenue Jean Poulhes, BP 84225, 31432 Toulouse Cedex 04, France; Tel: (33) 531224106; E-mail: pierre.cordelier@inserm.fr

Christian Neri (Ed)

In the newborn, KLF4 level is higher in colon than in small intestine although the levels in both organs rise with increasing age [1]. In Human, KLF4 protein is prominent at the epithelium surface in normal colon, and its expression gradually decreases toward the crypt [2]. In RKO cells (human colon cancer-derived cells), KLF4 stimulates intestinal alkaline phosphatase (IAP) gene expression, through a critical region including the IF-III cis element located within the proximal IAP promoter region [3]. In HCT-116 cells (human colon cancer-derived cells), KLF4 increases urokinase-type plasminogen activator receptor (u-PAR) [4]. Inversely, level of KLF4 transcript is significantly decreased in intestine during tumor formation in Min mice, a model of intestinal tumorogenesis, suggesting that KLF4 may play a role in gut development and/or tumor burden [1, 5]. KLF4 mRNA and protein levels are significantly and gradually decreased in the dysplasic epithelium of colon, as in adenomatous polyp and colorectal cancer (CRC) [2, 6]. In familial adenomatous polyposis (FAP) patients, KLF4 transcript level is lower in adenomas compared to paired normal-appearing mucosa [5].

The molecular mechanisms involved in KLF4 loss of expression are still unclear. The gene encoding KLF4 is localized on chromosome 9q, previously shown to exhibit loss of heterozigosity (LOH) in CRC [7]. LOH in KLF4 is found in 83% of CRC-derived cell lines, and in 20% of CRC[7]. Using methylation-specific PCR, Dang DT *et al* showed that KLF4 DNA is not methylated in either normal or tumor tissues from adenomas of Min mice [5]. However, KLF4 expression is stimulated by treating CRC-derived cell lines with 5-Aza-2-deoxycytidine (5-Aza-dC), implicating underlying aberrant epigenetic modifications [2b]. Again, KLF4 is methylated in a subset of tumors and cell lines [7]. Other chemicals are used to modulate KLF expression : 5,5'-DibromoDIM up-regulates expression of KLF4 in CRC-derived cell lines, and may represent a novel class of mechanism-based therapeutic drug for CRC [8]. Several point mutations in KLF4 result in a diminished ability to transactivate antiproliferative genes [7]. In addition, KLF4 expression is stimulated following treatment of HT-29 colon cancer cells with IFN-γ and Stat-1 recruitment [9]. In FAP, KLF4 is induced upon activation of the adenomatous polyposis coli (APC) gene [10], probably *via* CDX2 [10a]. KLF4 protein is predominantly expressed in cytoplasm but not in nucleus of CRC-derived cells, suggesting that impaired nuclear translocation of

KLF4 contributes to cancer formation [11]. Last, KLF4 expression is inversely correlated to beta-catenin/Tcf pathway in HT29 cells [12].

KLF4 is required for the terminal differentiation of goblet cells in the mouse intestine. Notch signaling pathway suppresses goblet cell formation and is up-regulated in intestinal tumors. Interestingly, KLF4 promoter activity is inhibited by Notch, which controls goblet cell differentiation in mouse GI tract [13], when Notch signaling suppresses KLF4 expression to support intestinal tumors and CRC progression [14]. Recently, stimulation of PPARgamma-dependent pathways resulted in KLF4 induction in colon cancer cells [15].

Because KLF4 expression is extensively lost in CRC and a downstream target of tumor suppressor genes, suggesting that its re-expression may counteract CRC progression. In cultured cells, expression of KLF4 is associated with growth arrest [16]. Constitutive expression of KLF4 in human CRC-derived cells decreases colony formation, cell proliferation, cell migration and invasion, and *in vivo* tumorogenesis [17]. As a negative regulator of gene expression, KLF4 acts trough multiple pathways: (i) induction of $p21^{WAF1/CIP1}$ to maintain the integrity of both G_1/S [18] and G_2/M [19] cell cycle checkpoints following DNA damage, reduction of cyclin D1 activity by interaction with the Sp1 binding domain on its promoter [20], (ii) repression of beta-catenin expression [10b], and (iii) inhibition of ornithine decarboxylase promoter activity through a target, GC-rich, region [21]. Of importance, KLF4 exerts a global inhibitory effect on protein and cholesterol biosynthesis [22]. Interestingly, recent studies suggest that KLF4 functions as an activator or repressor of transcription depending on whether it interacts with co-activators such as p300 and CREB-binding protein or co-repressors such as histone deacetylase HDAC3 [23]. Taken together, these studies clearly demonstrate that KLF4 plays an essential role in the regulation of cell growth in the colon.

Additional KLF family members modulate colon-derived cells proliferation. Despite binding to similar, if not identical, cis-acting DNA sequences, KLF4 and KLF5 often exhibit opposite effects on regulation of gene transcription and cell proliferation[24]. KLF5 promotes proliferation of intestinal epithelia in response to lysophophatidic acid (LPA) [25]. KLF5 protein expression correlates with activating KRAS mutations in intestinal tumors *in vitro* and *in* vivo [26]. Also, KLF5 modulates p53-independent apoptosis through Pim1 survival kinase [27]. Targeting KLF5 results in cell proliferation, transformation and multi-cellular tumor spheroids (MCTS) inhibition [26, 28]. Also, decreasing KLF5 expression using all-trans retinoic acid inhibits proliferation of intestinal epithelial cells [29]. As a consequence, small-molecule compounds are screened on their ability to inhibit the activity of this KLF family member [30]. However, KLF5 becomes anti-proliferative upon TGFβ-mediated acetylation, to maintain epithelial homeostasis [31].

It is less clear how KLF6 expression is regulated in CRC samples. Several studies describe the frequent inactivation by mutation of this putative tumor-suppressor gene [32]. KLF6 mutants lose the ability to induce $p21^{WAF1/CIP1}$, and fail to inhibit CRC-derived cells growth [32a]. LOH of KLF6 locus at chromosome 10p15 contributes to the development of non-polypoid colorectal carcinoma [33], and in the invasion step from an intra-mucosal to an invasive carcinoma [34]. Recently, KLF6 mutations were found in an Egyptian population. Most of them were missense mutations located in exon 2, and different in type and location as previously described in the Western population [35]. On the other hand, these findings are challenged by the absence of somatic mutations in DNA extracted from frozen tissue samples from larger series of CRC [36]. Interestingly, IGF-1, a well-known proliferation promoter, induces KLF6 expression in a p53-dependent manner [37]. Interestingly, inhibition of the receptor tyrosine kinase MET inhibited tumor growth in various colon tumor xenograft models following the up-regulation of many tumor suppressors including KLF6 [38]. Last, a recent study demonstrate that KLF9 expression is deregulated in CRC, suggesting that this KLF family member may also be involved in the carcinogenesis of human colorectal cancer [39].

1.2. Expression and Function of KLF in Gastric, Oesophageal and Bladder Cancers

In addition to colon, KLFs exert opposite effects on cell proliferation in numerous examples; while KLF4 and KLF6 inhibit cell growth, KLF5 stimulates proliferation. In gastric cancer, loss of KLF4 expression in primary tumors is associated with poor survival [40]. In parallel, KLF4 mutant mice show increased proliferation and altered differentiation of their gastric epithelia [40b]. Recently, Lee *et al* demonstrated that KLF4 positively regulates ghrelin expression in human gastric cancer cells [41]. Interestingly, KLF5 expression is also reduced in gastric carcinomas [42]. Disrupted KLF4 expression contributes to Sp1 over-expression, and to the development and progression of human gastric cancer [43]. Mechanistic studies indicate that promoter hypermethylation and hemizygous deletion contribute to KLF4 down-regulation in cancer [40a]. However, epigenetic alterations of KLF4 in gastric oncogenesis is still debated [44]. Enforced expression of KLF4 results in cell proliferation and tumor growth inhibition [40a]. In addition, somatic missense mutations are identified in the activation domain of the KLF6 gene in gastric carcinoma [45]. KLF6 LOH is observed in 16 of 37 informative cases [45a]. These data suggest that genetic alterations of KLF6 gene play an important role in the development or the progression of sporadic gastric cancers [45a]. Narla's group found that 53% of gastric cancer samples display a KLF6 locus specific loss, a decreased KLF6 expression. Moreover he demonstrated that enforced expression of KLF6 suppressed proliferation [46]. In the same study, KLF6 splicing variants were demonstrated to increase tumorigenicity *in vivo* [46]. KLF6 variant SV1 seems to be an important regulator of growth, migration, invasion and survival of gastric cancer cells and as a consequence, a new potential gene therapy target for this disease [47]. However, a recent study demonstrates that the IVS1 -27G/A polymorphism of KLF6 is not associated with an increased risk of gastric cancer in Korean population [48].

Down-regulation of KLF4 expression is also reported in oesophageal cancer [49], while over-expression of KLF5 in oesophageal epithelia leads to an increase of proliferation in basal cells, *in vivo* [50]. KLF4 represses the transcription of survivin in oesophageal squamous cancer cells, which is in favor of an anti-proliferative effect [51]. Last, KLF5 transcription factor is described to play an oncogenic role in a human bladder cancer cell line, *via* increased G_1 to S phase transition, up-regulation of cyclin D1 expression, phosphorylation of MAPK and Akt proteins, and inhibition of p27 and p15 expression [52].

1.3. Expression and Function of KLF in Hepatocellular Carcinoma

Recently, KLF4 has been associated with steaminess [53]. Interestingly, KLF4 is one of the many stemness-promoting genes induced by the hepatitis B antigen HBx that may contribute to the appearance of cancer stem cells in hepatocarcinogenesis [54]. KLF6 expression and function have been extensively investigated in hepatocellular carcinoma (HCC). KLF6 was first shown to be frequently inactivated either by LOH and/or inactivating somatic mutations, and that its inactivation contributes to pathogenesis of HCC [55]. Whereas wild-type KLF6 decreases cellular proliferation of HCC-derived cells, patient-derived KLF6 mutants do not [55b, c]. Also, these mutants are unable to transactivate $p21^{WAF1/CIP1}$ promoter [55c]. DNA methylation may marginally participate to KLF6 down-regulation in HCC [55d, 56]. Recently, it was shown that E2F3a up-regulates KLF6 in HepG2 liver cancer cells and induces apoptosis [57]. In the same model, KLF6 repressed the expression of the oncogenic pituitary tumor transforming gene 1 (PTTG1) by direct promoter interaction [58]. However, deregulation of KLF6 stability following ubiquitination may alter its tumor suppressor function in UV-B irradiated cell lines [59]. Transgenic expression of KLF6 in the liver results in decreased body and liver size, with decreased hepatocyte proliferation, when KLF6 +/ mice display increased liver mass with reduced p21 expression [60] and hepatic tumors when challenged with chemical carcinogens [61]. In the later model, KLF6 reduced MDM2 levels and increased p53 protein and transcriptional activity.

However, KLF6 status in HCC is still largely debated. No somatic mutations are found in freshly isolated HCC samples [55e, 62], and no difference in KLF6 expression are detected between HCC and adjacent normal liver [55e]. However, KLF6 expression is inhibited in macronodules, as compared to cirrhosis and HCC [62b]. Finally, KLF6 is essential for HCC-derived cells to evade apoptosis [63]. Together with the results describing KLF6 as a tumor suppressor gene, the latter observations create a paradox and indicate that not only loss, but also enforced expression of KLF6 contribute to tumorigenesis. This phenomenon is still not totally understood, and can be attributed to 'cell-type' specificity. Further studies are needed to identify such cellular factors, for KLF6 to switch from a growth-inhibiting tumor suppressor to a growth-promoting oncogene. In fact,

KLF6 alternative splicing that yield a dominant negative, proliferation-prone, KLF6-SV1 isoform is described in HCC [64] and in non-alcoholic fatty liver disease [65]. KLF6 alternative splicing in HCC is promoted by oncogenic *KRAS* [64]. We recently demonstrated that KLF6-SV2 variant is down-regulated in hepatocellular carcinoma and displays anti-proliferative and pro-apoptotic functions [66]. We observed a reduced expression of the SV2 variant in HCC samples *versus* surrounding tissues and normal liver. Interestingly, our findings demonstrate that the over-expression of the SV2 variant in IHH and HepG2 cells leads to a significant reduction of proliferation associated with cell death by apoptosis. We further demonstrate that the SV2 expression leads to an induction of the cell-cycle-controlling $p21^{(CIP/WAF1)}$ and the pro-apoptotic Bax genes, mediated by the p53 protein. We show further that the SV2 expression in IHH and HepG2 cells induces their sensitivity to the anti-cancer drug, gemcitabine. Thus, the discrepancy considering KLF6 expression and function in HCC may reside in the identification of the molecular isoform of this transcription factor.

2. EXPRESSION AND FUNCTION OF KLF IN OTHER EPITHELIAL CANCERS

2.1. Expression and Function of KLF in Prostate Cancer

KLF6 expression is altered in prostate cancer [67]. LOH analysis reveals that KLF6 allele is deleted in 77% of primary prostate tumors [67]. Again, tumor-derived KLF6 mutants fail to transactivate $p21^{WAF1/CIP1}$, and to impede cell proliferation [67-68], when wt KLF6 induces apotosis in prostate cancer cells *via* the ATF3 transcription factor [69]. A germline KLF6 single nucleotide polymorphism (IVS1-27G>A), together with increased transcription of three alternatively spliced KLF6 isoforms is reported in prostate cancer. KLF6 protein variants KLF6-SV1 and KLF6-SV2 are mislocalized to the cytoplasm, antagonize KLF6 function, leading to a decreased p21 expression and to an increased cell growth. These variants are up-regulated in tumor *vs.* normal prostatic tissue [70]. KLF6 SV1 is demonstrated to be essential for prostate cancer cell growth and spread [71]. However, significant genetic alterations of KLF6 seem restricted to a minority of high-grade prostate cancers [72]. Again, the finding of a broad loss of KLF6 expression in prostate cancer is strongly challenged by independent studies reporting the infrequent germ-line mutation of KLF6, or the presence of KLF6

IVS1-27G>A polymorphism, in prostate cancer [73]. Last, this polymorphism predicts the metastatic behavior of the tumors [74]. Taken together, KLF6 SV1 is a novel therapeutic target for prostate cancer. Selenium supplementation is effective in reducing the incidence of prostate cancer. Interestingly, selenium induces KLF4 expression in two cancer cell lines, derived from either androgen-dependent or androgen-independent prostate cancer [75]. Over-expression of KLF4 not only leads to an induction of apoptosis in control cells, but also enhances the DNA-suppressive and pro-apoptotic activities of selenium [75]. Thus, KLF4 may also play a key role in prostate cancer control.

2.1. Expression and Function of KLF in Breast and Ovarian Cancers

In breast cancer-derived cell lines, KLF2 is strongly induced following treatment with the antiproliferative compound zoledronic acid [76]. On the other hand, the histone methyltransferase EZH2 represses its expression [77]. KLF4 expression is associated with breast cancer progression [78]. Increased KLF4 expression is observed in neoplasic cells compared to adjacent uninvolved epithelium [79]. In addition, KLF4 is highly and frequently expressed in canine mammary carcinoma and correlates with a more aggressive phenotype [80]. KLF4 is activated prior to invasion through the basal membrane [79]. Its nuclear localization is associated with an aggressive phenotype in early stage-breast cancer [81]. On the other hand, KLF5 is a potential tumor suppressor gene in breast cancer [82]. KLF5 undergoes haplo-insufficiency during breast carcinogenesis, without evidence of neither point mutation, nor promoter methylation nor homozygous deletion [78, 82]. Re-expression of KLF5 in breast cancer-derived cells inhibits colony formation [82]. Unfortunately, a recent study demonstrates that patients with a higher KLF5 expression have a shorter disease-free survival and overall survival than those observed in patients with a lower KLF5 expression [83]. Thus, KLF5 expression and/or function in breast cancer remain to be elucidated.

Other KLF family members are involved in breast cancer progression. Tissue factor pathway inhibitor-2 (TFPI-2) is a matrix-associated Kunitz inhibitor that inhibits plasmin and trypsin-mediated activation of zymogen matrix metalloproteinases involved in tumor progression, invasion and metastasis. Its expression is lost in highly invasive breast cancer caused by DNA hypermethylation, and as a consequence, a lack in KLF6 transactivation [84]. A recent study by Gehrau *et al* demonstrates that KLF6 is detected in both the nucleus and the cytoplasm of normal breast cells, while breast carcinoma cells express KLF6 mainly in the cytoplasm. However, nuclear expression of KLF6 is found in early stage and small HER2-ERBB2 overexpressing ductal breast tumor cells [85]. In addition; KLF6 counteracts the oestrogen receptor oncogenic pathways in MCF-7 cells [86]. Following a genome-wide RNA interference strategy, Gumireddy K *et al* recently identified KLF17 as a key metastasis suppressor gene for breast cancer [87]. KLF17 functions as a direct inhibitor of Id1, and is significantly down-regulated in breast cancer [87].

Interestingly, KLF6 reduction in an ovarian cancer-derived cell line results in a reduction of E-cadherin expression. Conversely, KLF6-SV1 silencing strongly up-regulates E-cadherin, which alters ovarian tumor invasion and metastasis [88]. KLF8 plays a critical role in oncogenic transformation and in epithelial to mesenchymal transition [89]. KLF8 negatively regulates E-cadherin expression which promotes human breast cancer invasion and metastasis [90]. In ovarian cancer cells, KLF8 expression is induced by FAK, by activating the PI3K-Akt signaling pathway [91].

2.3. Expression and Function of KLF in Pancreatic Cancer

KLF4 is up-regulated in pancreatic intraepithelial neoplasia, a known precursor lesion of pancreatic cancer [92]. Ectopic expression of KLF4 results in cell cycle arrest of pancreatic carcinoma-derived BxPC-3 cells. In addition, cell growth is inhibited, as tumor growth and metastasis in an orthotopic mouse model. This anti-proliferative effect is due to the interaction of KLF4 with p27(Kip1) promoter [93]. KLF6 is neither lost nor mutated in pancreatic cancer, but accumulates in cytoplasm of the cancer cells caused by an alternative splicing [94]. Interestingly, KLF6 splicing correlates with tumor stage and survival, and may be useful as a prognostic factor.

2.4. Expression and Function of KLFs in Lung Carcinoma

Lung KLF (LKLF) is identified as a transactivator of cytosolic phospholipase A2 (cPLA2) in lung epithelial and non-small-cell lung cancer (NSLC) cells [95].

Interestingly, expression of a dominant negative form of LKLF inhibits the induction of cPLA2 promoter activity by *RAS* [95]. Because cPLA2 is critical for increased eicosanoid production associated with lung tumorigenesis, targeting LKLF may be of interest to treat NSLC. In addition, NSLC exhibit frequent allelic loss at chromosome 10p15 where KLF6 is located. Although somatic mutations are not detected in the coding sequence of KLF6, expression of KLF6 mRNA is down-regulated in 85% of primary tumors [96]. Thirty-four percent of informative samples had LOH in the KLF6 gene locus, without evidence of promoter hypermethylation [96]. Enforced expression of KLF6 resulted in NSLC-derived cells apoptosis [96], while PKC activates KLF6 to mediate lung cancer-derived cell lines growth inhibition [97]. In addition, a potential involvement of KLF6 IVS1-27G>A polymorphism is associated with lung cancer risk [98]. Again, an increased expression of KLF6-SV1 is observed in lung adenocarcinoma, and is associated with a decreased survival in patients. In lung cancer cell lines, KLF6-SV1 protects cells from apoptosis, and its targeted reduction induces apoptosis [96, 99]. Last KLF17, a well-characterized negative regulator of EMT and metastasis in breast cancer, has been recently reported to be down-regulated in lung adenocarcinoma [100]. In addition, transfection of KLF17 cDNA inhibited lung cancer cells proliferation [100].

2.5. Expression and Function of KLF in Skin Cancers

In melanoma cell lines, KLF4 expression is down-regulated as a result of homozygous deletion of the CDKN2A locus [101]. A similar study identifies an up-regulation of KLF5 in three melanoma cell lines with the most common and potent V600E mutation in B-RAF gene [102]. Several studies illustrate the function of KLF family members in skin squamous cell carcinoma (SCC). When induced in basal keratinocytes *in vitro*, KLF4 abolishes the distinctive properties of basal and parabasal epithelial cells [103]. Nuclear KLF4 causes a transitory apoptotic response and the skin progresses through phases of hyperplasia and dysplasia [103]. By 6 weeks, lesions highly resemble SCC [103]. KLF4 expression is later demonstrated to be associated with human skin SCC progression, as constitutive nuclear KLF4 staining pattern is detected in moderately and poorly differentiated tumors and metastasis [104]. Interestingly, retinoid receptors including RXRα can be used to prevent KLF4-mediated transformation and tumorogenesis, *in vitro* and *in vivo* [105].

3. EXPRESSION AND FUNCTION OF KLFs IN MISCELLANEOUS CANCERS

3.1. Expression and Function of KLFs in Central Nervous System Tumors

Dysfunction of KLF family members can contribute to oncogenesis in the brain. Gliomas are frequent tumors of the central nervous system with a poor prognosis. Because deletion of chromosome 10p15 is one of the most common chromosomal alterations in gliomas, KLF6 LOH is detected in 90% of glioblastoma samples [106], without evidences of KLF6 somatic mutations [106-107]. Mutations in KLF6 sequence are absent in astrocytic tumors [108], in meningiomas [107a], and in sporadic pituitary tumors [109]. In addition, mutations of KLF6 are infrequent (5.5% to 11.8%) in anaplastic astrocytomas, in low-grade diffuse astrocytomas and in oligodendrogliomas [107c]. Still, KLF6 expression is attenuated in human glioblastomas, when compared to primary astrocytes [106, 110]. KLF6 mutants loose their ability to reduce growth rate of glioma cells [110], whereas KLF6 SV1 expression is increased in glioblastomas to sustain cell proliferation [106].

3.2. Expression and Function of KLFs in Hematological Malignancies

KLF4, KLF9 and promyelocytic leukemia zinc finger (PLZF) are involved in maintaining cellular quiescence of naive B cells [111]. However, the accelerated response of memory B cells after activation through CD40 and the B cell receptor correlates with reduced expression of these genes, and the subsequent regulatory effects they exert on the cell cycle [111]. In accordance, KLF4 is a putative tumor suppressor in B-cell malignancies [112], whereas KLF5 confers drug resistance to acute lymphoblastic leukemia by promoting survivin expression [113]. Yasunaga *et al* identified aberrantly hypermethylated DNA sequences, including KLF4 gene, in adult T-cell leukemia (ATL) [114]. Restoring KLF4 expression in ATL-derived cell lines induces cell death by apoptosis [114]. The expression of KLF family members has been investigated in acute myeloid leukemia (AML). KLF2, KLF3, KLF5 and KLF6 are significantly lower expressed in AML blasts and CD34+ progenitor cells as compared to normal granulocytes [115]. In addition, KLF5 and KLF6 are increased and functionally involved in acute promyelocytic leukemia patients who received ATRA [115]. PLZF implication in hematological malignancies is more complex. PLZF is involved in the production of chimeric proteins formed by chromosomal translocations with retinoic acid receptor alpha to produce a dominant negative transcription factor that can decrease the expression of tumor suppressor genes[116]. In acute promyelocytic leukemia, RARα-PLZF overcomes PLZF to contribute to retinoic resistance [117]. In addition, increased expression of PLZF is a predictor of long-term survival in malignant melanoma patients [118]. The silencing of PLZF in melanoma-derived cells unblocks miR-221/-222 expression which in turn leads to enhanced proliferation [119]. Interestingly, CDK2 phosphorylates PLZF, which impairs PLZF transcriptional repression ability and antagonizes its growth inhibitory effects [120].

CONCLUSION

KLF family member proteins play a critical role in the growth and metastasis of numerous tumor types, at least in part by regulating the expression of cell cycle genes. Globally, KLF4 and KLF6 are considered as tumor suppressor gene, whereas KLF5 promotes cell proliferation. Family members have different transcriptional properties and can modulate each other's activity by a variety of mechanisms. Since cells can express multiple KLFs, KLF transcription factors build likely a transcriptional network to control cell proliferation. Effects of changes in KLF factors are context-dependent and can appear contradictory, considering differences in the expression profile of family members in various cells. In addition, KLF can be both pro-proliferative and anti-proliferative in the same cellular context, depending on signalling-mediated modifications [31] Last, KLF variants may antagonize the function of wild type proteins from which they derive. In conclusion, the study of KLF function in carcinogenesis is both complex and exciting and opens new avenues for the therapy of cancer.

ACKNOWLEDGEMENT

Declared none.

CONFLICT OF INTEREST

The author(s) confirm that this chapter content has no conflict of interest.

DISCOLUSURE

"This is an updated report of the earlier review article By Bureau C, Hanoun N, Torrisani J, Vinel JP, Buscail L, and Cordelier P." Expression and Function of

Kruppel Like-Factors (KLF) in Carcinogenesis". Current Genomics, 2009 Aug;10(5):353-360".

REFERENCES

[1] Ton-That, H.; Kaestner, K. H.; Shields, J. M.; Mahatanankoon, C. S.; Yang, V. W., Expression of the gut-enriched Kruppel-like factor gene during development and intestinal tumorigenesis. *FEBS Lett* **1997,** *419* (2-3), 239-43.

[2] (a) Shie, J. L.; Chen, Z. Y.; O'Brien, M. J.; Pestell, R. G.; Lee, M. E.; Tseng, C. C., Role of gut-enriched Kruppel-like factor in colonic cell growth and differentiation. *Am J Physiol Gastrointest Liver Physiol* **2000,** *279* (4), G806-14;

[2] (b) Xu, J.; Lu, B.; Xu, F.; Gu, H.; Fang, Y.; Huang, Q.; Lai, M., Dynamic down-regulation of Kruppel-like factor 4 in colorectal adenoma-carcinoma sequence. *J Cancer Res Clin Oncol* **2008,** *134* (8), 891-898.

[3] Hinnebusch, B. F.; Siddique, A.; Henderson, J. W.; Malo, M. S.; Zhang, W.; Athaide, C. P.; Abedrapo, M. A.; Chen, X.; Yang, V. W.; Hodin, R. A., Enterocyte differentiation marker intestinal alkaline phosphatase is a target gene of the gut-enriched Kruppel-like factor. *Am J Physiol Gastrointest Liver Physiol* **2004,** *286* (1), G23-30.

[4] Ai, W.; Liu, Y.; Langlois, M.; Wang, T. C., Kruppel-like factor 4 (KLF4) represses histidine decarboxylase gene expression through an upstream Sp1 site and downstream gastrin responsive elements. *J Biol Chem* **2004,** *279* (10), 8684-93.

[5] Dang, D. T.; Bachman, K. E.; Mahatan, C. S.; Dang, L. H.; Giardiello, F. M.; Yang, V. W., Decreased expression of the gut-enriched Kruppel-like factor gene in intestinal adenomas of multiple intestinal neoplasia mice and in colonic adenomas of familial adenomatous polyposis patients. *FEBS Lett* **2000,** *476* (3), 203-7.

[6] Choi, B. J.; Cho, Y. G.; Song, J. W.; Kim, C. J.; Kim, S. Y.; Nam, S. W.; Yoo, N. J.; Lee, J. Y.; Park, W. S., Altered expression of the KLF4 in colorectal cancers. *Pathol Res Pract* **2006,** *202* (8), 585-9.

[7] Zhao, W.; Hisamuddin, I. M.; Nandan, M. O.; Babbin, B. A.; Lamb, N. E.; Yang, V. W., Identification of Kruppel-like factor 4 as a potential tumor suppressor gene in colorectal cancer. *Oncogene* **2004,** *23* (2), 395-402.

[8] Cho, S. D.; Chintharlapalli, S.; Abdelrahim, M.; Papineni, S.; Liu, S.; Guo, J.; Lei, P.; Abudayyeh, A.; Safe, S., 5,5'-Dibromo-bis(3'-indolyl)methane induces Kruppel-like factor 4 and p21 in colon cancer cells. *Mol Cancer Ther* **2008,** *7* (7), 2109-20.

[9] Chen, Z. Y.; Shie, J. L.; Tseng, C. C., STAT1 is required for IFN-gamma-mediated gut-enriched Kruppel-like factor expression. *Exp Cell Res* **2002,** *281* (1), 19-27.

[10] (a) Dang, D. T.; Mahatan, C. S.; Dang, L. H.; Agboola, I. A.; Yang, V. W., Expression of the gut-enriched Kruppel-like factor (Kruppel-like factor 4) gene in the human colon cancer cell line RKO is dependent on CDX2. *Oncogene* **2001,** *20* (35), 4884-90;

[10] (b) Stone, C. D.; Chen, Z. Y.; Tseng, C. C., Gut-enriched Kruppel-like factor regulates colonic cell growth through APC/beta-catenin pathway. *FEBS Lett* **2002,** *530* (1-3), 147-52

[10] (c) Ghaleb, A. M.; McConnell, B. B.; Nandan, M. O.; Katz, J. P.; Kaestner, K. H.; Yang, V. W., Haploinsufficiency of Kruppel-like factor 4 promotes adenomatous polyposis coli dependent intestinal tumorigenesis. *Cancer Res* **2007,** *67* (15), 7147-54.

[11] Shie, J. L.; Tseng, C. C., A nucleus-localization-deficient mutant serves as a dominant-negative inhibitor of gut-enriched Kruppel-like factor function. *Biochem Biophys Res Commun* **2001,** *283* (1), 205-8.

[12] Flandez, M.; Guilmeau, S.; Blache, P.; Augenlicht, L. H., KLF4 regulation in intestinal epithelial cell maturation. *Exp Cell Res* **2008,** *314* (20), 3712-23.

[13] Zheng, H.; Pritchard, D. M.; Yang, X.; Bennett, E.; Liu, G.; Liu, C.; Ai, W., KLF4 gene expression is inhibited by the notch signaling pathway that controls goblet cell differentiation in mouse gastrointestinal tract. *Am J Physiol Gastrointest Liver Physiol* **2009,** *296* (3), G490-8.

[14] Ghaleb, A. M.; Aggarwal, G.; Bialkowska, A. B.; Nandan, M. O.; Yang, V. W., Notch inhibits expression of the Kruppel-like factor 4 tumor suppressor in the intestinal epithelium. *Mol Cancer Res* **2008,** *6* (12), 1920-7.

[15] Guo, J.; Chintharlapalli, S.; Lee, S. O.; Cho, S. D.; Lei, P.; Papineni, S.; Safe, S., Peroxisome proliferator-activated receptor gamma-dependent activity of indole ring-substituted 1,1-bis(3'-indolyl)-1-(p-biphenyl)methanes in cancer cells. *Cancer Chemother Pharmacol* *66* (1), 141-50.

[16] Shields, J. M.; Christy, R. J.; Yang, V. W., Identification and characterization of a gene encoding a gut-enriched Kruppel-like factor expressed during growth arrest. *J Biol Chem* **1996,** *271* (33), 20009-17.

[17] Dang, D. T.; Chen, X.; Feng, J.; Torbenson, M.; Dang, L. H.; Yang, V. W., Overexpression of Kruppel-like factor 4 in the human colon cancer cell line RKO leads to reduced tumorigenecity. *Oncogene* **2003,** *22* (22), 3424-30.

[18] Chen, X.; Johns, D. C.; Geiman, D. E.; Marban, E.; Dang, D. T.; Hamlin, G.; Sun, R.; Yang, V. W., Kruppel-like factor 4 (gut-enriched Kruppel-like factor) inhibits cell proliferation by blocking G1/S progression of the cell cycle. *J Biol Chem* **2001,** *276* (32), 30423-8.

[19] Yoon, H. S.; Yang, V. W., Requirement of Kruppel-like factor 4 in preventing entry into mitosis following DNA damage. *J Biol Chem* **2004,** *279* (6), 5035-41.

[20] Shie, J. L.; Chen, Z. Y.; Fu, M.; Pestell, R. G.; Tseng, C. C., Gut-enriched Kruppel-like factor represses cyclin D1 promoter activity through Sp1 motif. *Nucleic Acids Res* **2000,** *28* (15), 2969-76.

[21] Chen, Z. Y.; Shie, J. L.; Tseng, C. C., Gut-enriched Kruppel-like factor represses ornithine decarboxylase gene expression and functions as checkpoint regulator in colonic cancer cells. *J Biol Chem* **2002,** *277* (48), 46831-9.

[22] Whitney, E. M.; Ghaleb, A. M.; Chen, X.; Yang, V. W., Transcriptional profiling of the cell cycle checkpoint gene kruppel-like factor 4 reveals a global inhibitory function in macromolecular biosynthesis. *Gene Expr* **2006,** *13* (2), 85-96.

[23] Evans, P. M.; Zhang, W.; Chen, X.; Yang, J.; Bhakat, K. K.; Liu, C., Kruppel-like factor 4 is acetylated by p300 and regulates gene transcription *via* modulation of histone acetylation. *J Biol Chem* **2007,** *282* (47), 33994-4002.

[24] Ghaleb, A. M.; Nandan, M. O.; Chanchevalap, S.; Dalton, W. B.; Hisamuddin, I. M.; Yang, V. W., Kruppel-like factors 4 and 5: the yin and yang regulators of cellular proliferation. *Cell Res* **2005,** *15* (2), 92-6.

[25] Zhang, H.; Bialkowska, A.; Rusovici, R.; Chanchevalap, S.; Shim, H.; Katz, J. P.; Yang, V. W.; Yun, C. C., Lysophosphatidic acid facilitates proliferation of colon cancer cells *via* induction of Kruppel-like factor 5. *J Biol Chem* **2007,** *282* (21), 15541-9.

[26] Nandan, M. O.; McConnell, B. B.; Ghaleb, A. M.; Bialkowska, A. B.; Sheng, H.; Shao, J.; Babbin, B. A.; Robine, S.; Yang, V. W., Kruppel-like factor 5 mediates cellular transformation during oncogenic KRAS-induced intestinal tumorigenesis. *Gastroenterology* **2008,** *134* (1), 120-30.

[27] Zhao, Y.; Hamza, M. S.; Leong, H. S.; Lim, C. B.; Pan, Y. F.; Cheung, E.; Soo, K. C.; Iyer, N. G., Kruppel-like factor 5 modulates p53-independent apoptosis through Pim1 survival kinase in cancer cells. *Oncogene* **2008,** *27* (1), 1-8.

[28] Dardousis, K.; Voolstra, C.; Roengvoraphoj, M.; Sekandarzad, A.; Mesghenna, S.; Winkler, J.; Ko, Y.; Hescheler, J.; Sachinidis, A., Identification of differentially expressed genes involved in the formation of multicellular tumor spheroids by HT-29 colon carcinoma cells. *Mol Ther* **2007,** *15* (1), 94-102.

[29] Chanchevalap, S.; Nandan, M. O.; Merlin, D.; Yang, V. W., All-trans retinoic acid inhibits proliferation of intestinal epithelial cells by inhibiting expression of the gene encoding Kruppel-like factor 5. *FEBS Lett* **2004,** *578* (1-2), 99-105.

[30] Bialkowska, A. B.; Du, Y.; Fu, H.; Yang, V. W., Identification of novel small-molecule compounds that inhibit the proproliferative Kruppel-like factor 5 in colorectal cancer cells by high-throughput screening. *Mol Cancer Ther* **2009,** *8* (3), 563-70.

[31] Guo, P.; Dong, X. Y.; Zhang, X.; Zhao, K. W.; Sun, X.; Li, Q.; Dong, J. T., Pro-proliferative factor KLF5 becomes anti-proliferative in epithelial homeostasis upon signaling-mediated modification. *J Biol Chem* **2009,** *284* (10), 6071-8.

[32] (a) Reeves, H. L.; Narla, G.; Ogunbiyi, O.; Haq, A. I.; Katz, A.; Benzeno, S.; Hod, E.; Harpaz, N.; Goldberg, S.; Tal-Kremer, S.; Eng, F. J.; Arthur, M. J.; Martignetti, J. A.; Friedman, S. L., Kruppel-like factor 6 (KLF6) is a tumor-suppressor gene frequently inactivated in colorectal cancer. *Gastroenterology* **2004,** *126* (4), 1090-103;

[32] (b) Cho, Y. G.; Choi, B. J.; Song, J. W.; Kim, S. Y.; Nam, S. W.; Lee, S. H.; Yoo, N. J.; Lee, J. Y.; Park, W. S., Aberrant expression of krUppel-like factor 6 protein in colorectal cancers. *World J Gastroenterol* **2006,** *12* (14), 2250-3.

[33] Mukai, S.; Hiyama, T.; Tanaka, S.; Yoshihara, M.; Arihiro, K.; Chayama, K., Involvement of Kruppel-like factor 6 (KLF6) mutation in the development of nonpolypoid colorectal carcinoma. *World J Gastroenterol* **2007,** *13* (29), 3932-8.

[34] Miyaki, M.; Yamaguchi, T.; Iijima, T.; Funata, N.; Mori, T., Difference in the role of loss of heterozygosity at 10p15 (KLF6 locus) in colorectal carcinogenesis between sporadic and familial adenomatous polyposis and hereditary nonpolyposis colorectal cancer patients. *Oncology* **2006,** *71* (1-2), 131-5.

[35] Wahab, A. H.; Kassem, A. M.; Matter, S.; El Deen, A. F.; Helmy, A. S.; Ismaeil, M. M.; Zakaria, M. S., Role of KLF6 tumor suppressor gene mutations in the development of colorectal carcinoma in an Egyptian population. *Hepatogastroenterology 57* (104), 1405-10.

[36] Lievre, A.; Landi, B.; Cote, J. F.; Veyrie, N.; Zucman-Rossi, J.; Berger, A.; Laurent-Puig, P., Absence of mutation in the putative tumor-suppressor gene KLF6 in colorectal cancers. *Oncogene* **2005,** *24* (48), 7253-6.

[37] Bentov, I.; Narla, G.; Schayek, H.; Akita, K.; Plymate, S. R.; LeRoith, D.; Friedman, S. L.; Werner, H., Insulin-like growth factor-i regulates Kruppel-like factor-6 gene expression in a p53-dependent manner. *Endocrinology* **2008,** *149* (4), 1890-7.

[38] van der Horst, E. H.; Chinn, L.; Wang, M.; Velilla, T.; Tran, H.; Madrona, Y.; Lam, A.; Ji, M.; Hoey, T. C.; Sato, A. K., Discovery of fully human anti-MET monoclonal antibodies

with antitumor activity against colon cancer tumor models *in vivo*. *Neoplasia* **2009,** *11* (4), 355-64.

[39] Kang, L.; Lu, B.; Xu, J.; Hu, H.; Lai, M., Downregulation of Kruppel-like factor 9 in human colorectal cancer. *Pathol Int* **2008,** *58* (6), 334-8.

[40] (a) Wei, D.; Gong, W.; Kanai, M.; Schlunk, C.; Wang, L.; Yao, J. C.; Wu, T. T.; Huang, S.; Xie, K., Drastic down-regulation of Kruppel-like factor 4 expression is critical in human gastric cancer development and progression. *Cancer Res* **2005,** *65* (7), 2746-54;

[40] (b) Katz, J. P.; Perreault, N.; Goldstein, B. G.; Actman, L.; McNally, S. R.; Silberg, D. G.; Furth, E. E.; Kaestner, K. H., Loss of Klf4 in mice causes altered proliferation and differentiation and precancerous changes in the adult stomach. *Gastroenterology* **2005,** *128* (4), 935-45.

[41] Lee, H. J.; Kang, Y. M.; Moon, C. S.; Joe, M. K.; Lim, J. H.; Suh, Y. H.; Song, J.; Jung, M. H., KLF4 positively regulates human ghrelin expression. *Biochem J* **2009,** *420* (3), 403-11.

[42] (a) Kwak, M. K.; Lee, H. J.; Hur, K.; Park do, J.; Lee, H. S.; Kim, W. H.; Lee, K. U.; Choe, K. J.; Guilford, P.; Yang, H. K., Expression of Kruppel-like factor 5 in human gastric carcinomas. *J Cancer Res Clin Oncol* **2008,** *134* (2), 163-7; (b) Soon, M. S.; Hsu, L. S.; Chen, C. J.; Chu, P. Y.; Liou, J. H.; Lin, S. H.; Hsu, J. D.; Yeh, K. T., Expression of Kruppel-like factor 5 in gastric cancer and its clinical correlation in Taiwan. *Virchows Arch* *459* (2), 161-6.

[43] Kanai, M.; Wei, D.; Li, Q.; Jia, Z.; Ajani, J.; Le, X.; Yao, J.; Xie, K., Loss of Kruppel-like factor 4 expression contributes to Sp1 overexpression and human gastric cancer development and progression. *Clin Cancer Res* **2006,** *12* (21), 6395-402.

[44] Cho, Y. G.; Song, J. H.; Kim, C. J.; Nam, S. W.; Yoo, N. J.; Lee, J. Y.; Park, W. S., Genetic and epigenetic analysis of the KLF4 gene in gastric cancer. *Apmis* **2007,** *115* (7), 802-8.

[45] (a) Cho, Y. G.; Kim, C. J.; Park, C. H.; Yang, Y. M.; Kim, S. Y.; Nam, S. W.; Lee, S. H.; Yoo, N. J.; Lee, J. Y.; Park, W. S., Genetic alterations of the KLF6 gene in gastric cancer. *Oncogene* **2005,** *24* (28), 4588-90;

[45] (b) Chen, H.; Chen, L.; Zhang, Q. F., The Kruppel-like factor 6 genotype is associated with gastric cancer in a Chinese population. *J Int Med Res 38* (5), 1801-7;

[45] (c) Zhang, Q.; Tan, X. P.; Yuan, Y. S.; Hu, C. M.; He, C. H.; Wang, W. Z.; Li, J. C.; Zhao, Q.; Liu, N. Z., Decreased expression of KLF6 and its significance in gastric carcinoma. *Med Oncol 27* (4), 1295-302.

[46] Sangodkar, J.; Shi, J.; DiFeo, A.; Schwartz, R.; Bromberg, R.; Choudhri, A.; McClinch, K.; Hatami, R.; Scheer, E.; Kremer-Tal, S.; Martignetti, J. A.; Hui, A.; Leung, W. K.; Friedman, S. L.; Narla, G., Functional role of the KLF6 tumour suppressor gene in gastric cancer. *Eur J Cancer* **2009,** *45* (4), 666-76.

[47] Chen, H.; Chen, L.; Sun, L.; Zhen, H.; Li, X.; Zhang, Q., A small interfering RNA targeting the KLF6 splice variant, KLF6-SV1, as gene therapy for gastric cancer. *Gastric Cancer 14* (4), 339-52.

[48] Cho, Y. G.; Lee, H. S.; Song, J. H.; Kim, C. J.; Park, Y. K.; Nam, S. W.; Yoo, N. J.; Lee, J. Y.; Park, W. S., KLF6 IVS1 -27G/A polymorphism with susceptibility to gastric cancers in Korean. *Neoplasma* **2008,** *55* (1), 47-50.

[49] Wang, N.; Liu, Z. H.; Ding, F.; Wang, X. Q.; Zhou, C. N.; Wu, M., Down-regulation of gut-enriched Kruppel-like factor expression in esophageal cancer. *World J Gastroenterol* **2002,** *8* (6), 966-70.

[50] Goldstein, B. G.; Chao, H. H.; Yang, Y.; Yermolina, Y. A.; Tobias, J. W.; Katz, J. P., Overexpression of Kruppel-like factor 5 in esophageal epithelia *in vivo* leads to increased

proliferation in basal but not suprabasal cells. *Am J Physiol Gastrointest Liver Physiol* **2007,** *292* (6), G1784-92.

[51] Zhang, G.; Zhu, H.; Wang, Y.; Yang, S.; Liu, M.; Zhang, W.; Quan, L.; Bai, J.; Liu, Z.; Xu, N., Kruppel-like factor 4 represses the transcription of survivin gene in esophageal cancer cell lines. *Biol Chem* **2009**.

[52] Chen, C.; Benjamin, M. S.; Sun, X.; Otto, K. B.; Guo, P.; Dong, X. Y.; Bao, Y.; Zhou, Z.; Cheng, X.; Simons, J. W.; Dong, J. T., KLF5 promotes cell proliferation and tumorigenesis through gene regulation and the TSU-Pr1 human bladder cancer cell line. *Int J Cancer* **2006,** *118* (6), 1346-55.

[53] Li, H.; Collado, M.; Villasante, A.; Strati, K.; Ortega, S.; Canamero, M.; Blasco, M. A.; Serrano, M., The Ink4/Arf locus is a barrier for iPS cell reprogramming. *Nature* **2009,** *460* (7259), 1136-9.

[54] Arzumanyan, A.; Friedman, T.; Ng, I. O.; Clayton, M. M.; Lian, Z.; Feitelson, M. A., Does the hepatitis B antigen HBx promote the appearance of liver cancer stem cells? *Cancer Res* *71* (10), 3701-8.

[55] (a) Kremer-Tal, S.; Narla, G.; Chen, Y.; Hod, E.; DiFeo, A.; Yea, S.; Lee, J. S.; Schwartz, M.; Thung, S. N.; Fiel, I. M.; Banck, M.; Zimran, E.; Thorgeirsson, S. S.; Mazzaferro, V.; Bruix, J.; Martignetti, J. A.; Llovet, J. M.; Friedman, S. L., Downregulation of KLF6 is an early event in hepatocarcinogenesis, and stimulates proliferation while reducing differentiation. *J Hepatol* **2007,** *46* (4), 645-54;

[55] (b) Kremer-Tal, S.; Reeves, H. L.; Narla, G.; Thung, S. N.; Schwartz, M.; Difeo, A.; Katz, A.; Bruix, J.; Bioulac-Sage, P.; Martignetti, J. A.; Friedman, S. L., Frequent inactivation of the tumor suppressor Kruppel-like factor 6 (KLF6) in hepatocellular carcinoma. *Hepatology* **2004,** *40* (5), 1047-52;

[55] (c) Pan, X. C.; Chen, Z.; Chen, F.; Chen, X. H.; Jin, H. Y.; Xu, X. Y., Inactivation of the tumor suppressor Kruppel-like factor 6 (KLF6) by mutation or decreased expression in hepatocellular carcinomas. *J Zhejiang Univ Sci B* **2006,** *7* (10), 830-6;

[55] (d) Song, J.; Kim, C. J.; Cho, Y. G.; Kim, S. Y.; Nam, S. W.; Lee, S. H.; Yoo, N. J.; Lee, J. Y.; Park, W. S., Genetic and epigenetic alterations of the KLF6 gene in hepatocellular carcinoma. *J Gastroenterol Hepatol* **2006,** *21* (8), 1286-9;

[55] (e) Wang, S.; Chen, X.; Zhang, W.; Qiu, F., KLF6mRNA expression in primary hepatocellular carcinoma. *J Huazhong Univ Sci Technolog Med Sci* **2004,** *24* (6), 585-7.

[56] Hirasawa, Y.; Arai, M.; Imazeki, F.; Tada, M.; Mikata, R.; Fukai, K.; Miyazaki, M.; Ochiai, T.; Saisho, H.; Yokosuka, O., Methylation status of genes upregulated by demethylating agent 5-aza-2'-deoxycytidine in hepatocellular carcinoma. *Oncology* **2006,** *71* (1-2), 77-85.

[57] Gentner, B.; Giustacchini, A.; Visigalli, I.; Schira, G.; Biffi, A.; Naldini, L., Targeting Gene Transfer to Improve the Efficacy and Safety of Gene Therapy. *Human Gene Therapy* **2009,** *20* (5), 537-537.

[58] Lee, U. E.; Ghiassi-Nejad, Z.; Paris, A. J.; Yea, S.; Narla, G.; Walsh, M.; Friedman, S. L., Tumor suppressor activity of KLF6 mediated by downregulation of the PTTG1 oncogene. *FEBS Lett* *584* (5), 1006-10.

[59] Banck, M. S.; Beaven, S. W.; Narla, G.; Walsh, M. J.; Friedman, S. L.; Beutler, A. S., KLF6 degradation after apoptotic DNA damage. *FEBS Lett* **2006,** *580* (30), 6981-6.

[60] Narla, G.; Kremer-Tal, S.; Matsumoto, N.; Zhao, X.; Yao, S.; Kelley, K.; Tarocchi, M.; Friedman, S. L., *In vivo* regulation of p21 by the Kruppel-like factor 6 tumor-suppressor gene in mouse liver and human hepatocellular carcinoma. *Oncogene* **2007,** *26* (30), 4428-34.

[61] Tarocchi, M.; Hannivoort, R.; Hoshida, Y.; Lee, U. E.; Vetter, D.; Narla, G.; Villanueva, A.; Oren, M.; Llovet, J. M.; Friedman, S. L., Carcinogen-induced hepatic tumors in KLF6+/-

mice recapitulate aggressive human hepatocellular carcinoma associated with p53 pathway deregulation. *Hepatology 54* (2), 522-31.

[62] (a) Boyault, S.; Herault, A.; Balabaud, C.; Zucman-Rossi, J., Absence of KLF6 gene mutation in 71 hepatocellular carcinomas. *Hepatology* **2005,** *41* (3), 681-2; author reply 682-3;

[62] (b) Bureau, C.; Peron, J. M.; Bouisson, M.; Danjoux, M.; Selves, J.; Bioulac-Sage, P.; Balabaud, C.; Torrisani, J.; Cordelier, P.; Buscail, L.; Vinel, J. P., Expression of the transcription factor Klf6 in cirrhosis, macronodules, and hepatocellular carcinoma. *J Gastroenterol Hepatol* **2008,** *23* (1), 78-86.

[63] (a) D'Astolfo, D. S.; Gehrau, R. C.; Bocco, J. L.; Koritschoner, N. P., Silencing of the transcription factor KLF6 by siRNA leads to cell cycle arrest and sensitizes cells to apoptosis induced by DNA damage. *Cell Death Differ* **2008,** *15* (3), 613-6;

[63] (b) Sirach, E.; Bureau, C.; Peron, J. M.; Pradayrol, L.; Vinel, J. P.; Buscail, L.; Cordelier, P., KLF6 transcription factor protects hepatocellular carcinoma-derived cells from apoptosis. *Cell Death Differ* **2007,** *14* (6), 1202-10.

[64] Yea, S.; Narla, G.; Zhao, X.; Garg, R.; Tal-Kremer, S.; Hod, E.; Villanueva, A.; Loke, J.; Tarocchi, M.; Akita, K.; Shirasawa, S.; Sasazuki, T.; Martignetti, J. A.; Llovet, J. M.; Friedman, S. L., Ras promotes growth by alternative splicing-mediated inactivation of the KLF6 tumor suppressor in hepatocellular carcinoma. *Gastroenterology* **2008,** *134* (5), 1521-31.

[65] Miele, L.; Beale, G.; Patman, G.; Nobili, V.; Leathart, J.; Grieco, A.; Abate, M.; Friedman, S. L.; Narla, G.; Bugianesi, E.; Day, C. P.; Reeves, H. L., The Kruppel Like Factor 6 Genotype Is Associated With Fibrosis in Nonalcoholic Fatty Liver Disease. *Gastroenterology* **2008**.

[66] Hanoun, N.; Bureau, C.; Diab, T.; Gayet, O.; Dusetti, N.; Selves, J.; Vinel, J. P.; Buscail, L.; Cordelier, P.; Torrisani, J., The SV2 variant of KLF6 is down-regulated in hepatocellular carcinoma and displays anti-proliferative and pro-apoptotic functions. *J Hepatol 53* (5), 880-8.

[67] Narla, G.; Heath, K. E.; Reeves, H. L.; Li, D.; Giono, L. E.; Kimmelman, A. C.; Glucksman, M. J.; Narla, J.; Eng, F. J.; Chan, A. M.; Ferrari, A. C.; Martignetti, J. A.; Friedman, S. L., KLF6, a candidate tumor suppressor gene mutated in prostate cancer. *Science* **2001,** *294* (5551), 2563-6.

[68] Cheng, X. F.; Li, D.; Zhuang, M.; Chen, Z. Y.; Lu, D. X.; Hattori, T., Growth inhibitory effect of Kruppel-like factor 6 on human prostatic carcinoma and renal carcinoma cell lines. *Tohoku J Exp Med* **2008,** *216* (1), 35-45.

[69] Huang, X.; Li, X.; Guo, B., KLF6 induces apoptosis in prostate cancer cells through up-regulation of ATF3. *J Biol Chem* **2008,** *283* (44), 29795-801.

[70] Narla, G.; Difeo, A.; Reeves, H. L.; Schaid, D. J.; Hirshfeld, J.; Hod, E.; Katz, A.; Isaacs, W. B.; Hebbring, S.; Komiya, A.; McDonnell, S. K.; Wiley, K. E.; Jacobsen, S. J.; Isaacs, S. D.; Walsh, P. C.; Zheng, S. L.; Chang, B. L.; Friedrichsen, D. M.; Stanford, J. L.; Ostrander, E. A.; Chinnaiyan, A. M.; Rubin, M. A.; Xu, J.; Thibodeau, S. N.; Friedman, S. L.; Martignetti, J. A., A germline DNA polymorphism enhances alternative splicing of the KLF6 tumor suppressor gene and is associated with increased prostate cancer risk. *Cancer Res* **2005,** *65* (4), 1213-22.

[71] Narla, G.; DiFeo, A.; Yao, S.; Banno, A.; Hod, E.; Reeves, H. L.; Qiao, R. F.; Camacho-Vanegas, O.; Levine, A.; Kirschenbaum, A.; Chan, A. M.; Friedman, S. L.;

Martignetti, J. A., Targeted inhibition of the KLF6 splice variant, KLF6 SV1, suppresses prostate cancer cell growth and spread. *Cancer Res* **2005,** *65* (13), 5761-8.

[72] Chen, C.; Hyytinen, E. R.; Sun, X.; Helin, H. J.; Koivisto, P. A.; Frierson, H. F., Jr.; Vessella, R. L.; Dong, J. T., Deletion, mutation, and loss of expression of KLF6 in human prostate cancer. *Am J Pathol* **2003,** *162* (4), 1349-54.

[73] (a) Seppala, E. H.; Autio, V.; Duggal, P.; Ikonen, T.; Stenman, U. H.; Auvinen, A.; Bailey-Wilson, J. E.; Tammela, T. L.; Schleutker, J., KLF6 IVS1 -27G>A variant and the risk of prostate cancer in Finland. *Eur Urol* **2007,** *52* (4), 1076-81;

[73] (b) Agell, L.; Hernandez, S.; de Muga, S.; Lorente, J. A.; Juanpere, N.; Esgueva, R.; Serrano, S.; Gelabert, A.; Lloreta, J., KLF6 and TP53 mutations are a rare event in prostate cancer: distinguishing between Taq polymerase artifacts and true mutations. *Mod Pathol* **2008,** *21* (12), 1470-8;

[73] (c) Bar-Shira, A.; Matarasso, N.; Rosner, S.; Bercovich, D.; Matzkin, H.; Orr-Urtreger, A., Mutation screening and association study of the candidate prostate cancer susceptibility genes MSR1, PTEN, and KLF6. *Prostate* **2006,** *66* (10), 1052-60;

[73] (d) Koivisto, P. A.; Hyytinen, E. R.; Matikainen, M.; Tammela, T. L.; Ikonen, T.; Schleutker, J., Kruppel-like factor 6 germ-line mutations are infrequent in Finnish hereditary prostate cancer. *J Urol* **2004,** *172* (2), 506-7; (e) Muhlbauer, K. R.; Grone, H. J.; Ernst, T.; Grone, E.; Tschada, R.; Hergenhahn, M.; Hollstein, M., Analysis of human prostate cancers and cell lines for mutations in the TP53 and KLF6 tumour suppressor genes. *Br J Cancer* **2003,** *89* (4), 687-90.

[74] Narla, G.; DiFeo, A.; Fernandez, Y.; Dhanasekaran, S.; Huang, F.; Sangodkar, J.; Hod, E.; Leake, D.; Friedman, S. L.; Hall, S. J.; Chinnaiyan, A. M.; Gerald, W. L.; Rubin, M. A.; Martignetti, J. A., KLF6-SV1 overexpression accelerates human and mouse prostate cancer progression and metastasis. *J Clin Invest* **2008,** *118* (8), 2711-21.

[75] Liu, S.; Zhang, H.; Zhu, L.; Zhao, L.; Dong, Y., Kruppel-like factor 4 is a novel mediator of selenium in growth inhibition. *Mol Cancer Res* **2008,** *6* (2), 306-13.

[76] Ebert, R.; Zeck, S.; Meissner-Weigl, J.; Klotz, B.; Rachner, T. D.; Benad, P.; Klein-Hitpass, L.; Rudert, M.; Hofbauer, L. C.; Jakob, F., Kruppel-like factors KLF2 and 6 and Ki-67 are direct targets of zoledronic acid in MCF-7 cells. *Bone 50* (3), 723-32.

[77] Taniguchi, H.; Jacinto, F. V.; Villanueva, A.; Fernandez, A. F.; Yamamoto, H.; Carmona, F. J.; Puertas, S.; Marquez, V. E.; Shinomura, Y.; Imai, K.; Esteller, M., Silencing of Kruppel-like factor 2 by the histone methyltransferase EZH2 in human cancer. *Oncogene.*

[78] Chen, C. J.; Lin, S. E.; Lin, Y. M.; Lin, S. H.; Chen, D. R.; Chen, C. L., Association of expression of kruppel-like factor 4 and kruppel-like factor 5 with the clinical manifestations of breast cancer. *Pathol Oncol Res 18* (2), 161-8.

[79] Foster, K. W.; Frost, A. R.; McKie-Bell, P.; Lin, C. Y.; Engler, J. A.; Grizzle, W. E.; Ruppert, J. M., Increase of GKLF messenger RNA and protein expression during progression of breast cancer. *Cancer Res* **2000,** *60* (22), 6488-95.

[80] Chu, P. Y.; Hsu, N. C.; Liao, A. T.; Yeh, K. T.; Hou, M. F.; Liu, C. H., Elevated Kruppel-like factor 4 transcription factor in canine mammary carcinoma. *BMC Vet Res 7*, 58.

[81] Pandya, A. Y.; Talley, L. I.; Frost, A. R.; Fitzgerald, T. J.; Trivedi, V.; Chakravarthy, M.; Chhieng, D. C.; Grizzle, W. E.; Engler, J. A.; Krontiras, H.; Bland, K. I.; LoBuglio, A. F.; Lobo-Ruppert, S. M.; Ruppert, J. M., Nuclear localization of KLF4 is associated with an aggressive phenotype in early-stage breast cancer. *Clin Cancer Res* **2004,** *10* (8), 2709-19.

[82] Chen, C.; Bhalala, H. V.; Qiao, H.; Dong, J. T., A possible tumor suppressor role of the KLF5 transcription factor in human breast cancer. *Oncogene* **2002,** *21* (43), 6567-72.

[83] Tong, D.; Czerwenka, K.; Heinze, G.; Ryffel, M.; Schuster, E.; Witt, A.; Leodolter, S.; Zeillinger, R., Expression of KLF5 is a prognostic factor for disease-free survival and overall survival in patients with breast cancer. *Clin Cancer Res* **2006,** *12* (8), 2442-8.

[84] Guo, H.; Lin, Y.; Zhang, H.; Liu, J.; Zhang, N.; Li, Y.; Kong, D.; Tang, Q.; Ma, D., Tissue factor pathway inhibitor-2 was repressed by CpG hypermethylation through inhibition of KLF6 binding in highly invasive breast cancer cells. *BMC Mol Biol* **2007,** *8*, 110.

[85] Gehrau, R. C.; D'Astolfo, D. S.; Dumur, C. I.; Bocco, J. L.; Koritschoner, N. P., Nuclear expression of KLF6 tumor suppressor factor is highly associated with overexpression of ERBB2 oncoprotein in ductal breast carcinomas. *PLoS One 5* (1), e8929.

[86] Liu, J.; Du, T.; Yuan, Y.; He, Y.; Tan, Z.; Liu, Z., KLF6 inhibits estrogen receptor-mediated cell growth in breast cancer *via* a c-Src-mediated pathway. *Mol Cell Biochem 335* (1-2), 29-35.

[87] Gumireddy, K.; Li, A.; Gimotty, P. A.; Klein-Szanto, A. J.; Showe, L. C.; Katsaros, D.; Coukos, G.; Zhang, L.; Huang, Q., KLF17 is a negative regulator of epithelial-mesenchymal transition and metastasis in breast cancer. *Nat Cell Biol* **2009,** *11* (11), 1297-304.

[88] (a) DiFeo, A.; Narla, G.; Camacho-Vanegas, O.; Nishio, H.; Rose, S. L.; Buller, R. E.; Friedman, S. L.; Walsh, M. J.; Martignetti, J. A., E-cadherin is a novel transcriptional target of the KLF6 tumor suppressor. *Oncogene* **2006,** *25* (44), 6026-31;

[88] (b) DiFeo, A.; Narla, G.; Hirshfeld, J.; Camacho-Vanegas, O.; Narla, J.; Rose, S. L.; Kalir, T.; Yao, S.; Levine, A.; Birrer, M. J.; Bonome, T.; Friedman, S. L.; Buller, R. E.; Martignetti, J. A., Roles of KLF6 and KLF6-SV1 in ovarian cancer progression and intraperitoneal dissemination. *Clin Cancer Res* **2006,** *12* (12), 3730-9.

[89] Wang, X.; Zhao, J., KLF8 transcription factor participates in oncogenic transformation. *Oncogene* **2007,** *26* (3), 456-61.

[90] Wang, X.; Zheng, M.; Liu, G.; Xia, W.; McKeown-Longo, P. J.; Hung, M. C.; Zhao, J., Kruppel-like factor 8 induces epithelial to mesenchymal transition and epithelial cell invasion. *Cancer Res* **2007,** *67* (15), 7184-93.

[91] Wang, X.; Urvalek, A. M.; Liu, J.; Zhao, J., Activation of KLF8 Transcription by Focal Adhesion Kinase in Human Ovarian Epithelial and Cancer Cells. *J Biol Chem* **2008,** *283* (20), 13934-42.

[92] Prasad, N. B.; Biankin, A. V.; Fukushima, N.; Maitra, A.; Dhara, S.; Elkahloun, A. G.; Hruban, R. H.; Goggins, M.; Leach, S. D., Gene expression profiles in pancreatic intraepithelial neoplasia reflect the effects of Hedgehog signaling on pancreatic ductal epithelial cells. *Cancer Res* **2005,** *65* (5), 1619-26.

[93] Wei, D.; Kanai, M.; Jia, Z.; Le, X.; Xie, K., Kruppel-like factor 4 induces p27Kip1 expression in and suppresses the growth and metastasis of human pancreatic cancer cells. *Cancer Res* **2008,** *68* (12), 4631-9.

[94] Hartel, M.; Narla, G.; Wente, M. N.; Giese, N. A.; Martignoni, M. E.; Martignetti, J. A.; Friess, H.; Friedman, S. L., Increased alternative splicing of the KLF6 tumour suppressor gene correlates with prognosis and tumour grade in patients with pancreatic cancer. *Eur J Cancer* **2008,** *44* (13), 1895-903.

[95] Wick, M. J.; Blaine, S.; Van Putten, V.; Saavedra, M.; Nemenoff, R. A., Lung Kruppel-like factor (LKLF) is a transcriptional activator of the cytosolic phospholipase A2 alpha promoter. *Biochem J* **2005,** *387* (Pt 1), 239-46.

[96] Ito, G.; Uchiyama, M.; Kondo, M.; Mori, S.; Usami, N.; Maeda, O.; Kawabe, T.; Hasegawa, Y.; Shimokata, K.; Sekido, Y., Kruppel-like factor 6 is frequently down-regulated and induces apoptosis in non-small cell lung cancer cells. *Cancer Res* **2004,** *64* (11), 3838-43.

[97] Tahara, E.; Kadara, H.; Lacroix, L.; Lotan, D.; Lotan, R., Activation of protein kinase C by phorbol 12-myristate 13-acetate suppresses the growth of lung cancer cells through KLF6 induction. *Cancer Biol Ther* **2009,** *8* (9).

[98] Spinola, M.; Leoni, V. P.; Galvan, A.; Korsching, E.; Conti, B.; Pastorino, U.; Ravagnani, F.; Columbano, A.; Skaug, V.; Haugen, A.; Dragani, T. A., Genome-wide single nucleotide polymorphism analysis of lung cancer risk detects the KLF6 gene. *Cancer Lett* **2007,** *251* (2), 311-6.

[99] (a) DiFeo, A.; Feld, L.; Rodriguez, E.; Wang, C.; Beer, D. G.; Martignetti, J. A.; Narla, G., A functional role for KLF6-SV1 in lung adenocarcinoma prognosis and chemotherapy response. *Cancer Res* **2008,** *68* (4), 965-70;

[99] (b) Sangodkar, J.; Difeo, A.; Feld, L.; Bromberg, R.; Schwartz, R.; Huang, F.; Terzo, E. A.; Choudhri, A.; Narla, G., Targeted reduction of KLF6-SV1 restores chemotherapy sensitivity in resistant lung adenocarcinoma. *Lung Cancer* **2009**.

[100] Cai, X. D.; Zhou, Y. B.; Huang, L. X.; Zeng, Q. L.; Zhang, L. J.; Wang, Q. Q.; Li, S. L.; Feng, J. Q.; Han, A. J., Reduced expression of Kruppel-like factor 17 is related to tumor growth and poor prognosis in lung adenocarcinoma. *Biochem Biophys Res Commun 418* (1), 67-73.

[101] Bloethner, S.; Hemminki, K.; Thirumaran, R.; Chen, B.; Mueller-Berghaus, J.; Ugurel, S.; Schadendorf, D.; Kumar, R., Differences in global gene expression in melanoma cell lines with and without homozygous deletion of the CDKN2A locus genes. *Melanoma Res* **2006,** *4*, 297-307.

[102] Bloethner, S.; Chen, B.; Hemminki, K.; Müller-Berghaus, J.; Ugurel, S.; Schadendorf, D.; Kumar, R., Effect of common B-RAF and N-RAS mutations on global gene expression in melanoma cell lines. *Carcinogenesis* **2005,** *7*, 1224-1232.

[103] Foster, K.; Liu, Z.; Nail, C.; Li, X.; Fitzgerald, T.; Bailey, S.; Frost, A.; Louro, I.; Townes, T.; Paterson, A.; Kudlow, J.; Lobo-Ruppert, S.; Ruppert, J., Induction of KLF4 in basal keratinocytes blocks the proliferation-differentiation switch and initiates squamous epithelial dysplasia. *Oncogene* **2005,** *9*, 1491-15000.

[104] (a) Huang, C. C.; Liu, Z.; Li, X.; Bailey, S. K.; Nail, C. D.; Foster, K. W.; Frost, A. R.; Ruppert, J. M.; Lobo-Ruppert, S. M., KLF4 and PCNA identify stages of tumor initiation in a conditional model of cutaneous squamous epithelial neoplasia. *Cancer Biol Ther* **2005,** *4* (12), 1401-8;

[104] (b) Chen, Y. J.; Wu, C. Y.; Chang, C. C.; Ma, C. J.; Li, M. C.; Chen, C. M., Nuclear Kruppel-like factor 4 expression is associated with human skin squamous cell carcinoma progression and metastasis. *Cancer Biol Ther* **2008,** *7* (5).

[105] Jiang, W.; Deng, W.; Bailey, S. K.; Nail, C. D.; Frost, A. R.; Brouillette, W. J.; Muccio, D. D.; Grubbs, C. J.; Ruppert, J. M.; Lobo-Ruppert, S. M., Prevention of KLF4-mediated tumor initiation and malignant transformation by UAB30 rexinoid. *Cancer Biol Ther* **2009,** *8* (3).

[106] Camacho-Vanegas, O.; Narla, G.; Teixeira, M. S.; DiFeo, A.; Misra, A.; Singh, G.; Chan, A. M.; Friedman, S. L.; Feuerstein, B. G.; Martignetti, J. A., Functional inactivation of the KLF6 tumor suppressor gene by loss of heterozygosity and increased alternative splicing in glioblastoma. *Int J Cancer* **2007,** *121* (6), 1390-5.

[107] (a) Kohler, B.; Wolter, M.; Blaschke, B.; Reifenberger, G., Absence of mutations in the putative tumor suppressor gene KLF6 in glioblastomas and meningiomas. *Int J Cancer* **2004,** *111* (4), 644-5;

[107] (b) Montanini, L.; Bissola, L.; Finocchiaro, G., KLF6 is not the major target of chromosome 10p losses in glioblastomas. *Int J Cancer* **2004,** *111* (4), 640-1;

[107] (c) Jeng, Y. M.; Hsu, H. C., KLF6, a putative tumor suppressor gene, is mutated in astrocytic gliomas. *Int J Cancer* **2003,** *105* (5), 625-9.

[108] Koivisto, P. A.; Zhang, X.; Sallinen, S. L.; Sallinen, P.; Helin, H. J.; Dong, J. T.; Van Meir, E. G.; Haapasalo, H.; Hyytinen, E. R., Absence of KLF6 gene mutations in human astrocytic tumors and cell lines. *Int J Cancer* **2004,** *111* (4), 642-3.

[109] Vax, V. V.; Gueorguiev, M.; Dedov, II; Grossman, A. B.; Korbonits, M., The Kruppel-like transcription factor 6 gene in sporadic pituitary tumours. *Endocr Relat Cancer* **2003,** *10* (3), 397-402.

[110] (a) Yin, D.; Komatsu, N.; Miller, C. W.; Chumakov, A. M.; Marschesky, A.; McKenna, R.; Black, K. L.; Koeffler, H. P., KLF6: mutational analysis and effect on cancer cell proliferation. *Int J Oncol* **2007,** *30* (1), 65-72;

[110] (b) Kimmelman, A. C.; Qiao, R. F.; Narla, G.; Banno, A.; Lau, N.; Bos, P. D.; Nunez Rodriguez, N.; Liang, B. C.; Guha, A.; Martignetti, J. A.; Friedman, S. L.; Chan, A. M., Suppression of glioblastoma tumorigenicity by the Kruppel-like transcription factor KLF6. *Oncogene* **2004,** *23* (29), 5077-83.

[111] Good, K. L.; Tangye, S. G., Decreased expression of Kruppel-like factors in memory B cells induces the rapid response typical of secondary antibody responses. *Proc Natl Acad Sci U S A* **2007,** *104* (33), 13420-5.

[112] Kharas, M. G.; Yusuf, I.; Scarfone, V. M.; Yang, V. W.; Segre, J. A.; Huettner, C. S.; Fruman, D. A., KLF4 suppresses transformation of pre-B cells by ABL oncogenes. *Blood* **2007,** *109* (2), 747-55.

[113] Zhu, N.; Gu, L.; Findley, H.; Chen, C.; Dong, J.; Yang, L.; Zhou, M., KLF5 Interacts with p53 in regulating survivin expression in acute lymphoblastic leukemia. *J Biol Chem* **2006,** *21*, 14711-8.

[114] Yasunaga, J.; Taniguchi, Y.; Nosaka, K.; Yoshida, M.; Satou, Y.; Sakai, T.; Mitsuya, H.; Matsuoka, M., Identification of aberrantly methylated genes in association with adult T-cell leukemia. *Cancer Res* **2004,** *64* (17), 6002-9.

[115] Humbert, M.; Halter, V.; Shan, D.; Laedrach, J.; Leibundgut, E. O.; Baerlocher, G. M.; Tobler, A.; Fey, M. F.; Tschan, M. P., Deregulated expression of Kruppel-like factors in acute myeloid leukemia. *Leuk Res 35* (7), 909-13.

[116] Krug, U.; Ganser, A.; Koeffler, H., Tumor suppressor genes in normal and malignant hematopoiesis. *Oncogene.* **2002,** *21*, 3475-95.

[117] Guidez, F.; Parks, S.; Wong, H.; Jovanovic, J. V.; Mays, A.; Gilkes, A. F.; Mills, K. I.; Guillemin, M. C.; Hobbs, R. M.; Pandolfi, P. P.; de The, H.; Solomon, E.; Grimwade, D., RARalpha-PLZF overcomes PLZF-mediated repression of CRABPI, contributing to retinoid resistance in t(11;17) acute promyelocytic leukemia. *Proc Natl Acad Sci U S A* **2007,** *104* (47), 18694-9.

[118] Brunner, G.; Reitz, M.; Schwipper, V.; Tilkorn, H.; Lippold, A.; Biess, B.; Suter, L.; Atzpodien, J., Increased expression of the tumor suppressor PLZF is a continuous predictor of long-term survival in malignant melanoma patients. *Cancer Biother Radiopharm* **2008,** *23* (4), 451-9.

[119] Felicetti, F.; Errico, M. C.; Bottero, L.; Segnalini, P.; Stoppacciaro, A.; Biffoni, M.; Felli, N.; Mattia, G.; Petrini, M.; Colombo, M. P.; Peschle, C.; Care, A., The promyelocytic leukemia zinc finger-microRNA-221/-222 pathway controls melanoma progression through multiple oncogenic mechanisms. *Cancer Res* **2008,** *68* (8), 2745-54.

[120] Costoya, J. A.; Hobbs, R. M.; Pandolfi, P. P., Cyclin-dependent kinase antagonizes promyelocytic leukemia zinc-finger through phosphorylation. *Oncogene* **2008,** *27* (27), 3789-96.

Send Orders of Reprints at reprints@benthamscince.net
Advances in Genome Science, Vol. 1, **2013, 169-188** **169**

CHAPTER 7

Episcopic 3D Imaging Methods: Tools for Researching Gene Function

Wolfgang J. Weninger* **and Stefan H. Geyer**

IMG, Centre for Anatomy and Cell Biology, Medical University of Vienna, Waehringer Str. 13, A-1090 Vienna, Austria

Abstract: This work aims at describing episcopic 3D imaging methods and at discussing how these methods can contribute to researching the genetic mechanisms driving embryogenesis and tissue remodelling, and the genesis of pathologies. Several episcopic 3D imaging methods exist. The most advanced are capable of generating high-resolution volume data (voxel sizes from $0.5x0.5x1\mu m^3$ upwards) of small to large embryos of model organisms and tissue samples. Beside anatomy and tissue architecture, gene expression and gene product patterns can be three dimensionally analyzed in their precise anatomical and histological context with the aid of whole mount *in situ* hybridization or whole mount immunohistochemical staining techniques. Episcopic 3D imaging techniques were and are employed for analyzing the precise morphological phenotype of experimentally malformed, randomly produced, or genetically engineered embryos of biomedical model organisms. It has been shown that episcopic 3D imaging also fits for describing the spatial distribution of genes and gene products during embryogenesis, and that it can be used for analyzing tissue samples of adult model animals and humans. The latter offers the possibility to use episcopic 3D imaging techniques for researching the causality and treatment of pathologies or for staging cancer. Such applications, however, are not yet routine and currently only preliminary results are available. We conclude that, although episcopic 3D imaging is in its very beginnings, it represents an upcoming methodology, which in short terms will become an indispensable tool for researching the genetic regulation of embryo development as well as the genesis of malformations and diseases.

Keywords: 3D modelling, episcopic microscopy, imaging, embryo, development, gene expression.

BACKGROUND

Challenging hereditary diseases and pathologies requires profound knowledge of the genetic and epigenetic pathways regulating their genesis. Therefore modern biomedicine developed a number of different approaches for researching the role

***Address correspondence to Wolfgang J. Weninger:** IMG, Center for Anatomy and Cell Biology, Medical University of Vienna, Waehringerstr. 13, A-1090 Vienna, Austria, E.U.; Tel +43 1 40160 37560; Fax +43 1 40160 937560; E-mail: wolfgang.weninger@meduniwien.ac.at

Christian Neri (Ed)

and function of genes and gene products in normal and diseased embryos, organs, and tissues. The following study shall provide a very brief overview of techniques capable of researching gene function. It will then focus on imaging methods, which can quickly generate correctly registered digital image stacks, captured from the block surfaces of sequentially cut, histological processed, and embedded specimens (three dimensional (3D) episcopic imaging methods). These methods will be described in more detail and their value for researching gene function will be discussed.

ANALYZING GENE FUNCTION – A BRIEF OVERVIEW

Basically three main approaches exist for researching the activity and function of genes and gene products.

1. Gene Profiling

Tissue samples of normal or diseased model organisms or humans can be analyzed with the aid of microarray techniques [1-6]. Microarray studies provide information about the activity of thousands of genes in very short times. They reveal the patterns of gene activity in individual tissues and provide clues about gene function and their association with regulatory or signalling pathways. However, it is difficult to compare multiple micro-array experiments and array-derived data is only of limited value in defining gene function, because it frequently lacks spatial and/or temporal specificity. Thus microarray techniques are highly useful for describing and even quantifying the activity of multiple genes in tissue samples, but in order to properly understand gene function, gene-expression patterns need to be placed in the proper 3D context of tissue architecture.

2. Phenotyping Mutants of Model Organisms

Model organisms with randomly introduced or targeted gene disruptions are produced and their phenotype is comprehensively examined. This study will not discuss the methods, which are used for analyzing the cognitive or receptive function or the behaviour of animals. It will also not discuss methods designed for analyzing the morphological phenotype of adult animals or humans (*e.g.* computed tomography, magnetic resonance imaging, ultrasound, *etc.*). But it will discuss some methods capable of providing morphological descriptions of tissue samples and embryos of model organisms.

3. *In situ* Gene Expression Analysis

Expression patterns of genes and gene products are analysed in their precise anatomical and histological context. For conducting such analysis methods were developed, which permit analysis of gene or gene product patterns in the context of tissue architecture, tissue samples, or tissues and organs of model organisms and their embryos respectively.

With the exception of anatomical dissection, traditional techniques for analyzing microscopic and macroscopic anatomy and gene expression patterns, are two-dimensional (2D) (electron-microscopic, histological, and macroscopic sections, *etc.*). But tissues and organs are three-dimensional (3D) and gene products are distributed and act three dimensionally. Thus the last decades saw the development of a vast number of methods for creating 3D information of cell, tissue, and organ morphology, 3D information of gene expression and gene product patterns, or both. Examples for techniques capable of analyzing small specimens, such as tissue samples or embryos are: *in vivo* microscopy [7-10], micro-computed tomography (μCT) [11-15], micro-magnetic resonance imaging (μMRI) [9, 16-23], ultrasound biomicroscopy (UBM) [24-28], optical projection tomography (OPT) [29-33], confocal microscopy [34-39], atomic force microscopy [40-42], 3D electron tomography [43, 44], histological or macroscopic section based 3D reconstruction methods [45-49], and 3D episcopic imaging methods (see below).

This paper does not compare all the different methods for volume data generation and gene expression analysis. It will solely focus on the description, analysis and discussion of methods that permit volume data generation on the basis of episcopic images of physically sectioned specimens.

EPISCOPIC 3D IMAGING METHODS – DEFINITION

Under the term "episcopic 3D imaging methods" we summarize all 3D imaging techniques that create volume data by capturing images of subsequent surfaces of blocks, containing histological processed and embedded specimens, during their physical sectioning on microtomes. This includes methods such as "fast 3D serial reconstruction" [50], "Epi-3D" [51], "episcopic fluorescence image capturing"

[52], "surface imaging microscopy" [53, 54], "high resolution episcopic microscopy (HREM)" [55, 56], and serial block-face scanning electron microscopy (SBF-SEM) [57]. The term "surface imaging microscopy" was used twice. We will refer to the earlier method as SIM1 [53] and the later as SIM_2 [54]. We will not go into detail with SBF-SEM, which is capable of generating stacks of precisely aligned electron microscopy images and permits highly detailed 3D analysis of cell organelles, axons, and synapses. Although a very sophisticated method, it can neither be used for describing the morphological phenotype of mutants, nor for examining gene expression patterns in the morphological context.

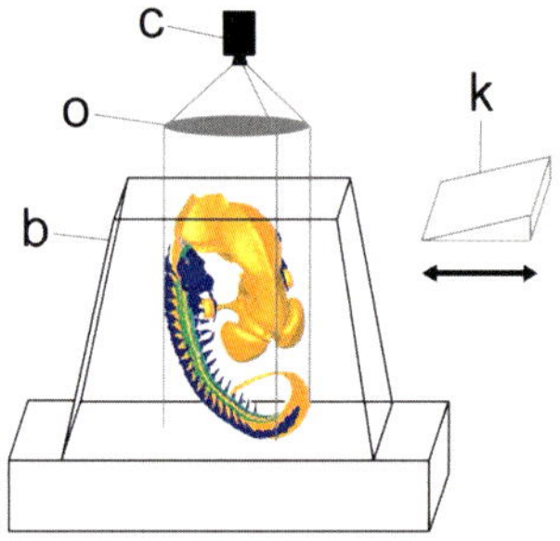

Figure 1: Set up used for episcopic 3D data generation. A camera (c), sitting on an optic (o) is aligned perpendicular to the surface of a block of embedding medium (b) containing a specimen. Note that the field of view (marked by grey lines) should be as narrow as possible. Its position cannot be changed during sectioning. k, knife.

EPISCOPIC 3D IMAGING METHODS – WORK FLOW

All episcopic 3D imaging techniques utilize sacrificed embryos and tissue samples. The specimens are fixed and sometimes pre-processed and whole mount stained for enhancing contrasts. Then they are embedded in histological embedding media and mounted on a microtome. Digital images of the tissues on the surfaces of the blocks of embedding medium containing the specimens are captured with a camera sitting on a magnifying optic. The optical pathway of the optic is aligned precisely perpendicular to the block surface (Fig. (**1**)). Depending on the technique the tissues on the block surface are either identified by their intrinsic contrast or by administering dyes to the block surface. After capturing an image of the block face, a small slice of the block is removed using either the microtome blade or micro-mills. Routinely this slice is thrown away, although some methods permit preservation of histological sections. Now, a digital image of the freshly cut - if necessary also freshly stained - block surface is captured and the next slice of embedding medium is removed. This procedure is repeated until the region of interest is sectioned and a stack of aligned digital images, showing subsequent block faces with tissue of the specimens is produced (Fig. (**2**)). In some methods, data generation is fully automated. This generates homogeneous data sets, and saves time and man power.

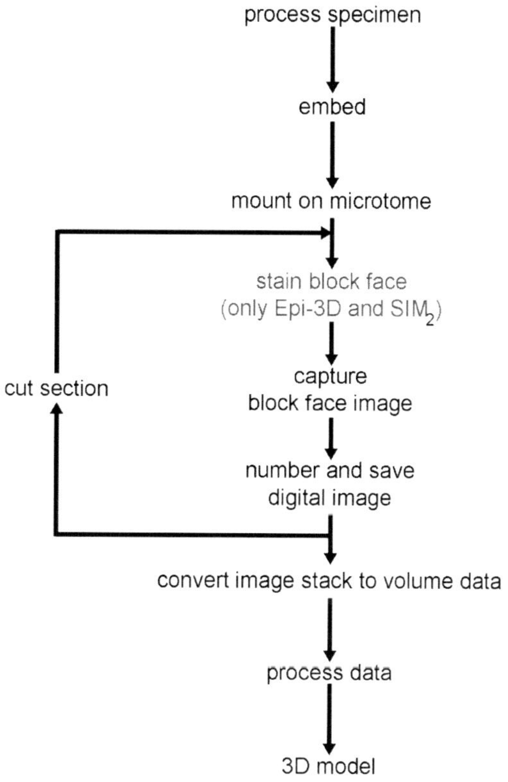

Figure 2: Flow chart of the episcopic 3D data generation protocol. Note that this flow chart provides only the main steps. Depending on the method, additional steps might be necessary for analysing gene expression patterns.

After data generation, the stack of aligned episcopic images is converted into a Volume data set. The dimension of one voxel is defined by the resolution of the captured block face image (its Pixel size) and the thickness of the slice of embedding medium removed (Fig. (**3**)). Virtual volume or surface rendered 3D models can be created and visualized with state of the art 3D visualization software packages (*e.g.*, Amira (Visage Imaging Inc.), Volocity (Improvision), SURFdriver (www.surfdriver.com), Osirix (www.osirix-viewer.com), arivis Browser (arivis GmbH)).

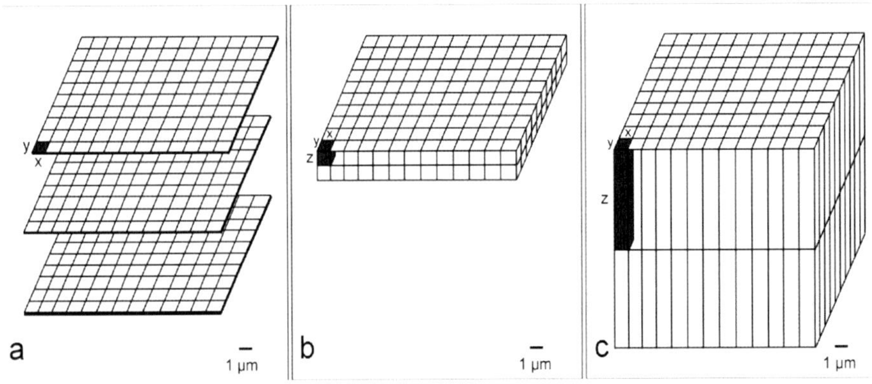

Figure 3: From pixel to voxel. a. Schematic drawing of three subsequent two-dimensional (2D) images. Each image consists of 10 x 13 pixels. b. Volume data created from images of 1 µm thick sections. The voxels are cubic, because the distance between the images is the same as the length of the x and y coordinate of the pixels of the 2D images. c. Volume data created from images of 7 µm thick sections. While the 2D resolution (resolution of the 2D images) is relatively high, the 3D resolution (resolution of the volume data set) is low.

EPISCOPIC 3D IMAGING METHODS – TECHNICAL DETAILS

The following chapter will shortly describe each method and discusses potential applications. The methods are ranked according to their year of publication. Table **1** highlights the characteristics of the different episcopic imaging methods.

Table 1: Characteristics of modern episcopic 3D imaging techniques. Note that all specimens have to become fixed and dehydrated prior to embedding

	Fast 3D serial reconstruction (1990)	**Epi-3D (1998)**	**EFIC (2002)**	**SIM$_1$ (2002)**	**HREM (2006)**	**SIM$_2$ (2008)**
specimens	bones	embryos adult material	embryos adult material biopsies	embryos adult material biopsies	embryos adult material biopsies	embryos adult material biopsies
specimen preprocessing	bleaching	lead acetate immersion during dehydration			eosin staining during dehydration	

Table 1: contd...

embedding medium	resin	paraffin mix or resin	paraffin mix	resin	resin	wax or resin
slice thickness	≥ 39 µm	≥ ~6 µm	≥ ~1 µm	≥ ~1 µm	≥ ~1 µm	≥0.2 µm (resin) ≥2.5 µm (wax)
visualization of cells, tissues and organs	bleaching	lead acetate / sodium-sulfide reaction	auto-fluorescence	auto-fluorescence	eosin	special mixtures for fast staining
visualization of gene expression	no	no	yes if whole mount stained	yes if whole mount stained	yes if whole mount stained	currently not

"Fast 3D serial reconstruction" [50] is designed and was used for obtaining information on the density of trabeculae in vertebrae [50, 58-60]. The bones are bleached and embedded in black resin, which results in excellent contrasts. Volume data and 3D models of osseous structures can be generated quickly and in high quality. However, the method is restricted to the analysis of the morphology of uniform structures. Due to the improvement of computed tomography, osseous structures can be visualized even faster in similar resolution with modern computer tomography apparatuses, wherefore, as far as we know, the original method is not longer in use.

"Epi-3D" [51] is designed and was applied for analyzing embryo morphology and the topology of neonate and adult organs respectively [61-64]. After fixation, the specimens are dehydrated and simultaneously pre-stained with lead acetate, before they are embedded in embedding media on wax basis. The histological blocks are sectioned on a conventional microtome. For contrasting the tissues on the block surface each freshly cut block surface is stained for 1 second with lead acetate. This method permits simple analysis of morphology, but does not offer the possibility for analyzing gene expression patterns.

"EFIC" [52] is designed and was applied for analyzing the morphology of the organ systems of normal and malformed embryos and for analysing the topology of diseased tissues [65, 66]. The specimens are embedded in reddish stained embedding medium on wax basis. Monochrome light of 460 nm wavelength is applied to the block surface in order to excite autofluorescence of the tissues. No block surface staining is required for achieving tissue contrast. Therefore the data generation process can be fully automated. In principle the system offers the possibility to visualise the expression patterns of transgenes and gene products, if specimens are whole mount stained with LacZ or NBT/BCIP prior to embedding. The bluish stain extincts tissue autofluorescence, wherefore dyed tissues appear black in the digital images. This "negative contrast" however, hampers decent gene expression analysis. If specifically stained tissues are located next to cavities, lumina, or surfaces, there is no possibility to distinguish between the true border of the cavity, lumen, or surface and the beginning of the stained tissue.

"SIM_1" [53] is designed and was applied for analysing embryos and tissue samples [67]. It was the first episcopic method that became commercialized. The specimens are embedded in a black embedding medium on resin basis and tissue autofluorescence is used for obtaining tissue information. In principle, SIM_1 is capable of visualizing multiple gene expression or gene product patterns in whole mount labelled specimens. However, it is bound to use fluorescent antibodies for detecting specifically labelled gene products. Unfortunately reliable whole mount fluorescence staining methods do not exist for many biomedical relevant model organisms. Furthermore the opacity of the embedding medium hampers positioning of the specimens and the definition of the correct field of view to become visualized.

"HREM" [56] is designed and was applied for analyzing embryos and tissue samples [68-75]. Specimens become immersed and embedded in resin dyed with eosin. The eosin causes an unspecific staining of the tissues. The intrinsic fluorescence of the resin covers shining through effects. The digital images of the block surfaces nearly resemble the quality of digital images of unspecifically stained histological sections (http://www.meduniwien.ac.at/3D-Rekonstr/HREM/). If whole mount LacZ or NBT/BCIP stained specimens are embedded, HREM is capable of visualizing specific gene expression or gene product patterns in the context of morphology and tissue architecture. In principle analysis of multiple gene expression patterns in one specimen are feasible. But they are not possible with the current technology.

"SIM_2" [54] is designed and was used for analyzing heart morphology. It uses waxes or resins as embedding mediums and an ultramiller for removing embedding medium from the blocks. The block surfaces are etched and then the freed tissues on the block face are stained for approximately 20 seconds. The staining solutions are similar to traditional histochemical stains, but are optimized for quick staining. SIM_2 is the youngest of the described methods (2007). Therefore the whole range of potential applications has not yet been explored. Basically it fits for analyzing the tissue architecture of all organ systems and, in principle, it has the potential for analyzing gene expression patterns.

EPISCOPIC 3D IMAGING METHODS – RESOLUTION OF VOLUME DATA

Most episcopic 3D imaging methods permit data generation of small to large specimens in high to low resolution. But, as in traditional microscopy, increasing the resolution means narrowing the field of view and thus the volume that can be analysed. This results in the simple relation; the larger the specimen to be visualized, the lower the resolution of the volume data. Two ways exist for overcoming this problem. One is to increase the resolution of the scanner or digital camera used for image capturing. However, this is limited by the numerical aperture of the optics, which the light coming from the specimen has to path to reach the chip of the scanner or camera. Increasing the target size of the scanner without decreasing the numerical aperture of the optics will result in larger digital images without additional information. The second approach to visualize large specimens in high resolution is to raster the block surface with a camera sitting on an optic with low numerical aperture. Multiple, overlapping high-resolution images of one block surface can be captured and digitally stitched together to form a single high resolution block face image. Although true high-resolution images of large specimens can be generated in this way, capturing multiple images from the same block surface is hampered by several problems. The most serious are: 1.) the block surface heats up, which results in artefacts or even breaking of the blocks, and 2.) complicated and expensive data generation apparatuses are required.

The resolution of volume data (voxel size) depends on both, the resolution (Pixel size) of the digital block surface images and on the thickness of the slices of embedding medium removed between capturing two subsequent images (Fig. (**3**)).

The 3D resolution usable for 3D analysis is defined by the largest of these two factors. As an example, images stacks derived from images with Pixel sizes of 0.5x0.5 μm^2, which were captured in 10 μm distances (voxel size 0.5x0.5x10 μm^3) are insufficient for analysing single cells, or the topology of capillaries and small embryonic blood vessels. Capillaries (~5μm diameter) might run for long distances, or might even branch "between" the images of two subsequent 10 μm sections. Such topological information is entirely lost. This is one of the main problems of physical section based 3D reconstruction methods [46-48, 76]. They have to use relatively thick sections in order to reduce the number of sections. By this the time expense for section processing and manual image capturing is kept feasible. Episcopic imaging methods do not have this problem. Capturing of digital images can be performed automatically during the sectioning process and takes only a few seconds. Therefore the time expense for the researcher is independent from the number of digital images captured. Episcopic 3D imaging methods typically generate image series of 500 – 2 000 single images, while producing sections of 1-3 μm thickness within a few hours operating time.

The dimension of one voxel (voxel size) of a data volume provides information about its numeric resolution. However, interpolation errors during data generation and data processing, and other artefacts must be taken into account. Therefore information resting in only one Voxel is unreliable under routine conditions. Furthermore, two cells (or structures) can only be discriminated, if at least one voxel, belonging to neither of them is located in between them. Therefore methods generating data volumes with voxels sizes of, for example, 10x10x10 μm^3, are numerically cellular, although they effectively are far from being capable of analyzing cells or even cell distributions. For analyzing the shape or for detecting the nature of a cell (or structure) the cell has to cover many voxels in sufficient contrast. Episcopic 3D image methods on average produce volume data of 2x2x2 μm^3 voxel size. Such data are not capable for analyzing cell organelles, but they are of a sufficient resolution to permit rough estimations of cell morphology, cell counts, and analysis of cell distributions.

EPISCOPIC 3D IMAGING METHODS – ADVANTAGES

The attractiveness of episcopic 3D imaging methods results from several advantages. The most important are:

1. Fast Data Generation

Most episcopic 3D imaging methods are capable of generating digital images of (in most cases) near to histological quality by capturing digital images directly from the surface of the embedding block during sectioning. In contrast to traditional histology and histological section based 3D reconstruction methods [45, 47, 48], no tedious section mounting, section realignment, section processing, and section image capturing is required. Volume data, *e.g.* consisting of 2 000 single images are available for analysis a few hours after starting the sectioning process. Since data generation can be automated, man-power is only necessary for setting up the system, starting sectioning, as well as subsequent data archiving, data visualisation, and data analysis.

2. High Data Quality

Episcopic 3D imaging methods capture images directly from the block surface. If the optic is not moved (Fig. (**1**)) and decent microtomes are used, the images of the resulting image stack are accurately aligned. Furthermore, similar to CT or MRI data, episcopically captured images do not show distortions introduced by sectioning, section mounting, or section processing. However, like with all post-mortem imaging techniques (including μCT, μMRI, *etc.*), artefacts introduced by specimen processing (fixation, embedding, *etc.*) do occur.

3. High Resolution

As discussed in the preceding chapter episcopically generated volume data are of high 3D resolution (Fig. (**3**)). This resolution is unmatched by other 3D methods, except for techniques optimized for analyzing cellular details, such as confocal microscopy, electron microscopy *etc.* But data generated with such techniques do not permit spatial analysis of cells in the 3D context of whole embryos or tissue samples.

4. Cheap Set Up

Currently no apparatus permitting episcopic volume data generation is commercially available. The single components however, are relatively cheap. They are standard equipment of most biomedical laboratories, and can be easily put together to function as an episcopic data generation apparatus. Nevertheless specialist knowledge is required for data interpretation and trouble shooting.

EPISCOPIC 3D IMAGING METHODS – TECHNICAL CHALLENGES

Episcopic imaging has three big and several small technical challenges. Here, we will only discuss the major ones.

1. Correct Positioning of the Specimens

Once automated data generation has started, the position of the optics in relation to the block must not be changed. Changes inevitably result in misaligned digital images and thus incorrect volume data. Therefore, prior to sectioning, the volume to visualize must be exactly positioned in respect to the optical pathway (Fig. (**1**)).

2. Shining Through Effect

Episcopic 3D imaging methods capture images directly from the surface of blocks of embedding material, which contain the specimen. Only tissue information from the very block face shall be visible in the digital images. However, tissues, which rest deep inside the block do "shine through" to and become visible at the block face. This diminishes contrasts and even obscures details in the block face images.

Epi-3D and SIM_2 try to overcome the shining through problem by contrasting only the tissues on each freshly cut block surface. However, this complicates automation of block face staining and thus automation of data generation. Unavoidable minimal differences in block face staining time result in different staining intensities between subsequent section images. Ultimately this hampers digital data processing and subsequent 3D modelling. Good section homogeneity and automation can be achieved with all episcopic imaging methods, which do not require block face staining (fast 3D serial reconstruction technique, SIM_1, EFIC, and HREM). These techniques use dyed embedding materials for eliminating, or at least reducing the shining through effect.

3. Block Heating

The illumination of the block surface must be sufficiently bright to achieve tissue contrast by light reflection or by exciting tissues to emit fluorescence signals. The required high intensities of light can cause heating of the block face and subsequently heating of the entire block. This heating may result in expansion of the block, melting of the embedding medium, and even breaking of the block.

Automation of data generation, shutter systems, and special compositions of the embedding medium help to reduce the negative effects caused by heating.

EPISCOPIC 3D IMAGING METHODS – ANALYZING TISSUE ARCHITECTURE AND MORPHOLOGICAL DETAIL

All episcopic imaging methods produce volume data of high resolution. Except for the fast 3D serial reconstruction technique, all methods permit the detection of organ and tissue details and even cells in the context of whole embryos, parts of embryos, or larger tissue samples.

Existing episcopic 3D imaging methods use different modalities for enhancing tissue contrasts. Consequently the volume data produced show dramatic differences in respect to image contrast and data quality. In brief: "Fast 3D serial reconstruction" provides excellent contrasts between bleached tissues and embedding medium. But the images do not permit analysis of tissue architecture. "Epi-3D" permits analysis of anatomical detail and differentiation of tissues, but in general, contrasts are relatively low. Furthermore, tissue contrast between subsequent images frequently differs, because each freshly cut block surface has to be stained individually prior to sectioning. "EFIC" and "SIM_1" use autofluorescence for detecting cells and tissues. The resulting digital images permit the detection of morphological details and the differentiation of tissues. But contrasts are low and borderlines of anatomical structures are often dubious.

"HREM" and "SIM_2" use traditional histological dyes for contrasting cells and tissues. Therefore the data quality is high and the appearance of the 2D images is similar to that of the digital images of histological sections, stained with overview dyes. In contrast to histological section series HREM images show delicate shining through artefacts (preceding chapter) and small scratches – the impressions of small irregularities of the edge of the blade. Their reflections appear in all images, because in contrast to histological sections, block faces are not cover-slipped. On the contrary, a series of HREM images is homogenous in respect to tissue contrast, while a series of images captured from subsequent histological sections is inhomogeneous. This inhomogenity is caused by different conditions during processing and staining of the individual sections. SIM_2 image series also may feature such inhomogeneous tissue contrasts between subsequent images, because each freshly cut block face has to become stained individually prior to image capturing.

EPISCOPIC 3D IMAGING METHODS – ANALYSING GENE EXPRESSION PATTERNS

EFIC, SIM_1, and HREM permit fast analysis of gene expression patterns in the context of morphological detail and tissue architecture [53, 77]. All kinds of samples (embryos, biopsies, tissue samples) can be analysed (http://www.meduniwien.ac.at/3D-Rekonstr/HREM/). However, for analyzing gene expression patterns the samples must be whole mount stained prior to embedding. The development of whole mount staining protocols is a major challenge. All antibodies and chemicals required must penetrate the entire sample. This is relatively easy with embryos of small size and small tissue samples, but it becomes complicated with larger embryos and bigger tissue samples.

While EFIC and HREM use a bluish colour reaction, SIM_1 uses fluorescence antibodies for detecting gene activity. That makes SIM_1 the only episcopic 3D imaging method that currently permits simultaneous analysis of multiple gene expression patterns in one specimen by using different fluorochromes. However, whole mount fluorescence *in situ* hybridization is still a major challenge.

Basically also SIM_2 might be expanded to permit 3D analysis of the patterns of specifically labelled gene products. However, a prerequisite must be the development of protocols and substances for fast on block *in situ* hybridization or immunohistochemical staining. Currently, neither adequate substances, nor protocols do exist.

All episcopic 3D imaging methods that permit the preservation of physical sections, bear the potential to collect the sections and to perform *in situ* hybridization or immunohistochemistry on the sections. After staining, the collected sections can be placed under a microscope and digital images can be captured with a digital camera sitting on the phototubus of the microscope. Proper alignment and equalization of the images of the sections can be achieved by matching them with the corresponding episcopically generated images (*e.g.* [78]).

However, with the exception of the alignment and distortion problems, such an approach has the same limitations as all physical section based 3D reconstruction methods. This is mainly the enormous human resources and the time expense necessary for generating the digital image series.

EPISCOPIC 3D IMAGING METHODS – CONCLUSIONS

Episcopic 3D imaging techniques can cheaply and quickly provide precisely registered high-resolution volume data in a highly automated fashion. The data can be used for detailed morphological analysis of biomedical model organisms and for 3D analysis of gene expression patterns in the context of tissue architecture and morphology of embryos, tissue samples and even biopsies. However even the advanced methods are still under construction. The main challenges of the next years will be 1) the development of fully automated, commercially available data generation apparatuses and 2) the modification and standardization of existing, and the development of new whole mount gene expression and gene product labelling protocols. We think that, within the next decades, episcopic 3D imaging will become an indispensable routine tool for researching the genetic regulation of embryo development as well as the genesis of malformations and diseases.

CONFLICT OF INTEREST

The author(s) confirm that this chapter content has no conflict of interest.

ACKNOWLEDGEMENTS

Declared none.

DISCLOSURE

This is an updated report of our earlier article W.J. Weninger and S.H. Geyer.Episcopic 3D Imaging Methods: Tools for Researching Gene Function. Curr Genomics 2009, 9:282-289.

REFERENCES

[1] Seltmann, M.; Horsch, M.; Drobyshev, A.; Chen, Y.; de Angelis, M. H.; Beckers, J. Assessment of a systematic expression profiling approach in ENU-induced mouse mutant lines. *Mamm Genome,* **2005,** *16*(1), 1-10.

[2] Reecy, J. M.; Spurlock, D. M.; Stahl, C. H. Gene expression profiling: insights into skeletal muscle growth and development. *J Anim Sci,* **2006,** *84 Suppl*, E150-4.

[3] Beckers, J.; Hoheisel, J.; Mewes, W.; Vingron, M.; Hrabe de Angelis, M. H. Molecular phenotyping of mouse mutant resources by RNA expression profiling. *Curr Genomics,* **2002,** *3*, 121-129.

[4] Mitiku, N.; Baker, J. C. Genomic analysis of gastrulation and organogenesis in the mouse. *Dev Cell,* **2007,** *13*(6), 897-907.

[5] Zhu, H.; Cabrera, R. M.; Wlodarczyk, B. J.; Bozinov, D.; Wang, D.; Schwartz, R. J.; Finnell, R. H. Differentially expressed genes in embryonic cardiac tissues of mice lacking Folr1 gene activity. *BMC Dev Biol,* **2007,** *7*, 128.

[6] Niemann, H.; Carnwath, J. W.; Kues, W. Application of DNA array technology to mammalian embryos. *Theriogenology,* **2007,** *68 Suppl 1*, S165-77.

[7] Shah, K. Current advances in molecular imaging of gene and cell therapy for cancer. *Cancer Biol Ther,* **2005,** *4*(5), 518-23.

[8] Wessels, J. T.; Busse, A. C.; Mahrt, J.; Dullin, C.; Grabbe, E.; Mueller, G. A. *In vivo* imaging in experimental preclinical tumor research--a review. *Cytometry A,* **2007,** *71*(8), 542-9.

[9] Li, X.; Liu, J.; Davey, M.; Duce, S.; Jaberi, N.; Liu, G.; Davidson, G.; Tenent, S.; Mahood, R.; Brown, P.; Cunningham, C.; Bain, A.; Beattie, K.; McDonald, L.; Schmidt, K.; Towers, M.; Tickle, C.; Chudek, S. Micro-magnetic resonance imaging of avian embryos. *J Anat,* **2007,** *211*(6), 798-809.

[10] Massoud, T. F.; Gambhir, S. S. Molecular imaging in living subjects: seeing fundamental biological processes in a new light. *Genes Dev,* **2003,** *17*(5), 545-80.

[11] Johnson, J. T.; Hansen, M. S.; Wu, I.; Healy, L. J.; Johnson, C. R.; Jones, G. M.; Capecchi, M. R.; Keller, C. Virtual histology of transgenic mouse embryos for high-throughput phenotyping. *PLoS Genet,* **2006,** *2*(4), e61.

[12] Borah, B.; Gross, G. J.; Dufresne, T. E.; Smith, T. S.; Cockman, M. D.; Chmielewski, P. A.; Lundy, M. W.; Hartke, J. R.; Sod, E. W. Three-dimensional microimaging (MRmicroI and microCT), finite element modeling, and rapid prototyping provide unique insights into bone architecture in osteoporosis. *Anat Rec,* **2001,** *265*(2), 101-10.

[13] Metscher, B. D. MicroCT for developmental biology: A versatile tool for high-contrast 3D imaging at histological resolutions. *Dev Dyn,* **2009,** *238*(3), 632-40.

[14] Degenhardt, K.; Wright, A. C.; Horng, D.; Padmanabhan, A.; Epstein, J. A. Rapid 3D phenotyping of cardiovascular development in mouse embryos by micro-CT with iodine staining. *Circ Cardiovasc Imaging,* **2010,** *3*(3), 314-22.

[15] Nagase, T.; Sasazaki, Y.; Kikuchi, T.; Machida, M. Rapid 3-dimensional imaging of embryonic craniofacial morphology using microscopic computed tomography. *Journal of Computer Assisted Tomography,* **2008,** *32*(5), 816-821.

[16] Bamforth, S. D.; Schneider, J. E.; Bhattacharya, S. High-throughput analysis of mouse embryos by magnetic resonance imaging. *Cold Spring Harbor Protocols,* **2012,** *2012*(1).

[17] Cleary, J. O.; Price, A. N.; Thomas, D. L.; Scambler, P. J.; Kyriakopoulou, V.; McCue, K.; Schneider, J. E.; Ordidge, R. J.; Lythgoe, M. F. Cardiac phenotyping in *ex vivo* murine embryos using microMRI. *NMR Biomed,* **2009,** *22*(8), 857-66.

[18] Dhenain, M.; Ruffins, S. W.; Jacobs, R. E. Three-dimensional digital mouse atlas using high-resolution MRI. *Dev Biol,* **2001,** *232*(2), 458-70.

[19] Schneider, J. E.; Bamforth, S. D.; Farthing, C. R.; Clarke, K.; Neubauer, S.; Bhattacharya, S. High-resolution imaging of normal anatomy, and neural and adrenal malformations in mouse embryos using magnetic resonance microscopy. *J Anat,* **2003,** *202*(2), 239-47.

[20] Schneider, J. E.; Bhattacharya, S. Making the mouse embryo transparent: identifying developmental malformations using magnetic resonance imaging. *Birth Defects Res C Embryo Today,* **2004,** *72*(3), 241-9.

[21] Schneider, J. E.; Bose, J.; Bamforth, S. D.; Gruber, A. D.; Broadbent, C.; Clarke, K.; Neubauer, S.; Lengeling, A.; Bhattacharya, S. Identification of cardiac malformations in mice lacking Ptdsr using a novel high-throughput magnetic resonance imaging technique. *BMC Dev Biol,* **2004,** *4*(1), 16.

[22] Bogani, D.; Willoughby, C.; Davies, J.; Kaur, K.; Mirza, G.; Paudyal, A.; Haines, H.; McKeone, R.; Cadman, M.; Pieles, G.; Schneider, J. E.; Bhattacharya, S.; Hardy, A.; Nolan, P. M.; Tripodis, N.; Depew, M. J.; Chandrasekara, R.; Duncan, G.; Sharpe, P. T.; Greenfield, A.; Denny, P.; Brown, S. D.; Ragoussis, J.; Arkell, R. M. Dissecting the genetic complexity of human 6p deletion syndromes by using a region-specific, phenotype-driven mouse screen. *Proc Natl Acad Sci U S A,* **2005,** *102*(35), 12477-82.

[23] Jacobs, R. E.; Fraser, S. E. Magnetic resonance microscopy of embryonic cell lineages and movements. *Science,* **1994,** *263*(5147), 681-4.

[24] Roellig, K.; Drews, B.; Goeritz, F.; Hildebrandt, T. B. The long gestation of the small naked mole-rat (Heterocephalus glaber Rüppell, 1842) studied with ultrasound biomicroscopy and 3D-ultrasonography. *PLoS One,* **2011,** *6*(3), e17744-e17744.

[25] Loveless, M. E.; Li, X.; Huamani, J.; Lyshchik, A.; Dawant, B.; Hallahan, D.; Gore, J. C.; Yankeelov, T. E. A method for assessing the microvasculature in a murine tumor model using contrast-enhanced ultrasonography. *Journal of Ultrasound in Medicine: Official Journal of the American Institute of Ultrasound in Medicine,* **2008,** *27*(12), 1699-1709.

[26] Phoon, C. K.; Ji, R. P.; Aristizabal, O.; Worrad, D. M.; Zhou, B.; Baldwin, H. S.; Turnbull, D. H. Embryonic heart failure in NFATc1-/- mice: novel mechanistic insights from in utero ultrasound biomicroscopy. *Circ Res,* **2004,** *95*(1), 92-9.

[27] Phoon, C. K. Imaging tools for the developmental biologist: ultrasound biomicroscopy of mouse embryonic development. *Pediatr Res,* **2006,** *60*(1), 14-21.

[28] Liang, H. D.; Blomley, M. J. The role of ultrasound in molecular imaging. *Br J Radiol,* **2003,** *76 Spec No 2*, S140-50.

[29] Walls, J. R.; Coultas, L.; Rossant, J.; Henkelman, R. M. Three-dimensional analysis of vascular development in the mouse embryo. *PLoS One,* **2008,** *3*(8), e2853.

[30] Breckenridge, R.; Kotecha, S.; Towers, N.; Bennett, M.; Mohun, T. Pan-myocardial expression of Cre recombinase throughout mouse development. *Genesis (New York, N.Y.: 2000),* **2007,** *45*(3), 135-144.

[31] Quintana, L.; Sharpe, J. Optical Projection Tomography of Vertebrate Embryo Development. *Cold Spring Harbor Protocols,* **2011,** *2011*, pdb.top116-pdb.top116-pdb.top116-pdb.top116.

[32] Sharpe, J. Optical projection tomography as a new tool for studying embryo anatomy. *J Anat,* **2003,** *202*(2), 175-81.

[33] Sharpe, J.; Ahlgren, U.; Perry, P.; Hill, B.; Ross, A.; Hecksher-Sorensen, J.; Baldock, R.; Davidson, D. Optical projection tomography as a tool for 3D microscopy and gene expression studies. *Science,* **2002,** *296*(5567), 541-5.

[34] Wanninger, A., The application of confocal microscopy and 3D imaging software in Functional, Evolutionary, and Developmental Zoology: reconstructing myo- and neurogenesis in space and time. In *Modern Research and Educational Topics in Microscopy*, Formatex, Bardajoz, Spain: 2007.

[35] Hecksher-Sorensen, J.; Sharpe, J. 3D confocal reconstruction of gene expression in mouse. *Mech Dev,* **2001,** *100*(1), 59-63.

[36] Chvatal, A.; Anderova, M.; Kirchhoff, F. Three-dimensional confocal morphometry - a new approach for studying dynamic changes in cell morphology in brain slices. *J Anat,* **2007,** *210*(6), 671-83.

[37] Kolker, S. J.; Tajchman, U.; Weeks, D. L. Confocal imaging of early heart development in Xenopus laevis. *Dev Biol,* **2000,** *218*(1), 64-73.

[38] Liebling, M.; Forouhar, A. S.; Wolleschensky, R.; Zimmermann, B.; Ankerhold, R.; Fraser, S. E.; Gharib, M.; Dickinson, M. E. Rapid three-dimensional imaging and analysis of the beating embryonic heart reveals functional changes during development. *Dev Dyn,* **2006,** *235*(11), 2940-8.

[39] Sands, G. B.; Gerneke, D. A.; Hooks, D. A.; Green, C. R.; Smaill, B. H.; Legrice, I. J. Automated imaging of extended tissue volumes using confocal microscopy. *Microsc Res Tech,* **2005,** *67*(5), 227-39.

[40] Matsko, N. B. Atomic force microscopy applied to study macromolecular content of embedded biological material. *Ultramicroscopy,* **2007,** *107*(2-3), 95-105.

[41] Matsko, N.; Mueller, M. AFM of biological material embedded in epoxy resin. *J Struct Biol,* **2004,** *146*(3), 334-43.

[42] Kedrov, A.; Ziegler, C.; Muller, D. J. Differentiating ligand and inhibitor interactions of a single antiporter. *J Mol Biol,* **2006,** *362*(5), 925-32.

[43] Frank, J., *Three dimensional imaging with the transmission electron microscope.* 1st ed.; Kluwer Academic / Plenum Publishers: 1992.

[44] Lucic, V.; Forster, F.; Baumeister, W. Structural studies by electron tomography: from cells to molecules. *Annu Rev Biochem,* **2005,** *74*, 833-65.

[45] Weninger, W. J.; Streicher, J.; Müller, G. B. [3-dimensional reconstruction of histological serial sections using a computer]. *Wien Klin Wochenschr,* **1996,** *108*(16), 515-20.

[46] Streicher, J.; Donat, M. A.; Strauss, B.; Spörle, R.; Schughart, K.; Müller, G. B. Computer-based three-dimensional visualization of developmental gene expression. *Nat Genet,* **2000,** *25*(2), 147-52.

[47] Streicher, J.; Weninger, W. J.; Müller, G. B. External marker-based automatic congruencing: a new method of 3D reconstruction from serial sections. *Anat Rec,* **1997,** *248*(4), 583-602.

[48] Soufan, A. T.; Ruijter, J. M.; van den Hoff, M. J.; de Boer, P. A.; Hagoort, J.; Moorman, A. F. Three-dimensional reconstruction of gene expression patterns during cardiac development. *Physiol Genomics,* **2003,** *13*(3), 187-95.

[49] Soufan, A. T.; van den Hoff, M. J.; Ruijter, J. M.; de Boer, P. A.; Hagoort, J.; Webb, S.; Anderson, R. H.; Moorman, A. F. Reconstruction of the patterns of gene expression in the developing mouse heart reveals an architectural arrangement that facilitates the understanding of atrial malformations and arrhythmias. *Circ Res,* **2004,** *95*(12), 1207-15.

[50] Odgaard, A.; Andersen, K.; Melsen, F.; Gundersen, H. J. A direct method for fast three-dimensional serial reconstruction. *J Microsc,* 1990, *159*(Pt 3), 335-42.

[51] Weninger, W. J.; Meng, S.; Streicher, J.; Müller, G. B. A new episcopic method for rapid 3-D reconstruction: applications in anatomy and embryology. *Anat Embryol (Berl),* **1998,** *197*(5), 341-8.

[52] Weninger, W. J.; Mohun, T. Phenotyping transgenic embryos: a rapid 3-D screening method based on episcopic fluorescence image capturing. *Nat Genet,* **2002,** *30*(1), 59-65.

[53] Ewald, A. J.; McBride, H.; Reddington, M.; Fraser, S. E.; Kerschmann, R. Surface imaging microscopy, an automated method for visualizing whole embryo samples in three dimensions at high resolution. *Dev Dyn,* **2002,** *225*(3), 369-75.

[54] Gerneke, D. A.; Sands, G. B.; Ganesalingam, R.; Joshi, P.; Caldwell, B. J.; Smaill, B. H.; Legrice, I. J. Surface imaging microscopy using an ultramiller for large volume 3D reconstruction of wax- and resin-embedded tissues. *Microsc Res Tech,* **2007,** *70*(10), 886-94.

[55] Mohun, T. J.; Weninger, W. J. Imaging heart development using high-resolution episcopic microscopy. *Curr Opin Genet Dev,* **2011,** *21*(5), 573-8.

[56] Weninger, W. J.; Geyer, S. H.; Mohun, T. J.; Rasskin-Gutman, D.; Matsui, T.; Ribeiro, I.; Costa Lda, F.; Izpisua-Belmonte, J. C.; Müller, G. B. High-resolution episcopic microscopy: a rapid technique for high detailed 3D analysis of gene activity in the context of tissue architecture and morphology. *Anat Embryol,* **2006,** *211*(3), 213-21.

[57] Denk, W.; Horstmann, H. Serial block-face scanning electron microscopy to reconstruct three-dimensional tissue nanostructure. *PLoS Biol,* **2004,** *2*(11), e329.

[58] Odgaard, A. Three-dimensional methods for quantification of cancellous bone architecture. *Bone,* **1997,** *20*(4), 315-28.

[59] Odgaard, A. Quantification of Connectivity in Cancellous Bone, with Special Emphasis on 3-D Reconstructions. *Bone,* **1993,** *14*, 173-82.

[60] Odgaard, A.; Andersen, K.; Ullerup, R.; Frich, L. H.; Melsen, F. Three-dimensional reconstruction of entire vertebral bodies. *Bone,* **1994,** *15*(3), 335-42.

[61] Mohun, T. J.; Leong, L. M.; Weninger, W. J.; Sparrow, D. B. The morphology of heart development in Xenopus laevis. *Dev Biol,* **2000,** *218*(1), 74-88.

[62] Weninger, W. J.; Streicher, J.; Müller, G. B. Anatomical compartments of the parasellar region: adipose tissue bodies represent intracranial continuations of extracranial spaces. *J Anat,* **1997,** *191 (Pt 2)*, 269-75.

[63] Weninger, W. J.; Müller, G. B.; Reiter, C.; Meng, S.; Rabl, S. U. Intimal hyperplasia of the infant parasellar carotid artery: a potential developmental factor in atherosclerosis and SIDS. *Circ Res,* **1999,** *85*(10), 970-5.

[64] Meng, S.; Costa Lda, F.; Geyer, S. H.; Viana, M. P.; Reiter, C.; Muller, G. B.; Weninger, W. J. Three-dimensional description and mathematical characterization of the parasellar internal carotid artery in human infants. *J Anat,* **2008,** *212*(5), 636-44.

[65] Weninger, W. J.; Floro, K. L.; Bennett, M. B.; Withington, S. L.; Preis, J. I.; Barbera, J. P.; Mohun, T. J.; Dunwoodie, S. L. Cited2 is required both for heart morphogenesis and establishment of the left-right axis in mouse development. *Development,* **2005,** *132*(6), 1337-48.

[66] Going, J. J.; Mohun, T. J. Human breast duct anatomy, the 'sick lobe' hypothesis and intraductal approaches to breast cancer. *Breast Cancer Res Treat,* **2006,** *97*(3), 285-91.

[67] Ewald, A. J.; Peyrot, S. M.; Tyszka, J. M.; Fraser, S. E.; Wallingford, J. B. Regional requirements for Dishevelled signaling during Xenopus gastrulation: separable effects on blastopore closure, mesendoderm internalization and archenteron formation. *Development,* **2004,** *131*(24), 6195-209.

[68] Gindes, L.; Matsui, H.; Achiron, R.; Mohun, T.; Yen Ho, S.; Gardiner, H. Comparison between ex-vivo HREM imaging and in-vivo 4DHRTV ultrasound of the first trimester fetal heart. *Ultrasound Obstet Gynecol,* **2011.**

[69] Dunlevy, L.; Bennett, M.; Slender, A.; Lana-Elola, E.; Tybulewicz, V. L.; Fisher, E. M.; Mohun, T. Down's syndrome-like cardiac developmental defects in embryos of the transchromosomic Tc1 mouse. *Cardiovasc Res,* **2010,** *88*(2), 287-95.

[70] Delaurier, A.; Burton, N.; Bennett, M.; Baldock, R.; Davidson, D.; Mohun, T. J.; Logan, M. P. The Mouse Limb Anatomy Atlas: an interactive 3D tool for studying embryonic limb patterning. *BMC Dev Biol,* **2008,** *8*, 83.

[71] Hasson, P.; DeLaurier, A.; Bennett, M.; Grigorieva, E.; Naiche, L. A.; Papaioannou, V. E.; Mohun, T. J.; Logan, M. P. Tbx4 and tbx5 acting in connective tissue are required for limb muscle and tendon patterning. *Dev Cell,* **2010,** *18*(1), 148-56.

[72] Geyer, S. H.; Weninger, W. J. Some Mice Feature 5th Pharyngeal Arch Arteries and Double-Lumen Aortic Arch Malformations. *Cells Tissues Organs,* **2012.**

[73] Geyer, S. H.; Maurer, B.; Potz, L.; Singh, J.; Weninger, W. J. High-Resolution Episcopic Microscopy Data-Based Measurements of the Arteries of Mouse Embryos: Evaluation of Significance and Reproducibility under Routine Conditions. *Cells Tissues Organs,* **2011**.

[74] Weninger, W. J.; Maurer, B.; Zendron, B.; Dorfmeister, K.; Geyer, S. H. Measurements of the diameters of the great arteries and semi-lunar valves of chick and mouse embryos. *J Microsc,* **2009,** *234*(2), 173-90.

[75] Pieles, G.; Geyer, S. H.; Szumska, D.; Schneider, J.; Neubauer, S.; Clarke, K.; Dorfmeister, K.; Franklyn, A.; Brown, S. D.; Bhattacharya, S.; Weninger, W. J. microMRI-HREM pipeline for high-throughput, high-resolution phenotyping of murine embryos. *J Anat,* **2007,** *211*(1), 132-7.

[76] Soufan, A. T.; van den Berg, G.; Ruijter, J. M.; de Boer, P. A.; van den Hoff, M. J.; Moorman, A. F. Regionalized sequence of myocardial cell growth and proliferation characterizes early chamber formation. *Circ Res,* **2006,** *99*(5), 545-52.

[77] Weninger, W. J.; Mohun, T. J. Three-dimensional analysis of molecular signals with episcopic imaging techniques. *Methods in molecular biology (Clifton, N.J,* **2007,** *411*, 35-46.

[78] Verbeek, F. J.; Huijsmans, D. P.; Baeten, R. J.; Schoutsen, N. J.; Lamers, W. H. Design and implementation of a database and program for 3D reconstruction from serial sections: a data-driven approach. *Microsc Res Tech,* **1995,** *30*(6), 496-512.

Send Orders of Reprints at reprints@benthamscince.net
Advances in Genome Science, Vol. 1, 2013, 189-209

CHAPTER 8

Drosophila Lethal Giant Larvae Neoplastic Mutant as a Genetic Tool for Cancer Modelling

Francesca Froldi, Marcello Ziosi, Annalisa Pession and Daniela Grifoni*

Cancer Modelling in Drosophila, Alma Mater Studiorum, Departments of Biologia Evoluzionistica Sperimentale, and Patologia Sperimentale, Bologna, Italy

Abstract: *Drosophila lethal giant larvae* (*lgl*) is a tumour suppressor gene whose function in establishing apical-basal cell polarity as well as in exerting proliferation control in epithelial tissues is conserved between flies and mammals. Individuals bearing *lgl* null mutations show a gradual loss of tissue architecture and an extended larval life in which cell proliferation never ceases and no differentiation occurs, resulting in prepupal lethality. When tissues from those individuals are transplanted into adult normal recipients, a subset of cells, possibly the cancer stem cells, are able to proliferate and migrate to distant sites forming metastases which eventually kill the host. This phenotype closely resembles that of mammalian epithelial cancers, in which loss of cell polarity is one of the hallmarks of a malignant, metastatic behaviour associated with poor prognosis. Lgl protein shares with its human counterpart Human giant larvae-1 (Hugl-1) significant stretches of sequence similarity that we demonstrated to translate into a complete functional conservation, pointing out a role in cell proliferation control and tumourigenesis also for the human homologue. The functional conservation and the power of fly genetics, that allows the researcher to manipulate the fly genome at a level of precision that exceeds that of any other multicellular genetic system, make this *Drosophila* mutant a very suitable model in which to investigate the mechanisms underlying epithelial tumour formation, progression and metastatisation. In this review, we will summarise the results obtained in the last years using this model for the study of cancer biology. Moreover, we will discuss how recent advances in developmental genetics techniques have succeeded in enhancing the similarities between fly and human tumourigenesis, giving *Drosophila* a pivotal role in the study of such a complex genetic disease.

Keywords: *lethal giant larvae*, tumour suppressor, *Drosophila*, Hugl-1, epithelial cancers, animal model.

GENERAL INTRODUCTION

Carcinomas are the most frequent form of cancer. They arise from epithelial cells and are often highly invasive and metastatic. It has extensively been demonstrated

***Address correspondence to Daniela Grifoni:** Departments of Biologia Evoluzionistica Sperimentale and Patologia Sperimentale, Bologna, Italy, to: Department of "Farmacia e Biotecnologie", Bologna, Italy; Tel: +39(0)512094179; Fax: +39(0)512094286; E-mail: daniela.grifoni@unibo.it

Christian Neri (Ed)

that a clonal accumulation of subsequent mutations is the cause of cellular escape from mechanisms controlling its fate [1, 2]. The first hit affecting a somatic cell in humans is however hardly distinguishable in the final lesion, and several decades may be required for the mutant cell to progress into a metastatic neoplasia, in a subtly tuned balance of forward and reverse signals. When this process leads to an *in situ* carcinoma, the reverse pathway is usually switched off and the lesion, sometimes confined within the basement membrane for years, proceeds invading the underlying connective tissue and seeding cells to distant sites, where secondary tumours may form [2, 3].

Several of the aspects of human carcinogenesis can be profitably investigated using *Drosophila* as an *in vivo* model since epithelial cancers have long been described in flies and, so far, more than 50 genes have been found whose mutation gives rise to uncontrolled proliferation acting in a recessive manner [4]. For these reasons they are called *Drosophila* tumour suppressor genes (TSG) and their mutations fall in two main categories. Genes whose loss of function (LOF) in the whole animal causes excessive proliferation of epithelial tissues without compromising monolayer organization, tissue architecture and differentiation, are defined hyperplastic tumour suppressors [4]. On the other hand, LOF mutations of neoplastic TSG cause overproliferation accompanied by loss of cell architecture and proper cell-cell contacts, 3D growth and failure to differentiate. Neoplastic potential is usually assayed through the ability of a tissue to give rise to a primary mass able to invade/metastasise following to transplantation into an adult host [4]. Among the fly neoplastic TSG so far identified, *lethal giant larvae* (*lgl*), *scribble* (*scrib*) and *discs large* (*dlg*) are of particular interest as their deprivation in the entire individual causes neoplastic growth of both the ectodermal derivatives: imaginal discs and larval brain [4, 5]. Imaginal discs are larval structures of epithelial origin whose primordia segregate in the embryo as quiescent cell clusters which begin to proliferate intensely starting from the third larval instar. Two days after, their cells cease dividing and during metamorphosis they differentiate into adult appendages. Imaginal discs are thus considered the system of election in which to study cell proliferation control and morphogenesis in *Drosophila* and their monolayered structure perfectly recapitulates many morphological, structural and molecular features of human epithelia coating internal organs, from which carcinomas arise [6].

lgl was the first neoplastic TSG described in *Drosophila* [7], but its exhaustive phenotypic characterization came only in 1978 with the isolation of a spontaneous *locus* deletion [8]. Its LOF causes neoplastic overproliferation of both imaginal epithelia and neuroblasts of the larval brain. These structures not only show excessive growth but also loss of apical-basal polarity and tissue architecture which result in an extended larval period and pre-pupal lethality. Cell growth appears to be slow in *lgl* mutants; 5 days-old *lgl* wing discs indeed contain about one third cells respect to the wild type counterpart [4]; despite this, since $lgl^{-/-}$ cells are unable to exit cell cycle, after many additional days of proliferation discs reach an enormous cell number and eventually form large amorphous masses which fail to form proper cellular contacts and do not differentiate, presenting all the main features of malignant carcinomas, Fig (**1**, Upper panel). They show loss of positional cues since, in *lgl* defective discs, cells straddle the anterior-posterior compartment boundary and intermix [9], Fig. (**1**, Lower panel). Moreover they are able to fuse with nearby tissues, so displaying local invasiveness, and show a high metastatic potential upon transplantation [8].

Two biochemical markers associated with aggressive human cancers have also been observed in *lgl* tumours: secretion of Collagenase IV [11] and Matrix Metallo-Proteinase 1 (MMP1) [12], both involved in basement membrane degradation; the latter has further been shown to be essential for *lgl* cells spreading because *lgl*; *MMP1* double mutant cells show a strong reduction in the ability to leave primary lesions [12]. The metastatic index of *lgl* mutant tissues was initially assessed by Woodhouse *et al.* in 1998 [13]. They defined it as the fraction of hosts with secondary tumours in various districts of the body divided by the fold increase in size of the primary tumour after transplantation. However, since the open circulatory system of the insect calls into question the possibility of a passive transport of tumour cells to distant sites, the authors developed a second and more powerful *in vivo* assay to test metastatic abilities of cancer cells [14]. They refined the criteria of index calculation and only those metastases occurring in the host's ovaries were taken into account. The *Drosophila* ovary consists of 15-20 ovarioles surrounded by a peritoneal sheath of cells. Each ovariole is in turn surrounded by a muscle layer allocated between two layers of extracellular matrix flanking the basement membrane of the follicular epithelium, an adult somatic

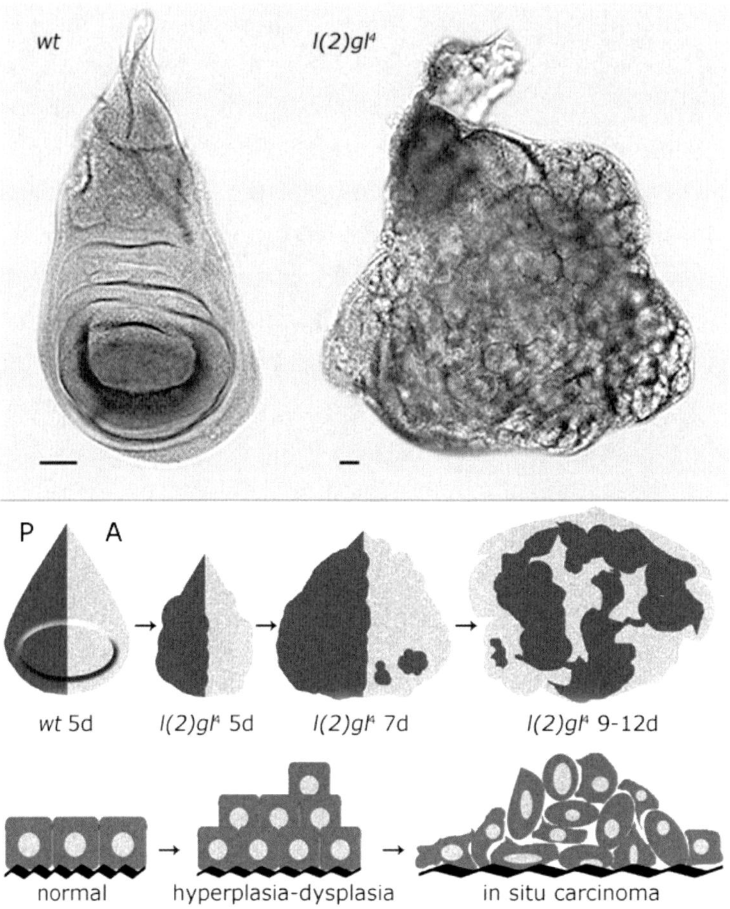

Figure 1: Upper panel: wild-type and *l(2)gl*^{4} wing imaginal discs. The wild type (wt) disc at 5 days after egg laying (AEL) ceases to proliferate and starts to differentiate into the adult wing; on the contrary, *l(2)gl*^{4} disc never exits cell cycle.
Lower panel: morphological parallelism between *lgl* mutant progression and a mammalian epithelial cancer development. The anterior (A) and posterior (P) compartments of the wing discs are independent developmental units and cells belonging to either of the two never intermix in a wild type disc [10].

columnar epithelium coating the oocyte. Since cancer cells have to pass through all these continuous layers to colonise the ovaries, the recovering of metastases within these organs is strict evidence of active invasion, thus seeming a very reliable indicator of metastatic potential [14].

Lineage analysis of these *lgl* tumours suggests that only a small fraction of the cancer cells (less than 2%) displays this metastatic behaviour [13]. Moreover, by extending the proliferation time of primary tumours through serial transplantations into multiple hosts, the frequency of metastases increases [14]. These two observations highlight the possible requirement of additional genetic lesions to promote invasion, further reinforcing the parallelism with mammalian tumourigenesis. Karyotypic instability was indeed observed in *lgl* mutant tumour cells after serial transplantations into adult hosts [8].

LGL IN CELL POLARITY AND PROLIFERATION

A correct apical-basal cell polarity is the main feature of a healthy epithelial tissue. From invertebrates to mammals, cells composing an epithelial sheath possess complex junctional structures that define apical and basolateral domains, favour cell-cell communication and cytoskeleton architecture, restrict motility and maintain tissue integrity, making the cell layer work as a functional unit [15, 16].

Indeed, loss of apical-basal polarity is a hallmark of malignant carcinomas. Often, as the tumour progresses, morphological alterations typical of the so-called EMT (Epithelial-to-Mesenchimal Transition) are observed: cells loosen their contacts, become round, grow in a disorganised multilayer and are prone to migration. From a molecular point of view ETM is characterised by the disruption of adherens junctions through functional inactivation of E-cadherins [17, 18].

lgl, *scrib* and *dlg* are clearly involved in epithelial cell polarity establishment in *Drosophila*. Given the similarities between their phenotypes, the strong genetic interaction shown and their mutual requirement for correct localisation, the three proteins are much likely part of the same genetic pathway and act as a complex in establishing the basolateral membrane domain identity as well as controlling proliferation [19].

Loss of apical-basal polarity caused by their LOF, due to failure to form proper adherens junctions [19], is peculiar as it results from lateral spreading of proteins normally restricted to the apical cell cortex.

lgl encodes a 127 kD protein with several WD40 repeats predicted to fold into a β-propeller domain involved in protein-protein contact [20]. It is able to form homo-oligomers and takes part to the multimeric cytoskeletal network that also contains non muscle type II myosin heavy chain [21]. Lgl is a highly stable protein; gene transcription is restricted to the two major phases of cell proliferation in the developing fly, namely, early embryogenesis and late third instar larvae [22]; moreover, *lgl* zygotes lacking maternal supply die as embryos, so maternal pulse is sufficient to sustain the earliest stages of development [23]. Also *dgl* and *scrib* encode for scaffolding proteins, rich in protein-protein interaction domains [4]. Scrib and Dgl are restricted to the lateral membrane domain, while Lgl is more widespread and enriched all along the basolateral cortex and it is also found in the cytoplasm. They colocalise at septate junctions (*Drosophila* homologue of the vertebrate tight junction), which lie basal to the adherens junctions, and act by antagonising the Par3/Par6/atypical Protein Kinase C (aPKC) complex, which localises at the adherens junctions and participates in the specification of the apical domain identity [4, 5, 24]. The mutually exclusive activity of Lgl and aPKC complexes is required for the correct formation and positioning of cell junctions which, in turn, provide cells with the correct apical-basal and cytoskeletal structure [25], Fig. (**2**).

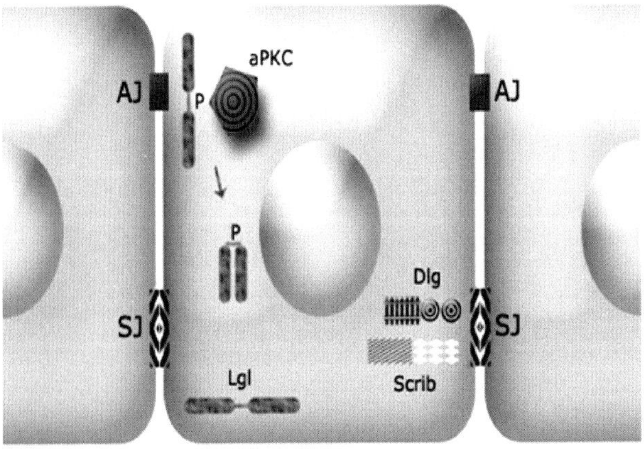

Figure 2: Schematic representation of proteins involved in the establishment of the apical-basal cell polarity in a *Drosophila* epithelial cell. The explanation is in the text. AJ: Adherens Junction; SJ: Septate Junction.

The mechanism through which these complexes regulate each-other was initially investigated in the stem cells of the *Drosophila* larval brain, called neuroblasts [25]. Although cell-cell adhesion structures are lacking in neuroblasts, the same protein complexes are involved in the apical and basal compartimentalisation of several cell fate determinants, necessary for the asymmetric cell division and the consequent self renewal and commitment of the stem population [26]. It was demonstrated that Lgl association with the plasma membrane, which seems to be essential for its function, is regulated through phosphorylation in conserved residues by aPKC [25]. Upon phosphorylation Lgl is released, and thus excluded from the apical plasma membrane, in an autoinhibited form [20], Fig. (**2**). Lgl in turn counteracts aPKC activity at the basolateral domain possibly sequestering it in an inactive form [25, 27]. The phenotype observed in *lgl*$^{-/-}$ tissues can indeed be reproduced by expanding aPKC activity domain in neuroblasts. The expression of a prenylated form, which is targeted to the entire cell cortex, results in an abnormal increase in neuroblast population and tumorous hyperproliferation due to an excess of self-renewal [28].

The same is true for epithelial tissues. The basolateral spreading of aPKC activity causes cytoplasmic release of Lgl from the basolateral membrane leading to hyperproliferation and massive overgrowth in the wing disc and deep morphogenetic alterations in the adult wing [29].

Understanding how loss of cell polarity and control of cell proliferation are linked would represent a relevant advance in the comprehension of the molecular bases of cancer development. Alleles of *lgl*, *scrib* and *dlg* were identified in a screening for dominant suppressors of a hypomorphic mutation of cyclin E, a key regulator of G1-S phase transition [30], implying a direct role in negative regulation of cell cycle.

Indeed *lgl* mutant tissue in the eye disc shows ectopic Cyclin E expression and thus ectopic S-phases without loss of apical-basal polarity [31], demonstrating that loss of proliferation control is not a direct consequence of polarity loss, at least in this context.

Recently it was demonstrated that the proliferative defects in the eye disc described above are due to the deregulation of the Hippo (Hpo) pathway [32], a very conserved signalling network that plays a central role in the control of epithelial tissues proliferation [33], establishing for the first time a direct link between *lgl* and cell proliferation control.

The Hpo pathway consists of several upstream components which function converges on Yorkie repression (Yki), a transcriptional activator normally sequestered in the cytoplasm. Upon pathway inactivation, Yki enters nucleus and activates the transcription of its target genes involved in cell growth, proliferation and resistance to apoptosis such as cycE, dMyc, dIAP1 (Drosophila Inhibitor of APoptosis 1) and the microRNA *bantam*. Yki is thus a growth promoter, whereas the upstream components act as tumour suppressors by antagonising the growth-promoting activity of Yki [33]. *lgl* mutant cells in the eye imaginal disc show upregulation of Hpo pathway target genes as cycE and dIAP1; consistently, in *lgl* mutant cells Yki was found to localise also in the nucleus. Removal of one copy of *yki* partially reduced the ectopic S-phases phenotype previously described [31, 32], indicating that Yki activity is required for *lgl*$^{-/-}$ proliferative defects. *lgl*-dependent regulation of Yki has also been observed in the wing imaginal disc [34].

MAMMALIAN LGL HOMOLOGUES

Establishment of epithelial cell polarity seems to be a very conserved process from flies to humans [24]. Homologues of *lgl* have been identified in many species including yeast, slime molds, worms, insects, mouse and man [35].

There are two *Drosophila* Lgl human homologues: Hugl-1 and Hugl-2 [36]. Hugl-1 is, so far, the best characterised. It shows the typical WD40 repeats, is mainly membrane associated, forms homo-oligomers and interacts with non muscle myosin type II, similar to *Drosophila* Lgl [35]. Fly and human proteins show a significant global sequence similarity of 62.5% Fig. (**3**), and we demonstrated that Hugl-1 is able to form heterocomplexes with Lgl *in vitro* [37]. However, we found conclusive evidence of the functional conservation between the two proteins in the full rescue of the null *lgl* mutant by the human Hugl-1 cDNA [37]. Analogous rescuing ability has been demonstrated also for rat Dlg [38] and human Scrib [39].

We and others showed that Hugl-1 is down-regulated or completely absent in a variety of human epithelial malignancies such as breast, lung, prostate, ovarian cancers and melanomas [37, 40] and it has also been implicated in colorectal cancer progression where it was found to be associated with advanced stages and lymph node metastases [41].

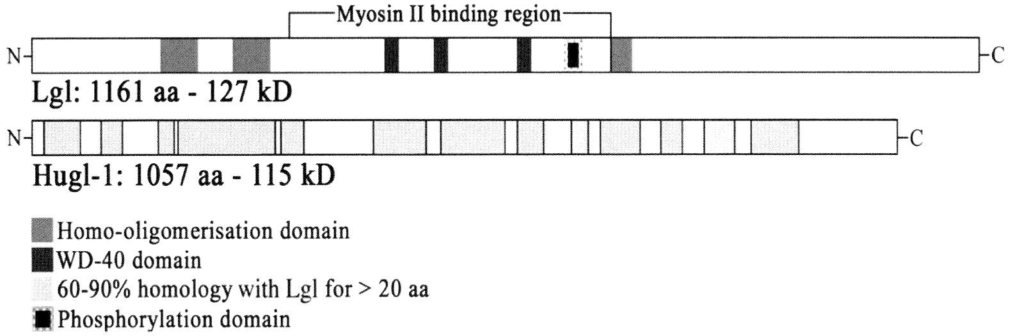

Figure 3: Structure of *Drosophila* Lgl and human Hugl-1 proteins.

In addition, we observed that a Hugl-1 cytoplasmic release, due to aPKC basolateral spreading, strictly correlates to cancer progression in ovarian carcinoma [29]. Its loss has been correlated with reduced survival in glioblastoma [42] and the occurrence of aberrant splicing variants with hepatocellular carcinoma progression [43].

Interestingly, pronounced similarities with *Drosophila lgl* mutant were also found in *Lgl* KO mice [44]. *Lgl-1* $^{-/-}$ individuals presented at birth severe brain dysplasia due to an abnormal expansion of progenitor cells, unable to exit cell cycle and to differentiate. These cells formed neuroepithelial rosette-like structures, similar to the neuroblastic rosettes found in human primitive neuroectodermal tumours (PNETs) occurring at pediatric age [44]. Noticeably, Hugl-1 gene maps to 17p11.2, a region often shown to undergo chromosomal breakage in human PNETs [35].

All these data strongly support the notion that also mammalian *lgl* homologues act as TSG so giving a major relevance to *Drosophila lgl* mutant in the study of human tumourigenesis.

LGL CLONAL PHENOTYPES

One of the most useful and broadly utilised investigation system offered by *Drosophila* genetics is undoubtedly the FLP (Flippase)/FRT (Flippase Recognition Target)-based mitotic clonal analysis technique that allows researchers to generate marked patches of homozygous mutant tissue in an otherwise heterozygous animal [45], Fig (**4a**).

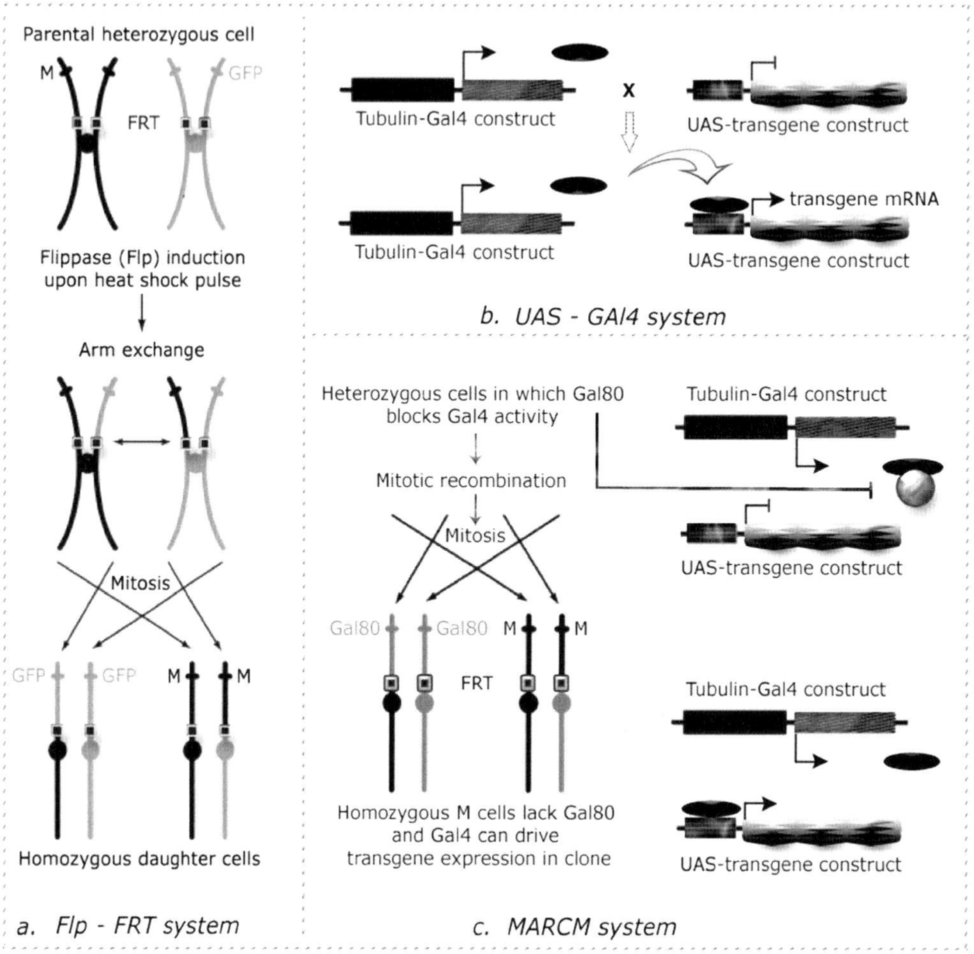

Figure 4: a: Flp-FRT system [38]: the site-specific yeast recombinase Flippase (FLP) mediates recombination at pericentromeric FRT sequences. These sequences are located at the same position on homologous chromosomes, one carrying a mutant allele and the other carrying a visible cell marker (*i.e.* GFP) on the same chromosomal arm. The consequence of the recombination event would be the formation of a mutant cell and a wild-type sibling. These cells and their progeny are clearly distinguishable as the first completely lacks the cell marker while the second is marked with twice the intensity with respect to the background. b: UAS-Gal4 system [46]: in this binary system a line contains the yeast Gal4 transactivator cDNA cloned downstream of a given promoter (in this case, *tubulin* promoter) and the second line contains the transgene of interest under the control of UAS cassettes, yeast enhancers specifically recognised by Gal4. In the progeny of the cross between those two lines, both elements are present in the cells so that the transgene will be transcribed under the control of the chosen promoter. c: MARCM system [47]: it is a combination of the Flp-FRT and UAS-Gal4 systems, with the addition of another yeast element: Gal80 repressor. Through this system, the presence of Gal80 either in one or two copies blocks Gal4 function, so that only in the homozygous mutant cell it will be possible to express any number of UAS-transgenes.

Most commonly the FLP is expressed under the control of a heat-shock promoter (hs-FLP) as it ensures the possibility to control either the developmental stage in which to induce recombination or the clone number that is related to heat pulse length; the distribution of clones is instead random throughout the animal.

Clonal analysis is an extremely versatile system in *Drosophila* and it is widely employed because pericentromeric FRT sequences have been introduced in both arms of each of the *Drosophila* chromosomes allowing to generate clones homozygous mutants for nearly every gene. This not only makes it possible to uncover additional stage-specific functions of pleiotropic genes whose mutations in the whole animal are lethal in earlier stages, but also to investigate potential cross-talking between mutant and wild type tissue and non cell-autonomous effects due to certain mutations.

Induction of mitotic clones with this technique is also extremely helpful for the study of proliferation control: it is indeed possible to precisely assess the proliferation rate/behaviour of a mutant clone comparing it to that of the wild type one, named the 'twin clone', generated in the same recombination event.

Mitotic recombination has begun revealing its potential also in mouse. For example, clones of $p53^{-/-}$ cells were generated in various organs mimicking the loss of heterozygosity (LOH) typical of patients affected by Li-Fraumeni syndrome [48]. However the system is far from being routinely applied, because technically laborious, and the current minimum level of genetic manipulation generally feasible in mouse is the whole organ through conditional knock-outs [49], confirming the pivotal role of *Drosophila* in this field.

While *lgl* mutant larvae show neoplastic overgrowth of the epithelial tissues, $lgl^{-/-}$ cells growing in mosaic larvae are usually eliminated by cell death; thus this opposite phenotype results from mechanisms triggered by the confrontation of lgl^{-} and lgl^{+} cells.

Our results on wing imaginal disc showed that *lgl* clones are invariantly smaller than wild type twins and are nearly completely disappeared by the time the adult tissue forms [50]. As discussed above, *lgl* cells are much slower growing than wild type cells [4] and it is not unlikely that *lgl* clones could fall prey of Cell Competition (CC). CC is an interesting phenomenon that has been characterised in *Drosophila* wing imaginal discs and that is involved in maintaining organ homeostasis during development [51]. Cells that for some reasons acquire the ability to proliferate faster than the surrounding ones, induce the slower growing cells to undergo apoptosis so ensuring a normal final size of the organ. CC was first described in 1975 where cells heterozygous for a *Minute* mutation were eliminated by surrounding wild type tissue [52]. The *Minute* mutations are dominant, homozygous lethal LOF mutations of different ribosomal proteins. Heterozygous *Minute* cells are viable in a homotypic context but display a slow growing phenotype.

dMyc oncoprotein, the fly unique homologue of the mammalian family of Myc factors which are often found deregulated in human cancers [53], has been shown to play a leading role in CC: cells bearing low levels of this protein die by apoptosis if growing in a high dMyc expressing context [54, 55].

There is suggestive evidence in literature regarding the relevance of CC to cancer biology [51], however a clear connection between CC and tumour development is yet to be made.

An alternative process has been shown to mediate the elimination of unfit/polarity-deficient cells from *Drosophila* imaginal epithelia: Intrinsic Tumour Suppression (ITS), which consists in the endocytic activation of Eiger, *Drosophila* TNF (Tumour Necrosis Factor) homologue, ultimately leading to the apoptotic death of the mutant cells [56].

We found that dMyc-induced CC is the main process through which *lgl* mutant clones are eliminated from imaginal wing epithelia [50]. Indeed $lgl^{-/-}$ clones expressed very low levels of dMyc protein relative to the background resulting in apoptotic death of mutant cells and compensatory proliferation of the surrounding wild-type tissue. Instead, $lgl^{-/-}$ clones in the slow-dividing and non-competitive *Minute* background were able to overgrow, upregulate dMyc and outcompete adjacent wild-type cells. Silencing dMyc inside those clones abolished tumorous growth. [50]. CC can therefore function as a tumour suppressive mechanism preserving epithelial integrity through the elimination of damaged and potentially harmful cells before they acquire additional oncogenic mutations.

Several reports have been published that confirmed that polarity-deficient cells are subject to CC, both in *Drosophila* [57-59] and mammalian epithelia [59, 60]. Indeed, inducible cell culture systems have been employed to demonstrate that such non-autonomous phenomenon is conserved. Tamori and colleagues knocked down the *Lgl* interactor Mahjong (that they proved to be involved in *lgl*-mediated CC in *Drosophila*) in Madin-Darby canine kidney epithelial cells and showed that Mahjong-deficient cells undergo apoptotic cell death only when surrounded by wild-type cells [59]. Similar results were obtained by Norman and colleagues upon *Scrib* knockdown in the same cell line: the authors demonstrated that while *Scrib*^{KD} cells were viable when surrounded by cells of the same kind, they died and were apically extruded in presence of *Scrib*^{WT} cells [60].

In addition, in a murine model of familial polyposis, *APC*$^{-}$ cells deprived of c-Myc were out-competed by surrounding, c-Myc-expressing wild-type cells, thereby reverting the malignant phenotype, suggesting that Myc-induced CC could play a role in mammalian carcinogenesis [61].

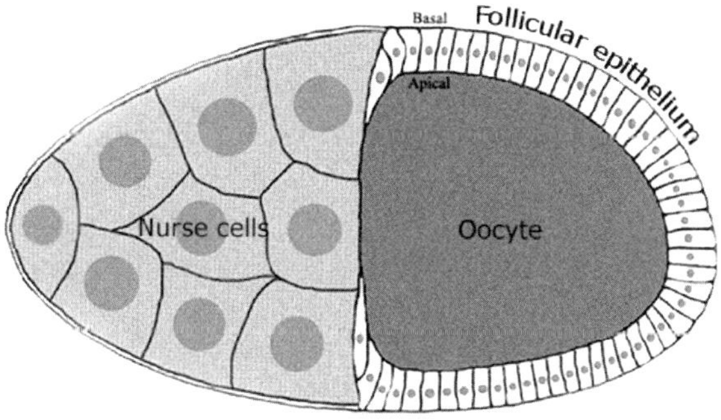

Figure 5: Scheme of a *Drosophila* egg chamber. The oocyte and the nurse cells are of germinal origin, while the follicle cells form an epithelial sheet of somatic origin coating the oocyte. The apical and basal polarity of the follicle cells is also indicated.

It is worth concluding this section discussing an adult tissue: the *Drosophila* follicular epithelium, Fig. (**5**), in which cells cease to divide during mid-oogenesis and it can be therefore compared to a non-proliferating tissue. Follicular epithelia carrying temperature-sensitive *lgl* and *dgl* alleles at non-permissive temperature displayed increased proliferation and loss of monolayered organisation with follicle cells accumulation at both poles of the egg chamber. In addition, groups of cells were seen to invade recurrently between the germ-line-derived nurse cells and the oocyte [23, 62]. Clonal induction of *lgl*, *scrib* and *dgl* LOF mutations in follicle cells also resulted in overgrowth and invasion [4, 19, 62].

lgl mutant clones, despite their slow growth, may have in this case a greater proliferative advantage since embedded in a context of a non-proliferating epithelium. Indeed, *lgl* mutant clones in this context upregulate dMyc and do not undergo cell death [50].

MODELS OF ONCOGENIC COOPERATION

An interesting improvement of the FRT/FLP based clonal analysis for its application in cancer research is the MARCM (Mosaic Analysis with a Repressible Cell Marker) technique [47]. It allows to generate GFP-marked mitotic clones of homozygous mutant tissue in which to express any transgene of interest, Fig (**4c**). It is thus possible to create groups of cells, where LOF mutations of TSG and gain-of-function mutations of proto-oncogenes are simultaneously present, surrounded by wild-type tissue, clearly mimicking very closely mammalian cancer onset.

This sophisticated genetic tool has enabled researchers to find out in *lgl, scrib* and *dgl* mutants a strong cooperation between different genetic lesions in tumour growth and metastatic activity [63, 64].

An *eyeless*-FLP (*ey*-FLP) construct was employed in order to confine clone induction to a specific territory: *ey* promoter is active throughout embryonic and larval development in the eye imaginal discs and optic lobes (those areas of the larval brain that will be dedicated to vision in the adult) [65].

scrib clones within an otherwise wild-type eye disc showed loss of cell polarity and overproliferation, however very little mutant tissue was recovered in the adult organ implying its obliteration during metamorphosis. But, when a second oncogenic mutation, such as a constitutively activated form of the onco-proteins Ras (Ras^{ACT}) or Notch ($Notch^{ACT}$), was introduced in the *scrib* background, a dramatic overgrowth was observed. The clonal tissue overwhelmed the wild-type tissue fusing with the nearby imaginal discs and brain and formed totally undifferentiated masses [63].

The same strong cooperation was also demonstrated for the metastatic phenotype [64]. In this case the MARCM technique was employed to develop a very useful *in vivo* assay: taking advantage of the possibility to generate GFP-marked eye disc clones, the authors were able to follow tumorous cells migration in the whole animal throughout development. Clones expressing Ras^{ACT} alone overgrew to form non-invasive tumours. However, when Ras^{ACT} expression was combined with *scrib, dlg* or *lgl* null mutations, these tumours became highly metastatic. They displayed a consistent invasion of the ventral nerve cord, an area of the larval brain not immediately adjacent to the territories of *eyeless* promoter expression. Cancer cells were also found to spread to other tissues. Very similarly to what is known to happen for human carcinomas, active basement membrane degradation was observed and tumorous cells also showed a lowered expression of E-cadherin in association to metastatic behaviour [64]. In both cases, however, the simultaneous expression of cell-cycle promoters and apoptosis inhibitors in *scrib* clones, with the intention to reproduce the possible consequences of Ras activation, failed to give rise to the neoplastic phenotype [63, 64]. This clearly implicates the requirement of specific targets downstream of Ras activation.

These papers were convincing evidence that for a tumour to arise, in *Drosophila* as in humans, simultaneous mutations of tumour suppressors and proto-oncogenes are required [63, 64].

Cell polarity-deficient cells grown in a clonal context are eliminated by non-autonomous JNK-induced cell death [50, 56, 63]. JNK signal transduction pathway is an evolutionarily conserved stress-induced cascade that has roles in multiple cellular processes including proliferation, differentiation, morphogenesis, and most of all, apoptosis, and has been implicated in several aspects of tumour development. A few papers were published in these latter years which demonstrated that JNK signalling activation due to polarity genes loss switches from pro-apoptotic, and thus tumour suppressive, to tumour promoting in presence of oncogenic Ras signalling [66, 67].

Several lines of evidence seem to indicate that JNK signalling in the context of oncogenic cooperation between cell polarity loss and activated Ras promotes invasion by upregulating MMP1 expression and thus basement membrane degradation: inhibition of MMP1 activity either by genetic inactivation or by expression of MMP1 inhibitors also abolished invasiveness of mutant clones [68, 69].

The molecular events leading to transformation in these oncogenic cooperation models have been extensively analysed in recent years and several other pathways, also implicated in human tumourigenesis, have proved to contribute to the acquisition of malignant traits by mutant cells [57, 70, 71].

CONCLUSIONS

Cell cultures and mammalian models have undoubtedly provided a large amount of information on how mutations affecting various aspects of cell physiology, such us proliferation, apoptosis, differentiation and migration can eventually lead to cancer emergence and metastasis. It is however becoming increasingly clear that the understanding of these complex processes unavoidably needs to go through the comprehension of the cross-talks between cancer cells and the environment in which they develop.

In this review we presented *Drosophila lethal(2)giant larvae* mutant as a powerful *in vivo* system where many of the aspects of carcinogenesis, including those of tumour-microenvironment interactions, can be fruitfully recapitulated.

All the major pathways involved in tumourigenesis are extremely conserved between flies and humans [72] and the employment of an easily manipulatable *in vivo* model is therefore a golden opportunity for cancer research.

CONFLICT OF INTEREST

The author(s) confirm that this chapter content has no conflict of interest.

ACKNOWLEDGEMENT

This work was supported by grants from the CaRisBO and AIRC (IG-12093) Foundations.

DISCLOSURE

"This is an updated report of the earlier review article By Francesca Froldi, Marcello Ziosi, Annalisa Pession and Daniela Grifoni "*Drosophila lethal giant larvae* neoplastic mutant as a genetic tool for cancer modeling". Current Genomics, 2008, 9, 147-154.

ABBREVIATIONS

TSG	=	Tumour Suppressor Genes
LOF	=	Loss Of Function
lgl	=	*lethal giant larvae*
scrib	=	*scribble*
dlg	=	*discs large*
AEL	=	After Egg Laying
MMP-1	=	Matrix Metallo Proteinase-1
EMT	=	Epithelial-to-Mesenchimal Transition
aPKC	=	atypical Protein Kinase C
Hpo	=	Hippo
Yki	=	Yorkie
dIAP1	=	*drosophila* Inhibitor of Apoptosis 1
Hugl-1/2	=	Human giant larvae 1/2
PNETs	=	Primitive Neuro-Ectodermal Tumours
Flp	=	Flippase

FRT = Flippase Recognition Targets

LOH = Loss Of Heterozygosity

UAS = Upstream Activating Sequences

GFP = Green Fluorescent Protein

CC = Cell Competition

ITS = Intrinsic Tumour Suppression

MARCM = Mitotic Analysis with a Repressible Cell Marker

JNK = c-Jun N-terminal Kinase

REFERENCES

[1] Hanahan, D.; Weiberg, R. A. The hallmarks of cancer. *Cell*, **2000**, *100*, 57-70.

[2] Mareel, M.; Leroy, A. Clinical, cellular and molecular aspects of cancer invasion. *Phys. Rev.*, **2003**, *83*, 337-376.

[3] Bissel, M.J.; Radisky, D. Putting tumours in context. *Nat. Rev. Cancer*, **2001**, *1*, 46-54.

[4] Bilder, D. Epithelial polarity and proliferation control: links from the *Drosophila* neoplastic tumor suppressors. *Genes Dev.*, **2004**, *18*, 1909-1925.

[5] Humbert, P.; Russel, S.; Richardson, H. Dgl, Scribble and Lgl in cell polarity, cell proliferation and cancer. *Bioessays*, **2003**, *25*, 542-553.

[6] Weinkove, D.; Leevers, S. J. The genetic control of organ growth: insights from *Drosophila*. *Curr. Opin. Genet. Dev.*, **2000**, *10*, 75-80.

[7] Bridges, C. B.; Brehme, K. S. *The mutants of Drosophila melanogaster*. Carnegie Institute of Washington, Washington DC, **1944**.

[8] Gateff, E. The genetics and epigenetics of neoplasms in *Drosophila*. *Biol. Rev.*, **1978**, *53*, 123-168.

[9] Agrawal, N.; Kango, M.; Mishra, A.; Sinha, P. Neoplastic transformation and aberrant cell-cell interactions in genetic mosaics of *lethal(2)giant larvae* (*lgl*), a tumor suppressor gene of *Drosophila*. *Dev. Biol.*, **1995**, *172*, 218-229.

[10] Garcia-Bellido, A.; Ripoll, P.; Morata, G. Developmental compartimentalisation of the wing disk of *Drosophila*. *Nat. New. Biol.*, **1973**, *245*, 251-253.

[11] Woodhouse, E.; Hersperger, E.; Stetler-Stevenson, W. G.; Liotta, L. A.; Shearn, A. Increased type IV collagenase in lgl-induced invasive tumours of *Drosophila*. *Cell Growth Differ.*, **1994**, *5*, 151-159.

[12] Beaucher, M.; Hersperger, E.; Page-McCaw, A.; Shearn, A. Metastatic ability of *Drosophila* tumors depends on MMP activity. *Dev. Biol.*, **2007**, *303*, 625-634.

[13] Woodhouse, E.; Hersperger, E; Shearn, A. Growth, metastasis, and invasiveness of *Drosophila* tumors caused by mutations in specific tumor suppressor genes. *Dev. Genes Evol.*, **1998**, *207*, 542-550.

[14] Beaucher, M.; Goodliffe, J.; Hersperger, E.; Trunova, S.; Frydman, H.; Shearn, A. *Drosophila* brain tumor metastases express both neuronal and glial cell type markers. *Dev. Biol.*, **2007**, *301*, 287-297.

[15] Nelson, J. W. Adaptation of core mechanisms to generate cell polarity. *Nature,* **2003**, *422*, 766-774.

[16] Pilot, F.; Lecuit, T. Compartmentalized morphogenesis in epithelia: from cell to tissue shape. *Dev. Dyn.,* **2005**, *232*, 685-694.

[17] Huber, M. A.; Kraut, N.; Beug, H. Molecular requirements for epithelial-mesenchymal transition during tumor progression. *Curr. Opin. Cell Biol.,* **2005**, *17*, 548-558.

[18] Guarino, M. Epithelial-mesenchymal transition and tumour invasion. *Int. J. Biochem. Cell. Biol.,* **2007**, *39*, 2153-2160.

[19] Bilder, D.; Li M.; Perrimon, N. Cooperative regulation of cell polarity and growth by *Drosophila* tumor suppressors. *Science,* **2000**, *289*, 113-116.

[20] Betschinger, J.; Eisenhaber, F; Knoblich, J. A. Phosphorylation-induced autoinhibition regulates the cytoskeletal protein Lethal (2) giant larvae. *Curr. Biol.,* **2005**, *15*, 276-282.

[21] Strand, D.; Jakobs, R.; Merdes, G.; Neumann, B.; Kalmes, A.; Heid, H. W.; Husmann, I.; Mechler, B. M. The *Drosophila lethal(2)giant larvae* tumor suppressor protein forms homo-oligomers and is associated with nonmuscle myosin II heavy chain. *J. Cell Biol.,* **1994**, *127,* 1361-1373.

[22] Mechler, B. M.; McGinnis, W.; Gehring, W. J. Molecular cloning of *lethal(2)giant larvae*, a recessive oncogene of *Drosophila melanogaster. EMBO J.,* **1985**, *4*, 1551-1557.

[23] Manfruelli, P.; Arquier, N.; Hanratty, W. P.; Sémériva, M. The tumor suppressor gene, *lethal(2)giant larvae* (*1(2)gl*), is required for cell shape change of epithelial cells during *Drosophila* development. *Development,* **1996**, *122*, 2283-2294.

[24] Dow, L.; Humbert, P.O. Polarity regulators and the control of epithelial architecture, cell migration, and tumorigenesis. *Int. Rev. Cytol.,* **2007**, *262*, 253-302.

[25] Betschinger, J.; Mechtler, K.; Knoblich, J. A. The Par complex directs asymmetric cell division by phosphorylating the cytoscheletal protein Lgl. *Nature,* **2003**, *422*, 326-330.

[26] Albertson, R.; Doe, C. Q. Dgl, Scrib and Lgl regulate neuroblast cell size and mitotic spindle asymmetry. *Nat. Cell Biol.,* **2003**, *5*, 166-170.

[27] Rolls, M. M.; Albertson, R.; Shih, H. P.; Lee, C. Y.; Doe, C. Q. *Drosophila* aPKC regulates cell polarity and cell proliferation in neuroblasts and epithelia. *J. Cell Biol.,* **2003**, *163*, 1089-1098.

[28] Lee, C.Y.; Robinson, K. J.; Doe, C. Q. Lgl, Pins and aPKC regulate neuroblast self-renewal versus differentiation. *Nature,* **2006**, *439*, 594-598.

[29] Grifoni, D.; Garoia, F.; Bellosta, P.; De Biase, D.; Collina, G.; Strand, D.; Cavicchi, S.; Pession, A. aPKCζ cortical loading is associated with Lgl cytoplasmic release and tumor growth in *Drosophila* and human epithelia. *Oncogene,* **2007**, *26*, 5960-5965.

[30] Brumby, A.; Secombe, J.; Horsfield, J.; Coombe, M.; Amin, N., Coates, D.; Saint, R.; Richardson, H. A genetic screen for dominant modifiers of a cyclin E hypomorphic mutation identifies novel regulators of S-phase entry in *Drosophila. Genetics,* **2004**, *168*, 227-251.

[31] Grzeschik, N. A.; Amin, N.; Secombe, J.; Brumby, A. M.; Richardson, H. E. Abnormalities in cell proliferation and apico-basal cell polarity are separable in *Drosophila lgl* mutant clones in the developing eye. *Dev. Biol.,* **2007**, *311*, 106-123.

[32] Grzeschik, N. A.; Parsons, L.; Allot, M.; Harvey, K. F.; Richardson H. E. Lgl, aPKC and Crumbs regulate the Salvador/Warts/Hippo pathway through two distinct mechanisms. *Curr. Biol.,* **2010**, *20*, 573-581.

[33] Halder, G.; Johnson, R. L. Hippo signaling: growth control and beyond. *Development,* **2011**, *130*, 9-22.

[34] Sun, G.; Irvine, K. Regulation of Hippo signaling by Jun kinase signaling during compensatory cell proliferation and regeneration, and in neoplastic tumors. *Dev. Biol.,* **2011**, *350*, 139-151.

[35] Strand, D.; Unger, S.; Corvi, R.; Hartenstein, K.; Schenkel, H.; Kalmes, A.; Merdes, G.; Neumann, B.; Krieg-Schneider, F.; Coy, J. F. *et al.* A human homologue of the *Drosophila* tumour suppressor gene *l(2)gl* maps to 17p11.2-12 and codes for a cytoskeletal protein that associates with nonmuscle myosin II heavy chain. *Oncogene,* **1995**, *11*, 291-301.

[36] Zimmerman, T.; Kashyap, A.; Hartmann, U.; Otto, G.; Galle, P. R.; Strand, S.; Strand D. Cloning and characterization of the promoter of Hugl-2, the human homologue of *Drosophila lethal giant larvae*, (*lgl*) polarity gene. *Biochem. Biophys. Res. Commun.*, **2008**, *366*, 1067-1073.

[37] Grifoni, D.; Garoia, F.; Schimanski, C. C.; Schmitz, G.; Laurenti, E.; Galle, P. R.; Pession, A.; Cavicchi, S.; Strand, D. The human protein Hugl-1 substitutes for *Drosophila* Lethal giant larvae tumour suppressor function *in vivo*. *Oncogene,* **2004**, *23*, 8688-8694.

[38] Thomas, U.; Phannavong, B.; Muller, B.; Garner, C. C.; Gundelfinger, E.D. Functional expression of rat synapse-associated proteins SAP97 and SAP102 in *Drosophila* dgl-1 mutants: effects on tumor suppression and synaptic bouton structure. *Mech. Dev.*, **1997**, *62*, 161-174.

[39] Dow, L. E.; Brumby, A. M.; Muratore, R.; Coombe M. L.; Sedelies, K. A.; Trapani, J. A.; Russell, S. M.; Richardson, H. E.; Humbert, P. O. hScrib is a functional homologue of the *Drosophila* tumour suppressor Scribble. *Oncogene,* **2003**, *22*, 9225-9230.

[40] Kuphal, S.; Wallner, S.; Schimanski, C. C.; Bataille, F.; Hofer, P.; Strand, S.; Strand, D.; Bosserhoff, A. K. Expression of Hugl-1 is strongly reduced in malignant melanoma. *Oncogene,* **2006**, *25*, 103-110.

[41] Schimanski, C. C.; Schmitz, G.; Kashyap, A.; Bosserhoff, A. K.; Bataille, F.; Schäfer, S. C.; Lehr, H. A.; Berger, M. R.; Galle, P. R.; Strand, S.; Strand, D. Reduced expression oh Hugl-1, the human homologue of *Drosophila* tumour suppressor gene *lgl*, contributes to progression of colorectal cancer. *Oncogene,* **2005**, *24*, 3100-3109.

[42] Korshunov, A.; Sycheva, R.; Golanov, A. Genetically distinct and clinically relevant subtypes of glioblastoma defined by array-based comparative genomic hybridization (array-CGH). *Acta Neuropathol.*, **2006**, *111*, 465-474.

[43] Lu, X.; Feng, X.; Man, X.; Yang, G.; Tang, L.; Du, D.; Zhang, F.; Yuan, H.; Huang, Q.; Zhang, Z.; Liu, Y.; Strand, D.; Chen, Z. Aberrant splicing of Hugl-1 is associated with epatocellular carcinoma progression. *Clin. Cancer Res.*, **2009**, *15*, 3287-3296.

[44] Klezovitch, O.; Fernandez, T. E.; Tapscott, S. J.; Vasioukhin, V. Loss of cell polarity causes severe brain dysplasia in *Lgl1* knockout mice. *Genes Dev.*, **2004**, *18*, 559-571.

[45] Golic, K. G. Site-specific recombination between homologous chromosomes in *Drosophila. Science,* **1991**, *252*, 958-961.

[46] Brand, H. A.; Perrimon, N. Targeted gene expression as a means of altering cell fates and generating dominant phenotypes. *Development,* **1993**, *118*, 401-415.

[47] Lee, T.; Luo, L. Mosaic analysis with a repressible cell marker (MARCM) for *Drosophila* neural development. *TRENDS Neurosci,*. **2001**, *24*, 251-254.

[48] Wang, W.; Warren, M.; Bradley, A. Induced mitotic recombination of *p53 in vivo*. *PNAS,* **2007**, *104*, 4501-4505.

[49] Giuriato, S.; Rabin, K.; Fan, A. C.; Shachaf, C. M.; Felsher, D. W. Conditional animal models: a strategy to define when oncogenes will be effective targets to treat cancer. *Semin Cancer Biol.*, **2004**, *14*, 3-11.

[50] Froldi, F.; Ziosi, M.; Garoia, F.; Pession, A.; Grzeschik, N. A.; Bellosta, P.; Strand, D.; Richardson, H. E.; Pession, A.; Grifoni, D. The *lethal giant larvae* tumour suppressor mutation requires dMyc oncoprotein to promote clonal malignancy. *BMC Biol.*, **2010**, *8*, 33.

[51] Moreno, E. Is cell competition relevant to cancer? *Nat. Rev. Cancer,* **2008**, *8*, 141-147.

[52] Morata, G.; Ripoll, P. Minutes: mutants of *Drosophila* autonomously affecting cell division rate. *Dev. Biol.,* **1975**, *42*, 211-221.
[53] Meyer, N.; Penn, L. Z. Reflecting on 25 years with MYC. *Nat. Rev. Cancer,* **2008**, *8*, 976-990.
[54] de la Cova, C.; Abril, M.; Bellosta, P.; Gallant, P.; Johnston, L. A. Drosophila Myc regulates organ size by inducing cell competition. *Cell,* **2004**, *117*, 107-116.
[55] Moreno, E.; Basler, K. dMyc transforms cells into super-competitors. *Cell,* **2004**, *117*, 117-129.
[56] Igaki, T.; Pastor-Pareja, J. C.; Aonuma, H.; Miura, M.; Xu, T. Intrinsic Tumor Suppression and Epithelial Maintenance by Endocytic Activation of Eiger/TNF signaling in *Drosophila. Dev. Cell,* **2009**, *16*, 458-465.
[57] Menéndez, J.; Pérez-Garijo, A.; Calleja, M.; Morata, G. A tumor-suppressing mechanism in *Drosophila* involving cell competition and the Hippo pathway. *PNAS,* **2010**, *107*, 14651-14656.
[58] Chen, C.; Schroeder, M. C.; Kango-Singh, M.; Tao, C.; Halder, G. Tumor suppression by cell competition through regulation of the Hippo pathway. *PNAS,* **2011**, *109*, 484-489.
[59] Tamori, Y.; Bialucha, C. U.; Tian, A. G.; Kajita, M.; Huang, Y. C.; Norman, M.; Poulton, J.; Ivanovitch, K.,;Disch, L.; Liu, T.; Deng, W. M.; Fujita, Y. Involvement of Lgl and Mahjong/VprBP in cell competition. *PLoS Biol.*, **2010**, *8*, e1000422.
[60] Norman, M.; Wisniewska, K. A.; Lawrenson, K.; Garcia-Miranda, P.; Tada, M.; Kajita, M.; Mano, H.; Ishikawa, S.; Ikegawa, M.; Shimada, T.; Fujita, Y. Loss of Scribble causes cell competition in mammalian cells. *J. Cell Sci.*, **2012**, *125*, 59-66.
[61] Sansom, O. J.; Meniel, V. S.; Muncan, V.; Phesse, T. J.; Wilkins, J. A.; Reed, K. R.; Vass, J. K.; Athineos, D.; Clevers, H.; Clarke, A. R. Myc deletion rescues Apc deficiency in the small intestine. *Nature,* **2007**, *446*, 676-679.
[62] Goode, S.; Perrimon, N. Inhibition of patterned cell shape change and cell invasion by Discs large during *Drosophila* oogenesis. *Genes Dev.,* **1997**, *11*, 2532-2544.
[63] Brumby, A. M.; Richardson, H. E. *scribble* mutants cooperate with oncogenic Ras or Notch to cause neoplastic overgrowth in *Drosophila. EMBO J.,* **2003**, *22*, 5769-5779.
[64] Pagliarini, R. A.; Xu, T. A genetic screen in *Drosophila* for metastatic behaviour. *Science,* **2003**, *302*, 1227-1231.
[65] Newsome, T. P.; Asling, B.; Dickson B. J. Analysis of *Drosophila* photoreceptor axon guidance in eye-specific mosaics. *Development,* **2000**, *127*, 851-860.
[66] Igaki, T.; Pagliarini, R. A.; Xu, T. Loss of Cell Polarity Drives Tumor Growth and Invasion through JNK Activation in *Drosophila. Curr. Biol.*, **2006**, *16*, 1139-1146.
[67] Leong, G. R.; Goulding, K. R.; Amin, N.; Richardson, H. E; Brumby, A. M. Scribble mutants promote aPKC and JNK-dependent epithelial neoplasia independently of Crumbs. *BMC Biol.*, **2009**, *7*, 62.
[68] Uhlirova, M.; Bohmann, D. JNK- and Fos-regulated Mmp1 expression cooperates with Ras to induce invasive tumors in *Drosophila. EMBO J.*, **2006**, *25*, 5294-5304.
[69] Srivastava, A.; Pastor-Pareja, J. C; Igaki, T.; Pagliarini, R.; Xu, T. Basement membrane remodelling is essential for *Drosophila* disc eversion and tumor invasion. *PNAS,* **2007**, *104*, 2721-2726.
[70] Wu, M.; Pastor-Pareja, J. C.; Xu, T. Interaction between RasV12 and scribbled clones induces tumour growth and invasion. *Nature,* **2010**, *463*, 545-549.
[71] Willecke, M.; Toggweiler, J.; Basler, K. Loss of PI3K blocks cell-cycle progression in a *Drosophila* tumor model. *Oncogene,* **2011**, *30*, 4067-4074.
[72] Brumby, A. M. Richardson, H.E. Using *Drosophila melanogaster* to map human cancer pathways. *Nat. Rev. Cancer,* **2005**, *5*, 626-639.

Send Orders of Reprints at reprints@benthamscince.net
 Advances in Genome Science, Vol. 1, 2013, 210-240

CHAPTER 9

Genetic Insights Through Genome Wide Association Studies in Type 2 Diabetes Mellitus will Lead to New Therapeutics

Marcel G. M. Wolfs[1,‡], Naishi Li[1,4,‡], Jingyuan Fu[2], Cisca Wijmenga[2], Timon W. van Haeften[1,3], Marten H. Hofker[1,*]

[1]Department of Pathology and Medical Biology, Medical Biology Section, Molecular Genetics and [2]Department of Genetics, University Medical Center Groningen, University of Groningen, PO Box 30001, 9700 RB Groningen, The Netherlands; [3]Department of Internal Medicine, G 02-228, University Medical Center Utrecht, PO Box 85500, 3508 GA Utrecht, The Netherlands; [4]Department of Endocrinology, Key Laboratory of Endocrinology of Ministry of Health, Peking Union Medical College Hospital, Peking Union Medical College, Chinese Academy of Medical Science, 100730, Beijing, China

Abstract: Type 2 diabetes is a disorder of dysregulated glucose homeostasis. Normal glucose homeostasis is a complex process involving several interacting mechanisms, such as insulin secretion, insulin sensitivity, glucose production, and glucose uptake. The dysregulation of one or more of these mechanisms due to environmental and/or genetic factors, can lead to a defective glucose homeostasis. Hyperglycemia is managed by augmenting insulin secretion and/or interaction with hepatic glucose production, as well as by decreasing dietary caloric intake and raising glucose metabolism through exercise. Although these interventions can delay disease progression and correct blood glucose levels, they are not able to cure the disease or stop its progression entirely. Better management of type 2 diabetes is sorely needed. Advances in genotyping techniques and the availability of large patient cohorts have made it possible to identify common genetic variants associated with type 2 diabetes through genome-wide association studies (GWAS). So far, genetic variants on 50 loci have been identified. Most of these loci contain or lie close to genes that were not previously linked to diabetes and they may thus harbor targets for new drugs. It is also hoped that further genetic studies will pave the way for predictive genetic screening. The newly discovered candidate genes for type 2 diabetes can be classified based on their presumed molecular function, and we discuss the relation between these gene classes and current treatments. We go on to consider whether the new genes provide opportunities for developing alternative drug therapies.

Keywords: Type 2 diabetes, drug targets, genetics, personalized medicine.

[*]**Address correspondence to Marten H. Hofker:** Department of Pathology and Medical Biology, Medical Biology Section, Molecular Genetics, University Medical Center Groningen, University of Groningen, Hanzeplein 1, 9713 GZ Groningen, The Netherlands; Tel: +31-(0)50-363 5777; Fax: +31-(0)50-363 8971; E-mail: M.H.Hofker01@umcg.nl
‡These authors contributed equally.

INTRODUCTION

The last few decades have witnessed a dramatic increase in the prevalence of type 2 diabetes mellitus, due to changes in food intake combined with less physical exercise, a lifestyle often referred to as Western. The long-term consequences of type 2 diabetes are severe and include cardiovascular disease, retinopathy, neuropathy, nephropathy, and diabetic foot disease. While it has been estimated that worldwide around one billion people are obese, over 180 million people suffer from type 2 diabetes and this number is expected to double over the next 25 years. The number of annual deaths due to type 2 diabetes is difficult to estimate because they are often hidden under cardiovascular disease, but the WHO has estimated it to be between 1 and 3 million in 2006. However, since mortality from this disease often occurs many years after its onset, even 3 million deaths is probably an underestimate of the death toll in the near future. Although external factors such as food-intake-related obesity have attracted much attention, the genetic predisposition for diabetes is also important. While the life-time risk for type 2 diabetes in the Western world is around 10%, first-degree relatives of patients have a 20–40% risk for the disease, and concordance rates for identical twins have been estimated to be 57% or higher (up to 90%) for type 2 diabetes in male twins [1]. The endeavor to find the underlying genes was unsuccessful for many years although hundreds of genetic associations have been described, based largely on candidate genes. Unfortunately, replication was only possible for a very few variants. However, the recent advent of genome-wide association studies (GWAS) holds promise, since such studies have now uncovered a number of common genetic variants related to diabetes for which replication is also possible. These genes had not previously been linked to diabetes and these loci are therefore expected to lead to new insights into the disease mechanisms.

GLUCOSE HOMEOSTASIS AND DIABETES

Diabetes is a state of persistent hyperglycemia, leading to irreversible damage in a number of tissues, especially the retina, the kidney glomeruli, neural tissue and blood vessels. Normally plasma glucose levels are kept within a narrow range, a process referred to as glucose homeostasis. This homeostasis is regulated in a complex way and involves several axes.

After eating a meal, a person's plasma glucose rises temporarily, after which glucose is rapidly taken up by liver and muscle and to a lesser degree by fat tissue, due to the effects of insulin released from the pancreatic β-cells. Under these circumstances, glucose is converted into glycogen, which is then stored in the liver. When plasma glucose levels have dropped, insulin secretion gradually diminishes. In the post-prandial state, a minimum level of plasma glucose is assured by glucose production in the liver, partly due to release from glycogen and partly due to new formation from precursors such as lactate and the amino acids alanine and glutamine. Insulin has a number of other effects, including lipid storage in adipocytes, and it even increases DNA replication and protein synthesis. The regulation of insulin release is complex; apart from the prevailing plasma glucose level, release from various peptides from the intestine (*e.g.*, glucagon-like peptide 1, GLP1) after a meal and feedback mechanisms ensure adequate insulin release.

Type 2 diabetes is characterized by the combination of disturbances in insulin secretion, and an impairment of the effects of insulin, so-called insulin resistance, which is often related to obesity. Insulin resistance is caused by defects in the signaling pathways that process the insulin signal in its target tissues. It has often been stated that the disease starts with insulin resistance accompanied by raised insulin and glucose levels. At a later stage, we presume β-cells undergo further damage and apoptosis, and are not able to keep up with the demand for insulin release, which then results in higher glucose levels. At least one of the pathways involved in insulin secretion, β-cell regeneration, β-cell survival, or β-cell development are important in determining the vulnerability of the β-cell pool in insulin-resistant conditions. Studies have shown decreases in β-cell function prior to the augmentation of plasma glucose in normal glucose-tolerant first-degree relatives of type 2 diabetes subjects; in these studies these relatives had normal insulin sensitivity [2, 3]. While obesity is a major risk factor for diabetes – around 50% of obese subjects will develop type 2 diabetes at some stage (depending on the age when they became obesity) –, it should be noted that 20% of type 2 diabetes subjects are not obese. It thus seems that obesity is a major risk factor for developing type 2 diabetes, but that it is the vulnerability of the β-cell pool which determines whether obesity in fact triggers type 2 diabetes.

GENETICS OF TYPE 2 DIABETES

Table 1: Genes Associated with Diabetes: Overview of Their loci, Risk SNPs, Candidate genes, Expressed Tissue, Functions, and Related Medication

Chr	Locus and Risk SNPs	Candidate Gene for type 2 diabetes	Expressed tissue (up tissue database)	Proposed function(s) for gene product	Drug(s) affecting the same pathway as the diabetes gene
1	1p12 rs10923931-T	NOTCH2	Brain, Epithelium, Mammary tumor, Placenta, T-cell, Testis	Growth and development; transcription factor; membrane receptor	Unknown
		ADAM30	Testis	Membrane-anchored protein	
2	2p16.1 rs243021-A	BCL11A	Brain, Embryo, Fetal brain, Lymph, Tonsil	Unknown	Unknown
	2p21 rs7578597-T	THADA	Amygdala, Brain, Pancreas, Skin, Testis, Uterus	Apoptosis	Unknown
	2p25 rs35927125-G	KLF11†	Brain, Erythroid cell, Hepatoma, Pancreas,Placenta	regulation of Pdx-1 transcription to influence beta cells activity	Unknown
	2q23.3 rs7560163-C	RBM43	Tongue	Unknown	Unknown
		RND3	Fetal brain, Kidney, Placenta, Spleen	Binding GTP but has no GTPase activity	
	2q24.2 rs7593730-?	RBMS1	Muscle, Placenta	Unknown	Unknown
		ITGB6	Keratinocyte, Pancreas, Skeletal muscle	Unknown	
	2q24.3 rs3923113-A	GRB14	Skin	Interacting with insulin receptors to produce an inhibitory effect on insulin receptor signaling	Unknown
	2q32	NEUROD1‡	Eye, Retina, Rhabdomyosarcoma	Growth and development; transcription factor that activates several genes including insulin and is important for early β-cell development	Unknown
	2q36.3 rs7578326-A	IRS1	Epithelium, Eye, Skeletal muscle	Being phosphorylated by insulin receptor tyrosine kinase, an important component in insulin signaling pathway	Unknown
3	3p14.1 rs831571-C	PSMD6	Bone marrow, Lymph, Mammary cancer, Ovary, Placenta, Testis	Unknown	Unknown
	3p14.1 rs4607103-C	ADAMTS9	Brain, Fetus	Cleavage of proteoglycans	Unknown

Table 1: cont…

3	3q27.2 rs1470579-C	IGF2BP2	Colon adenocarcinoma, Pancreas, Pancreatic islet	Growth and development	Unknown
	3q27.3 rs16861329-G	ST6GAL1	Liver, Lymph, Placenta, Skin, Spleen, Thymus	Catalyze the transfer of sialic acid from CMP-sialic acid to galactose-containing substrates normally found in the Golgi	Unknown
	3p25 rs1801282	PPARG	Adipose, Adipose tissue, Bone marrow, Colon carcinoma, Heart, Placenta	Nuclear receptor (transcription factor) that regulates adipocyte differentiation	Thiazolidinediones
4	4p16.1 rs1801214-T	WFS1†	Amygdala, Brain	Apoptosis; Endoplasmic Reticulum stress pathway activation	Unknown
	4p16.3 rs6815464-C	MAEA	Embryo, Lung, Spleen, Thymus	Mediating the attachment of erythroblasts to macrophages	Unknown
5	5q13.3 rs4457053-G	ZBED3	Skin, Spleen	Unknown	Unknown
6	6p21.2 rs9470794-C	ZFAND3	Pancreas, PCR rescued clones	Unknown	Unknown
	6p21.2 rs1535500-T	KCNK16	Kidney	Potassium channel proteins	Sulfonylurea derivatives
	6p22.3 rs10440833-A	CDKAL1	Placenta,Testis,Trachea	Growth and development/Proinsulin to insulin conversion	Unknown
	6q13 rs1048886-G	C6orf57	Whole body	Unknown	Unknown
7	7p15.1 rs849134-A	JAZF1	Amygdala, Brain, Testis	Cell cycle regulation; transcriptional repressor	Unknown
	7p15.3-p15.1	GCK‡	Lung, Pancreas, Placenta	Catalyzes reaction from glucose to glucose-6-phosphate	Unknown
	7q32.1 rs6467136-G	GCC1	Brain, Cervix carcinoma, Hepatoma, Skin, Uterus	Peripheral membrane protein	Unknown
		PAX4†	Colon, Insulinoma, PCR rescued clones, Placenta	Pancreatic islet development	
	7q32.3 rs972283-G	KLF14	Epithelium	Part of a transcriptional co-repressor complex	Unknown
8	8q22.1 rs896854-T	TP53INP1	Lung, Thymus	interact physically with p53 and regulatep53 transcriptional activity	Unknown
	8q24.11 rs3802177-G	SLC30A8	Amygdala, Brain, Lung, Pancreas	B-cell ion homeostasis and insulin secretion; cellular efflux of Zn^{2+} ions/Proinsulin to insulin conversion	Sulfonylurea derivatives

Table 1: cont…

9	9p21.3 rs10965250-G	CDKN2A-2B	Hematopoietic, Skin, Thyroid carcinoma/Heart, Skin	Cell cycle regulation	Unknown
	9p24.1 rs17584499-T	PTPRD	Brain, PCR rescued clones, Placenta	A receptor-type of the protein tyrosine phosphatase (PTP) family associated with insulin signaling	Unknown
	9p24.2 rs7041847-A	GLIS3†	Brain,Testis	Transcription factor, regulation of the development of pancreaticβ-cells and insulin gene expression	Unknown
	9q21.31 rs13292136-C	CHCHD9 (CHCHD2P9)	Whole body	Unknown	Unknown
	9q34.3	CEL‡	Colon, Mammary gland, Milk, Pancreas	Glycoprotein that is important in regulation of cholesterol metabolism	Unknown
10	10p13 rs12779790-G	CDC123	Bone marrow, Fibrosarcoma, Foreskin, Lymph	Cell cycle regulation	Unknown
		CAMK1D	Pancreas	Regulation of granulocyte function	
	10q22.1 rs1802295-A	VPS26A	Brain, Colon, Umbilical cord blood	A component of the retromer complex, involved in retrograde transport of proteins from endosomes to the trans-Golgi network	Unknown
	10q23.33 rs5015480-C	HHEX	Bone marrow, Breast, Leukemia, Myeloid leukemia cell, Peripheral blood monocyte, Uterus	Growth and development; transcription factor	Unknown
		IDE	Brain, Testis	Termination of the response to insulin	
	10q25.2 rs7903146-T	TCF7L2	Fetus, Gastric carcinoma, Uterus	Wnt signaling/Proinsulin to insulin conversion	Unknown
11	11p15.1 rs5215-C	KCNJ11†	Brain, Breast, Ovary, Placenta, Spleen	B-cell ion homeostasis and insulin secretion	Sulfonylurea derivatives
	11p15.1	ABCC8‡	Brain, Foreskin, Pancreas, Pancreatic islet	B-cell ion homeostasis and insulin secretion; ATP-binding cassette transporter that modulates ATP-sensitive potassium channels and insulin release	Sulfonylurea derivatives
	11p15.5 rs231362-G	KCNQ1	Colon, Heart, Kidney, Pancreas, Pooled	B-cell ion homeostasis and insulin secretion	Sulfonylurea derivatives
	11q13.4 rs1552224-A	CENTD2 (ARAP1)	Brain, Epithelium, Pancreas, Spleen	Regulate the cell-specific trafficking of a receptor protein involved in apoptosis	Unknown
	11q14.3 rs1387153-T	MTNR1B	Ovary, Retina	An integral membrane GPCR for melatonin	Unknown

Table 1: cont…

12	12q14.3 rs1531343-C	HMGA2	Aorta endothelial cell, Hepatoma	A transcriptional regulating factor in adipogenesis and mesenchymal differentiation	Unknown
	12q21.1 rs7961581-C	TSPAN8	Colon carcinoma, Human small intestine, Liver	Glycoprotein involved in the mediation of signal transduction	Unknown
		LGR5	Placenta	Unknown	
	12q24.31 rs7957197-T	TCF1† (HNF1A)	Liver	Growth and development; A transcription factor required for the expression of several liver-specific genes, important for Wnt signaling	Unknown
13	13q12.1	PDX1‡	Pancreatic islet	Growth and development; nuclear protein that acts as a transcriptional activator of several genes including insulin and is important for early β-cell development	Unknown
	13q31.1 rs1359790-G	SPRY2	Brain,Muscle,Skin	Inhibits growth factor-mediated, receptor tyrosine kinase-induced, MAPK signaling	Unknown
15	15q22.2 rs7172432-?	C2CD4A-4B	Colon/ Mammary gland	Nuclear factor related with inflammation	Unknown
	15q24.3 rs7178572-G	HMG20A	Uterus	Unknown	Unknown
	15q25.1 rs11634397-G	ZFAND6	Hypothalamus, Kidney, Uterus	Unknown	Unknown
	15q26.1 rs2028299-C	AP3S2	Brain, Colon, Muscle, Skin	A clathrin-associated adaptor complex that may be involved in vesicle transport and sorting	Unknown
	15q26.1 rs8042680-A	PRC1	Epithelium, Kidney, Placenta	Cytokinesis	Unknown
16	16q12.2 rs11642841-A	FTO	Brain, Cervix, Eye, Lung	Associated to obesity	Unknown
17	17p13.3 rs391300-G	SRR	Brain, Fetal brain cortex, Melanoma	Alter glutamate signaling and affect insulin or glucagon secretion	Unknown
	17cen-q21.3	TCF2† (HNF1B)	Colon, Liver, Thalamus	Growth and development; transcription factor that forms a complex with the product of TCF1 important for Wnt signaling/Cell cycle regulation	Unknown
19	19q13.11 rs3786897-A	PEPD	Epithelium, Hepatocyte, Kidney, Liver, Mammary cancer, Placenta, Skin, Uterus	Enzyme serves an important role in the recycling of proline	Unknown
20	20q13.12 rs6017317-G	FITM2	Unknown	An evolutionarily conserved family of proteins involved in fat storage	Unknown
		R3HDML	Unknown	Unknown	

Table 1: cont...

20	20q13.12 rs4812829-A	HNF4A†	Kidney, Liver	Growth and development; transcription factor	Unknown
X	Xq28 rs5945326-A	DUSP9	Kidney, Liver, Lung, Placenta	Specificity protein phosphatase for members of the ERK family of MAP kinases	Unknown

Genes included in the list are involved in type 2 diabetes, Maturity Onset Diabetes of the Young (MODY), Permanent Neonatal Diabetes Mellitus (PNDM) or selected syndromic forms of diabetes [4-8, 11-39]. The cut-off p-value for the inclusion of type 2 diabetes genes identified by GWAS is 5×10^{-8}. †Genes involved in both monogenic diabetes and type 2 diabetes. ‡Genes merely involved in monogenic diabetes. Candidate genes for type 2 diabetes and their loci and risk SNPs are derived from Catalog of Published Genome-Wide Association Studies of the National Human Genome Research Institute (http://www.genome.gov/gwastudies/index.cfm?pageid=26525384#searchForm). The data of expression of genes in up tissues are derived from David Bioinformatics Resources 6.7 of National Institute of Allergy and Infectious Diseases (NIAID), NIH (http://david.abcc.ncifcrf.gov/home.jsp). And the proposed functions for gene products are derived from a few functional research literatures and the relationship with pathways about type 2 diabetes, most of which are still needed to be further investigated.

Although monogenic forms of diabetes have been found (Table **1**) (reviewed in [4]), the majority of cases of type 2 diabetes do not show inheritance as a Mendelian trait, but rather as a genetically complex disorder in which genetic variants predispose individuals to develop the disease. Twin studies have revealed concordance rates of over 50 % (and reported up to 90%) for type 2 diabetes in identical twins [1], which still leaves a substantial role for environmental factors, such as excess food and limited physical activity. The rapid rise in diabetes prevalence over the last few decades strongly suggests that genetic variants involved in type 2 diabetes are interacting with environmental factors.

CANDIDATE-BASED ASSOCIATION STUDIES

Several candidate gene association studies have been performed to identify genes involved in type 2 diabetes. These studies test some of the genetic variants in a gene that is a strong candidate for being involved in the disease. Although a few genes have been identified in this way, in general these studies have not been very successful because the results could not be replicated and the p-values for association of the genetic variants were moderate.

Obviously, the few diabetes genes that were identified and replicated in candidate-based association studies had already been linked to a diabetic phenotype, since the tested genes were mainly selected on the basis of their function and their potential relationship with diabetes. The genes identified by candidate gene approaches were found to be involved in rare, monogenic forms of diabetes. The TCF2 gene was linked to maturity-onset diabetes of the young (MODY) 5 [5], while WFS1 gene mutations led to Wolfram (or DIDMOAD) syndrome, a rare, lethal, neurodegenerative disorder that includes diabetes insipidus, diabetes mellitus, optic nerve degeneration, and inner ear deafness [6]. The KCNJ11 gene is related to permanent neonatal diabetes mellitus (PNDM), a rare form of diabetes starting before the age of 6 months [7]. Finally, mutations in PPARγ can lead to insulin resistance, hypertension, and lipodystrophy [8]. Although the molecular impact of mutations in these genes is not yet fully understood, it is likely that mutations in them will have differing effects and that mutations involved in type 2 diabetes will have a milder impact on human physiology than those involved in monogenic diabetes.

GENOME-WIDE ASSOCIATION STUDIES (GWAS)

Recent advances in genotyping techniques and the collection of large, type 2 diabetes patient cohorts have made it possible to perform hypothesis-free genome-wide association studies (GWAS) to identify common genetic variants that increase susceptibility to type 2 diabetes. The human genome harbors around 3 billion base-pairs, which contain at least 3 million common single nucleotide polymorphisms (SNPs) according to the International HapMap Project. However, alleles represented by SNPs that are close together have often stayed on the same chromosome in further generations, forming a so-called haplotype. This implies that when one variant has been typed, we know the genotype of a set of other variants that surround the initial genetic marker; in genetic terminology it is said that the SNPs that stayed together on a chromosome are in high linkage disequilibrium. It has been estimated that, in a Caucasian population, assessing 500,000 SNPs will detect around 80% of the common genetic variation. GWAS typically involve the assessment of such numbers of SNPs determined in large case-control studies for association with a disease or with a so-called “phenotypic trait”, *i.e.*, type 2 diabetes. It should be noted that GWAS identify association of a genetic locus, and not directly of a gene. It is likely that the most associated SNPs in a genomic region are not the causal variant, but that the disease-producing SNP is in high linkage disequilibrium with the “associated” SNPs. It is also possible that a SNP, even when located in a gene, can influence the expression of a nearby gene located several thousand base-pairs or more away [9]. It is therefore difficult to determine which gene is responsible for the association signal in a GWAS with full certainty. A detailed description of a GWAS is reviewed by McCarthy et al [10].

The genome-wide approach has been very successful for type 2 diabetes, leading to the identification of quite a few common genetic variants associated with the disease lying near genes that had not previously been associated with a diabetic phenotype (Table **1**) [11-39]. In the near future this number will probably further increase and a large part of the genetic variation that confers susceptibility to type 2 diabetes will be unveiled. At present, the total effect of the identified genetic variants only explains 10 to 15% of the total type 2 diabetes heritability [40-42]. This small explained percentage of the total type 2 diabetes heritability indicates that the largest proportion of genetic variation involved in type 2 diabetes still await their discovery. Unveiling this so-called missing heritability [43, 44] likely will lead to more interesting discoveries about genetics of type 2 diabetes.

FUNCTIONS OF CANDIDATE GENES FOR TYPE 2 DIABETES

As discussed above, type 2 diabetes is a disease characterized by impaired β-cell secretion of insulin, in combination with resistance to insulin in its target tissues. Both insulin secretion and insulin sensitivity are influenced by genetic and environmental factors [45]. Genes that harbor variants associated with diabetes can thus be expected to exert their effect through one of these pathways.

Most genes that play a role in monogenic diabetes are important for pancreatic β-cell growth and/or function (Table **1**). Genetic variants in three (KCNJ11, TCF2, WFS1) out of four replicated genes for type 2 diabetes found by candidate-based genetic association studies are also related to decreased β-cell function. The best characterized gene is KCNJ11 which encodes a member of a β-cell potassium channel involved in insulin secretion [46]. Mutations in TCF2 are associated with insulin secretion [47], while knock-out mutations in WFS1 have been found to lead to Endoplasmic Reticulum (ER) stress and subsequent apoptosis of pancreatic β-cells [48]. The protein product of the fourth gene identified by a candidate gene approach, PPARγ, is involved in adipocyte differentiation and function (reviewed in [49]). Since rare mutations in this gene lead to insulin resistance and lipodystrophy, variations in this gene are likely to contribute to type 2 diabetes susceptibility through altered insulin effects in adipose tissue.

To explore the functions of the newly discovered genes, various studies have investigated their roles in determining sub-phenotypes of type 2 diabetes, such as peripheral insulin sensitivity and β-cell insulin secretion, and the genetic variants identified in GWAS (except KCNQ1 because variants in this gene have only recently been identified) [15, 50-89]. In general, these studies indicate that the genetic variants in the genes identified by association studies mainly act through interference with insulin secretion instead of peripheral insulin sensitivity, and are thus similar to the genes identified through candidate gene approaches and in monogenic diabetes. However, insulin secretion and insulin sensitivity are related through complex and poorly understood mechanisms. Expression studies have revealed that most of the newly discovered candidate genes for type 2 diabetes are expressed in multiple cell types throughout the body and that the expression of some of these genes under diabetic conditions is unaltered in the pancreatic islets whereas it is different in other tissues (reviewed in [90]). Functional studies of the molecular mechanisms by which most genetic variants lead to impaired insulin secretion and the use of animal models have been quite limited but are sorely needed [91].

In the past five years, GWASs have led to the discovery of 50 new candidate genes related to type 2 diabetes. However, it is still possible that there are mutations in genes that affect other axes of the underlying mechanisms of diabetes. It has been proposed that multiple hits in several diabetes pathways generally occur in subjects destined to develop type 2 diabetes. Some genes related to insulin resistance have been found by GWAS, although the number of genes related to β-cell failure is much larger. Thus, genetic variation associated with type 2 diabetes seems to be more prominently involved in β-cell failure than in insulin resistance.

GENETIC SUBGROUPS

Genes Related to β-Cell Failure

It can be proposed that some of the candidate genes for type 2 diabetes related to β-cell failure can be grouped into four subgroups based on what is known about their molecular function, see Fig. (**1**) (Some candidate genes are not included in this figure

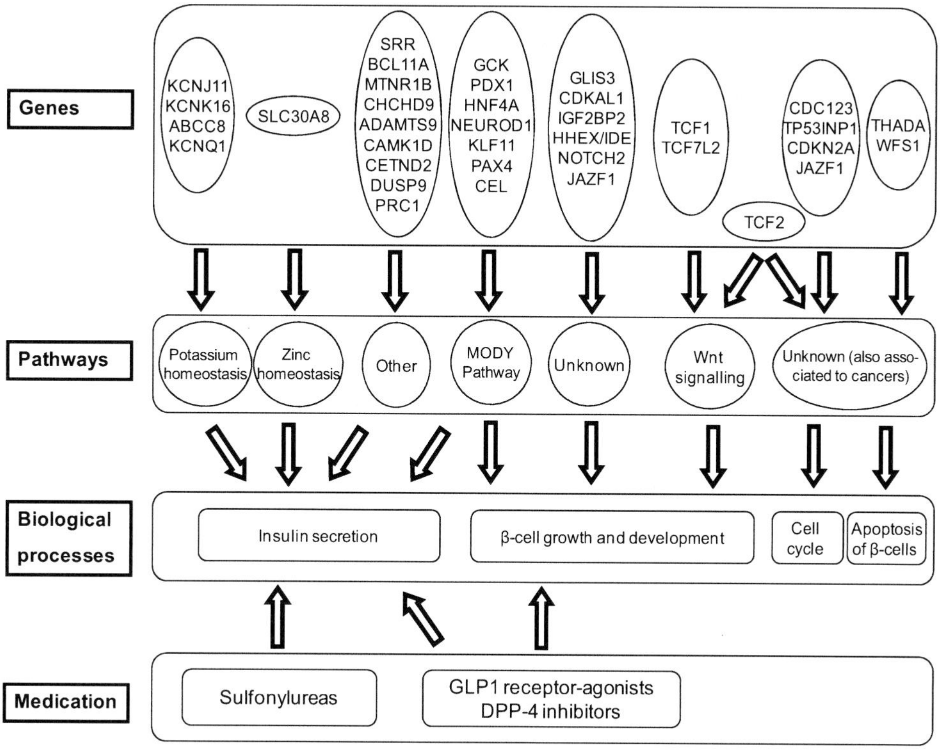

Figure 1: Classification of candidate genes for type 2 diabetes with a potential role in β-cell function.

because their functional relation to diabetes is totally elusive). First, KCNJ11, KCNQ1, ABCC8, KCNK16, SLC30A8, SRR, ADAMTS9, MTNR1B, CAMK1D, CENTD2, DUSP9,BCL11A, PRC1 and CHCHD9 gene products are probably involved in cellular ion homeostasis and insulin secretion. A second group of genes – TCF7L2, TCF1, TCF2, HHEX / IDE, IGF2BP2, CDKAL1, GLIS3,and NOTCH2 –is likely to be involved in the growth and development of the pancreas, which likely influences the overall capacity of the pancreas to secrete insulin. Of this group TCF7L2, TCF1, and TCF2 are important in Wnt signaling (reviewed in [92]). In line with the presence of this group of type 2 diabetes candidate genes, some genes whose deficiency can induce MODY, such as NEUROD1, PDX1, HNF4A, and PAX4, are also related to pancreas growth and development. A third group is related with cell cycle events: JAZF1 and TCF2 have both been associated with prostate cancer and these genes might play a role in the regulation of the cell cycle [93-95]. CDKN2A, TP53INP1, and CDC123 also play a role in the cell cycle. Finally,

THADA and WFS1 are thought to be involved in the apoptosis of β-cells. Genes involved in the cell cycle and apoptosis of β-cells might play a role in diabetes by dysregulating the response of β-cells when a higher insulin output is called for when insulin resistance is present.

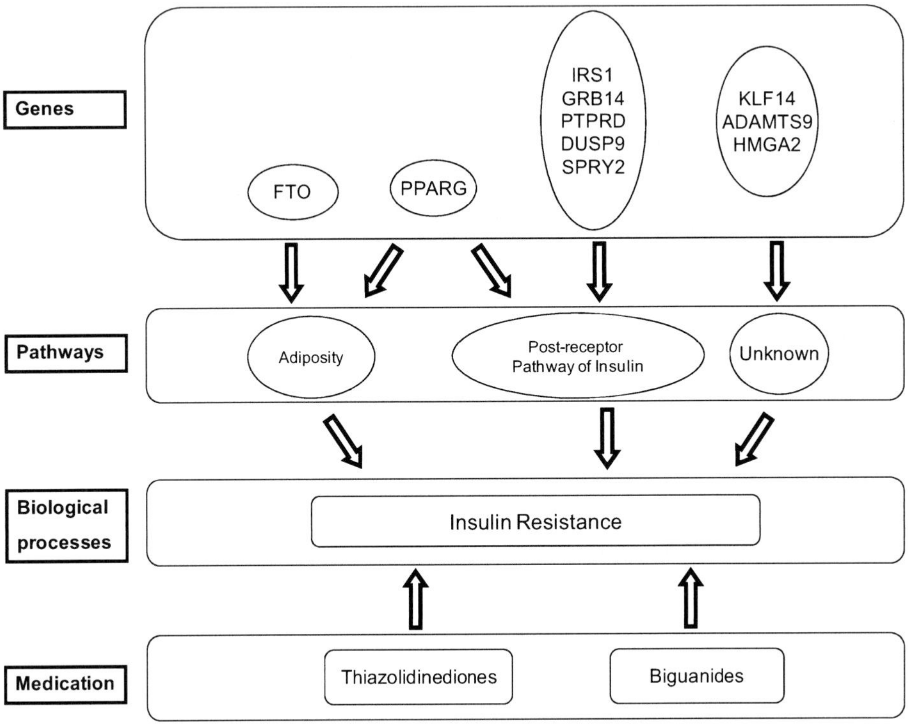

Figure 2: Classification of diabetes genes according to the mechanism of insulin resistance.

Genes Related to Insulin Resistance

Several genes associated to type 2 diabetes in GWAS seem to be related to insulin resistance, which can be classified into four subgroups. First, FTO, can be classified as a separate group amongst the type 2 diabetes candidate genes since it is the only gene that is associated with both obesity and type 2 diabetes [96, 97]. In line with this, Fto deficient mice (Fto-/-) have a significant reductionin adipose tissue and lean body mass. Thus, the association between FTO and type 2 diabetes might be driven by obesity. PPARG may be the only member of the second subgroup, which is believed to influence insulin sensitivity by affecting both adiposity and post-receptor pathway of insulin. A third group includes IRS1,

GRB14, PTPRD, DUSP9 and SPRY2, are closely associated with the insulin receptor signaling pathway, and are thought to be related to type 2 diabetes by this mechanism. For example, the protein product encoded by IRS1 is phosphorylated by insulin receptor tyrosine kinase directly. Finally, KLF14, ADAMTS9 and HMGA2 may be also related to insulin resistance [28]; however, knowledge about their mechanism is still limited.

MANAGEMENT OF TYPE 2 DIABETES AND HOW IT IS RELATED TO GENETICS

The corner stone for diabetes management still lies in diet and exercise [98, 99]. There is also a slowly expanding list of drugs being used to treat type 2 diabetes, all of which act through one of the pathways important in diabetes pathophysiology. However, neither changes in lifestyle nor the use of medication are sufficient to cure diabetes, although both interventions can delay the progression of disease. There is therefore an urgent need to develop new medications or strategies to counter the huge increase in cases expected in the future. Since the management of type 2 diabetes with either lifestyle changes, medication or both, is more effective when started at an early stage, improving the techniques for early diagnosis and the opportunities for early intervention will greatly improve the effects of current ways of managing type 2 diabetes.

Pathways important for the function, growth and development of pancreatic β-cells provide obvious drug targets and are already being used, since defects in insulin secretion play a central role in diabetes. Sulfonylurea derivatives are widely used and act through improving β-cell function by closing β-cell potassium channels and thereby enhancing insulin secretion. Some of the genes associated with both monogenic (ABCC8 and KCNJ11) and complex diabetes (KCNJ11, KCNQ1) encode subunits of these potassium channels and accordingly, monogenic diabetes of this type can be well managed with sulfonylureas. SLC30A8 encodes a zinc transporter protein in the β-cell and the function of this gene product might be related to that of the potassium channel genes by regulating ion homeostasis in the β-cell [100].

The actions of various other drugs involve provoking different effects rather than augmenting insulin secretory function: thiazolidinediones enhance insulin activity by acting on adipose tissue, metformin lowers hepatic glucose output, α-glucosidase inhibitors delay digestion and absorption of intestinal carbohydrate, sodium-glucose cotransporter 2 inhibitors increase urinary glucose excretion and pramlintidine delays gastric emptying and inhibits the release of glucagons. GLP1 receptor-agonists or DPP-4 inhibitors have combined actions on food intake by regulating satiety and enhancing insulin secretion in the short term, and β-cell neogenesis and proliferation in the long term (reviewed in [101]). The exact physiological mechanisms that underlie these improvements are, however, unclear. Except for PPARγ and genes involved in pancreas growth and survival, there is no evidence that any of the other genes identified by GWAS are involved in any of these processes. The protein product of PPARγ is likely to be involved in the mechanism targeted by thiazolidinediones because these compounds are ligands of PPARγ receptors in spite that physiological ligand has not been found [102]. Metformin, the only representative of biguanides, is now believed to be the most widely prescribed oral antidiabetic drug in the world, but its mechanism is too complex to be completely understood. There is some evidence that PPARγ can inhibit hepatic glucose production through the AMPK pathway [103]. However recent research showed that metformin was equally effective to the AMPK deficient mice (AMPK-/-) as to wild type mice, which suggests that metformin could reduce hepatic glucose production through AMPK-independent pathways [104, 105]. Although it seems likely that some of the newly discovered candidate genes for type 2 diabetes act in a pathway that is targeted by GLP1 receptor-agonists or DPP-4 inhibitors, more research into both the mechanisms of the medication and the function of the genes is needed.

IMPLICATIONS FOR PREVENTION AND TREATMENT

Genetic Screening for Prediction and Prevention

The effectiveness of current type 2 diabetes management is greatly improved when it is started at an early stage of the disease. If genetic testing could be used to predict type 2 diabetes, preventive measures could be taken and diabetes could potentially be managed more easily. However, the variants associated with type 2 diabetes that have been identified so far only explain a small percentage of the total genetic variation that is thought to be present [40-42]. It is therefore not yet possible to perform accurate predictive genetic testing but, in the near future, research should provide more insight into the opportunities for such testing. Firstly, it is expected that many more common genetic variants will be identified by performing GWAS in other populations and by improving their power and genomic coverage (*i.e.*, more patients included in studies and more SNPs tested). Secondly, performing thorough analyses of the genomic regions that show association to the disease by resequencing large numbers of patients and controls may identify genetic variants that have higher odds ratios than the common genetic variants identified so far. In a GWAS it is unlikely that the functional variant will be identified, but rather one or more common variants to which the disease variant is in linkage disequilibrium, *i.e.*, that the actual causative mutation(s) is in close vicinity to the tested variant. If the frequency of the causal disease variant is lower than the tested common SNP, we can expect the odds ratio of the causal SNP to be higher and this would make the causal SNP a better candidate to use for genetic testing. One drawback of genetic testing is that many variants have to be tested for each subject, which is laborious and costly at present, but improved genotyping and resequencing technologies will make screening for such variants feasible in the near future.

New Opportunities for Intervention

Even if the exact functions of the majority of the genes associated with type 2 diabetes are still elusive, they can be broadly grouped into several classes, see Figs. (**1**) and (**2**). When certain genes are involved in the same molecular pathway or physiological process, not only these genes but the entire pathway or process would be an obvious target for new anti-diabetes drugs. In combination with genetic screening, such information could be used to optimize diabetes management by prescribing drugs that act on those pathways that are affected in a patient, and *vice versa*, it is also possible that altering the activity, be it stimulation or inhibition, of undisturbed processes could improve glucose homeostasis in selected patients. Improved drug treatment in diabetes can be seen in the current management of various MODY subtypes. As described above, mutations in genes involved in β-cell signaling and/or growth in the pancreas have been found to be responsible for MODY subtypes. After these mutations were identified, the management of these diseases was greatly improved by the use of sulfonylurea derivatives, which enhances insulin secretion instead of exogenous insulin [100]. The good response of MODY3 patients to sulfonylurea treatment serves as an excellent example of personalized therapy based on genetic screening.

Evidence that genes identified by GWAS can represent targets for managing type 2 diabetes is provided by the association to this disease of variants in the KCNQ1 and KCNJ11 genes. Both these genes encode proteins that are members of β-cell potassium channels important for insulin secretion and these channels are targeted by sulfonylurea derivates which are already widely used anti-diabetic drugs [100].

Another gene, found through a candidate-based association study and likely to act in the pathway targeted by thiazolidinediones, is PPARγ. On the physiological level no relationships are known between current medications and all the other genetic variants identified in either GWAS or candidate gene association studies. Therefore all these genes represent new potential targets for intervention.

However, most of the candidate genes for type 2 diabetes have not yet been studied well enough to predict whether interference with their products could be used in managing diabetes. Studies are needed in order to better classify these genes.

Common variants in candidate genes for type 2 diabetes relating to cell cycle events and apoptosis, and representing different alleles than those associated to type 2 diabetes, are also associated with various cancers. In addition, the risks of developing diabetes and prostate cancer are correlated in a complex way: overall diabetes risk and prostate cancer risk are inversely correlated, but while diabetes risk and the risk of developing aggressive prostate cancer show no correlation, these two risks *are* positively correlated in lean men and in men who undertake vigorous physical activity [106]. Systemic administration of agents interfering with the gene products of diabetes genes that are also associated with cancers could, therefore, be beneficial in treating diabetes, but they might be potentially carcinogenic. Tissue-specific drug targeting or additive medication to overcome the carcinogenic side-effects of such medication needs to be developed to overcome this problem. Hence, cell-cycle related genes are probably not the most promising targets for developing new diabetes treatments, unless cell-type-specific targeting of drugs becomes a reality.

Although the genes containing variants associated with type 2 diabetes that seem to be involved in pancreatic growth and development have not been well studied yet, we can speculate on several possibilities for intervention with such genes. Because most genes in this category are not associated with other diseases, they represent promising intervention targets. Variants in these genes might affect the regenerative or proliferative potential of the β-cell population when there is an increased demand for insulin secretion. In this case it would be possible to use such genes, or the molecular pathways they are involved in, as a target for intervention, in an attempt to correct the poor response of the β-cells and ultimately aiming to improve insulin secretion.

Another possibility is that variants in these genes cause pancreatic damage, which would lead to endocrine malfunction of the pancreas at an early stage of the diabetic development. The possibilities for therapeutic intervention would largely depend on the severity and reversibility of such a malfunction.

Genetic variants in three genes identified (or confirmed) by GWAS (TCF7L2, CDKAL1, and SLC30A8) are associated with impairment of proinsulin-to-insulin conversion, whereas variants in other genes identified by GWAS are not associated to this [107]. Even if the affected proinsulin-to-insulin conversion is not a primary but a secondary effect of genetic variants in one of the three genes, reconstituting this process would enhance insulin secretion and thereby enhance β-cell function. Although the conversion of proinsulin-to-insulin has long been known to be involved in type 2 diabetes [108], the mechanisms that regulate this process are still undetermined and there is no current therapy that acts by interfering with this process. Association of TCF7L2, CDKAL1, and SLC30A8 to impaired proinsulin-to-insulin conversion has provided evidence that these genes are somehow related to this process and this clears the way for functional research and development of new medication.

CONCLUSIONS

GWAS are an important tool for identifying genetic variation and they have been very successful in finding 50 genetic loci involved in type 2 diabetes. Genotyping techniques will continue to improve and patient cohorts will become larger, which will lead to more genetic variants being identified in the near future. Although the true causal SNPs are often not known, it is anticipated they will be uncovered by deep sequencing of the genomic regions containing association signals. These advances will help elucidate a higher percentage of the total genetic variation, and improve opportunities for genetic screening and personalized diabetes care. Functional studies and animal models harboring specific gene deletions will be needed to study the role of such mutations, while genetic studies using sharply defined endophenotypes will greatly enhance our knowledge about the candidate genes for type 2 diabetes identified by GWAS.

At present, it is too early to expect results from GWAS to lead to new therapies. Nevertheless, it is clear that genetic studies are crucial to dissecting the mechanisms underlying the biological processes and to finding ways to intervene with them, as they provide important clues for directing the focus of functional research. The ongoing avalanche of genetic information about type 2 diabetes will thus pave the way to further research in the field and has already yielded important information to aid the development of new therapeutic strategies.

ACKNOWLEDGEMENTS

We thank the Dutch Diabetes Foundation (project grant 541741) for financial support. We thank J.L. Senior for critically reading the manuscript.

CONFLICT OF INTEREST

The author(s) confirm that this chapter content has no conflict of interest.

DISCLOSURE

This chapter is an updated version of the following manuscript: Wolfs MG, Hofker MH, Wijmenga C, van Haeften TW. Type 2 Diabetes Mellitus: New Genetic Insights will Lead to New Therapeutics. Curr Genomics. 2009 Apr; 10(2):110-8.

REFERENCES

[1] Newman, B.; Selby, J. V.; King, M. C.; Slemenda, C.; Fabsitz, R.; Friedman, G. D. Concordance for type 2 (non-insulin-dependent) diabetes mellitus in male twins. *Diabetologia* **1987**, *30* (10), 763-8.

[2] van Haeften, T. W.; Dubbeldam, S.; Zonderland, M. L.; Erkelens, D. W. Insulin secretion in normal glucose-tolerant relatives of type 2 diabetic subjects. Assessments using hyperglycemic glucose clamps and oral glucose tolerance tests. *Diabetes Care* **1998**, *21* (2), 278-82.

[3] Pimenta, W.; Korytkowski, M.; Mitrakou, A.; Jenssen, T.; Yki-Jarvinen, H.; Evron, W.; Dailey, G.; Gerich, J. Pancreatic beta-cell dysfunction as the primary genetic lesion in

NIDDM. Evidence from studies in normal glucose-tolerant individuals with a first-degree NIDDM relative. *JAMA* **1995**, *273* (23), 1855-61.

[4] Vaxillaire, M.; Froguel, P. Monogenic diabetes in the young, pharmacogenetics and relevance to multifactorial forms of type 2 diabetes. *Endocr Rev* **2008**, *29* (3), 254-64.

[5] Horikawa, Y.; Iwasaki, N.; Hara, M.; Furuta, H.; Hinokio, Y.; Cockburn, B. N.; Lindner, T.; Yamagata, K.; Ogata, M.; Tomonaga, O.; Kuroki, H.; Kasahara, T.; Iwamoto, Y.; Bell, G. I. Mutation in hepatocyte nuclear factor-1 beta gene (TCF2) associated with MODY. *Nat Genet* **1997**, *17* (4), 384-5.

[6] Strom, T. M.; Hortnagel, K.; Hofmann, S.; Gekeler, F.; Scharfe, C.; Rabl, W.; Gerbitz, K. D.; Meitinger, T. Diabetes insipidus, diabetes mellitus, optic atrophy and deafness (DIDMOAD) caused by mutations in a novel gene (wolframin) coding for a predicted transmembrane protein. *Hum Mol Genet* **1998**, *7* (13), 2021-8.

[7] Gloyn, A. L.; Pearson, E. R.; Antcliff, J. F.; Proks, P.; Bruining, G. J.; Slingerland, A. S.; Howard, N.; Srinivasan, S.; Silva, J. M.; Molnes, J.; Edghill, E. L.; Frayling, T. M.; Temple, I. K.; Mackay, D.; Shield, J. P.; Sumnik, Z.; van Rhijn, A.; Wales, J. K.; Clark, P.; Gorman, S.; Aisenberg, J.; Ellard, S.; Njolstad, P. R.; Ashcroft, F. M.; Hattersley, A. T. Activating mutations in the gene encoding the ATP-sensitive potassium-channel subunit Kir6.2 and permanent neonatal diabetes. *N Engl J Med* **2004**, *350* (18), 1838-49.

[8] Barroso, I.; Gurnell, M.; Crowley, V. E.; Agostini, M.; Schwabe, J. W.; Soos, M. A.; Maslen, G. L.; Williams, T. D.; Lewis, H.; Schafer, A. J.; Chatterjee, V. K.; O'Rahilly, S. Dominant negative mutations in human PPARgamma associated with severe insulin resistance, diabetes mellitus and hypertension. *Nature* **1999**, *402* (6764), 880-3.

[9] Schadt, E. E.; Molony, C.; Chudin, E.; Hao, K.; Yang, X.; Lum, P. Y.; Kasarskis, A.; Zhang, B.; Wang, S.; Suver, C.; Zhu, J.; Millstein, J.; Sieberts, S.; Lamb, J.; GuhaThakurta, D.; Derry, J.; Storey, J. D.; Avila-Campillo, I.; Kruger, M. J.; Johnson, J. M.; Rohl, C. A.; van Nas, A.; Mehrabian, M.; Drake, T. A.; Lusis, A. J.; Smith, R. C.; Guengerich, F. P.; Strom, S. C.; Schuetz, E.; Rushmore, T. H.; Ulrich, R. Mapping the genetic architecture of gene expression in human liver. *PLoS Biol* **2008**, *6* (5), e107.

[10] McCarthy, M. I.; Abecasis, G. R.; Cardon, L. R.; Goldstein, D. B.; Little, J.; Ioannidis, J. P.; Hirschhorn, J. N. Genome-wide association studies for complex traits: consensus, uncertainty and challenges. *Nat Rev Genet* **2008**, *9* (5), 356-69.

[11] Zeggini, E.; Weedon, M. N.; Lindgren, C. M.; Frayling, T. M.; Elliott, K. S.; Lango, H.; Timpson, N. J.; Perry, J. R.; Rayner, N. W.; Freathy, R. M.; Barrett, J. C.; Shields, B.; Morris, A. P.; Ellard, S.; Groves, C. J.; Harries, L. W.; Marchini, J. L.; Owen, K. R.; Knight, B.; Cardon, L. R.; Walker, M.; Hitman, G. A.; Morris, A. D.; Doney, A. S.; McCarthy, M. I.; Hattersley, A. T. Replication of genome-wide association signals in UK samples reveals risk loci for type 2 diabetes. *Science* **2007**, *316* (5829), 1336-41.

[12] Florez, J. C.; Manning, A. K.; Dupuis, J.; McAteer, J.; Irenze, K.; Gianniny, L.; Mirel, D. B.; Fox, C. S.; Cupples, L. A.; Meigs, J. B. A 100K genome-wide association scan for diabetes and related traits in the Framingham Heart Study: replication and integration with other genome-wide datasets. *Diabetes* **2007**, *56* (12), 3063-74.

[13] Frayling, T. M. Genome-wide association studies provide new insights into type 2 diabetes aetiology. *Nat Rev Genet* **2007**, *8* (9), 657-62.

[14] Hanson, R. L.; Bogardus, C.; Duggan, D.; Kobes, S.; Knowlton, M.; Infante, A. M.; Marovich, L.; Benitez, D.; Baier, L. J.; Knowler, W. C. A search for variants associated with young-onset type 2 diabetes in American Indians in a 100K genotyping array. *Diabetes* **2007**, *56* (12), 3045-52.

[15] Rampersaud, E.; Damcott, C. M.; Fu, M.; Shen, H.; McArdle, P.; Shi, X.; Shelton, J.; Yin, J.; Chang, Y. P.; Ott, S. H.; Zhang, L.; Zhao, Y.; Mitchell, B. D.; O'Connell, J.; Shuldiner, A. R. Identification of novel candidate genes for type 2 diabetes from a genome-wide association scan in the Old Order Amish: evidence for replication from diabetes-related quantitative traits and from independent populations. *Diabetes* **2007**, *56*, 3053-62.

[16] Salonen, J. T.; Uimari, P.; Aalto, J. M.; Pirskanen, M.; Kaikkonen, J.; Todorova, B.; Hypponen, J.; Korhonen, V. P.; Asikainen, J.; Devine, C.; Tuomainen, T. P.; Luedemann, J.; Nauck, M.; Kerner, W.; Stephens, R. H.; New, J. P.; Ollier, W. E.; Gibson, J. M.; Payton, A.; Horan, M. A.; Pendleton, N.; Mahoney, W.; Meyre, D.; Delplanque, J.; Froguel, P.; Luzzatto, O.; Yakir, B.; Darvasi, A. Type 2 diabetes whole-genome association study in four populations: the DiaGen consortium. *Am J Hum Genet* **2007**, *81* (2), 338-45.

[17] Genome-wide association study of 14,000 cases of seven common diseases and 3,000 shared controls. *Nature* **2007**, *447* (7145), 661-78.

[18] Scott, L. J.; Mohlke, K. L.; Bonnycastle, L. L.; Willer, C. J.; Li, Y.; Duren, W. L.; Erdos, M. R.; Stringham, H. M.; Chines, P. S.; Jackson, A. U.; Prokunina-Olsson, L.; Ding, C. J.; Swift, A. J.; Narisu, N.; Hu, T.; Pruim, R.; Xiao, R.; Li, X. Y.; Conneely, K. N.; Riebow, N. L.; Sprau, A. G.; Tong, M.; White, P. P.; Hetrick, K. N.; Barnhart, M. W.; Bark, C. W.; Goldstein, J. L.; Watkins, L.; Xiang, F.; Saramies, J.; Buchanan, T. A.; Watanabe, R. M.; Valle, T. T.; Kinnunen, L.; Abecasis, G. R.; Pugh, E. W.; Doheny, K. F.; Bergman, R. N.; Tuomilehto, J.; Collins, F. S.; Boehnke, M. A genome-wide association study of type 2 diabetes in Finns detects multiple susceptibility variants. *Science* **2007**, *316* (5829), 1341-5.

[19] Sladek, R.; Rocheleau, G.; Rung, J.; Dina, C.; Shen, L.; Serre, D.; Boutin, P.; Vincent, D.; Belisle, A.; Hadjadj, S.; Balkau, B.; Heude, B.; Charpentier, G.; Hudson, T. J.; Montpetit, A.; Pshezhetsky, A. V.; Prentki, M.; Posner, B. I.; Balding, D. J.; Meyre, D.; Polychronakos, C.; Froguel, P. A genome-wide association study identifies novel risk loci for type 2 diabetes. *Nature* **2007**, *445* (7130), 881-5.

[20] Steinthorsdottir, V.; Thorleifsson, G.; Reynisdottir, I.; Benediktsson, R.; Jonsdottir, T.; Walters, G. B.; Styrkarsdottir, U.; Gretarsdottir, S.; Emilsson, V.; Ghosh, S.; Baker, A.; Snorradottir, S.; Bjarnason, H.; Ng, M. C.; Hansen, T.; Bagger, Y.; Wilensky, R. L.; Reilly, M. P.; Adeyemo, A.; Chen, Y.; Zhou, J.; Gudnason, V.; Chen, G.; Huang, H.; Lashley, K.; Doumatey, A.; So, W. Y.; Ma, R. C.; Andersen, G.; Borch-Johnsen, K.; Jorgensen, T.; van Vliet-Ostaptchouk, J. V.; Hofker, M. H.; Wijmenga, C.; Christiansen, C.; Rader, D. J.; Rotimi, C.; Gurney, M.; Chan, J. C.; Pedersen, O.; Sigurdsson, G.; Gulcher, J. R.; Thorsteinsdottir, U.; Kong, A.; Stefansson, K. A variant in CDKAL1 influences insulin response and risk of type 2 diabetes. *Nat Genet* **2007**, *39* (6), 770-5.

[21] Unoki, H.; Takahashi, A.; Kawaguchi, T.; Hara, K.; Horikoshi, M.; Andersen, G.; Ng, D. P.; Holmkvist, J.; Borch-Johnsen, K.; Jorgensen, T.; Sandbaek, A.; Lauritzen, T.; Hansen, T.; Nurbaya, S.; Tsunoda, T.; Kubo, M.; Babazono, T.; Hirose, H.; Hayashi, M.; Iwamoto, Y.; Kashiwagi, A.; Kaku, K.; Kawamori, R.; Tai, E. S.; Pedersen, O.; Kamatani, N.; Kadowaki, T.; Kikkawa, R.; Nakamura, Y.; Maeda, S. SNPs in KCNQ1 are associated with susceptibility to type 2 diabetes in East Asian and European populations. *Nat Genet* **2008**, *40* (9), 1098-102.

[22] Yasuda, K.; Miyake, K.; Horikawa, Y.; Hara, K.; Osawa, H.; Furuta, H.; Hirota, Y.; Mori, H.; Jonsson, A.; Sato, Y.; Yamagata, K.; Hinokio, Y.; Wang, H. Y.; Tanahashi, T.; Nakamura, N.; Oka, Y.; Iwasaki, N.; Iwamoto, Y.; Yamada, Y.; Seino, Y.; Maegawa, H.; Kashiwagi, A.; Takeda, J.; Maeda, E.; Shin, H. D.; Cho, Y. M.; Park, K. S.; Lee, H. K.; Ng,

M. C.; Ma, R. C.; So, W. Y.; Chan, J. C.; Lyssenko, V.; Tuomi, T.; Nilsson, P.; Groop, L.; Kamatani, N.; Sekine, A.; Nakamura, Y.; Yamamoto, K.; Yoshida, T.; Tokunaga, K.; Itakura, M.; Makino, H.; Nanjo, K.; Kadowaki, T.; Kasuga, M. Variants in KCNQ1 are associated with susceptibility to type 2 diabetes mellitus. *Nat Genet* **2008**, *40* (9), 1092-7.

[23] Zeggini, E.; Scott, L. J.; Saxena, R.; Voight, B. F.; Marchini, J. L.; Hu, T.; de Bakker, P. I.; Abecasis, G. R.; Almgren, P.; Andersen, G.; Ardlie, K.; Bostrom, K. B.; Bergman, R. N.; Bonnycastle, L. L.; Borch-Johnsen, K.; Burtt, N. P.; Chen, H.; Chines, P. S.; Daly, M. J.; Deodhar, P.; Ding, C. J.; Doney, A. S.; Duren, W. L.; Elliott, K. S.; Erdos, M. R.; Frayling, T. M.; Freathy, R. M.; Gianniny, L.; Grallert, H.; Grarup, N.; Groves, C. J.; Guiducci, C.; Hansen, T.; Herder, C.; Hitman, G. A.; Hughes, T. E.; Isomaa, B.; Jackson, A. U.; Jorgensen, T.; Kong, A.; Kubalanza, K.; Kuruvilla, F. G.; Kuusisto, J.; Langenberg, C.; Lango, H.; Lauritzen, T.; Li, Y.; Lindgren, C. M.; Lyssenko, V.; Marvelle, A. F.; Meisinger, C.; Midthjell, K.; Mohlke, K. L.; Morken, M. A.; Morris, A. D.; Narisu, N.; Nilsson, P.; Owen, K. R.; Palmer, C. N.; Payne, F.; Perry, J. R.; Pettersen, E.; Platou, C.; Prokopenko, I.; Qi, L.; Qin, L.; Rayner, N. W.; Rees, M.; Roix, J. J.; Sandbaek, A.; Shields, B.; Sjogren, M.; Steinthorsdottir, V.; Stringham, H. M.; Swift, A. J.; Thorleifsson, G.; Thorsteinsdottir, U.; Timpson, N. J.; Tuomi, T.; Tuomilehto, J.; Walker, M.; Watanabe, R. M.; Weedon, M. N.; Willer, C. J.; Illig, T.; Hveem, K.; Hu, F. B.; Laakso, M.; Stefansson, K.; Pedersen, O.; Wareham, N. J.; Barroso, I.; Hattersley, A. T.; Collins, F. S.; Groop, L.; McCarthy, M. I.; Boehnke, M.; Altshuler, D. Meta-analysis of genome-wide association data and large-scale replication identifies additional susceptibility loci for type 2 diabetes. *Nat Genet* **2008**, *40* (5), 638-45.

[24] Timpson, N. J.; Lindgren, C. M.; Weedon, M. N.; Randall, J.; Ouwehand, W. H.; Strachan, D. P.; Rayner, N. W.; Walker, M.; Hitman, G. A.; Doney, A. S.; Palmer, C. N.; Morris, A. D.; Hattersley, A. T.; Zeggini, E.; Frayling, T. M.; McCarthy, M. I. Adiposity-related heterogeneity in patterns of type 2 diabetes susceptibility observed in genome-wide association data. *Diabetes* **2009**, *58* (2), 505-10.

[25] Takeuchi, F.; Serizawa, M.; Yamamoto, K.; Fujisawa, T.; Nakashima, E.; Ohnaka, K.; Ikegami, H.; Sugiyama, T.; Katsuya, T.; Miyagishi, M.; Nakashima, N.; Nawata, H.; Nakamura, J.; Kono, S.; Takayanagi, R.; Kato, N. Confirmation of multiple risk Loci and genetic impacts by a genome-wide association study of type 2 diabetes in the Japanese population. *Diabetes* **2009**, *58* (7), 1690-9.

[26] Tsai, F. J.; Yang, C. F.; Chen, C. C.; Chuang, L. M.; Lu, C. H.; Chang, C. T.; Wang, T. Y.; Chen, R. H.; Shiu, C. F.; Liu, Y. M.; Chang, C. C.; Chen, P.; Chen, C. H.; Fann, C. S.; Chen, Y. T.; Wu, J. Y. A genome-wide association study identifies susceptibility variants for type 2 diabetes in Han Chinese. *PLoS Genet* **2010**, *6* (2), e1000847.

[27] Voight, B. F.; Scott, L. J.; Steinthorsdottir, V.; Morris, A. P.; Dina, C.; Welch, R. P.; Zeggini, E.; Huth, C.; Aulchenko, Y. S.; Thorleifsson, G.; McCulloch, L. J.; Ferreira, T.; Grallert, H.; Amin, N.; Wu, G.; Willer, C. J.; Raychaudhuri, S.; McCarroll, S. A.; Langenberg, C.; Hofmann, O. M.; Dupuis, J.; Qi, L.; Segre, A. V.; van Hoek, M.; Navarro, P.; Ardlie, K.; Balkau, B.; Benediktsson, R.; Bennett, A. J.; Blagieva, R.; Boerwinkle, E.; Bonnycastle, L. L.; Bengtsson Bostrom, K.; Bravenboer, B.; Bumpstead, S.; Burtt, N. P.; Charpentier, G.; Chines, P. S.; Cornelis, M.; Couper, D. J.; Crawford, G.; Doney, A. S.; Elliott, K. S.; Elliott, A. L.; Erdos, M. R.; Fox, C. S.; Franklin, C. S.; Ganser, M.; Gieger, C.; Grarup, N.; Green, T.; Griffin, S.; Groves, C. J.; Guiducci, C.; Hadjadj, S.; Hassanali, N.; Herder, C.; Isomaa, B.; Jackson, A. U.; Johnson, P. R.; Jorgensen, T.; Kao, W. H.; Klopp, N.; Kong, A.; Kraft, P.;

Kuusisto, J.; Lauritzen, T.; Li, M.; Lieverse, A.; Lindgren, C. M.; Lyssenko, V.; Marre, M.; Meitinger, T.; Midthjell, K.; Morken, M. A.; Narisu, N.; Nilsson, P.; Owen, K. R.; Payne, F.; Perry, J. R.; Petersen, A. K.; Platou, C.; Proenca, C.; Prokopenko, I.; Rathmann, W.; Rayner, N. W.; Robertson, N. R.; Rocheleau, G.; Roden, M.; Sampson, M. J.; Saxena, R.; Shields, B. M.; Shrader, P.; Sigurdsson, G.; Sparso, T.; Strassburger, K.; Stringham, H. M.; Sun, Q.; Swift, A. J.; Thorand, B. Twelve type 2 diabetes susceptibility loci identified through large-scale association analysis. *Nat Genet* **2010**, *42* (7), 579-89.

[28] Yamauchi, T.; Hara, K.; Maeda, S.; Yasuda, K.; Takahashi, A.; Horikoshi, M.; Nakamura, M.; Fujita, H.; Grarup, N.; Cauchi, S.; Ng, D. P.; Ma, R. C.; Tsunoda, T.; Kubo, M.; Watada, H.; Maegawa, H.; Okada-Iwabu, M.; Iwabu, M.; Shojima, N.; Shin, H. D.; Andersen, G.; Witte, D. R.; Jorgensen, T.; Lauritzen, T.; Sandbaek, A.; Hansen, T.; Ohshige, T.; Omori, S.; Saito, I.; Kaku, K.; Hirose, H.; So, W. Y.; Beury, D.; Chan, J. C.; Park, K. S.; Tai, E. S.; Ito, C.; Tanaka, Y.; Kashiwagi, A.; Kawamori, R.; Kasuga, M.; Froguel, P.; Pedersen, O.; Kamatani, N.; Nakamura, Y.; Kadowaki, T. A genome-wide association study in the Japanese population identifies susceptibility loci for type 2 diabetes at UBE2E2 and C2CD4A-C2CD4B. *Nat Genet* **2010**, *42* (10), 864-8.

[29] Parra, E. J.; Below, J. E.; Krithika, S.; Valladares, A.; Barta, J. L.; Cox, N. J.; Hanis, C. L.; Wacher, N.; Garcia-Mena, J.; Hu, P.; Shriver, M. D.; Kumate, J.; McKeigue, P. M.; Escobedo, J.; Cruz, M. Genome-wide association study of type 2 diabetes in a sample from Mexico City and a meta-analysis of a Mexican-American sample from Starr County, Texas. *Diabetologia* **2011**, *54* (8), 2038-46.

[30] Cui, B.; Zhu, X.; Xu, M.; Guo, T.; Zhu, D.; Chen, G.; Li, X.; Xu, L.; Bi, Y.; Chen, Y.; Xu, Y.; Wang, W.; Wang, H.; Huang, W.; Ning, G. A genome-wide association study confirms previously reported loci for type 2 diabetes in Han Chinese. *PLoS One* **2011**, *6* (7), e22353.

[31] Kooner, J. S.; Saleheen, D.; Sim, X.; Sehmi, J.; Zhang, W.; Frossard, P.; Been, L. F.; Chia, K. S.; Dimas, A. S.; Hassanali, N.; Jafar, T.; Jowett, J. B.; Li, X.; Radha, V.; Rees, S. D.; Takeuchi, F.; Young, R.; Aung, T.; Basit, A.; Chidambaram, M.; Das, D.; Grundberg, E.; Hedman, A. K.; Hydrie, Z. I.; Islam, M.; Khor, C. C.; Kowlessur, S.; Kristensen, M. M.; Liju, S.; Lim, W. Y.; Matthews, D. R.; Liu, J.; Morris, A. P.; Nica, A. C.; Pinidiyapathirage, J. M.; Prokopenko, I.; Rasheed, A.; Samuel, M.; Shah, N.; Shera, A. S.; Small, K. S.; Suo, C.; Wickremasinghe, A. R.; Wong, T. Y.; Yang, M.; Zhang, F.; Abecasis, G. R.; Barnett, A. H.; Caulfield, M.; Deloukas, P.; Frayling, T. M.; Froguel, P.; Kato, N.; Katulanda, P.; Kelly, M. A.; Liang, J.; Mohan, V.; Sanghera, D. K.; Scott, J.; Seielstad, M.; Zimmet, P. Z.; Elliott, P.; Teo, Y. Y.; McCarthy, M. I.; Danesh, J.; Tai, E. S.; Chambers, J. C. Genome-wide association study in individuals of South Asian ancestry identifies six new type 2 diabetes susceptibility loci. *Nat Genet* **2011**, *43* (10), 984-9.

[32] Kho, A. N.; Hayes, M. G.; Rasmussen-Torvik, L.; Pacheco, J. A.; Thompson, W. K.; Armstrong, L. L.; Denny, J. C.; Peissig, P. L.; Miller, A. W.; Wei, W. Q.; Bielinski, S. J.; Chute, C. G.; Leibson, C. L.; Jarvik, G. P.; Crosslin, D. R.; Carlson, C. S.; Newton, K. M.; Wolf, W. A.; Chisholm, R. L.; Lowe, W. L. Use of diverse electronic medical record systems to identify genetic risk for type 2 diabetes within a genome-wide association study. *J Am Med Inform Assoc* **2012**, *19* (2), 212-8.

[33] Cho, Y. S.; Chen, C. H.; Hu, C.; Long, J.; Ong, R. T.; Sim, X.; Takeuchi, F.; Wu, Y.; Go, M. J.; Yamauchi, T.; Chang, Y. C.; Kwak, S. H.; Ma, R. C.; Yamamoto, K.; Adair, L. S.; Aung, T.; Cai, Q.; Chang, L. C.; Chen, Y. T.; Gao, Y.; Hu, F. B.; Kim, H. L.; Kim, S.; Kim, Y. J.; Lee, J. J.; Lee, N. R.; Li, Y.; Liu, J. J.; Lu, W.; Nakamura, J.; Nakashima, E.; Ng, D. P.; Tay,

W. T.; Tsai, F. J.; Wong, T. Y.; Yokota, M.; Zheng, W.; Zhang, R.; Wang, C.; So, W. Y.; Ohnaka, K.; Ikegami, H.; Hara, K.; Cho, Y. M.; Cho, N. H.; Chang, T. J.; Bao, Y.; Hedman, A. K.; Morris, A. P.; McCarthy, M. I.; Takayanagi, R.; Park, K. S.; Jia, W.; Chuang, L. M.; Chan, J. C.; Maeda, S.; Kadowaki, T.; Lee, J. Y.; Wu, J. Y.; Teo, Y. Y.; Tai, E. S.; Shu, X. O.; Mohlke, K. L.; Kato, N.; Han, B. G.; Seielstad, M. Meta-analysis of genome-wide association studies identifies eight new loci for type 2 diabetes in east Asians. *Nat Genet* **2012**, *44* (1), 67-72.

[34] Palmer, N. D.; McDonough, C. W.; Hicks, P. J.; Roh, B. H.; Wing, M. R.; An, S. S.; Hester, J. M.; Cooke, J. N.; Bostrom, M. A.; Rudock, M. E.; Talbert, M. E.; Lewis, J. P.; Ferrara, A.; Lu, L.; Ziegler, J. T.; Sale, M. M.; Divers, J.; Shriner, D.; Adeyemo, A.; Rotimi, C. N.; Ng, M. C.; Langefeld, C. D.; Freedman, B. I.; Bowden, D. W.; Voight, B. F.; Scott, L. J.; Steinthorsdottir, V.; Morris, A. P.; Dina, C.; Welch, R. P.; Zeggini, E.; Huth, C.; Aulchenko, Y. S.; Thorleifsson, G.; McCulloch, L. J.; Ferreira, T.; Grallert, H.; Amin, N.; Wu, G.; Willer, C. J.; Raychaudhuri, S.; McCarroll, S. A.; Langenberg, C.; Hofmann, O. M.; Dupuis, J.; Qi, L.; Segre, A. V.; van Hoek, M.; Navarro, P.; Ardlie, K.; Balkau, B.; Benediktsson, R.; Bennett, A. J.; Blagieva, R.; Boerwinkle, E.; Bonnycastle, L. L.; Bostrom, K. B.; Bravenboer, B.; Bumpstead, S.; Burtt, N. P.; Charpentier, G.; Chines, P. S.; Cornelis, M.; Couper, D. J.; Crawford, G.; Doney, A. S.; Elliott, K. S.; Elliott, A. L.; Erdos, M. R.; Fox, C. S.; Franklin, C. S.; Ganser, M.; Gieger, C.; Grarup, N.; Green, T.; Griffin, S.; Groves, C. J.; Guiducci, C.; Hadjadj, S.; Hassanali, N.; Herder, C.; Isomaa, B.; Jackson, A. U.; Johnson, P. R.; Jorgensen, T.; Kao, W. H.; Klopp, N.; Kong, A.; Kraft, P.; Kuusisto, J.; Lauritzen, T.; Li, M.; Lieverse, A.; Lindgren, C. M.; Lyssenko, V.; Marre, M.; Meitinger, T.; Midthjell, K.; Morken, M. A.; Narisu, N. A genome-wide association search for type 2 diabetes genes in African Americans. *PLoS One* **2012**, *7* (1), e29202.

[35] Qi, L.; Cornelis, M. C.; Kraft, P.; Stanya, K. J.; Linda Kao, W. H.; Pankow, J. S.; Dupuis, J.; Florez, J. C.; Fox, C. S.; Pare, G.; Sun, Q.; Girman, C. J.; Laurie, C. C.; Mirel, D. B.; Manolio, T. A.; Chasman, D. I.; Boerwinkle, E.; Ridker, P. M.; Hunter, D. J.; Meigs, J. B.; Lee, C. H.; Hu, F. B.; van Dam, R. M. Genetic variants at 2q24 are associated with susceptibility to type 2 diabetes. *Hum Mol Genet* **2010**, *19* (13), 2706-15.

[36] Sim, X.; Ong, R. T.; Suo, C.; Tay, W. T.; Liu, J.; Ng, D. P.; Boehnke, M.; Chia, K. S.; Wong, T. Y.; Seielstad, M.; Teo, Y. Y.; Tai, E. S. Transferability of type 2 diabetes implicated loci in multi-ethnic cohorts from Southeast Asia. *PLoS Genet* **2011**, *7* (4), e1001363.

[37] Huang, J.; Ellinghaus, D.; Franke, A.; Howie, B.; Li, Y. 1000 Genomes-based imputation identifies novel and refined associations for the Wellcome Trust Case Control Consortium phase 1 Data. *Eur J Hum Genet* **2012**, *20* (7), 801-5.

[38] Saxena, R.; Voight, B. F.; Lyssenko, V.; Burtt, N. P.; de Bakker, P. I.; Chen, H.; Roix, J. J.; Kathiresan, S.; Hirschhorn, J. N.; Daly, M. J.; Hughes, T. E.; Groop, L.; Altshuler, D.; Almgren, P.; Florez, J. C.; Meyer, J.; Ardlie, K.; Bengtsson Bostrom, K.; Isomaa, B.; Lettre, G.; Lindblad, U.; Lyon, H. N.; Melander, O.; Newton-Cheh, C.; Nilsson, P.; Orho-Melander, M.; Rastam, L.; Speliotes, E. K.; Taskinen, M. R.; Tuomi, T.; Guiducci, C.; Berglund, A.; Carlson, J.; Gianniny, L.; Hackett, R.; Hall, L.; Holmkvist, J.; Laurila, E.; Sjogren, M.; Sterner, M.; Surti, A.; Svensson, M.; Tewhey, R.; Blumenstiel, B.; Parkin, M.; Defelice, M.; Barry, R.; Brodeur, W.; Camarata, J.; Chia, N.; Fava, M.; Gibbons, J.; Handsaker, B.; Healy, C.; Nguyen, K.; Gates, C.; Sougnez, C.; Gage, D.; Nizzari, M.; Gabriel, S. B.; Chirn, G. W.; Ma, Q.; Parikh, H.; Richardson, D.; Ricke, D.; Purcell, S. Genome-wide association analysis identifies loci for type 2 diabetes and triglyceride levels. *Science* **2007**, *316* (5829), 1331-6.

[39] Lango, H.; Palmer, C. N.; Morris, A. D.; Zeggini, E.; Hattersley, A. T.; McCarthy, M. I.; Frayling, T. M.; Weedon, M. N. Assessing the combined impact of 18 common genetic variants of modest effect sizes on type 2 diabetes risk. *Diabetes* **2008**, *57* (11), 3129-35.

[40] van Hoek, M.; Dehghan, A.; Witteman, J. C.; van Duijn, C. M.; Uitterlinden, A. G.; Oostra, B. A.; Hofman, A.; Sijbrands, E. J.; Janssens, A. C. Predicting type 2 diabetes based on polymorphisms from genome-wide association studies: a population-based study. *Diabetes* **2008**, *57* (11), 3122-8.

[41] Manolio, T. A.; Collins, F. S.; Cox, N. J.; Goldstein, D. B.; Hindorff, L. A.; Hunter, D. J.; McCarthy, M. I.; Ramos, E. M.; Cardon, L. R.; Chakravarti, A.; Cho, J. H.; Guttmacher, A. E.; Kong, A.; Kruglyak, L.; Mardis, E.; Rotimi, C. N.; Slatkin, M.; Valle, D.; Whittemore, A. S.; Boehnke, M.; Clark, A. G.; Eichler, E. E.; Gibson, G.; Haines, J. L.; Mackay, T. F.; McCarroll, S. A.; Visscher, P. M. Finding the missing heritability of complex diseases. *Nature* **2009**, *461* (7265), 747-53.

[42] Maher, B. Personal genomes: The case of the missing heritability. *Nature* **2008**, *456* (7218), 18-21.

[43] Eichler, E. E.; Flint, J.; Gibson, G.; Kong, A.; Leal, S. M.; Moore, J. H.; Nadeau, J. H. Missing heritability and strategies for finding the underlying causes of complex disease. *Nat Rev Genet* **2010**, *11* (6), 446-50.

[44] Stumvoll, M.; Goldstein, B. J.; van Haeften, T. W. Type 2 diabetes: principles of pathogenesis and therapy. *Lancet* **2005**, *365* (9467), 1333-46.

[45] Inagaki, N.; Gonoi, T.; Clement, J. P. t.; Namba, N.; Inazawa, J.; Gonzalez, G.; Aguilar-Bryan, L.; Seino, S.; Bryan, J. Reconstitution of IKATP: an inward rectifier subunit plus the sulfonylurea receptor. *Science* **1995**, *270* (5239), 1166-70.

[46] Wang, L.; Coffinier, C.; Thomas, M. K.; Gresh, L.; Eddu, G.; Manor, T.; Levitsky, L. L.; Yaniv, M.; Rhoads, D. B. Selective deletion of the Hnf1beta (MODY5) gene in beta-cells leads to altered gene expression and defective insulin release. *Endocrinology* **2004**, *145* (8), 3941-9.

[47] Yamada, T.; Ishihara, H.; Tamura, A.; Takahashi, R.; Yamaguchi, S.; Takei, D.; Tokita, A.; Satake, C.; Tashiro, F.; Katagiri, H.; Aburatani, H.; Miyazaki, J.; Oka, Y. WFS1-deficiency increases endoplasmic reticulum stress, impairs cell cycle progression and triggers the apoptotic pathway specifically in pancreatic beta-cells. *Hum Mol Genet* **2006**, *15* (10), 1600-9.

[48] Stumvoll, M.; Haring, H. The peroxisome proliferator-activated receptor-gamma2 Pro12Ala polymorphism. *Diabetes* **2002**, *51* (8), 2341-7.

[49] Grarup, N.; Andersen, G.; Krarup, N. T.; Albrechtsen, A.; Schmitz, O.; Jorgensen, T.; Borch-Johnsen, K.; Hansen, T.; Pedersen, O. Association testing of novel type 2 diabetes risk alleles in the JAZF1, CDC123/CAMK1D, TSPAN8, THADA, ADAMTS9, and NOTCH2 loci with insulin release, insulin sensitivity, and obesity in a population-based sample of 4,516 glucose-tolerant middle-aged Danes. *Diabetes* **2008**, *57* (9), 2534-40.

[50] Grarup, N.; Rose, C. S.; Andersson, E. A.; Andersen, G.; Nielsen, A. L.; Albrechtsen, A.; Clausen, J. O.; Rasmussen, S. S.; Jorgensen, T.; Sandbaek, A.; Lauritzen, T.; Schmitz, O.; Hansen, T.; Pedersen, O. Studies of association of variants near the HHEX, CDKN2A/B, and IGF2BP2 genes with type 2 diabetes and impaired insulin release in 10,705 Danish subjects: validation and extension of genome-wide association studies. *Diabetes* **2007**, *56* (12), 3105-11.

[51] Pascoe, L.; Tura, A.; Patel, S. K.; Ibrahim, I. M.; Ferrannini, E.; Zeggini, E.; Weedon, M. N.; Mari, A.; Hattersley, A. T.; McCarthy, M. I.; Frayling, T. M.; Walker, M. Common variants of the novel type 2 diabetes genes CDKAL1 and HHEX/IDE are associated with decreased pancreatic beta-cell function. *Diabetes* **2007**, *56* (12), 3101-4.

[52] Staiger, H.; Machicao, F.; Kantartzis, K.; Schafer, S. A.; Kirchhoff, K.; Guthoff, M.; Silbernagel, G.; Stefan, N.; Fritsche, A.; Haring, H. U. Novel meta-analysis-derived type 2 diabetes risk loci do not determine prediabetic phenotypes. *PLoS One* **2008**, *3* (8), e3019.

[53] Staiger, H.; Machicao, F.; Stefan, N.; Tschritter, O.; Thamer, C.; Kantartzis, K.; Schafer, S. A.; Kirchhoff, K.; Fritsche, A.; Haring, H. U. Polymorphisms within novel risk loci for type 2 diabetes determine beta-cell function. *PLoS One* **2007**, *2* (9), e832.

[54] Bouatia-Naji, N.; Bonnefond, A.; Cavalcanti-Proenca, C.; Sparso, T.; Holmkvist, J.; Marchand, M.; Delplanque, J.; Lobbens, S.; Rocheleau, G.; Durand, E.; De Graeve, F.; Chevre, J. C.; Borch-Johnsen, K.; Hartikainen, A. L.; Ruokonen, A.; Tichet, J.; Marre, M.; Weill, J.; Heude, B.; Tauber, M.; Lemaire, K.; Schuit, F.; Elliott, P.; Jorgensen, T.; Charpentier, G.; Hadjadj, S.; Cauchi, S.; Vaxillaire, M.; Sladek, R.; Visvikis-Siest, S.; Balkau, B.; Levy-Marchal, C.; Pattou, F.; Meyre, D.; Blakemore, A. I.; Jarvelin, M. R.; Walley, A. J.; Hansen, T.; Dina, C.; Pedersen, O.; Froguel, P. A variant near MTNR1B is associated with increased fasting plasma glucose levels and type 2 diabetes risk. *Nat Genet* **2009**, *41* (1), 89-94.

[55] Rung, J.; Cauchi, S.; Albrechtsen, A.; Shen, L.; Rocheleau, G.; Cavalcanti-Proenca, C.; Bacot, F.; Balkau, B.; Belisle, A.; Borch-Johnsen, K.; Charpentier, G.; Dina, C.; Durand, E.; Elliott, P.; Hadjadj, S.; Jarvelin, M. R.; Laitinen, J.; Lauritzen, T.; Marre, M.; Mazur, A.; Meyre, D.; Montpetit, A.; Pisinger, C.; Posner, B.; Poulsen, P.; Pouta, A.; Prentki, M.; Ribel-Madsen, R.; Ruokonen, A.; Sandbaek, A.; Serre, D.; Tichet, J.; Vaxillaire, M.; Wojtaszewski, J. F.; Vaag, A.; Hansen, T.; Polychronakos, C.; Pedersen, O.; Froguel, P.; Sladek, R. Genetic variant near IRS1 is associated with type 2 diabetes, insulin resistance and hyperinsulinemia. *Nat Genet* **2009**, *41* (10), 1110-5.

[56] Saxena, R.; Elbers, C. C.; Guo, Y.; Peter, I.; Gaunt, T. R.; Mega, J. L.; Lanktree, M. B.; Tare, A.; Castillo, B. A.; Li, Y. R.; Johnson, T.; Bruinenberg, M.; Gilbert-Diamond, D.; Rajagopalan, R.; Voight, B. F.; Balasubramanyam, A.; Barnard, J.; Bauer, F.; Baumert, J.; Bhangale, T.; Bohm, B. O.; Braund, P. S.; Burton, P. R.; Chandrupatla, H. R.; Clarke, R.; Cooper-DeHoff, R. M.; Crook, E. D.; Davey-Smith, G.; Day, I. N.; de Boer, A.; de Groot, M. C.; Drenos, F.; Ferguson, J.; Fox, C. S.; Furlong, C. E.; Gibson, Q.; Gieger, C.; Gilhuijs-Pederson, L. A.; Glessner, J. T.; Goel, A.; Gong, Y.; Grant, S. F.; Grobbee, D. E.; Hastie, C.; Humphries, S. E.; Kim, C. E.; Kivimaki, M.; Kleber, M.; Meisinger, C.; Kumari, M.; Langaee, T. Y.; Lawlor, D. A.; Li, M.; Lobmeyer, M. T.; Maitland-van der Zee, A. H.; Meijs, M. F.; Molony, C. M.; Morrow, D. A.; Murugesan, G.; Musani, S. K.; Nelson, C. P.; Newhouse, S. J.; O'Connell, J. R.; Padmanabhan, S.; Palmen, J.; Patel, S. R.; Pepine, C. J.; Pettinger, M.; Price, T. S.; Rafelt, S.; Ranchalis, J.; Rasheed, A.; Rosenthal, E.; Ruczinski, I.; Shah, S.; Shen, H.; Silbernagel, G.; Smith, E. N.; Spijkerman, A. W.; Stanton, A.; Steffes, M. W.; Thorand, B.; Trip, M.; van der Harst, P.; van der, A. D.; van Iperen, E. P.; van Setten, J.; van Vliet-Ostaptchouk, J. V.; Verweij, N.; Wolffenbuttel, B. H.; Young, T.; Zafarmand, M. H.; Zmuda, J. M.; Boehnke, M.; Altshuler, D.; McCarthy, M.; Kao, W. H.; Pankow, J. S.; Cappola, T. P.; Sever, P. Large-scale gene-centric meta-analysis across 39 studies identifies type 2 diabetes loci. *Am J Hum Genet* **2012**, *90* (3), 410-25.

[57] van Vliet-Ostaptchouk, J. V.; van Haeften, T. W.; Landman, G. W.; Reiling, E.; Kleefstra, N.; Bilo, H. J.; Klungel, O. H.; de Boer, A.; van Diemen, C. C.; Wijmenga, C.; Boezen, H. M.; Dekker, J. M.; van 't Riet, E.; Nijpels, G.; Welschen, L. M.; Zavrelova, H.; Bruin, E. J.; Elbers, C. C.; Bauer, F.; Onland-Moret, N. C.; van der Schouw, Y. T.; Grobbee, D. E.; Spijkerman, A. M.; van der, A. D.; Simonis-Bik, A. M.; Eekhoff, E. M.; Diamant, M.; Kramer, M. H.; Boomsma, D. I.; de Geus, E. J.; Willemsen, G.; Slagboom, P. E.; Hofker, M. H.; t Hart, L. M. Common variants in the type 2 diabetes KCNQ1 gene are associated with impairments in insulin secretion during hyperglycaemic glucose clamp. *PLoS One* **2012**, *7* (3), e32148.

[58] Lyssenko, V.; Nagorny, C. L.; Erdos, M. R.; Wierup, N.; Jonsson, A.; Spegel, P.; Bugliani, M.; Saxena, R.; Fex, M.; Pulizzi, N.; Isomaa, B.; Tuomi, T.; Nilsson, P.; Kuusisto, J.; Tuomilehto, J.; Boehnke, M.; Altshuler, D.; Sundler, F.; Eriksson, J. G.; Jackson, A. U.; Laakso, M.; Marchetti, P.; Watanabe, R. M.; Mulder, H.; Groop, L. Common variant in MTNR1B associated with increased risk of type 2 diabetes and impaired early insulin secretion. *Nat Genet* **2009**, *41* (1), 82-8.

[59] Prokopenko, I.; Langenberg, C.; Florez, J. C.; Saxena, R.; Soranzo, N.; Thorleifsson, G.; Loos, R. J.; Manning, A. K.; Jackson, A. U.; Aulchenko, Y.; Potter, S. C.; Erdos, M. R.; Sanna, S.; Hottenga, J. J.; Wheeler, E.; Kaakinen, M.; Lyssenko, V.; Chen, W. M.; Ahmadi, K.; Beckmann, J. S.; Bergman, R. N.; Bochud, M.; Bonnycastle, L. L.; Buchanan, T. A.; Cao, A.; Cervino, A.; Coin, L.; Collins, F. S.; Crisponi, L.; de Geus, E. J.; Dehghan, A.; Deloukas, P.; Doney, A. S.; Elliott, P.; Freimer, N.; Gateva, V.; Herder, C.; Hofman, A.; Hughes, T. E.; Hunt, S.; Illig, T.; Inouye, M.; Isomaa, B.; Johnson, T.; Kong, A.; Krestyaninova, M.; Kuusisto, J.; Laakso, M.; Lim, N.; Lindblad, U.; Lindgren, C. M.; McCann, O. T.; Mohlke, K. L.; Morris, A. D.; Naitza, S.; Orru, M.; Palmer, C. N.; Pouta, A.; Randall, J.; Rathmann, W.; Saramies, J.; Scheet, P.; Scott, L. J.; Scuteri, A.; Sharp, S.; Sijbrands, E.; Smit, J. H.; Song, K.; Steinthorsdottir, V.; Stringham, H. M.; Tuomi, T.; Tuomilehto, J.; Uitterlinden, A. G.; Voight, B. F.; Waterworth, D.; Wichmann, H. E.; Willemsen, G.; Witteman, J. C.; Yuan, X.; Zhao, J. H.; Zeggini, E.; Schlessinger, D.; Sandhu, M.; Boomsma, D. I.; Uda, M.; Spector, T. D.; Penninx, B. W.; Altshuler, D.; Vollenweider, P.; Jarvelin, M. R.; Lakatta, E.; Waeber, G.; Fox, C. S.; Peltonen, L.; Groop, L. C.; Mooser, V.; Cupples, L. A.; Thorsteinsdottir, U.; Boehnke, M.; Barroso, I. Variants in MTNR1B influence fasting glucose levels. *Nat Genet* **2009**, *41* (1), 77-81.

[60] Kilpelainen, T. O.; Zillikens, M. C.; Stancakova, A.; Finucane, F. M.; Ried, J. S.; Langenberg, C.; Zhang, W.; Beckmann, J. S.; Luan, J.; Vandenput, L.; Styrkarsdottir, U.; Zhou, Y.; Smith, A. V.; Zhao, J. H.; Amin, N.; Vedantam, S.; Shin, S. Y.; Haritunians, T.; Fu, M.; Feitosa, M. F.; Kumari, M.; Halldorsson, B. V.; Tikkanen, E.; Mangino, M.; Hayward, C.; Song, C.; Arnold, A. M.; Aulchenko, Y. S.; Oostra, B. A.; Campbell, H.; Cupples, L. A.; Davis, K. E.; Doring, A.; Eiriksdottir, G.; Estrada, K.; Fernandez-Real, J. M.; Garcia, M.; Gieger, C.; Glazer, N. L.; Guiducci, C.; Hofman, A.; Humphries, S. E.; Isomaa, B.; Jacobs, L. C.; Jula, A.; Karasik, D.; Karlsson, M. K.; Khaw, K. T.; Kim, L. J.; Kivimaki, M.; Klopp, N.; Kuhnel, B.; Kuusisto, J.; Liu, Y.; Ljunggren, O.; Lorentzon, M.; Luben, R. N.; McKnight, B.; Mellstrom, D.; Mitchell, B. D.; Mooser, V.; Moreno, J. M.; Mannisto, S.; O'Connell, J. R.; Pascoe, L.; Peltonen, L.; Peral, B.; Perola, M.; Psaty, B. M.; Salomaa, V.; Savage, D. B.; Semple, R. K.; Skaric-Juric, T.; Sigurdsson, G.; Song, K. S.; Spector, T. D.; Syvanen, A. C.; Talmud, P. J.; Thorleifsson, G.; Thorsteinsdottir, U.; Uitterlinden, A. G.; van Duijn, C. M.; Vidal-Puig, A.; Wild, S. H.; Wright, A. F.; Clegg, D. J.; Schadt, E.; Wilson, J.

F.; Rudan, I.; Ripatti, S.; Borecki, I. B.; Shuldiner, A. R.; Ingelsson, E.; Jansson, J. O.; Kaplan, R. C.; Gudnason, V.; Harris, T. B.; Groop, L.; Kiel, D. P.; Rivadeneira, F. Genetic variation near IRS1 associates with reduced adiposity and an impaired metabolic profile. *Nat Genet* **2011**, *43* (8), 753-60.

[61] Langlais, P.; Yi, Z.; Finlayson, J.; Luo, M.; Mapes, R.; De Filippis, E.; Meyer, C.; Plummer, E.; Tongchinsub, P.; Mattern, M.; Mandarino, L. J. Global IRS-1 phosphorylation analysis in insulin resistance. *Diabetologia* **2011**, *54* (11), 2878-89.

[62] de Miguel-Yanes, J. M.; Shrader, P.; Pencina, M. J.; Fox, C. S.; Manning, A. K.; Grant, R. W.; Dupuis, J.; Florez, J. C.; D'Agostino, R. B., Sr.; Cupples, L. A.; Meigs, J. B. Genetic risk reclassification for type 2 diabetes by age below or above 50 years using 40 type 2 diabetes risk single nucleotide polymorphisms. *Diabetes Care* **2011**, *34* (1), 121-5.

[63] Goenaga, D.; Hampe, C.; Carre, N.; Cailliau, K.; Browaeys-Poly, E.; Perdereau, D.; Holt, L. J.; Daly, R. J.; Girard, J.; Broutin, I.; Issad, T.; Burnol, A. F. Molecular determinants of Grb14-mediated inhibition of insulin signaling. *Mol Endocrinol* **2009**, *23* (7), 1043-51.

[64] Cariou, B.; Capitaine, N.; Le Marcis, V.; Vega, N.; Bereziat, V.; Kergoat, M.; Laville, M.; Girard, J.; Vidal, H.; Burnol, A. F. Increased adipose tissue expression of Grb14 in several models of insulin resistance. *FASEB J* **2004**, *18* (9), 965-7.

[65] Cooney, G. J.; Lyons, R. J.; Crew, A. J.; Jensen, T. E.; Molero, J. C.; Mitchell, C. J.; Biden, T. J.; Ormandy, C. J.; James, D. E.; Daly, R. J. Improved glucose homeostasis and enhanced insulin signalling in Grb14-deficient mice. *EMBO J* **2004**, *23* (3), 582-93.

[66] Fernandez-Zapico, M. E.; van Velkinburgh, J. C.; Gutierrez-Aguilar, R.; Neve, B.; Froguel, P.; Urrutia, R.; Stein, R. MODY7 gene, KLF11, is a novel p300-dependent regulator of Pdx-1 (MODY4) transcription in pancreatic islet beta cells. *J Biol Chem* **2009**, *284* (52), 36482-90.

[67] McConnell, B. B.; Yang, V. W. Mammalian Kruppel-like factors in health and diseases. *Physiol Rev* **2010**, *90* (4), 1337-81.

[68] Winckler, W.; Weedon, M. N.; Graham, R. R.; McCarroll, S. A.; Purcell, S.; Almgren, P.; Tuomi, T.; Gaudet, D.; Bostrom, K. B.; Walker, M.; Hitman, G.; Hattersley, A. T.; McCarthy, M. I.; Ardlie, K. G.; Hirschhorn, J. N.; Daly, M. J.; Frayling, T. M.; Groop, L.; Altshuler, D. Evaluation of common variants in the six known maturity-onset diabetes of the young (MODY) genes for association with type 2 diabetes. *Diabetes* **2007**, *56* (3), 685-93.

[69] Gong, Z. C.; Huang, Q.; Dai, X. P.; Lei, G. H.; Lu, H. B.; Yin, J. Y.; Xu, X. J.; Qu, J.; Pei, Q.; Dong, M.; Zhou, B. T.; Shen, J.; Zhou, G.; Zhou, H. H.; Liu, Z. Q. NeuroD1 A45T and PAX4 R121W polymorphisms are associated with plasma glucose level of repaglinide monotherapy in Chinese patients with type 2 diabetes. *Br J Clin Pharmacol* **2012**, *74* (3), 501-9.

[70] Malecki, M. T.; Jhala, U. S.; Antonellis, A.; Fields, L.; Doria, A.; Orban, T.; Saad, M.; Warram, J. H.; Montminy, M.; Krolewski, A. S. Mutations in NEUROD1 are associated with the development of type 2 diabetes mellitus. *Nat Genet* **1999**, *23* (3), 323-8.

[71] Barroso, I.; Luan, J.; Middelberg, R. P.; Harding, A. H.; Franks, P. W.; Jakes, R. W.; Clayton, D.; Schafer, A. J.; O'Rahilly, S.; Wareham, N. J. Candidate gene association study in type 2 diabetes indicates a role for genes involved in beta-cell function as well as insulin action. *PLoS Biol* **2003**, *1* (1), E20.

[72] Chen, T.; Li, M.; Ding, Y.; Zhang, L. S.; Xi, Y.; Pan, W. J.; Tao, D. L.; Wang, J. Y.; Li, L. Identification of zinc-finger BED domain-containing 3 (Zbed3) as a novel Axin-interacting protein that activates Wnt/beta-catenin signaling. *J Biol Chem* **2009**, *284*, 6683-9.

[73] Duprat, F.; Girard, C.; Jarretou, G.; Lazdunski, M. Pancreatic two P domain K+ channels TALK-1 and TALK-2 are activated by nitric oxide and reactive oxygen species. *J Physiol* **2005**, *562* (Pt 1), 235-44.

[74] Brey, C. W.; Nelder, M. P.; Hailemariam, T.; Gaugler, R.; Hashmi, S. Kruppel-like family of transcription factors: an emerging new frontier in fat biology. *Int J Biol Sci* **2009**, *5*, 622-36.

[75] Small, K. S.; Hedman, A. K.; Grundberg, E.; Nica, A. C.; Thorleifsson, G.; Kong, A.; Thorsteindottir, U.; Shin, S. Y.; Richards, H. B.; Soranzo, N.; Ahmadi, K. R.; Lindgren, C. M.; Stefansson, K.; Dermitzakis, E. T.; Deloukas, P.; Spector, T. D.; McCarthy, M. I. Identification of an imprinted master trans regulator at the KLF14 locus related to multiple metabolic phenotypes. *Nat Genet* **2011**, *43*, 561-4.

[76] Savkovic, V.; Gaiser, S.; Iovanna, J. L.; Bodeker, H. The stress response of the exocrine pancreas. *Dig Dis* **2004**, *22* (3), 239-46.

[77] Johansson, B. B.; Torsvik, J.; Bjorkhaug, L.; Vesterhus, M.; Ragvin, A.; Tjora, E.; Fjeld, K.; Hoem, D.; Johansson, S.; Raeder, H.; Lindquist, S.; Hernell, O.; Cnop, M.; Saraste, J.; Flatmark, T.; Molven, A.; Njolstad, P. R. Diabetes and pancreatic exocrine dysfunction due to mutations in the carboxyl ester lipase gene-maturity onset diabetes of the young (CEL-MODY): a protein misfolding disease. *J Biol Chem* **2011**, *286* (40), 34593-605.

[78] Nielsen, T.; Sparso, T.; Grarup, N.; Jorgensen, T.; Pisinger, C.; Witte, D. R.; Hansen, T.; Pedersen, O. Type 2 diabetes risk allele near CENTD2 is associated with decreased glucose-stimulated insulin release. *Diabetologia* **2011**, *54* (5), 1052-6.

[79] Schafer, S. A.; Machicao, F.; Fritsche, A.; Haring, H. U.; Kantartzis, K. New type 2 diabetes risk genes provide new insights in insulin secretion mechanisms. *Diabetes Res Clin Pract* **2011**, *93 Suppl 1*, S9-24.

[80] Dietrich, K.; Birkmeier, S.; Schleinitz, D.; Breitfeld, J.; Enigk, B.; Muller, I.; Bottcher, Y.; Lindner, T.; Stumvoll, M.; Tonjes, A.; Kovacs, P. Association and evolutionary studies of the melatonin receptor 1B gene (MTNR1B) in the self-contained population of Sorbs from Germany. *Diabet Med* **2011**, *28*, 1373-80.

[81] Hammond, S. M.; Sharpless, N. E. HMGA2, microRNAs, and stem cell aging. *Cell* **2008**, *135* (6), 1013-6.

[82] Cleynen, I.; Van de Ven, W. J. The HMGA proteins: a myriad of functions (Review). *Int J Oncol* **2008**, *32* (2), 289-305.

[83] Champy, M. F.; Le Voci, L.; Selloum, M.; Peterson, L. B.; Cumiskey, A. M.; Blom, D. Reduced body weight in male Tspan8-deficient mice. *Int J Obes (Lond)* **2011**, *35* (4), 605-17.

[84] Warton, K.; Foster, N. C.; Gold, W. A.; Stanley, K. K. A novel gene family induced by acute inflammation in endothelial cells. *Gene* **2004**, *342* (1), 85-95.

[85] Ashcroft, F. M. ATP-sensitive potassium channelopathies: focus on insulin secretion. *J Clin Invest* **2005**, *115* (8), 2047-58.

[86] McCarthy, M. I. Genomics, type 2 diabetes, and obesity. *N Engl J Med* **2010**, *363* (24), 2339-50.

[87] Grarup, N.; Sparso, T.; Hansen, T. Physiologic characterization of type 2 diabetes-related loci. *Curr Diab Rep* **2010**, *10* (6), 485-97.

[88] Petrie, J. R.; Pearson, E. R.; Sutherland, C. Implications of genome wide association studies for the understanding of type 2 diabetes pathophysiology. *Biochem Pharmacol* **2011**, *81* (4), 471-7.

[89] Doria, A.; Patti, M. E.; Kahn, C. R. The emerging genetic architecture of type 2 diabetes. *Cell Metab* **2008**, *8* (3), 186-200.

[90] Cox, R. D.; Church, C. D. Mouse models and the interpretation of human GWAS in type 2 diabetes and obesity. *Dis Model Mech* **2011**, *4* (2), 155-64.

[91] Jin, T. The WNT signalling pathway and diabetes mellitus. *Diabetologia* **2008**, *51* (10), 1771-80.

[92] Gudmundsson, J.; Sulem, P.; Steinthorsdottir, V.; Bergthorsson, J. T.; Thorleifsson, G.; Manolescu, A.; Rafnar, T.; Gudbjartsson, D.; Agnarsson, B. A.; Baker, A.; Sigurdsson, A.; Benediktsdottir, K. R.; Jakobsdottir, M.; Blondal, T.; Stacey, S. N.; Helgason, A.; Gunnarsdottir, S.; Olafsdottir, A.; Kristinsson, K. T.; Birgisdottir, B.; Ghosh, S.; Thorlacius, S.; Magnusdottir, D.; Stefansdottir, G.; Kristjansson, K.; Bagger, Y.; Wilensky, R. L.; Reilly, M. P.; Morris, A. D.; Kimber, C. H.; Adeyemo, A.; Chen, Y.; Zhou, J.; So, W. Y.; Tong, P. C.; Ng, M. C.; Hansen, T.; Andersen, G.; Borch-Johnsen, K.; Jorgensen, T.; Tres, A.; Fuertes, F.; Ruiz-Echarri, M.; Asin, L.; Saez, B.; van Boven, E.; Klaver, S.; Swinkels, D. W.; Aben, K. K.; Graif, T.; Cashy, J.; Suarez, B. K.; van Vierssen Trip, O.; Frigge, M. L.; Ober, C.; Hofker, M. H.; Wijmenga, C.; Christiansen, C.; Rader, D. J.; Palmer, C. N.; Rotimi, C.; Chan, J. C.; Pedersen, O.; Sigurdsson, G.; Benediktsson, R.; Jonsson, E.; Einarsson, G. V.; Mayordomo, J. I.; Catalona, W. J.; Kiemeney, L. A.; Barkardottir, R. B.; Gulcher, J. R.; Thorsteinsdottir, U.; Kong, A.; Stefansson, K. Two variants on chromosome 17 confer prostate cancer risk, and the one in TCF2 protects against type 2 diabetes. *Nat Genet* **2007**, *39* (8), 977-83.

[93] Thomas, G.; Jacobs, K. B.; Yeager, M.; Kraft, P.; Wacholder, S.; Orr, N.; Yu, K.; Chatterjee, N.; Welch, R.; Hutchinson, A.; Crenshaw, A.; Cancel-Tassin, G.; Staats, B. J.; Wang, Z.; Gonzalez-Bosquet, J.; Fang, J.; Deng, X.; Berndt, S. I.; Calle, E. E.; Feigelson, H. S.; Thun, M. J.; Rodriguez, C.; Albanes, D.; Virtamo, J.; Weinstein, S.; Schumacher, F. R.; Giovannucci, E.; Willett, W. C.; Cussenot, O.; Valeri, A.; Andriole, G. L.; Crawford, E. D.; Tucker, M.; Gerhard, D. S.; Fraumeni, J. F., Jr.; Hoover, R.; Hayes, R. B.; Hunter, D. J.; Chanock, S. J. Multiple loci identified in a genome-wide association study of prostate cancer. *Nat Genet* **2008**, *40* (3), 310-5.

[94] Eeles, R. A.; Kote-Jarai, Z.; Giles, G. G.; Olama, A. A.; Guy, M.; Jugurnauth, S. K.; Mulholland, S.; Leongamornlert, D. A.; Edwards, S. M.; Morrison, J.; Field, H. I.; Southey, M. C.; Severi, G.; Donovan, J. L.; Hamdy, F. C.; Dearnaley, D. P.; Muir, K. R.; Smith, C.; Bagnato, M.; Ardern-Jones, A. T.; Hall, A. L.; O'Brien, L. T.; Gehr-Swain, B. N.; Wilkinson, R. A.; Cox, A.; Lewis, S.; Brown, P. M.; Jhavar, S. G.; Tymrakiewicz, M.; Lophatananon, A.; Bryant, S. L.; Horwich, A.; Huddart, R. A.; Khoo, V. S.; Parker, C. C.; Woodhouse, C. J.; Thompson, A.; Christmas, T.; Ogden, C.; Fisher, C.; Jamieson, C.; Cooper, C. S.; English, D. R.; Hopper, J. L.; Neal, D. E.; Easton, D. F. Multiple newly identified loci associated with prostate cancer susceptibility. *Nat Genet* **2008**, *40* (3), 316-21.

[95] Tung, Y. C.; Yeo, G. S. From GWAS to biology: lessons from FTO. *Ann N Y Acad Sci* **2011**, *1220*, 162-71.

[96] Hofker, M.; Wijmenga, C. A supersized list of obesity genes. *Nat Genet* **2009**, *41* (2), 139-40.

[97] Tuomilehto, J.; Lindstrom, J.; Eriksson, J. G.; Valle, T. T.; Hamalainen, H.; Ilanne-Parikka, P.; Keinanen-Kiukaanniemi, S.; Laakso, M.; Louheranta, A.; Rastas, M.; Salminen, V.; Uusitupa, M. Prevention of type 2 diabetes mellitus by changes in lifestyle among subjects with impaired glucose tolerance. *N Engl J Med* **2001**, *344* (18), 1343-50.

[98] Knowler, W. C.; Barrett-Connor, E.; Fowler, S. E.; Hamman, R. F.; Lachin, J. M.; Walker, E. A.; Nathan, D. M. Reduction in the incidence of type 2 diabetes with lifestyle intervention or metformin. *N Engl J Med* **2002**, *346* (6), 393-403.

[99] Hattersley, A. T.; Pearson, E. R. Minireview: pharmacogenetics and beyond: the interaction of therapeutic response, beta-cell physiology, and genetics in diabetes. *Endocrinology* **2006**, *147* (6), 2657-63.

[100] Drucker, D. J. Glucagon-like peptides: regulators of cell proliferation, differentiation, and apoptosis. *Mol Endocrinol* **2003**, *17* (2), 161-71.

[101] Elbrecht, A.; Chen, Y.; Cullinan, C. A.; Hayes, N.; Leibowitz, M.; Moller, D. E.; Berger, J. Molecular cloning, expression and characterization of human peroxisome proliferator activated receptors gamma 1 and gamma 2. *Biochem Biophys Res Commun* **1996**, *224* (2), 431-7.

[102] Zhou, G.; Myers, R.; Li, Y.; Chen, Y.; Shen, X.; Fenyk-Melody, J.; Wu, M.; Ventre, J.; Doebber, T.; Fujii, N.; Musi, N.; Hirshman, M. F.; Goodyear, L. J.; Moller, D. E. Role of AMP-activated protein kinase in mechanism of metformin action. *J Clin Invest* **2001**, *108* (8), 1167-74.

[103] Foretz, M.; Hebrard, S.; Leclerc, J.; Zarrinpashneh, E.; Soty, M.; Mithieux, G.; Sakamoto, K.; Andreelli, F.; Viollet, B. Metformin inhibits hepatic gluconeogenesis in mice independently of the LKB1/AMPK pathway via a decrease in hepatic energy state. *J Clin Invest* **2010**, *120* (7), 2355-69.

[104] Miller, R. A.; Birnbaum, M. J. An energetic tale of AMPK-independent effects of metformin. *J Clin Invest* **2010**, *120* (7), 2267-70.

[105] Leitzmann, M. F.; Ahn, J.; Albanes, D.; Hsing, A. W.; Schatzkin, A.; Chang, S. C.; Huang, W. Y.; Weiss, J. M.; Danforth, K. N.; Grubb, R. L., 3rd; Andriole, G. L. Diabetes mellitus and prostate cancer risk in the Prostate, Lung, Colorectal, and Ovarian Cancer Screening Trial. *Cancer Causes Control* **2008**, *19* (10), 1267-76.

[106] Kirchhoff, K.; Machicao, F.; Haupt, A.; Schafer, S. A.; Tschritter, O.; Staiger, H.; Stefan, N.; Haring, H. U.; Fritsche, A. Polymorphisms in the TCF7L2, CDKAL1 and SLC30A8 genes are associated with impaired proinsulin conversion. *Diabetologia* **2008**, *51* (4), 597-601.

[107] Jain, K.; Logothetopoulos, J. Secretion of insulin in a perifusion system and conversion of proinsulin to insulin by pancreatic islets from hyperglycemic rats. *Diabetes* **1977**, *26* (7), 650-6.

Send Orders of Reprints at reprints@benthamscince.net
Advances in Genome Science, Vol. 1, 2013, 241-260

CHAPTER 10

Indispensable Roles of Plastids in *Arabidopsis thaliana* Embryogenesis – Update

Mehdi Nafati and Kentaro Inoue*

Department of Plant Sciences, University of California, Davis, CA, USA

Abstract: The plastid is an organelle vital to all photosynthetic and some non-photosynthetic eukaryotes. Although the plastid has its own genome, more than 95% of proteins in this organelle are encoded in the nucleus, representing slightly lower than 10% of all the nuclear-encoded proteins in the model plant *Arabidopsis thaliana*. Recent functional genomic and molecular genetic studies have shown that *ca.* 400 non-redundant nuclear genes are required for proper embryo formation of *A. thaliana*, and that more than one third of them encode plastid-localized proteins. Classification of these genes based on terminal phenotypes of knockout mutants has revealed that non-photosynthetic metabolic activity of plastids is a prerequisite for the transition of preglobular to globular embryos, and that plastid gene expression becomes significant at or after the globular stage. Functions of the products of 28 out of 105 genes shown to be required for embryo development at the globular and later stages remain unknown. Several approaches to address this issue are discussed.

Keywords: *Arabidopsis thaliana*, embryogenesis, globular embryo, plastid, preglobular embryo, SeedGenes.

BACKGROUND

Plastids are organelles derived from an ancient form of cyanobacteria by endosymbiosis [1] and are vital for all photosynthetic and some nonphotosynthetic eukaryotes. In higher plants, plastids are present in all cell types except male gametophytes of certain species [2, 3]. Plastids exist in several distinct forms, such as chloroplasts in photosynthetic tissues, chromoplasts in some yellow, orange and red reproductive tissues, amyloplasts in non-colored storage tissues, and undifferentiated proplastids in meristematic cells. Most of these plastids are inter-convertible, and their development is closely associated with plant growth and development [4]. In addition to the oxygenic photosynthetic activity of chloroplasts

***Address correspondence to Kentaro Inoue:** Department of Plant Sciences, University of California, One Shields Avenue, Davis, CA 95616, USA; Tel 1-530-752-7931; Fax 1-530-752-9659; Email kinoue@ucdavis.edu which leads to carbon fixation, numerous metabolic processes such as the biosynthesis of fatty acids, amino acids, tetrapyrroles and various isoprenoids, including carotenoids and precursors to gibberellins, take place in plastids [5-8]. Hence, properly functioning plastids are essential for the viability of plants and this idea has been supported by recent genetic and functional genomic studies [9, 10].

During the evolution of plastids, most of the genes in the cyanobacterial endosymbiont were transferred to the host nuclear genome [1]. The resultant plastid still contains its own genome, which encodes about 100 proteins including major components of the photosynthetic electron transport machinery such as the D1 subunit of photosystem II, the large subunit of ribulose 1,5-bisphosphate carboxylase/oxygenase (RUBISCO), which directly catalyzes CO_2 fixation, and the ACCD subunit of the multi-subunit acetyl-CoA carboxylase (MS-ACC), which catalyzes the committed step of fatty acid biosynthesis [11]. However, the vast majority of plastid-localized proteins, including the small subunit of RUBISCO and most MS-ACC subunits, are encoded in the nuclear genome, and most of these proteins are synthesized on cytoplasmic ribosomes as precursors with N-terminal extensions called the transit peptides, which are cleaved off the precursors by stromal processing peptidase (SPP) [12]. Transit peptide-dependent protein import across the double-membrane envelope of plastids is catalyzed by two distinct protein complexes in the outer and inner membranes, called TOC and TIC (translocon at the outer- and inner-envelope-membrane of chloroplasts), respectively [13]. Based on extensive evaluation of several prediction programs that identify proteins with transit peptides, a total of 2,100 nuclear genes were estimated to encode plastid proteins in the model plant *Arabidopsis thaliana* [14]. Furthermore, no more than 100 nuclear genes encode transit peptide-less proteins, which include most outer envelope proteins [15], a few inner envelope proteins [16, 17] and α-carbonic anhydrase that is targeted to the chloroplast through a secretory pathway [18].

In the life cycle of flowering plants, embryogenesis is a crucial developmental period, which can be divided into morphogenesis and maturation phases [19]. In *A. thaliana*, embryo morphogenesis begins with the single-celled zygote, which undergoes a stereotypical cell division pattern giving rise to preglobular, globular, heart, torpedo, linear cotyledon, bent-cotyledon, and mature green stage embryos.

Undifferentiated plastids begin to develop into chloroplasts and increase their numbers at the torpedo stage before embryos enter into the maturation phase [20]. Our understanding of the molecular basis for *A. thaliana* embryo development has been rapidly advanced by results of molecular genetic and functional genomic studies, such as the SeedGenes project led by Dr. David W. Meinke (http://www.seedgenes.org), which presents comprehensive information about *A. thaliana* genes that are essential for seed development [21, 22], the Chloroplast Function Database by RIKEN (http://rarge.psc.riken.jp/chloroplast/), which lists a collection of transposon and T-DNA insertion lines in genes encoding chloroplast-localized proteins [23], and the Gene Networks in Seed Development project led by Drs Robert B. Goldberg and John J. Harada (http://seedgenenetwork.net), which utilizes laser capture microdissection, microarray and high-throughput sequencing technologies to profile the mRNA sets present in different seed regions and compartments throughout development [24].

A cytological study showed that plastids in *A. thaliana* embryonic cells remain as undifferentiated non-photosynthetic forms without detectable starch accumulation until the late globular stage when grana become visible [20]. Although the exact roles of these plastids remain unclear, a number of nuclear genes encoding plastid-targeted proteins have been found to be required for embryogenesis. Previously we made use of publicly available datasets and literature to elucidate the relevance of plastid activity to plant embryogenesis [9]. Three hundred thirty-nine non-redundant genes were found to be required for proper embryo formation in *A. thaliana*. Among them, 108 appeared to encode plastid-localized proteins and were further categorized based on the terminal phenotype of the knockout mutants. The outcome of this *in silico* analysis, taken together with the gene expression data, suggested that i) metabolic activities are required from the preglobular stage throughout all stages of embryo development, and ii) plastid gene expression becomes necessary during and/or after the globular stage to sustain various activities of the organelle including photosynthesis [9]. Soon after this study, an elegant work by Dr. Meinke's group presented a comprehensive list of nuclear genes that encode plastid-localized proteins necessary for proper seed development [10]. Furthermore, several additional genes encoding plastid proteins have been reported to be required for embryo development [25-27]. In this article, we provide an updated list of genes encoding plastid proteins required for *A. thaliana* embryogenesis which are categorized based on the terminal phenotypes of knockout mutants.

RECENT STUDIES ON NUCLEAR GENES ENCODING PLASTID PROTEINS NECESSARY FOR EMBRYOGENESIS IN *ARABIDOPSIS THALIANA*

The SeedGenes database Release 8, December, 2010 [22] lists 481 genes that give abnormal seed phenotypes when disrupted by mutation. Among them, 402 genes cause arrests at various stages of embryo development. It was previously estimated that 500-750 genes are essential for seed development [28, 29]. Therefore, the SeedGenes data may cover 55-80% of the entire complement of genes necessary for seed development.

Bryant *et al.* examined mutants present in the RIKEN collection [23] and identified a total of 381 nuclear genes encoding plastid-localized proteins associated with various phenotypes related to seed development [10]. Their data indicate that while genes encoding proteins involved in photosynthesis are not essential for embryo development, their disruption can lead to several phenotypic defects during seed development, such as the loss of pigmentation. They also showed that gametophyte lethals are underrepresented among knockouts of plastid protein-encoding genes. Among the genes necessary for proper seed development are 119 genes whose disruption causes embryo lethality. The authors' findings agree with those by Hsu *et al.* in that the proper embryo development requires the biosynthesis of amino acids and nucleotides, as well as expression of plastid-encoded genes. Bryant *et al.* further noted that the chloroplast-located *accD* gene is essential for embryo development in *A. thaliana* but not in barley (*Hordeum vulgare*) and *Brassica napus*, and discussed that the latter two species contain duplicated nuclear genes whose products are imported into the chloroplasts and can compensate for the lack of the chloroplast-encoded protein [10].

More recently, three additional nuclear genes encoding plastid-localized proteins have been reported to be essential for *A. thaliana* embryo development. The first one is *SECA2*, which together with another essential gene, *SCY2*, is proposed to encode components of the second chloroplastic SEC (cpSEC) system [26]. The SEC system catalyzes protein export in prokaryotic cells [30]. In chloroplasts, the first cpSEC system is believed to sort proteins from the stroma to the thylakoid lumen to assist assembly of photosynthetic machinery [31]. Indeed, disruption of genes encoding the first cpSEC machinery, namely *SCY1* (*AT2G18710*), *SECA1* (*AT4G01800*), and *SECE1* (*AT4G14870*), resulted in albino seedlings and sucrose-dependent heterotrophic growth, but did not cause embryo arrest [26]. These findings led the authors to suggest that the second cpSEC machinery, which consists of SECA2 and SCY2, is responsible for protein sorting to the inner envelope membrane, the process required for the biogenesis of embryonic plastids [26].

The second new gene for a plastid-localized protein essential for embryo development is *BELAYA SMERT'* (*BSM*), which encodes a plastidic mitochondrial transcription termination factor (mTERF) protein [27]. The mTERF family proteins are found in Metazoa and plants, and include DNA-binding regulators of mitochondrial transcription in animals [32]. BSM is one of 11 *A. thaliana* mTERF proteins that were shown to be targeted to plastids [27]. Babiychuk *et al.* showed that BSM is involved in plastid translation and that the *bsm*-null seeds, which were arrested at the globular stage of embryo development, could be germinated *in vitro*. Interestingly, further growth of the *bsm*-null mutant, albeit quite slow, could be achieved by external application of cytokinin and auxin, which appeared to promote cell proliferation and *de novo* organogenesis despite the lack of functional plastids. By characterizing mosaic plants, which were generated by a heat-inducible gene knockout system using the complemented *bsm*-null mutant and thus contained both BSM+ and *bsm*-null cells in a single plant, a non-cell-autonomous function of plastids during plant development was demonstrated. That is, the *bsm*-null cells turned albino, while BSM+ cells remained green because they contained functional plastids. The *bsm*-null tissues could develop to a certain extent apparently assisted by growth-promoting compounds supplied by the neighboring BSM+ cells. Interestingly, the *bsm*-null tissues suffered disruption in proper cell identity and patterning [27]. These results indicate the presence of several layers of complexity in the function of plastids in plant development.

The third gene encodes SPP, the enzyme that catalyzes removal of the stromal-targeting transit peptide from nuclear-encoded precursor proteins [12]. SPP is encoded by a single gene in *A. thaliana*, and a previous study using the antisense strategy demonstrated its importance in plant viability [33]. However, no study had shown the effect of the knockout of *SPP* until Trosch and Jarvis demonstrated in 2011 by careful analysis that *spp*-null resulted in embryo arrest at the 16-cell stage, which is intermediate between the stages arrested by *toc75* (2-cell stage) and by *tic110* (globular-stage) [25]. Toc75 is a protein import channel in the outer envelope membrane and Tic110 functions in protein import at the inner envelope membrane by recruiting molecular chaperone and/or forming the protein-conducting channel [13]. Based on this, Trosch and Jarvis (2011) discussed multiple versions of very insightful hypotheses that explain the different phenotypes of the *spp*, *toc75*, and *tic110* mutants [25].

UPDATED LIST OF NUCLEAR GENES ENCODING PLASTID PROTEINS NECESSARY FOR *A. THALIANA* EMBRYOGENESIS

Before this work, two lists of *A. thaliana* nuclear genes encoding plastid-located proteins necessary for embryo development had been published [9, 10]. The Hsu list consists of 108 genes and the Bryant list includes 119 genes, with 88 genes being found in both lists. Of the 108 genes in the Hsu list, two were found to be no longer suitable: the study reporting the embryo-lethality of the *TIC32* (*AT4G23430*) knockout was retracted [34], and the putative DEVH RNA helicase (*AT1G70070*) was localized to cytoplasmic granules often seen in close proximity to but distinct from chloroplasts [35]. The Bryan list contained *TIC32*, but not *AT1G70070*. Thirty-one genes in the Bryant list are absent in the Hsu list. Conversely, 19 genes in the Hsu list are not found in the Bryant list. Together with the three additional genes, the updated list consists of a total of 140 genes (Table **1**). The SeedGenes list of 402 genes includes *TIC32* but does not contain *SECA2*, *BSM* and *SPP*. The SeedGenes website also provides a supplemental list of pending additions, which lists *SECA2* as one of seven genes in the embryo-defective class (as of April 2nd, 2012). Hence, the new list represents 34% of non-redundant genes known to be essential for proper embryo development, similar to what was reported previously [9]. This fraction is close to four times larger than the proportion of genes encoding plastid-localized proteins (*ca.* 9%) in the entire *A. thaliana* nuclear genome [9]. This overrepresentation of plastid protein-coding genes appears to support the importance of plastid functions for plant viability from the early stages of development. In the meantime, it is also possible that duplication of genes encoding plastid-localized proteins may occur less frequently than that for non-chloroplast proteins, and thus availability of embryo-defective mutants may be skewed toward genes encoding plastid proteins.

As had been done in the Hsu list, the identified genes in the updated list were categorized based on two criteria. The first criterion was timing, which used the reported terminal phenotype of the null mutants arrested at the preglobular (I), globular (II), transition of globular to heart (III), and cotyledon stages (IV), respectively. For a gene with a single mutant allele showing heterogeneous seed phenotypes, or one with multiple alleles showing different phenotypes, the earliest stage was used for grouping because we considered that it should be the stage when the gene is first required. The second criterion was function, which was assigned to one of six categories: metabolism (M), plastid gene maintenance and expression (PGME), protein trafficking (PT), protein homeostasis (PH), membrane transport (MT), and others including unknown (U) (Table **1**).

The Group I genes are required for proper development of preglobular embryos and/or their conversion to globular embryos. Thirteen genes were added to the Hsu list of 19 in this group. Among the total of 32 genes, 16 genes encode various enzymes including those responsible for the biosynthesis of histidine and nicotinamide adenine dinucleotide, 12 code for proteins involved in plastid DNA replication, transcription, and translation, two are for transit peptide-dependent protein import (Toc75 and SPP), and two others are categorized into PH as they encode a molecular chaperone (GrpE) and a chloroplast protease (DegP), respectively (Table **1**). Compared to the Hsu list, updated Group I includes eight additional genes in the PGME category although genes in the M category are still overrepresented. Eight genes were added to Group II, resulting in a total of 46 members that are dispensable for globular embryo formation but become necessary for further embryo morphogenesis and maturation. Consistent with the Hsu list, Group II is enriched with genes in the PGME category, including two types of regulatory proteins, one containing pentatricopeptide repeat (PPR) domains and another with mTERF motifs (Table **1**). Four genes were added to Group III, three genes in the PGME category and one in the U category, while one gene each for PGME and PH and four U genes were added to Group IV.

Consequently, Group III is over-represented by PGME and appears more similar to Group II, whereas Group IV is more diverse than the former three in functional categories as in the Hsu list (Fig. **1**). Together, the updated list is overall consistent with the previous finding [9], showing a shift in functionalities necessary at the two early stages of embryo development. This list also supports the previous finding about the importance of plastid gene expression regulators for embryo development [10], as it includes a total of 22 genes encoding proteins containing PPR or mTERF domains (Table **1**).

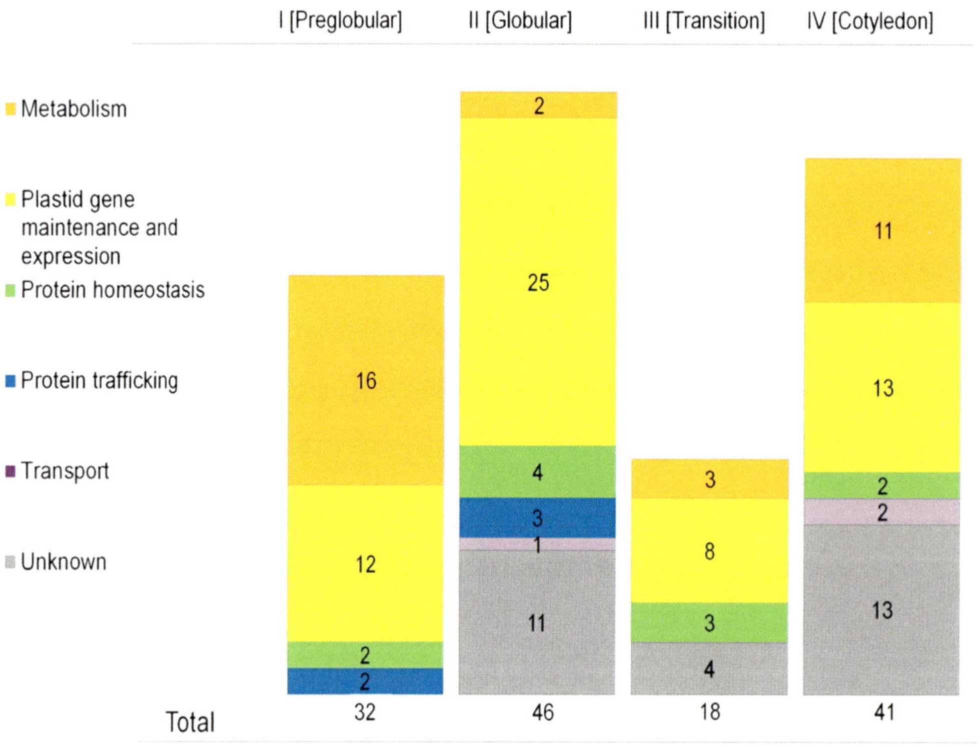

Figure 1: Functional grouping of genes encoding plastid proteins essential for *A. thaliana* embryogenesis. Genes essential for *A. thaliana* embryogenesis and encoding plastidic proteins are grouped by their mutant phenotypes and predicted functions. Predicted functions are based on sequence comparison and/or available experimental data, and divided into six categories: metabolism (orange blocks), plastid gene maintenance and expression (yellow blocks), protein homeostasis (green blocks), protein trafficking (blue blocks), transport (purple blocks), and others including unknown (gray blocks).

DOES PLASTID DIVISION PLAY A ROLE IN EMBRYO DEVELOPMENT?

As an organelle of endosymbiotic origin, the plastid depends for its biogenesis on posttranslational protein import and binary fission. Table **1** lists three genes encoding proteins necessary for the protein import (*TOC75*, *TIC110* and *SPP*), and available data clearly show that protein import is an essential process for embryo development [36-39]. By contrast, the machinery for the binary fission of the organelle appears to be dispensable for plant viability *e.g.*, [40, 41]. Nonetheless, during normal embryogenesis, plastid number increases, particularly from the heart to linear cotyledon stages [20]. Interestingly, this change coincides with increasing levels of *FTSZ* transcripts, which encode stromal components of the plastid division machinery [40], but not of transcripts for other division components (Fig. **2**). Indeed, a few genes required for embryo development in Table **1** were reported to be related to plastid division. The first one is *AtYLMG1-1* (*AT3G07430*) whose knockout led at the cytological level to both disruption of proper nucleoids distribution and arrest of chloroplast divisions, although causal relationships of these two responses remain elusive [42]. Another gene present in our list is *ARC2*, which encodes chaperonin 60α and its knockout caused altered plastid division [43]. Whether embryo-lethality of the *arc2* mutant is due to a chloroplast division defect or another yet to be discovered mechanism is unknown [44].

HOW CAN WE ELUCIDATE THE UNKNOWN FUNCTIONS OF THE PRODUCTS OF GENES ESSENTIAL FOR EMBRYOGENESIS?

It is obvious that the genes listed in Table **1** are important because they are essential for the viability of plants from the early stages of development. In order to address how plastids play essential roles in embryogenesis, one needs to know the functions of the proteins encoded by these essential genes. Fortunately, products of most of the genes listed in Table **1** have been characterized by biochemical assays or their functions could be deduced based on similarities in their primary structures to known proteins from different organisms. Indeed this information allowed us to make the functional categorizations (Fig. **1** and Table **1**). Nonetheless, the list includes quite a few genes whose products have unknown functions, but which are essential at the globular and later stages (Fig. **1**). How can their functions be studied? Biochemical and physiological characterization of the aborted embryos of the null mutants may be hindered due to the small size of the organs. Based on recent advancement in various methods and increasing knowledge, molecular genetic approaches other than a simple gene knockout would be obvious choices. One approach would be to down-regulate the expression of a gene by using an inducible or a development-stage specific

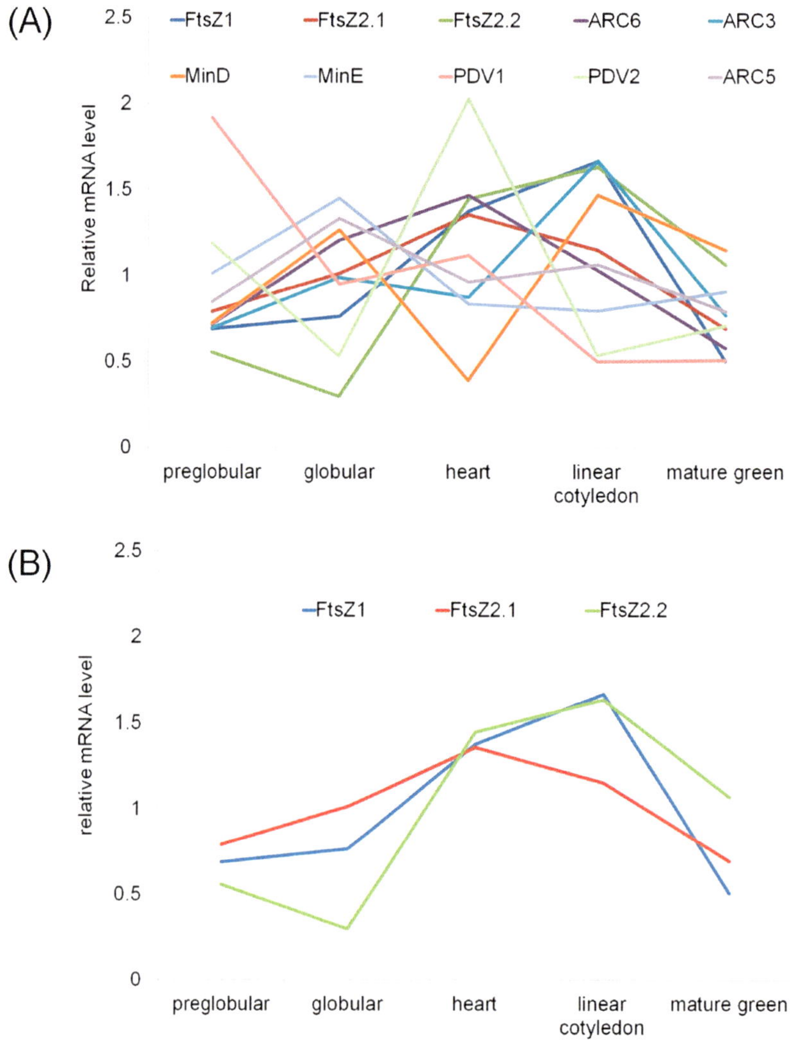

Figure 2: Expression pattern of FtsZ genes during *A. thaliana* embryogenesis. Histogram showing the expression values of genes coding for ARC6 (At5g42480), ARC3 (At1g75010), MinD (At5g24020), MinE (At5g10490), PDV1 (At5g53280), PDV2 (At2g16070), ARC5 (At3g19720),

FtsZ1 (At5g55280), FtsZ2.1 (At2g36250), and FtsZ2.2 (At3g52750) in *Arabidopsis thaliana* embryo at different stages of development (preglobular, globular, heart, linear cotyledon, mature green). Panel A shows expression of all the genes, and panel B shows expression of genes encoding FtsZ homologs. Data were extracted from the Gene Networks in Seed Development project (http://seedgene-network.net), and normalized to an average value.

promoter, and to examine the phenotypes of the resultant plants. Indeed, Huang *et al.* successfully used an inducible RNA interference strategy to demonstrate that *OEP80*, whose knockout causes an embryo arrest at the globular stage [45], is important for chloroplast development at postembryonic stages [46]. However, this approach may not always work. For example, if a low level of expression or a slow rate of protein turnover is sufficient to maintain the functionality of the gene product, phenotypic defects may not be observed unless the gene is completely knocked out. Another approach would be to test the effects of partial loss-of-function of the gene product. This can be done by examining EMS-mutagenized lines with point mutations in the gene of interest [47] or a genetic complementation assay. An example of the latter assay is found in a recent report by Hsu *et al.*, who successfully used a constitutive promoter (Cauliflower Mosaic Virus 35S promoter) to rescue the embryo-lethal *oep80*-null mutant, and also to show that the nucleotide sequence encoding the 52 N-terminal amino acids was not required for functional expression of the *OEP80* gene and that an additional C terminal tag remained intact without disrupting the functionality of OEP80 [48]. Apparently, this strategy may not provide useful information if the targeted residues are essential for the functionality of the gene product, or vice versa. For example, a recent study showed by a complementation assay that certain residues of the protein encoded by the *TOC75* gene, whose knockout results in embryo lethality at the preglobular stage [36, 39], are not essential for the function of the protein although these residues are required for cleavage of an extension signal [49]. An alternative approach would be to use dominant negative mutants, as was demonstrated for the *TIC110* gene, whose knockout causes an embryo-lethality at the globular stage [38].

Table 1: Nuclear Genes Encoding Plastid Proteins Required for *Arabidopsis thaliana* Embryogenesis

Gene[a]	ID[b]	Function (Cat)[c]
I. Genes whose disruption causes embryo abortion at preglobular stage (32)		

AT1G31860	C2'	phosphoribosyl-ATP pyrophosphohydrolase/phosphoribosyl-AMP cyclohydrolase (His biosynthesis)	(M)
AT1G34430	NC	dihydrolipoamide S-acetyltransferase	(M)
AT1G74960	C	ketoacyl-acyl carrier protein synthase	(M)
AT2G01350	C	ouinolinic acid phosphoribosyl transferase (NAD biosynthesis)	(M)
AT2G28880	NC	4-amino-4-deoxychorismate synthase (folate biosynthesis)	(M)
AT2G36230	C	N'-5'-phosphoribosyl-formimino-5-aminoimidazole-4-carboxamide ribonucleotide isomerase (His biosynthesis)	(M)

Table 1: cont…

*AT3G04790**	C	ribulose-5-phosphate isomerase	(M)
AT3G54660	C2	glutathione reductase	(M)
*AT3G55010**	C	phosphoribosylformylglycinamidine cyclo-ligase (PUR5)	(M)
*AT4G00620**	C	folic acid biosynthesis	(M)
AT4G26500	C	activator of plastidic and mitochondrial desulfurases (AtSufE)	(M)
AT4G26900	C	imidazole glycerol phosphate synthase (His biosynthesis)	(M)
AT4G33680	C	Lys-LL-diaminopimelate amino transferase	(M)
AT5G10330	C	histidinol phosphate aminotransferase (His biosynthesis)	(M)
AT5G14760	C	L-Asp oxidase (NAD biosynthesis)	(M)
AT5G50210	C	ouinolinate synthase (NAD biosynthesis)	(M)
*AT1G08840^*	C2	DNA replication helicase	(PGME)
*AT1G48350**	NC	chloroplast 50S Ribosomal Protein L18	(PGME)
*AT1G59990**	NC	putative RNA helicase	(PGME)
*AT2G33800**	C2	chloroplast 30S Ribosomal Protein S5	(PGME)
AT3G02660	C	Tyr-tRNA synthetase	(PGME)
AT3G18290	C2	zinc finger protein-related	(PGME)
*AT3G63490**	C	chloroplast 50S Ribosomal Protein L1	(PGME)
*AT5G13510**	NC	chloroplast 50S Ribosomal Protein L10	(PGME)
*AT5G14320**	C2	30S ribosomal protein S13	(PGME)
AT5G24120	C	RNA polymerase sigma subunit SigE	(PGME)
*AT5G39980^*	C2	PPR protein	(PGME)
*AT5G50390^*	NC	PPR protein	(PGME)
*AT2G47940^*	NC	DegP2	(PH)
AT5G17710	C	co-chaperone GrpE family protein	(PH)
AT3G46740	C	translocon at the outer-envelope-membrane of chloroplasts 75 (Toc75)	(PT)
AT5G42390#	**C**	**stromal processing peptidase (SPP)**	(PT)
II. Genes whose disruption causes embryo abortion at globular stage (46)			
*AT4G31780**	C	MGDG synthase 1 (MGD1)	(M)
AT4G39120	C	myoinositol monophosphatase like 2 (His biosynthesis)	(M)
AT1G05190	NC	ribosomal protein L6 family protein	(PGME)

*AT1G07320**	NC	plastid ribosolam protein L4	(PGME)
AT1G30610	C	PPR-containing protein	(PGME)
AT1G62750	C	elongation factor Tu family protein	(PGME)
AT1G78630	NC	ribosomal protein L13 family protein	(PGME)
*AT1G79490^*	NC	PPR-containing protein	(PGME)
AT2G01860	C	PPR-containing protein	(PGME)
AT2G04530	NC	RNase Z	(PGME)
*AT2G04842^*	NC	Thr-tRNA synthetase	(PGME)
*AT2G36000**	NC	mitochondrial transcription termination factor family (mTERF) domain protein	(PGME)

Table 1: cont…

*AT2G38770^*	C	U5 associated protein	(PGME)
AT3G18390	NC	chloroplast splicing factor (CRS1)	(PGME)
AT3G48110	C2	Gly-tRNA synthetase	(PGME)
AT3G49240	C	PPR-containing protein	(PGME)
*AT4G02790**	C	putative ribosome biogenesis GTPase	(PGME)
AT4G02990#	**C**	**plastidic mTERF, BELAYA SMERT (BSM)**	(PGME)
AT4G04350	C	Leu-tRNA synthetase	(PGME)
AT4G26300	NC	Arg-tRNA synthetase	(PGME)
AT4G29060	C	elongation factor Ts family protein	(PGME)
*AT4G39620^*	C	PPR-containing protein	(PGME)
AT5G03800	C	similar to PPR-containing protein	(PGME)
AT5G16715	C	Val-tRNA synthetase	(PGME)
AT5G18570	C	GTP1/OBG family protein	(PGME)
AT5G26742	NC	DEAD/DEAH box RNA helicase	(PGME)
AT5G50280	NC	PPR-containing protein	(PGME)
AT1G02560	C	ATP-dependent Clp protease proteolytic subunit (ClpP5)	(PH)
AT3G04340	NC	FtsH protease family protein	(PH)
AT3G10670	C	plastidic SufC-like protein (Fe-S cluster biogenesis)	(PH)
AT5G18820	NC	RuBisCO subunit binding-protein alpha subunit (Cpn60a)	(PH)
AT1G06950	C	translocon at the inner-envelope-membrane of chloroplasts 110 (Tic110)	(PT)
AT1G21650#	**C**	**SecA homolog (SECA2)**	(PT)
*AT2G31530^*	C	SecY homolog (SCY2)	(PT)
AT4G32400	C	nucleotide export	(T)
AT1G10510	C	Leu-rich repeat family protein, similar to ribonuclease inhibitor	(U)
*AT1G10910**	C	unknown	(U)
AT2G25660	C	unknown	(U)
AT3G12080	C	GTP-binding family protein	(U)
*AT3G24560^*	C	ATP binding	(U)
AT5G19620	C	outer envelope protein 80	(U)
AT5G40160	C	ankyrin repeat protein	(U)

AT5G57930	C	Fe-S cluster related	(U)
AT5G63420	C	RNA-metabolising metallo-beta-lactamase family protein l RNase J homolog.	(U)
*AT5G64580**	C2	unknown	(U)
AT5G66055	C	ankyrin repeat protein (AKRP); essential for plastid differentiation	(U)
III. Genes whose disruption causes embryo abortion at the transition of globular to cotyledon stage (18)			
AT3G20440	C2	Branching enzyme 1BE1 (Glycoside Hydrolase family 13)	(M)
AT3G25860	C	dihydrolipoamide S-acetyltransferase	(M)
AT5G16390	C	biotin carboxyl-carrier protein 1 (BCCP1, also known as CAC1A)	(M)
AT1G23400	C2	chloroplast intron splicing factor	(PGME)

Table 1: cont…

*AT1G64790**	NC	translational activator (GCN1)	(PGME)
AT1G75350	NC	ribosomal protein L31 family protein	(PGME)
AT2G03050^$	C	similar to the mTERF	(PGME)
AT3G29290^	NC	PPR-containing protein	(PGME)
*AT4G34620**	C	mitochondrial ribosomal protein S16	(PGME)
AT5G22800	C2	aminoacyl-tRNA synthetase	(PGME)
AT5G67570	C4	PPR-containing protein	(PGME)
AT1G79560	C2'	AAA and metalloprotease (FtsH12)	(PH)
AT2G04030	C	heat shock protein (Hsp90)	(PH)
AT3G16290	C2	FtsH protease (AAA ATPase)	(PH)
AT2G37920^	NC	copper transporter related	(U)
*AT3G14900**	C	unknown	(U)
AT3G61780	C2	unknown	(U)
AT5G62990	C	unknown	(U)
IV. Genes whose disruption causes embryo abortion at cotyledon stage (41)			
AT1G08510	C4	acyl-acyl carrier protein thioesterase	(M)
AT1G48850	NC	chorismate synthase/5-enolpyruvylshikimate-3-phosphate phospholyase	(M)
AT1G78580^	C	trehalose-6-phosphate synthase 1	(M)
AT2G19450^	C4	diacylglycerol O-acyltransferase / acyl CoA:diacylglycerol acyltransferase	(M)
AT3G06350	C	dehydroquinate dehydratase; shikimate dehydrogenase	(M)
AT3G55610	C	delta 1-pyrroline-5-carboxylate synthetase B (Pro synthesis)	(M)
AT4G23100	C	gamma-glutamylcysteine synthetase	(M)
AT4G30580	C	2-acylglycerophosphoethanolamine acyltransferase	(M)
AT5G24400	C	6-phosphogluconolactonase	(M)
AT5G52920	C4	similar to pyruvate kinase isozyme G	(M)
AT5G61410	C	ribulose-5-phosphate-3-epimerase	(M)
AT1G06145^	UC	similar to PPR-containing protein	(PGME)
AT1G14610^	C	Val-tRNA synthetase	(PGME)
AT1G74970	NC	ribosomal protein S9	(PGME)
AT1G79350^	NC	DNA-binding protein	(PGME)

AT3G06430	NC	PPR-containing protein (AtPPR2)	(PGME)
AT3G10690	C3	DNA gyrase subunit A family protein	(PGME)
*AT3G18110^*	C	PPR-containing protein	(PGME)
*AT3G27750**	C2	VPS9 domain protein	(PGME)
AT3G49170	C	PPR-containing protein	(PGME)
AT4G20090	C	PPR-containing protein	(PGME)
*AT4G20740**	C	PPR protein	(PGME)
AT5G02250	C4	ribonuclease II family protein	(PGME)
AT5G27270	NC	PPR-containing protein	(PGME)

Table 1: cont...

*AT1G12410**	C	ClpR2	(PH)
AT2G28000	C3	Cpn60α (also known as ARC2)	(PH)
AT1G19800	C4	permease (TGD1, trigalactosyldiacylglycerol 1)	(T)
AT4G33460	NC	ABC-type transport protein	(T)
*AT1G21390^*	NC	unknown	(U)
AT1G49510	C2	unknown	(U)
AT1G56200	C3	unknown	(U)
*AT1G58210^*	C	kinase interacting family protein, similar to kinase interacting protein 1	(U)
*AT2G01735^*	C	RIE1 (zinc finger (C3HC4-type RING finger) family protein)	(U)
*AT2G39080**	NC	unknown	(U)
AT3G07430	C	YGGT family protein (AtYLMG1-1)	(U)
AT4G28210	C	unknown	(U)
*AT5G11890**	NC	unknown	(U)
AT5G22640	C	MORN (Membrane Occupation and Recognition Nexus) repeat-containing protein	(U)
*AT5G51020**	C	outer envelope protein involved in plastid division, CRUMPLED LEAF (CRL)	(U)
AT5G53860	C	unknown	(U)
*AT5G58250**	NC	unknown	(U)
Genes whose disruption causes embryo lethality, but its terminal phenotype unknown (3)			
*AT1G17220**	NC	chloroplast translation Initiation factor (cpIF2)	(PGME)
*AT3G01370**	C2	chloroplast intron splicing factor	(PGME)
*AT5G04810***	C	PPR protein	(PGME)

[a]Five types of genes are indicated.

*Thirty genes, which are not listed in Hsu et al. (2010), but found in Bryant et al. (2011) and in SeedGenes.

**One gene, which are not found in Hsu et al. (2010) or in the 481 genes required for seed development in SeedGenes, but listed in Bryant *et al.* (2011). This gene is included in the supplemental list of pending additions to SeedGenes (updated on January 13th, 2011).

^Nineteen genes, which are not in Bryant et al.(2011), but in Hsu et al. (2010) and in SeedGenes.

#Three genes, which are not in Hsu et al. (2010), in Bryant et al. (2011), and in the 481 seed-development essential genes in SeedGenes. *SECA2* is included in the supplemental list of pending additions to SeedGenes (updated on January 13th, 2011). Knockout of *SPP* gene results in embryo arrest at 16-cell stage (Trosch and Jarvis 2011), which we categorize into preglobular.

$One gene, which is listed as terminal phenotype unknown in Hsu et al. (2010).

[b]ID indicates identity confidence as defined in SeedGenes database. C, confirmed by the presence of multiple alleles causing an embryo arrest or by the genetic complementation assay; C2, having multiple null-lines with insertions in different portions of exons showing different terminal phenotypes; C2', having multiple alleles including the ones with 5'UTR insertion causing a phenotype different from those with coding region insertions; C3, having null-mutant seeds that can germinate and develop into seedlings but not beyond; C4, having null-mutant seeds that can germinate and develop into mature plants; NC, not confirmed (only a single mutant allele with sequence information is available); UC, uncertain (insertion or mutation site not within coding region or 5' UTR

and either downstream of stop codon or more than 250 bp upstream of start codon. The information of identity confidence extracted from SeedGenes database has been further updated with recent reports.
[c]Functions are assigned based on annotation in public database (GreenPhylDB http://greenphyl.cirad.fr/cgi-bin/greenphyl.cgi, individual publications or manually. Cat, functional categories: M, metabolism; PGME, plastid gene maintenance and expression; PH, protein homeostasis; PT, protein trafficking; T, transport; U, unknown and others.

CONCLUSION

The plastids of higher plants have gained the ability to develop into a variety of non-photosynthetic types and play vital roles in the growth and development of the organisms. However, except for chloroplasts in photosynthetic tissues, the plastid functions essential at each developmental stage are not completely understood yet. The current study updates our previous *in silico* study [9] and complements others [10] to shed light on the functions of plastids during embryogenesis, the earliest stage of plant development following zygote formation.

From the updated list, we can make the following hypotheses: The key processes from the onset of embryogenesis at the preglobular stage include the biosynthesis of electron carriers and amino acids, and disruption of either process can lead to embryo lethality. From the globular stage, genes coding for ribosomal proteins, t-RNA synthases, as well as a large number of pentatricopeptides proteins, putatively associated with protein biosynthesis, are necessary. Disruption of those genes can also be embryo-lethal, as recently exemplified by the disruption of AtPPR2, which binds to plastid 23S rRNA [50], and of the *ASD* gene, which encodes the 50S ribosomal protein L21 [51]. Developing seeds turn green only after the heart stage. Hence, it seems logical that mutation of genes associated with photosynthesis may lead to loss of pigmentation and later to seedling auxotrophy rather than early embryo lethality.

ACKNOWLEDGEMENTS

We thank Professor Daniel Potter and Dr. Shih-Chi Hsu for their helpful comments on the manuscript. The work on the essential plastid proteins Toc75 and OEP80 in the Inoue laboratory has been supported by the Division of Molecular and Cellular Biosciences at the US National Science Foundation (Grant no. 1050602).

CONFLICT OF INTEREST

The authors confirm that this chapter content has no conflict of interest.

DISCLOSURE

This is an updated report of our earlier review article, "Indispensable roles of plastids in Arabidopsis thaliana embryogenesis". Current Genomics, 2010, 11, 338-349. Shih-Chi Hsu, Mark F Belmore, John J Harada, Kentaro Inoue.

REFERENCES

[1] Keeling, P. J., The endosymbiotic origin, diversification and fate of plastids. *Philosophical transactions of the Royal Society of London. Series B, Biological sciences* **2010,** *365* (1541), 729-748.

[2] Corriveau, J. L.; Coleman, A. W., Rapid Screening Method to Detect Potential Biparental Inheritance of Plastid DNA and Results for over 200 Angiosperm Species. *Am J Bot* **1988,** *75* (10), 1443-1458.

[3] Zhang, Q.; Liu, Y.; Sodmergen, Examination of the cytoplasmic DNA in male reproductive cells to determine the potential for cytoplasmic inheritance in 295 angiosperm species. *Plant & cell physiology* **2003,** *44* (9), 941-951.

[4] Waters, M.; Pyke, K., Plastid development and differentiation. In *Annual Plant Reviews. Plastids*, Møller, S. G., Ed. Blackwell: Oxford, 2005; Vol. 13, pp 30-59.

[5] Seo, M.; Koshiba, T., Complex regulation of ABA biosynthesis in plants. *Trends in plant science* **2002,** *7* (1), 41-48.

[6] DellaPenna, D.; Pogson, B. J., Vitamin synthesis in plants: Tocopherols and carotenoids. *Annu Rev Plant Biol* **2006,** *57*, 711-738.

[7] Neuhaus, H. E.; Emes, M. J., Nonphotosynthetic metabolism in plastids. *Annu Rev Plant Phys* **2000,** *51*, 111-140.

[8] Yamaguchi, S.; Kamiya, Y., Gibberellin biosynthesis: Its regulation by endogenous and environmental signals. *Plant and Cell Physiology* **2000,** *41* (3), 251-257.

[9] Hsu, S. C.; Belmonte, M. F.; Harada, J. J.; Inoue, K., Indispensable Roles of Plastids in Arabidopsis thaliana Embryogenesis. *Current genomics* **2010,** *11* (5), 338-349.

[10] Bryant, N.; Lloyd, J.; Sweeney, C.; Myouga, F.; Meinke, D., Identification of nuclear genes encoding chloroplast-localized proteins required for embryo development in Arabidopsis. *Plant physiology* **2011,** *155* (4), 1678-1689.

[11] Sato, S.; Nakamura, Y.; Kaneko, T.; Asamizu, E.; Tabata, S., Complete structure of the chloroplast genome of Arabidopsis thaliana. *DNA research : an international journal for rapid publication of reports on genes and genomes* **1999,** *6* (5), 283-290.

[12] Richter, S.; Zhong, R.; Lamppa, G., Function of the stromal processing peptidase in the chloroplast import pathway. *Physiologia plantarum* **2005,** *123* (4), 362-368.

[13] Li, H. M.; Chiu, C. C., Protein transport into chloroplasts. *Annu Rev Plant Biol* **2010,** *61*, 157-180.

[14] Richly, E.; Leister, D., An improved prediction of chloroplast proteins reveals diversities and commonalities in the chloroplast proteomes of Arabidopsis and rice. *Gene* **2004,** *329*, 11-16.

[15] Inoue, K., The chloroplast outer envelope membrane: The edge of light and excitement. *J Integr Plant Biol* **2007,** *49* (8), 1100-1111.

[16] Miras, S.; Salvi, D.; Ferro, M.; Grunwald, D.; Garin, J.; Joyard, J.; Rolland, N., Non-canonical transit peptide for import into the chloroplast. *Journal of Biological Chemistry* **2002,** *277* (49), 47770-47778.

[17] Nada, A.; Soll, J., Inner envelope protein 32 is imported into chloroplasts by a novel pathway. *J Cell Sci* **2004,** *117* (17), 3975-3982.

[18] Villarejo, A.; Buren, S.; Larsson, S.; Dejardin, A.; Monne, M.; Rudhe, C.; Karlsson, J.; Jansson, S.; Lerouge, P.; Rolland, N.; von Heijne, G.; Grebe, M.; Bako, L.; Samuelsson, G., Evidence for a protein transported through the secretory pathway en route to the higher plant chloroplast. *Nat Cell Biol* **2005,** *7* (12), 1224-1231.

[19] Goldberg, R. B.; de Paiva, G.; Yadegari, R., Plant embryogenesis: zygote to seed. *Science* **1994,** *266* (5185), 605-614.

[20] Mansfield, S.; Briarty, L., Early embryogenesis in Arabidopsis thaliana. II. The developing embryo. . *Can. J. Bot.* **1991,** *69*, 461–476.

[21] Tzafrir, I.; Dickerman, A.; Brazhnik, O.; Nguyen, Q.; McElver, J.; Frye, C.; Patton, D.; Meinke, D., The Arabidopsis SeedGenes Project. *Nucleic acids research* **2003,** *31* (1), 90-93.

[22] Meinke, D.; Muralla, R.; Sweeney, C.; Dickerman, A., Identifying essential genes in Arabidopsis thaliana. *Trends in plant science* **2008,** *13* (9), 483-491.

[23] Myouga, F.; Akiyama, K.; Motohashi, R.; Kuromori, T.; Ito, T.; Iizumi, H.; Ryusui, R.; Sakurai, T.; Shinozaki, K., The Chloroplast Function Database: a large-scale collection of Arabidopsis Ds/Spm- or T-DNA-tagged homozygous lines for nuclear-encoded chloroplast proteins, and their systematic phenotype analysis. *The Plant journal : for cell and molecular biology* **2010,** *61* (3), 529-542.

[24] Le, B. H.; Cheng, C.; Bui, A. Q.; Wagmaister, J. A.; Henry, K. F.; Pelletier, J.; Kwong, L.; Belmonte, M.; Kirkbride, R.; Horvath, S.; Drews, G. N.; Fischer, R. L.; Okamuro, J. K.; Harada, J. J.; Goldberg, R. B., Global analysis of gene activity during Arabidopsis seed development and identification of seed-specific transcription factors. *Proceedings of the National Academy of Sciences of the United States of America* **2010,** *107* (18), 8063-8070.

[25] Trosch, R.; Jarvis, P., The stromal processing peptidase of chloroplasts is essential in Arabidopsis, with knockout mutations causing embryo arrest after the 16-cell stage. *PloS one* **2011,** *6* (8), e23039.

[26] Skalitzky, C. A.; Martin, J. R.; Harwood, J. H.; Beirne, J. J.; Adamczyk, B. J.; Heck, G. R.; Cline, K.; Fernandez, D. E., Plastids contain a second sec translocase system with essential functions. *Plant physiology* **2011,** *155* (1), 354-369.

[27] Babiychuk, E.; Vandepoele, K.; Wissing, J.; Garcia-Diaz, M.; De Rycke, R.; Akbari, H.; Joubes, J.; Beeckman, T.; Jansch, L.; Frentzen, M.; Van Montagu, M. C.; Kushnir, S., Plastid gene expression and plant development require a plastidic protein of the mitochondrial transcription termination factor family. *Proceedings of the National Academy of Sciences of the United States of America* **2011,** *108* (16), 6674-6679.

[28] McElver, J.; Tzafrir, I.; Aux, G.; Rogers, R.; Ashby, C.; Smith, K.; Thomas, C.; Schetter, A.; Zhou, Q.; Cushman, M. A.; Tossberg, J.; Nickle, T.; Levin, J. Z.; Law, M.; Meinke, D.; Patton, D., Insertional mutagenesis of genes required for seed development in Arabidopsis thaliana. *Genetics* **2001,** *159* (4), 1751-1763.

[29] Franzmann, L. H.; Yoon, E. S.; Meinke, D. W., Saturating the Genetic-Map of Arabidopsis-Thaliana with Embryonic Mutations. *Plant Journal* **1995,** *7* (2), 341-350.

[30] du Plessis, D. J.; Nouwen, N.; Driessen, A. J., The Sec translocase. *Biochimica et biophysica acta* **2011,** *1808* (3), 851-865.

[31] Albiniak, A. M.; Baglieri, J.; Robinson, C., Targeting of lumenal proteins across the thylakoid membrane. *Journal of experimental botany* **2012,** *63* (4), 1689-1698.

[32] Roberti, M.; Polosa, P. L.; Bruni, F.; Manzari, C.; Deceglie, S.; Gadaleta, M. N.; Cantatore, P., The MTERF family proteins: mitochondrial transcription regulators and beyond. *Biochimica et biophysica acta* **2009,** *1787* (5), 303-311.

[33] Zhong, R.; Wan, J.; Jin, R.; Lamppa, G., A pea antisense gene for the chloroplast stromal processing peptidase yields seedling lethals in Arabidopsis: survivors show defective GFP import in vivo. *The Plant journal : for cell and molecular biology* **2003,** *34* (6), 802-812.

[34] Hormann, F.; Kuchler, M.; Sveshnikov, D.; Oppermann, U.; Li, Y.; Soll, J., Tic32, an essential component in chloroplast biogenesis. (vol 279, pg 34756, 2004). *Journal of Biological Chemistry* **2009,** *284* (42), 29240.

[35] Kobayashi, K.; Otegui, M. S.; Krishnakumar, S.; Mindrinos, M.; Zambryski, P., INCREASED SIZE EXCLUSION LIMIT 2 encodes a putative DEVH box RNA helicase involved in plasmodesmata function during Arabidopsis embryogenesis. *The Plant cell* **2007,** *19* (6), 1885-1897.

[36] Baldwin, A.; Wardle, A.; Patel, R.; Dudley, P.; Park, S. K.; Twell, D.; Inoue, K.; Jarvis, P., A molecular-genetic study of the Arabidopsis Toc75 gene family. *Plant physiology* **2005,** *138* (2), 715-733.

[37] Kovacheva, S.; Bedard, J.; Patel, R.; Dudley, P.; Twell, D.; Rios, G.; Koncz, C.; Jarvis, P., In vivo studies on the roles of Tic110, Tic40 and Hsp93 during chloroplast protein import. *The Plant journal : for cell and molecular biology* **2005,** *41* (3), 412-428.

[38] Inaba, T.; Alvarez-Huerta, M.; Li, M.; Bauer, J.; Ewers, C.; Kessler, F.; Schnell, D. J., Arabidopsis tic110 is essential for the assembly and function of the protein import machinery of plastids. *The Plant cell* **2005,** *17* (5), 1482-1496.

[39] Hust, B.; Gutensohn, M., Deletion of core components of the plastid protein import machinery causes differential arrest of embryo development in Arabidopsis thaliana. *Plant Biol (Stuttg)* **2006,** *8* (1), 18-30.

[40] Schmitz, A. J.; Glynn, J. M.; Olson, B. J.; Stokes, K. D.; Osteryoung, K. W., Arabidopsis FtsZ2-1 and FtsZ2-2 are functionally redundant, but FtsZ-based plastid division is not essential for chloroplast partitioning or plant growth and development. *Molecular plant* **2009,** *2* (6), 1211-1222.

[41] Vitha, S.; Froehlich, J. E.; Koksharova, O.; Pyke, K. A.; van Erp, H.; Osteryoung, K. W., ARC6 is a J-domain plastid division protein and an evolutionary descendant of the cyanobacterial cell division protein Ftn2. *The Plant cell* **2003,** *15* (8), 1918-1933.

[42] Kabeya, Y.; Nakanishi, H.; Suzuki, K.; Ichikawa, T.; Kondou, Y.; Matsui, M.; Miyagishima, S. Y., The YlmG protein has a conserved function related to the distribution of nucleoids in chloroplasts and cyanobacteria. *BMC plant biology* **2010,** *10*, 57.

[43] Apuya, N. R.; Yadegari, R.; Fischer, R. L.; Harada, J. J.; Zimmerman, J. L.; Goldberg, R. B., The Arabidopsis embryo mutant schlepperless has a defect in the chaperonin-60alpha gene. *Plant physiology* **2001,** *126* (2), 717-730.
[44] Suzuki, K.; Nakanishi, H.; Bower, J.; Yoder, D. W.; Osteryoung, K. W.; Miyagishima, S. Y., Plastid chaperonin proteins Cpn60 alpha and Cpn60 beta are required for plastid division in Arabidopsis thaliana. *BMC plant biology* **2009,** *9*, 38.
[45] Patel, R.; Hsu, S. C.; Bedard, J.; Inoue, K.; Jarvis, P., The Omp85-related chloroplast outer envelope protein OEP80 is essential for viability in Arabidopsis. *Plant physiology* **2008,** *148* (1), 235-245.
[46] Huang, W.; Ling, Q.; Bedard, J.; Lilley, K.; Jarvis, P., In vivo analyses of the roles of essential Omp85-related proteins in the chloroplast outer envelope membrane. *Plant physiology* **2011,** *157* (1), 147-159.
[47] Till, B. J.; Reynolds, S. H.; Greene, E. A.; Codomo, C. A.; Enns, L. C.; Johnson, J. E.; Burtner, C.; Odden, A. R.; Young, K.; Taylor, N. E.; Henikoff, J. G.; Comai, L.; Henikoff, S., Large-scale discovery of induced point mutations with high-throughput TILLING. *Genome research* **2003,** *13* (3), 524-530.
[48] Hsu, S. C.; Nafati, M.; Inoue, K., OEP80, an essential protein paralogous to the chloroplast protein translocation channel Toc75, exists as a 70-kD protein in the Arabidopsis thaliana chloroplast outer envelope. *Plant molecular biology* **2012,** *78* (1-2), 147-158.
[49] Shipman-Roston, R. L.; Ruppel, N. J.; Damoc, C.; Phinney, B. S.; Inoue, K., The significance of protein maturation by plastidic type I signal peptidase 1 for thylakoid development in Arabidopsis chloroplasts. *Plant physiology* **2010,** *152* (3), 1297-1308.
[50] Lu, Y.; Li, C.; Wang, H.; Chen, H.; Berg, H.; Xia, Y., AtPPR2, an Arabidopsis pentatricopeptide repeat protein, binds to plastid 23S rRNA and plays an important role in the first mitotic division during gametogenesis and in cell proliferation during embryogenesis. *The Plant journal : for cell and molecular biology* **2011,** *67* (1), 13-25.
[51] Yin, T.; Pan, G.; Liu, H.; Wu, J.; Li, Y.; Zhao, Z.; Fu, T.; Zhou, Y., The chloroplast ribosomal protein L21 gene is essential for plastid development and embryogenesis in Arabidopsis. *Planta* **2011,** *235*, 907-921.

Send Orders of Reprints at reprints@benthamscince.net
Advances in Genome Science, Vol. 1, 2013, 261-292

CHAPTER 11

Mass Spectrometry-Based Approaches Toward Absolute Quantitative Proteomics

Keiji Kito[1] and Takashi Ito[2,*]

[1]Department of Life Science, School of Agriculture, Meiji University, Kawasaki 214-8571, Japan and [2]Department of Biophysics and Biochemistry, Graduate School of Science, University of Tokyo, Tokyo 113-0032, Japan

Abstract: Mass spectrometry has served as a major tool for the discipline of proteomics to catalogue proteins in an unprecedented scale. With chemical and metabolic techniques for stable isotope labeling developed over the past decade, it is now routinely used as a method for relative quantification to provide valuable information on alteration of protein abundance in a proteome-wide scale. Recently, absolute or stoichiometric quantification of proteome is becoming feasible, in particular, with the development of strategies with isotope-labeled standards composed of concatenated peptides. On the other hand, remarkable progress has been also made in label-free quantification methods based on the number of identified peptides. Here we review these mass spectrometry-based approaches for absolute quantification of proteome and discuss their implications.

Keywords: Quantitative proteomics, Mass spectrometry, Absolute quantification, Stable isotope labeling, Label-free.

1. INTRODUCTION

Proteomics is a promising discipline in the aim of profiling of whole proteins, identifying their modifications and interactions, and providing abundance information. In so doing, it contributes to reveal molecular systems underlying various biological phenomena and provide valuable information on disease mechanisms as well as biomarkers for diagnosis and prognosis. Mass spectrometry (MS) enables protein identification and quantification in a large-scale, and hence serves as the most powerful tool to address these purposes of proteomics [1-6].

In general, proteins are digested with specific proteases, such as trypsin, into a

***Address correspondence to Takashi Ito:** Department of Biophysics and Biochemistry, Graduate School of Science, University of Tokyo, Tokyo 113-0032, Japan; 2-11-16 Yayoi, Bunkyo-ku, Tokyo 113-0032 Japan; Tel: 81-3-5841-3047; Fax: 81-3-5841-4691; E-mail: ito@bi.s.u-tokyo.ac.jp

Christian Neri (Ed)

distinct set of peptides. These peptides are then ionized and introduced into MS instruments. Tandem mass spectrometry (MS/MS) has been widely used in proteome analysis, where a peptide ion to be analyzed is selectively isolated and fragmented to obtain an MS/MS spectrum. Since the pattern of these fragment ions is dependent on the amino acid sequence of its precursor ion, it can be compared with theoretical ones calculated from protein sequences in the database. If the MS/MS spectrum measured for a peptide shows a reliable matching with one of the theoretical spectra, the peptide is identified as the one used for the calculation of the theoretical spectrum.

Since the MS/MS analysis identifies individual peptides, it is amenable to be combined with liquid chromatography (LC) as a peptide separation step that precedes MS. LC-MS/MS is a commonly used technical platform, where peptides are separated with reverse-phase LC, directly ionized *via* electrospray ionization, and introduced into mass spectrometer. Alternatively, peptides eluted from LC are separately collected and spotted on a plate for matrix assisted laser desorption ionization (MALDI). These systems especially have an impact on the analysis of highly complex samples composed of a great number of proteins. Their power can be further enhanced by the use of multidimensional separation with cation-exchange and reverse-phase LC to identify more than 1,000 proteins at once [7-10]. Thus, MS is a high throughput and powerful protein identification system.

MS has also an ability to provide quantitative information in proteome analysis. Various strategies have been developed to detect relative changes in protein abundance between the samples to be compared [11-13]. They can be divided into two categories; one is based on stable isotope labeling and the other is the so-called label-free method.

In the former, samples to be analyzed for relative quantification are differentially labeled with stable isotope, combined, and simultaneously subjected to MS. Ratio of peak intensity between the ions of an isotope pair (*i.e.,* light and heavy peptide ions) gives relative difference in abundance of the protein from which the peptide is derived. Various methods have been developed for stable isotope labeling of proteome, including chemical, proteolytic, and metabolic labeling ones [14-18].

By contrast, in the latter or label-free methods, samples to be compared are individually introduced into mass spectrometer, and peak intensities or frequencies of identification measured in separate runs are compared to calculate relative change in protein abundance [11, 12]. Accordingly, label-free strategies are much simpler but more error-prone than isotope labeling ones, due to systematic variations among individual runs and stochastic nature of the indices used for calculation.

Although these techniques are widely used for relative quantification in proteomics studies, the ultimate goal of quantitative proteomics is definitely the absolute measurement of protein abundance. Absolute quantification provides a far more precise description of molecular events in the biological processes than relative quantification. Furthermore, absolute quantification data can be readily exchanged among different studies to facilitate data integration. Recent advent in proteomics has enabled MS-based absolute quantification by extending the technologies originally developed for relative quantification [19]. Thus, we review both relative and absolute quantification techniques either with or without stable isotope labeling.

2. QUANTITATIVE TECHNIQUES BASED ON STABLE ISOTOPE LABELING

In stable isotope labeling methods, quantitative values were calculated based on ratio of peak intensity between isotope pair ions, one of which contains only native isotopes whereas the other bears a number of heavy stable isotopes. Having the same chemical properties, two peptide ions of an isotope pair can be simultaneously introduced into mass spectrometer but clearly distinguished by their mass difference. Simultaneous measurement of ion intensities in the same analysis eliminates not only run-to-run variations in performance of LC and MS, amounts of injected sample, and ion-suppression effect of co-eluting ions, but also limitations in intrinsic dynamic range of each MS, thereby enabling more accurate and reliable quantification.

2.1. Relative Quantification

In relative quantification, samples to be compared are differentially labeled with stable isotopes. These samples are then combined and subjected to quantitative MS.

Peak intensity ratio between heavy and light peptides is measured to learn relative change in protein abundance. Various labeling methods have been developed, including chemical, proteolytic, isobaric, and metabolic labeling techniques.

The most popular method of chemical labeling would be the isotope-coded affinity tags (ICAT) approach, in which a compound containing stable isotope is coupled to Cys residues in proteins [20, 21]. Different isotopomers of the compound, each having a unique mass, are used for labeling of different samples. Following this differential labeling procedure, the samples are mixed and subjected to protease digestion followed by affinity-purification of Cys-containing peptides. Besides the original ICAT methods, strategies have been reported for chemical labeling of carboxyl, amino, or thiol moieties [14-18, 22]. In the labeling method coupled with hydrolysis, proteins are digested with protease in the presence of ^{18}O-labeled water so that ^{18}O is incorporated at the carboxyl end of each peptide [23, 24]. Another *in vitro* labeling method is an isobaric tagging strategy, in which each tag has an identical mass but contains stable isotopes at unique atomic positions to generate a reporter ion with a unique mass-charge ratio upon fragmentation [25, 26].

An obvious advantage of these *in vitro* labeling methods is that they can be applied to tissue samples, for which *in vivo* labeling is difficult or practically impossible. On the other hand, they require tangled procedures for sample handling and labeling. Accordingly, the samples to be compared are forced to be combined at later stages of the procedure, allowing variations in earlier steps to affect accuracy of quantification.

An alternative labeling method is the *in vivo* metabolic incorporation of stable isotopes, where cells are cultivated in a medium supplemented with an appropriate stable isotope-labeled nutrient that is essential for growth to achieve labeling of whole proteome [27-30]. Amino acids are usually used as the labeled essential nutrient, and such a procedure is often called SILAC for stable isotope labeling by amino acids in cell culture [28]. Proteome from multicellular and mammalian organisms, such as worm, fly, rat, and mouse, can be also metabolically labeled by feeding stable isotope-labeled microorganisms or nutrient-defined diet [31-33].

An obvious advantage of these metabolic labeling methods over the chemical and hydrolytic ones is that protein samples can be combined at much earlier step in the procedure, for instance, at the stage of cell harvest. Accordingly, the effect of experimental errors can be minimized. On the other hand, it is difficult, or even impossible, to complete stable isotope labeling of animal tissues and metabolically inactive cells. To partly circumvent these difficulties, an interesting method termed culture-derived isotope tags (CDITs) or super-SILAC was developed, in which relative abundance of proteins in the tissues refractory to metabolic labeling are quantified using the isotope-labeled proteome of the cell line derived from the tissue [34-36].

These methods have been widely implemented in relative quantification in proteomics studies. With the remarkable analytical power of multidimensional LC, relative differences of several hundred to thousand proteins were successfully quantified among yeast cells grown in different culture conditions [37-41]. The strategy was also applied to distinguish contaminants in purification; quantitative comparison between purified and mock-purified samples revealed specific components in a protein complex and a cellular compartment [42, 43]. It can be combined with specific purification techniques for protein complex and post-translationally modified peptides to grasp dynamics of protein interactions [42, 44-50] and phosphorylation in stimulated cells and cell cycle progression [46, 49, 51-57].

2.2. Absolute Quantification

In MS-based absolute quantification, a known amount of isotope-labeled authentic standard is mixed with the analyte, and the mixture is introduced into mass spectrometer. The absolute amount of the analyte is calculated from the ratio of ion intensity between the analyte and its standard. Accordingly, known amounts of stable isotope-labeled synthetic peptides, proteins, or peptide concatemers have been used as a standard for absolute or stoichiometric quantification of proteins. Different types of standard are added to the samples at the different stages of the procedure, and have distinct pros and cons (Fig. (**1**), Table (**1**)). Accordingly, the most suitable standard should be selected, depending on the purpose of the experiment, or on whether it intends to quantify a small number of targets including their post-translational modifications, obtain highly accurate data for a single unique protein, or measure absolute or stoichiometric abundance of many proteins.

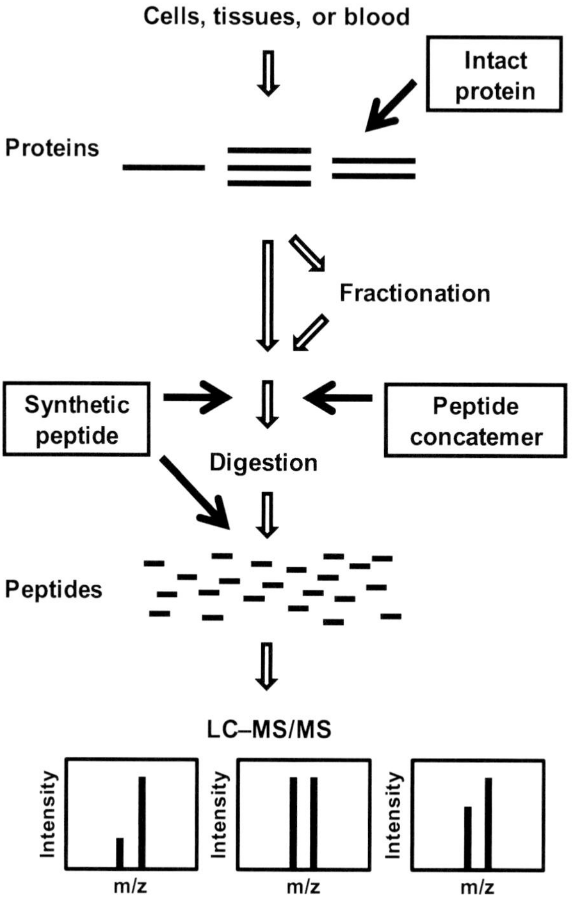

Figure 1: Strategies for absolute quantification using stable isotope-labeled standards. Different types of stable isotope standard are spiked at different steps of the sample preparation procedure. Intact protein standard can be spiked as soon as proteins are extracted from cells, tissues or bloods, even if subsequent fractionation steps (*e.g.,* SDS-PAGE, gel filtration) are included in the procedure. While synthetic peptide standard is spiked before or after digestion with protease, peptide-concatenated standard has to be spiked prior to digestion to allow co-proteolysis of target and standard. Note that synthetic peptide and peptide concatenated standard have to be spiked after protein fractionation steps.

2.2.1. Quantification with Stable Isotope-Labeled Peptide Standard

In the original report that uses a stable isotope-labeled peptide as an internal standard for MS-based absolute quantification [58], the amount of endogenous enkephalin in thalamus extract was measured with an ^{18}O-incorporated standard peptide. To reduce background noises and enhance the specificity of measurement, multiple reaction monitoring (MRM) mode, in which intensity of multiple fragment ions but not the peptide ion are monitored, was used for absolute quantification of enkaphalin in human tissue [59].

Table 1: Summary of approaches for absolute quantification of protein abundance with mass spectrometry

	Standard types	**Measured values**	**Spiking time point**	**Normalization**	**Accuracy**	**Coverage**	**Applicability to post-translational modification**	**Noise origin**
Isotope labeling	Synthetic peptide	Ratio to standard	Before or after digestion	—	Medium	Low	Applicable	S/N of ion peak, Missed cleavage
	Intact protein	Ratio to standard	Just after protein extraction	—	Very High	Low	Not	S/N of ion peak
	Peptide concatemer (QconCAT, PCS)	Ratio to standard	Prior to digestion	—	High	Medium	Not	S/N of ion peak, Missed cleavage (QconCAT)
Label-free	—	Peak intensity	—	Average of most intense three peaks	Low	Medium	Not	Variation of ionization efficiency
	—	Spectral count	—	Observable peptides	Low	High	Not	Stochastically calculated index

Stable isotope-labeled peptides were initially applied to absolute quantification of a specific protein [60]; three peptides from apolipoprotein A-I were selected, ^{2}H- and ^{13}C-labeled, and used as standard for quantification of its abundance in human serum. To increase specificity and sensitivity, measurement of difference in intensities of fragment ion(s) [SRM (selected reaction monitoring) or MRM mode] between analyte and standard peptide was implemented in absolute quantification of protein. Expression level of G protein-coupled receptor rhodopsin in rod outer segment membrane was quantified using a ^{2}H-labeled synthetic peptide [61]. For quantification of low abundance proteins, enrichment of the target protein is necessary. For instance, following SDS-PAGE separation of yeast extract, a gel slice containing the target protein was excised, mixed with ^{13}C- and ^{15}N-labeled standard peptides, and subjected to trypsin digestion followed by MS [62]. In this study, abundance of Sir2 and Sir4 were determined as ~10^{3} copies per cell. Absolute abundance of proteins in blood or tissue was also quantified using synthetic peptides as isotope-labeled standards in the MRM mode. For instance, absolute amount of C-reactive protein, a well-known diagnostic marker for rheumatoid arthritis, was measured in human serum depleted of abundant proteins (serum albumin and immunoglobulin G) [63]. Similarly, GST-α in human liver, a marker for acute hepatocyte damage, was quantified distinctly from other isozymes [64].

Stable isotope-labeled synthetic peptides have been applied to simultaneous quantification of multiple proteins. For instance, the stoichiometry among the 10 components of human spliceosomal U1 small ribonucleoprotein complex was determined using chemical labeling of sample-derived and synthetic standard peptides with isotope-coded reagents after trypsin digestion [65]. Similarly, concentrations of 8 endogenous proteins in human serum were quantified by spiking isotope-labeled standard peptides [66]. In another study, isotope-labeled peptides were applied to absolute quantification of 32 key proteins in the postsynaptic density of rat, including calmodulin-dependent protein kinase, synaptic GTPase-activating protein, glutamate receptors, and scaffold proteins [67]; the absolute and stoichiometric abundance obtained from this study provided valuable information on abundance of receptor subtypes and protein interactions.

Taken together, isotope-labeled synthetic peptide is definitely a powerful tool for stoichiometric quantification for protein complexes [68-71], but also for absolute quantification in tissue or blood samples from animals and human [72-76]. However, it is too expensive to synthesize multiple stable isotope-labeled peptides for quantification of multiple proteins. In this context, it is interesting to note that the CDIT strategy described above allows quantification of tissue proteins without any labeled synthetic peptides as follows [34]. Absolute amounts of the proteins in the cultured cell line labeled *in vivo* are determined with unlabeled synthetic peptides. These quantified labeled proteins are, in turn, used as the standards for absolute quantification of unlabeled tissue proteins. Indeed, this strategy was successfully used to quantify 103 proteins in the mouse brain.

Peptide standard can be used to quantify not only protein abundance but also post-translational modifications. Isotope-labeled unphosphorylated and phosphorylated peptides can be used to quantify phosphorylation stoichiometry. For instance, stoichiometry of phosphorylation at Ser-1126 of separase, which negatively regulates its activity, was measured in different stages of the cell cycle either in a peptide ion scanning mode [77] or in SRM mode [62]. Simultaneous quantification of multiple phosphoryation sites was also reported. Two phosphorylated sites of cyclin-dependent kinases, which inhibit their activity, are regulated in a cell cycle-specific manner. Stoichiometry of four possible patterns of these phosphorylation sites was determined to reveal that isoforms unphosphorylated and phosphorylated at both sites were dominant in M-phase and S-phase, respectively, whereas those phosphorylated at either site were minorities [78]. Phosphorylation stoichiometry for signaling kinase involved in cell growth were successfully quantified, providing monitoring techniques for evaluation of response to kinase inhibitors as anticancer drugs [79, 80].

Modifications other than phosphorylation can be quantified. For instance, abundance of polyubiquitin chain branched at the Lys-48 was measured in mammalian cells treated with or without the proteasome inhibitor MG132 [81]. Ubiquitination is mediated *via* an isopeptide bond between the carboxy-terminal end of ubiquitin and the ε-amino group of a Lys residue in the substrate. Since the amino acid sequence at the C-terminal end of ubiquitin is Arg-Gly-Gly, tryptic digestion of ubiquitinated proteins produces peptides containing Lys conjugated with Gly-Gly or diglycine at its ε-amino group. In a study aiming at absolute quantification of ubiquitin conjugated sites, a peptide bearing diglycine-conjugated Lys residue was synthesized and used as an isotope-labeled standard. Topology of polyubiquitin chain of an *in vitro* ubiquitinated protein, or which of the seven Lys residues is used for branching, was also analyzed using ubiquitin-standard peptides [82]. Beside phosphorylation and ubiquitination, absolute quantification of farnesylation was reported for H-Ras using a ^{2}H-labeled, farnesylated carboxy-terminal peptide [83].

2.2.2. Quantification with Stable Isotope-Labeled Intact Protein

When a proteolytic standard peptide is used for absolute quantification, efficiency of protease digestion is critically important for accuracy, because incomplete digestion of the analyte leads to underestime its amount. While some studies optimized and monitored cleavage efficiency for specific peptides [60, 61], it is difficult to assess a large number of peptides in terms of cleavage efficiency. Notably, measured abundances of an analyte protein can differ substantially, depending on the tryptic peptides used as the standard, presumably because efficiency of trypsin digestion is different from one site to another [63]. In particular, when SDS-PAGE is used as a pre-fractionation step, recovery of peptides is affected not only by the efficiency of in-gel digestion but also by that of peptide extraction from gel slices [84].

An ideal standard for absolute quantification of a protein is obviously the protein *per se* that is labeled with stable isotope, because it can be spiked at the earliest stage of sample preparation to minimize experimental errors and shares exactly the same efficiency of protease digestion with the target proteins in the sample. For instance, concentrations of insulin in sera of normal individuals and diabetic patients were quantified with an isotope-labeled recombinant standard protein expressed in and purified from *E. coli* [85]. Similarly, ^{15}N-labeled recombinant standard proteins were used for absolute quantification of 6 proteins localized in postsynaptic density [86], and expression level of alcohol dehydrogenase isozyme ADH1C1 was quantified in human liver tissue using a ^{13}C- and ^{15}N-labeled recombinant intact protein [87].

An intriguing study was reported to directly compare two strategies, one with a synthetic peptide standard and the other with an intact protein standard, in absolute quantification of *Staphylococcus aureus* superantigenic toxins spiked into drinking water and urine samples [88]. The amount quantified with the synthetic peptide standards was smaller than that obtained with the intact protein standard, presumably because of incomplete digestion of the toxins in sample. This result indicates the power of the strategy using an intact protein as a stable isotope standard.

Isotope-labeled intact standard protein can be also used for finding out and quantification of post-translational modifications. Isotope-labeled free- and conjugated-ubiquitin allowed quantification of status of ubiquitin pool in cultured cells and tissues [89]. Peptides with modification site would have decreased peak intensities among others in a whole protein, depending on the degree of modification. Based on this principle, modification levels of anaphase promoting complex were tracked in different stages of cell cycle [90].

2.2.3. Peptide-Concatenated Standard

Although the strategies using synthetic peptides or intact proteins as stable isotope-labeled standards allow us to learn absolute amounts of specific proteins, a large scale analysis requires preparation and handling of many standard peptides/proteins, thereby raising many concerns. First, it is expensive to prepare many stable isotope-labeled synthetic peptides. Second, the purity of synthetic peptides is variable from one to another and is often unsatisfactory for accurate quantification. Third, it is also a daunting task to express and purify many recombinant proteins as stable isotope-labeled standards. Forth, since individual standards are differentially lost during the course of experiments, one cannot guarantee their precise amounts or even their equimolarity.

To overcome these bottlenecks, two groups independently conceived a strategy that uses a peptide concatemer as a standard, namely QconCAT [91] and PCS for peptide-concatenated standard [92]. In both strategies, tryptic peptides used for quantification are concatenated into a single artificial protein. This protein was metabolically labeled with stable isotope in *E. coli*, purified, and mixed with a protein sample to obtain absolute or stoichiometric quantities of multiple proteins (Fig. (**2**)). This peptide concatenation strategy eliminates the daunting task for the preparation of many standard peptides/proteins. Furthermore, since all isotope-labeled peptides are contained in a single protein, they are always added to the sample at exactly the same molarity. The QconCAT approach provided absolute quantitative data of more than 10 proteins in chick skeletal muscle of different developmental stages [91, 93]. It was also used for quantification of plasma proteins [94]; absolute amounts of 13 proteins in human plasma of 20-10,000 fmol/μl concentrations were successfully measured in an MRM mode.

However, it should be noted that, as discussed above, one of the critical keys for accurate quantification is to equalize cleavage efficiency between the analyte and its standard. It was shown that the amino acid context around a trypsin cleavage site substantially affects the efficiency of its digestion [95, 96]. In contrast to QconCAT, PCS contains each standard peptide with its natural flanking sequences on both sides to faithfully recapitulate the efficiency of tryptic cleavage of parental proteins or analytes (Fig. (**2**)). The involvement of flanking sequences was demonstrated to improve the accuracy of quantification, and led to accurate quantification of stoichiometry among 5 subunits in eIF2B stable complex of yeast within 5% measurement error [92]. Similarly, others successfully quantified absolute and stoichiometric abundance of each subunit of transducin, a heterotrimeric G-protein, using a PCS incorporating the flanking sequences of each tryptic peptide [97]. Notably, the QconCAT strategy, which lacks flanking sequences, resulted in an underestimation of the amounts of *Staphylococcus* toxins spiked into drinking water and urine samples when compared to the strategy using an intact protein standard [88], presumably because the target protein was less efficiently digested than the QconCAT. These studies highlight the importance of incorporation of flanking sequences into the standard to ensure highly accurate absolute or stoichiometric quantification.

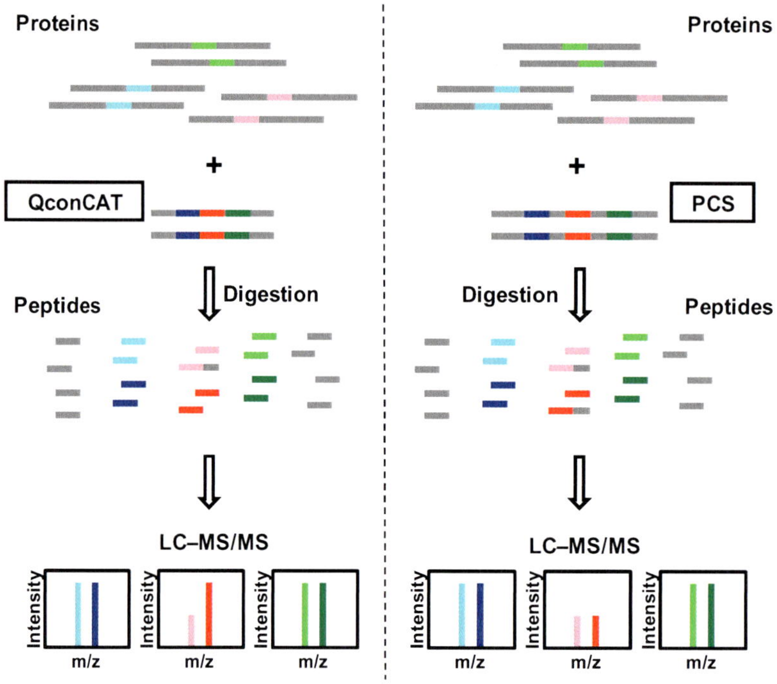

Figure 2: Stable isotope-labeled and peptide-concatenated standard for absolute or stoichiometric quantification. In strategy using peptide-concatenated standard, tryptic peptides to be quantified are selected from target proteins and concatenated to generate an artificial protein, which serves as a stable isotope-labeled standard. Colored fragments denote the selected tryptic peptides. Target and standard peptides that form an isotope pair are indicated by similar colors (blue and light blue, red and pink, and green and light green). Peptide-concatenation not only guarantees equimolar spiking of every standard peptide but also eliminates the need to prepare many standard molecules. Two types of peptide-concatenated standard, namely QconCAT [91] and PCS [92], have been reported. While each peptide included in QconCAT is the tryptic peptide *per se*, that in PCS carries its natural flanking sequences at both sides to faithfully recapitulate the efficiency of proteolytic cleavage in the target protein. Even a peptide excised less efficiently from the target protein (colored in pink) can be precisely quantified with PCS, because its standard (colored in red) is also excised from the PCS at a similar efficiency.

Peptide-concatenated artificial proteins are, in most cases, recovered in insoluble fraction. Since the order of the peptides affect the success rate of production in an *in vitro* translation system [98], an efficient algorithm is needed to optimize the peptide order for improvement of the solubility of peptide-concatenated standard proteins.

The insoluble-prone nature of peptide-concatenated artificial proteins not only makes their handling difficult but also limits their sizes. Thus, many standards have to be used concurrently in a large scale analysis. It then becomes important to know precise amounts of the standards to integrate quantitative data obtained from each of them. To solve this issue and to expand the scale of analysis, we proposed a hierarchical PCS strategy, in which each primary PCS includes a unique "bar-code" peptide and is quantified by a secondary PCS composed of the bar-code peptides [92]. It is also possible to adjust the amount of each primary PCS to extend the dynamic range of quantification.

3. QUANTITATIVE TECHNIQUES BASED ON LABEL-FREE STRATEGIES

Stable isotope-labeling strategies described above have enabled quantitative MS-based proteomics. However, they inevitably require additional steps for isotope labeling and/or preparation of the standards. By contrast, the so-called label-free (standard-free) method is simple and requires no additional experimental steps; it just exploits peak intensity of peptide ion or identification frequency for a particular protein to obtain quantitative data (Fig. (**3**), Table (**1**)). While absolute quantification with stable-isotope labeling can quantify only the proteins with corresponding isotope standards, label-free strategies can, in principle, quantify any protein from which a peptide is unambiguously identified. This implies that the label-free methods are amenable to a large scale analysis. On the other hand, they provide less accurate quantitative values than those by label-based ones, due to run-to-run variations and a stochastic nature of the measurement.

3.1. Relative Quantification

3.1.1. Quantification Based on Peak Intensity

In relative quantification based on peak intensity, each sample is separately subjected to MS. Peptide peak intensity is measured in individual runs and change

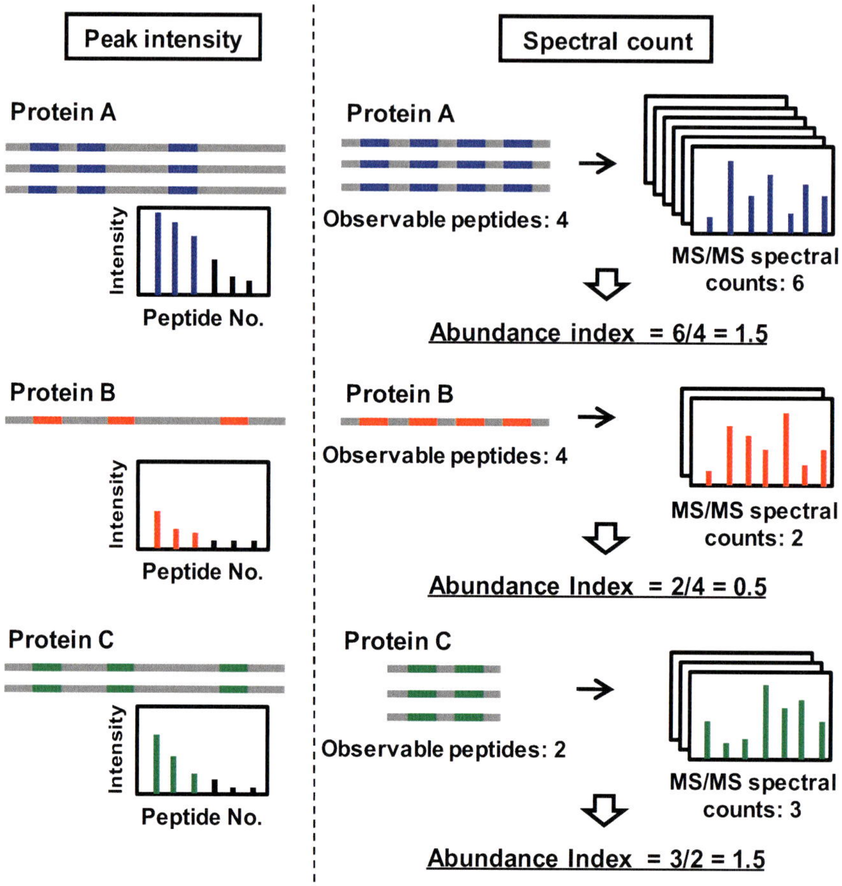

Figure 3: Label-free methods for absolute quantitative proteomics. In the strategy based on intensities of individual peptides, average of intensities of the three most intense ions is used to generate reliable absolute quantitative data. In the strategy based on identification frequency, spectral count (*i.e.,* number of identified MS/MS spectra) is the most useful indicator to estimate protein abundance. Since larger proteins tend to provide more observable peptides than smaller ones, each spectral count is divided by observable peptide number for normalization to ensure accuracy of quantification. For instance, while proteins A and C present at the same abundance have different spectral counts (*i.e.,* 6 and 3, respectively), they share the same normalized spectral count (*i.e.,* 1.5).

in protein abundance is calculated *via* a comparison among different analyses. This approach has been applied to quantification of relative change in protein expression [99, 100] and quantitative profiling of purified proteins to identify *bona fide* components of a protein complex and reveal dynamics of protein-protein interactions [101, 102]. Recently, sum of the fragment ion intensity of identified peptide was calculated for each protein and successfully used for relative protein abundance with significant reproducibility [103, 104].

In contrast to stable isotope labeling methods, label-free approach based on peak intensity is error-prone [105], due to run-to-run variations in performance of LC and MS, amounts of injected samples, and ion-suppression effect of co-detected ions, and also due to limited dynamic range of each mass spectrometry. Systematic errors induced by the first two factors can be normalized by spiking an identical amount of standard protein into every sample to be compared [106, 107], the total ion intensities over the entire analysis [99, 103], or using the average intensity ratios between target peptide and co-eluting peptides as a pseudo internal standard [108]. Furthermore, high reproducibility of retention time for each peptide is required to extract an ion pair from different runs, when either, but not both, of the pair is identified in MS/MS analysis. Otherwise, an algorithm to align peptide ion maps of different analyses has to be developed to compare peak intensity of each peptide ion [109].

3.1.2. Quantitative Approaches Based on Identification Frequency

An increase in protein abundance usually results in an increase in the number of identifications of its tryptic peptides, and *vice versa*. Thus, identification frequency, which is the number of identified peptides, precursor ions, or MS/MS spectra (spectral count) for each protein, can be used to estimate relative difference in protein abundance. Peptide identification number was originally applied to quantitative proteomics analysis of urine sample from healthy donors and patients [110]. Among the factors of identification frequencies, spectral count showed the highest correlation with relative protein abundance, suggesting it to be the best index for relative quantification [9]. Relative quantitative approaches were also taken to compare protein expression in yeast and mammalian cells under different culture conditions [99, 111, 112], and the screening of phosphotyrosine-binding proteins in mammalian cells [113].

An intriguing study compared relative abundance calculated from the spectral count with that obtained using a stable isotope-labeling method [112]. When only the peptides with high signal-to-noise ratio in the extracted ion chromatogram were included in the calculation, the two methods showed a positive correlation within 1.5-fold error. Quantitative data determined by isotope labeling and spectral count would compensate each other to improve accuracy of quantification.

3.2. Absolute Quantification

3.2.1. Quantification Based on Peak Intensity

Although difference in peak intensity has been exploited for relative quantification, individual peptides differ in propensities to be ionized and in efficiencies for isolation and detection, and they are also dependent on MS equipments. Accordingly, ion intensities can be different even among the peptides present at the same molarity (*e.g.*, peptides derived from a single protein). Thus, these differences should be corrected for absolute quantification.

To reduce the effect of variance in individual peak intensities, average of ion intensities for multiple peptides in a particular protein was used as a quantitative value to estimate protein abundance [114, 115]. A relationship was observed between protein abundance and average of intensities of the three most intense peptide ions. The average for the three most intense ions had an ability to predict the abundance of known amount of proteins with less than ~15% error. Stoichiometry of GroEL and GroES of *E.coli* was successfully quantified to be 2:1, consistent with known structure of this molecular chaperone [114]. While proven useful by an evaluation using a mixture of known amounts of proteins, this strategy may generate a large error for low abundance proteins, for which only a small number of peptides can be identified, and fails to quantify proteins from which only one or two peptides are identified.

3.2.2. Quantitative Approaches Based on Identification Frequency

Label-free approaches based on identification frequency, which had been applied to relative quantification, was also modified to estimate absolute protein abundance. Larger proteins have more peptides that are detectable by MS than smaller ones. Accordingly, the number of identified peptides, precursor ions, and MS/MS spectra (spectra count) may be different between the two proteins that exist at the same abundance but differ in sizes. Thus, to know absolute or stoichiometric quantity, one should use the percentile fraction of the protein sequence covered by identified peptides or normalize the number of identifications by either protein size or observable peptide kinds.

3.2.2.1. Normalization with Protein Size

It was observed that the number of identified peptides correlates with the codon adaptaion index of the protein, which serves as an indicator of protein abundance [7]. The number of identified peptides per protein molecular weight provided stoichiometric abundance for clathrin and its adaptor proteins in clathrin-coated vesicles [116]. Clustering analysis using abundance index, or the number of identified precursor ions per protein molecular weight, successfully identified proteins associated with SAGA, a histone acetyltransferase complex, as those having a similar pattern of purification abundance [117]. Spectral count per protein length was further optimized to give normalized spectral abundance factor (NSAF); the latter index is calculated by dividing the former by the sum of all spectral count per protein length in each MS run to eliminate variation of each analysis [118]. Quantitative analysis with NSAF revealed the subunit stoichiometry of yeast Mediator, a transcriptional coactivator complex [119]. A study examined correlations between relative protein abundance and three indices (*i.e.,* sequence coverage, identified peptide number, and spectral count) by spiking known amounts of 6 different proteins into yeast cell extract [9]. While sequence coverage and peptide numbers failed to linearly correlate with relative abundance, a strong correlation was observed between relative abundance and spectral count with 2-order of magnitude. Furthermore, spectral count per molecular weight of each protein had a linear correlation with stoichiometry of 6 different proteins [9], suggesting that spectral count is the most useful index for absolute or stoichiometric quantification based on the frequency of identifications. Normalization strategy using protein size was recently implemented into intensity based quantitative approach [103]. Total intensity of fragment ions for identified peptides is divided by protein length to calculate absolute protein abundance.

3.2.2.2. Normalization with Observable Peptides

The number of observable peptides in each protein has been used as an alternative normalization factor. For instance, the protein abundance index (PAI) is calculated by dividing the number of identified precursor ions by the number of theoretically observable tryptic peptides for each protein, to roughly estimate protein abundance [120]. This index was later refined to be emPAI or exponentially modified PAI (*i.e.,*

10^{PAI}-1), where theoretically observable peptides were defined as those within a range of mass-to-charge ratio of scanning in mass spectrometer [121]. The emPAI demonstrated its ability by successfully estimating absolute abundance of 46 proteins, which had been measured using synthetic peptides, with 2-3 of average deviation factor: more than 2- to 3-fold difference in absolute abundance can be detected with this index. The values of emPAI can be calculated so easily that it is quite useful in obtaining an approximation of absolute protein abundance in a large-scale analysis. More recently, normalization with observable peptide was also used for quantitative method based on peak intensity, in which sum of all identified peptide intensity was divided by observable peptide number with detectable length, to measure protein copies in mammalian cells [122].

Sequence coverage rate and the identification frequency per protein length or theoretically observable tryptic peptide kinds have a correlation with protein abundance and hence serve as useful indices to obtain absolute quantification data. Meanwhile, individual peptides have different propensities to be detected and identified by MS/MS analysis, not only because they differ in efficiency of proteolytic digestion, ionization, and detection in mass spectrometer, but because qualities and patterns of fragment ions are variable depending on their amino acid composition.

To sophisticate the approaches based on peptide and spectral count, observability of each peptide, or a probability that the peptide is identified with MS analysis, was introduced as a novel index [123-125]. Absolute protein expression profiling approach, termed APEX, was developed where peptide observability was predicted from 4,023 tryptic peptides of 40 abundant proteins identified in a shotgun analysis of the yeast proteome [124]. Using this dataset containing 714 observed and 3,309 not-observed peptides, a probability for identification of each peptide from the yeast proteome was calculated based on the frequencies of each amino acid, peptide length, and molecular weight. Redundant spectral count was normalized by the sum of observability for each peptide from a corresponding protein and by the probability of protein identification calculated by ProteinProphet [126]. Resultant values were divided by the sum of the values of all identified proteins to generate an APEX score that would provide an absolute value of protein abundance. APEX successfully measured the abundance of 10 proteins, known amounts of which had been spiked into yeast cell extract, with mean difference of approximately 2-fold at 2-order magnitude: this approach had an ability to detect more than 2-fold difference in protein abundance.

Another group computationally predicted observable peptides, termed 'proteotypic peptides', using a much larger dataset composed of more than 600,000 peptides from yeast proteins identified on four different experimental methods [127]. A proteotypic peptide is defined as the peptide that was detected in more than half of the proteomics studies in which the protein was detected. Approximately 500 physiochemical properties of more than 16,000 proteotypic peptides from 4,030 yeast proteins were used to develop a classifier that distinguishes between proteotypic and non-proteotypic peptides. Proteotypic peptides were successfully predicted with 65-80% coverage and less than 10% error. As originally suggested [128], a dataset of proteotypic peptides would be of particular use in selecting peptides to be used as standards in quantification with stable isotope-labeling [129, 130]. It would be also useful to generate a library of selected information-rich peptides for the reduction of time for database search and the improvement of accuracy of identification [131].

Prediction of peptide observability is an important factor for normalization of identified peptide and spectral count to generate a more accurate index for estimating protein abundance. However, it should be noted that the probability of peptide identification would be dependent on the experimental designs, types of MS instruments, and analytical conditions, as suggested by the studies of prediction for different data sets [132]. Indeed, propensity of a peptide to be proteotypic was shown to depend on experimental methods [127]. Thus, application of classification values generated from a dataset to other datasets may substantially compromise both accuracy of prediction and coverage. Customized score for peptide observability may help generate more accurate estimation of absolute protein abundance.

4. COVERAGE AND DYNAMIC RANGE OF MASS SPECTROMETRY BASED PROTEOMICS

Current potential of MS-based proteomics still not attains to covering entire proteome. Fractionation of protein or peptide mixture prior to MS can improve the coverage of protein identification over proteome. For yeast proteome, protein separation with SDS-PAGE or peptide fractionation with liquid phase IEF prior to application to LC-MS analysis or peptide fractionations *via* two- or three-dimensional LC led to identification of 1,500-2,500 proteins [7-9, 133-135]. However, even the coverage of these analyses corresponds to ~30% of the yeast proteome. By contrast, western-blotting approach over yeast proteome, in which each yeast ORF tagged with TAP-tag was immunodetected, provided absolute quantity as copies per cell for 4,251 proteins or ~70% of the total proteome [136]. While MS is a general system for protein identification that does not require any specific strains such as the tagged strains, its coverage and sensitivity in detection is dependent on combination of multiple different types of fractionation procedures and a multitude of MS analyses [6]. Not only SDS-PAGE and IEF but also protein solubility-based fractionation and segmentation of MS scan range allowed identification of 4,399 yeast proteins with ~70% coverage of proteome [137]. A high coverage of fly proteome was achieved by combining multiple separation procedures including fractionation of cellular compartments, protein separation using gel filtration and isoelectric focusing, and peptide separation with multidimensional LC [138]. This study succeeded in cataloguing more than 9,000 proteins to cover ~60% of the fly proteome. For worm and plant, ~50% coverage were achieved by collection of protein identification for different developmental stages and organs [139, 140]. Furthermore, combination of different types of MS and experimental methods proved to increase the sensitivity of protein identification significantly [141, 142]. All possible means to increase coverage have to be combined to achieve a truly proteome-wide quantitative analysis.

Accurate quantification of absolute abundance requires high specificity and wide dynamic range. Background noise peak and co-detected irrelevant ions often interfere with detection of weak target peaks, thereby affecting both specificity and dynamic range. To reduce background noise and enhance specificity and dynamic range, SRM or MRM data acquisition mode, in which intensity of selected or multiple fragment ions other than peptide ion itself are monitored, have been used for measurement of absolute quantity [59, 61, 63, 67, 87, 94, 143, 144]. Combination of SRM mode and protein separation with SDS-PAGE was shown to allow quantification of low abundance proteins (~10^3 copies per cell) [62]. Quantification in MRM mode for peptide mixture pre-fractionated with strong cation-exchange LC allowed us to measure absolute amounts of proteins spiked in plasma at the concentration of 1-10 ng/ml [144, 145]. To achieve systematic SRM assay, many efforts have been made to generate the list of pairs of peptide and fragment ions in whole proteome, by screening the best one from proteomics MS/MS data repository [146-149] or synthesized proteins and peptides [150, 151]. High-resolution MS is an alternative technical basis to increase specificity and dynamic range, because it can better separate an analyte from co-eluting peptides with similar mass-to-charge ratio. High-resolution can also generate spectrum with low background noise leading to an increase in specificity and dynamic range [152, 153]. For instance, LTQ-Orbitrap mass spectrometer [154] has achieved a strong linearity in quantification of spiked proteins within 4-orders of dynamic range [155], providing analytical platform for absolute or stoichiometric quantification of protein abundance and complex architecture with accuracy and wide dynamic range [156, 157].

5. CONCLUSION

The pros and cons of stable isotope-labeling strategies and label-free approaches were summarized in terms of absolute quantification (Table (**1**)). Methods based on stable isotope standard would provide accurate quantitative data, because of calculation *via* the ratio of co-detected ion pair, one from the target and the other from the standard. In these methods, quality of quantified peak (*e.g.*, signal-to-noise ratio) is a critical factor to affect accuracy and dynamic range. By contrast, strategies for absolute quantification based on indices of normalized identification frequency are easy to implement with no additional experimental step and can be applied to a high throughput and comprehensive analysis. However, these methods have an intrinsic limitation in their accuracy, in particular, for low abundance proteins from which only a small number of peptides is identified. Assuming that both strategies would compensate with each other, we can propose a combined strategy where a proteome-wide estimation of abundance are achieved with label-free methods and corrected using a limited number of stable isotope-labeled standard spiked into the sample. Among the strategies using stable isotope labeling, the peptide concatenation, such as PCS, would be the most versatile one to achieve accurate quantification of absolute protein abundance in a large scale. These quantitative techniques along with the advance in detection coverage would eventually lead to absolute quantification over whole proteome, thereby significantly contributing to both basic and applied studies in various fields of biology and medicine.

ACKNOWLEDGEMENTS

This work was partly supported by Grant-in-Aid for Scientific Research on Priority Areas and the 21st Century COE Program from the Ministry of Education, Culture, Sports, Science and Technology, Japan, and Core Research for Evolutional Science and Technology (CREST) and Bioinformatics Research and Development (BIRD) from Japan Science and Technology Agency (JST).

CONFLICT OF INTEREST

The authors confirm that this chapter content has no conflict of interest.

DISCLOSURE

This is an updated report of our earlier review article, "Mass spectrometry-based approaches toward absolute quantitative proteomics" Current Genomics, 2008, 9, 263-274, Keiji Kito and Takashi Ito.

REFERENCES

[1] Pandey, A., Mann, M. Proteomics to study genes and genomes. *Nature* **2000**, *405*: 837-846.
[2] Aebersold, R., Goodlett, D.R. Mass spectrometry in proteomics. *Chem. Rev.* **2001**, *101*: 269-295.
[3] Aebersold, R., Mann, M. Mass spectrometry-based proteomics. *Nature* **2003**, *422*: 198-207.
[4] Smith, J.C., Lambert, J.P., Elisma, F., Figeys, D. Proteomics in 2005/*2006*: developments, applications and challenges. *Anal. Chem.* **2007**, *79*: 4325-4343.
[5] Cravatt, B.F., Simon, G.M., Yates, J.R., III. The biological impact of mass-spectrometry-based proteomics. *Nature* **2007**, *450*: 991-1000.
[6] Ahrens, C.H., Brunner, E., Qeli, E., Basler, K., Aebersold, R. Generating and navigating proteome maps using mass spectrometry. *Nat. Rev. Mol. Cell Biol.* **2010,** *11*: 789-801.
[7] Washburn, M.P., Wolters, D., Yates, J.R., III. Large-scale analysis of the yeast proteome by multidimensional protein identification technology. *Nat. Biotechnol.* **2001**, *19*: 242-247.
[8] Peng, J., Elias, J.E., Thoreen, C.C., Licklider, L.J., Gygi, S.P. Evaluation of multidimensional chromatography coupled with tandem mass spectrometry (LC/LC-MS/MS) for large-scale protein analysis: the yeast proteome. *J. Proteome Res.* **2003**, *2*: 43-50.
[9] Liu, H., Sadygov, R.G., Yates, J.R., III. A model for random sampling and estimation of relative protein abundance in shotgun proteomics. *Anal. Chem.* **2004**, *76*: 4193-4201.

[10] Motoyama, A., Yates, J.R. III. Multidimensional LC separations in shotgun proteomics. *Anal, Chem.* **2008,** *80*: 7187-7193.

[11] Bantscheff, M., Schirle, M., Sweetman, G., Rick, J., Kuster, B. Quantitative mass spectrometry in proteomics: a critical review. *Anal. Bioanal. Chem.* **2007**, *389*: 1017-1031.

[12] Mueller, L.N., Brusniak, M.Y., Mani, D.R., Aebersold, R. An assessment of software solutions for the analysis of mass spectrometry based quantitative proteomics data. *J. Proteome Res.* **2008,** *7*: 51-61.

[13] Bachi, A., Bonaldi, T. Quantitative proteomics as a new piece of the systems biology puzzle. *J. Proteomics* **2008,** *71*: 357-367.

[14] Ong, S.E., Foster, L.J., Mann, M. Mass spectrometric-based approaches in quantitative proteomics. *Methods* **2003**, *29*: 124-130.

[15] Sechi, S., Oda, Y. Quantitative proteomics using mass spectrometry. *Curr. Opin. Chem. Biol.* **2003**, *7*: 70-77.

[16] Goshe, M.B., Smith, R.D. Stable isotope-coded proteomic mass spectrometry. *Curr. Opin. Biotechnol.* **2003**, *14*: 101-109.

[17] Tao, W.A., Aebersold, R. Advances in quantitative proteomics *via* stable isotope tagging and mass spectrometry. *Curr. Opin. Biotechnol.* **2003**, *14*: 110-118.

[18] Ong, S.E., Mann, M. Mass spectrometry-based proteomics turns quantitative. *Nat. Chem. Biol.* **2005**, *1*: 252-262.

[19] Brun, V., Masselon, C., Garin, J., Dupuis, A. Isotope dilution strategies for absolute quantitative proteomics. *J. Proteomics* **2009,** *72*: 740-749.

[20] Gygi, S.P., Rist, B., Gerber, S.A., Turecek, F., Gelb, M.H., Aebersold, R. Quantitative analysis of complex protein mixtures using isotope-coded affinity tags. *Nat. Biotechnol.* **1999**, *17*: 994-999

[21] Han, D.K., Eng, J., Zhou, H., Aebersold, R. Quantitative profiling of differentiation-induced microsomal proteins using isotope-coded affinity tags and mass spectrometry. *Nat. Biotechnol.* **2001**, *19*: 946-951.

[22] Regnier, F.E., Julka, S. Primary amine coding as a path to comparative proteomics. *Proteomics* **2006**, *6*: 3968-3979.

[23] Yao, X., Freas, A., Ramirez, J., Demirev, P.A., Fenselau, C. Proteolytic ^{18}O labeling for comparative proteomics: model studies with two serotypes of adenovirus. *Anal. Chem.* **2001**, *73*: 2836-2842.

[24] Stewart, I.I., Thomson, T., Figeys, D. ^{18}O labeling: a tool for proteomics. *Rapid Commun. Mass Spectrom.* **2001**, *15*: 2456-2465.

[25] Thompson, A., Schäfer, J., Kuhn, K., Kienle, S., Schwarz, J., Schmidt, G., Neumann, T., Johnstone, R., Mohammed, A.K., Hamon, C. Tandem mass tags: a novel quantification strategy for comparative analysis of complex protein mixtures by MS/MS. *Anal. Chem.* **2003**, *75*: 1895-1904.

[26] Ross, P.L., Huang, Y.N., Marchese, J.N., Williamson, B., Parker, K., Hattan, S., Khainovski, N., Pillai, S., Dey, S., Daniels, S., Purkayastha, S., Juhasz, P., Martin, S., Bartlet-Jones, M., He, F., Jacobson, A., Pappin, D.J. Multiplexed protein quantitation in *Saccharomyces cerevisiae* using amine-reactive isobaric tagging reagents. *Mol. Cell. Proteomics* **2004**, *3*: 1154-1169.

[27] Oda, Y., Huang, K., Cross, F.R., Cowburn, D., Chait, B.T. Accurate quantitation of protein expression and site-specific phosphorylation. *Proc. Natl. Acad. Sci. U.S.A.* **1999**, *96*: 6591-6596.

[28] Ong, S.E., Blagoev, B., Kratchmarova, I., Kristensen, D.B., Steen, H., Pandey, A., Mann, M. Stable isotope labeling by amino acids in cell culture, SILAC, as a simple and accurate approach to expression proteomics. *Mol. Cell. Proteomics* **2002**, *1*: 376-386.

[29] Berger, S.J., Lee, S.W., Anderson, G.A., Pasa-Tolic, L., Tolic, N., Shen, Y., Zhao, R., Smith, R.D. High-throughput global peptide proteomic analysis by combining stable isotope amino acid labeling and data-dependent multiplexed-MS/MS. *Anal. Chem.* **2002**, *74*: 4994-5000.

[30] Zhu, H., Pan, S., Gu, S., Bradbury, E.M., Chen, X. Amino acid residue specific stable isotope labeling for quantitative proteomics. *Rapid Commun. Mass Spectrom.* **2002**, *16*: 2115-2123.

[31] Krijgsveld, J., Ketting, R.F., Mahmoudi, T., Johansen, J., Artal-Sanz, M., Verrijzer, C.P., Plasterk, R.H., Heck, A.J. Metabolic labeling of *C. elegans* and *D. melanogaster* for quantitative proteomics. *Nat. Biotechnol.* **2003**, *21*, 927-931.

[32] Wu, C.C., MacCoss, M.J., Howell, K.E., Matthews, D.E., Yates, J.R., III. Metabolic labeling of mammalian organisms with stable isotopes for quantitative proteomic analysis. *Anal. Chem.* **2004**; *76*, 4951-4959.

[33] Krüger, M., Moser, M., Ussar, S., Thievessen, I., Luber, C.A., Forner, F., Schmidt, S., Zanivan, S., Fässler, R., Mann, M. SILAC mouse for quantitative proteomics uncovers kindlin-3 as an essential factor for red blood cell function. *Cell* **2008,** *134*: 353-364.

[34] Ishihama, Y., Sato, T., Tabata, T., Miyamoto, N., Sagane, K., Nagasu, T., Oda, Y. Quantitative mouse brain proteomics using culture-derived isotope tags as internal standards. *Nat. Biotechnol.* **2005**, *23*: 617-621.

[35] Geiger, T., Cox, J., Ostasiewicz, P., Wisniewski, J.R., Mann, M. Super-SILAC mix for quantitative proteomics of human tumor tissue. *Nat. Methods* **2010,** *7*: 383-385.

[36] Geiger, T., Wisniewski, J.R., Cox, J., Zanivan, S., Kruger, M., Ishihama, Y., Mann, M. Use of stable isotope labeling by amino acids in cell culture as a spike-in standard in quantitative proteomics. *Nat. Protoc.* **2011,** *6*: 147-157.

[37] Washburn, M.P., Ulaszek, R., Deciu, C., Schieltz, D.M., Yates, J.R., III. Analysis of quantitative proteomic data generated *via* multidimensional protein identification technology. *Anal. Chem.* **2002**, *74*: 1650-1657.

[38] Washburn, M.P., Koller, A., Oshiro, G., Ulaszek, R.R., Plouffe, D., Deciu, C., Winzeler, E., Yates, J.R., III. Protein pathway and complex clustering of correlated mRNA and protein expression analyses in *Saccharomyces cerevisiae. Proc. Natl. Acad. Sci. U.S.A.* **2003**, *100*: 3107-3112.

[39] Kolkman, A., Olsthoorn, M.M., Heeremans, C.E., Heck, A.J., Slijper, M. Comparative proteome analysis of *Saccharomyces cerevisiae* grown in chemostat cultures limited for glucose or ethanol. *Mol. Cell. Proteomics* **2005**, *4*: 1-11.

[40] Kolkman, A., Daran-Lapujade, P., Fullaondo, A., Olsthoorn, M.M., Pronk, J.T., Slijper, M., Heck, A.J. Proteome analysis of yeast response to various nutrient limitations. *Mol. Syst. Biol.* **2006**, *2*: 2006.0026.

[41] Pham, T.K., Wright, P.C. The proteomic response of Saccharomyces cerevisiae in very high glucose conditions with amino acid supplementation. *J. Proteome Res.* **2008,** *7*: 4766-4774.

[42] Ranish, J.A., Yi, E.C., Leslie, D.M., Purvine, S.O., Goodlett, D.R., Eng, J., Aebersold, R. The study of macromolecular complexes by quantitative proteomics. *Nat. Genet.* **2003**, *33*: 349-355.

[43] Foster, L.J., de Hoog, C.L., Mann M. Unbiased quantitative proteomics of lipid rafts reveals high specificity for signaling factors. *Proc. Natl. Acad. Sci. U.S.A.* **2003**, *100*: 5813-5818.

[44] Blagoev, B., Kratchmarova, I., Ong, S.E., Nielsen, M., Foster, L.J., Mann, M. A proteomics strategy to elucidate functional protein-protein interactions applied to EGF signaling. *Nat. Biotechnol.* **2003**, *21*: 315-318.

[45] Foster, L.J., Rudich, A., Talior, I., Patel, N., Huang, X., Furtado, L.M., Bilan, P.J., Mann, M., Klip, A. Insulin-dependent interactions of proteins with GLUT4 revealed through stable isotope labeling by amino acids in cell culture (SILAC). *J. Proteome Res.* **2006**, *5*: 64-75.

[46] Pflieger, D., Jünger, M.A., Müller, M., Rinner, O., Lee, H., Gehrig, P.M., Gstaiger, M., Aebersold, R. Quantitative proteomic analysis of protein complexes: concurrent identification of interactors and their state of phosphorylation. *Mol. Cell. Proteomics* **2008**, *7*: 326-346.

[47] Gingras, A.C., Gstaiger, M., Raught, B., Aebersold, R. Analysis of protein complexes using mass spectrometry. *Nat. Rev. Mol. Cell Biol.* **2007**, *8*: 645-654.

[48] Yang, W., Steen, H., Freeman, M.R. Proteomic approaches to the analysis of multiprotein signaling complexes. *Proteomics* **2008**, *8*: 832-851.

[49] Pijnappel, W.P., Kolkman, A., Baltissen, M.P., Heck, A.J., Timmers, H.M. Quantitative mass spectrometry of TATA binding protein-containing complexes and subunit phosphorylations during the cell cycle. *Proteome Sci.* **2009,** *7*: 46.

[50] Kaake, R.M., Milenković, T., Przulj, N., Kaiser, P., Huang, L. Characterization of cell cycle specific protein interaction networks of the yeast 26S proteasome complex by the QTAX strategy. *J. Proteome Res.* **2010,** *9*: 2016-2029.

[51] Blagoev, B., Ong, S.E., Kratchmarova, I., Mann, M. Temporal analysis of phosphotyrosine-dependent signaling networks by quantitative proteomics. *Nat. Biotechnol.* **2004**, *22*: 1139-1145.

[52] Gruhler, A., Olsen, J.V., Mohammed, S., Mortensen, P., Faergeman, N.J., Mann, M., Jensen, O.N. Quantitative phosphoproteomics applied to the yeast pheromone signaling pathway. *Mol. Cell. Proteomics* **2005**, *4*: 310-327.

[53] Zhang, Y., Wolf-Yadlin, A., Ross, P.L., Pappin, D.J., Rush, J., Lauffenburger, D.A., White, F.M. Time-resolved mass spectrometry of tyrosine phosphorylation sites in the epidermal growth factor receptor signaling network reveals dynamic modules. *Mol. Cell. Proteomics* **2005**, *4*: 1240-1250.

[54] Olsen, J.V., Blagoev, B., Gnad, F., Macek, B., Kumar, C., Mortensen, P., Mann, M. Global, *in vivo*, and site-specific phosphorylation dynamics in signaling networks. *Cell* **2006**, *127*: 635-648.

[55] Li, X., Gerber, S.A., Rudner, A.D., Beausoleil, S.A., Haas, W., Villén, J., Elias, J.E., Gygi, S.P. Large-scale phosphorylation analysis of alpha-factor-arrested Saccharomyces cerevisiae. *J. Proteome Res.* **2007**, *6*: 1190-1197.

[56] Dephoure, N., Zhou, C., Villén, J., Beausoleil, S.A., Bakalarski, C.E., Elledge, S.J., Gygi, S.P. A quantitative atlas of mitotic phosphorylation. *Proc. Natl. Acad. Sci. U.S.A.* **2008,** *105*: 10762-10767.

[57] Monetti, M., Nagaraj, N., Sharma, K., Mann, M. Large-scale phosphosite quantification in tissues by a spike-in SILAC method. *Nat. Methods* **2011,** *8*: 655-658.

[58] Desiderio, D.M., Kai, M. Preparation of stable isotope-incorporated peptide internal standards for field desorption mass spectrometry quantification of peptides in biologic tissue. *Biomed. Mass Spectrom.* **1983**, *10*: 471-479.

[59] Kusmierz, J.J., Sumrada, R., Desiderio, D.M. Fast atom bombardment mass spectrometric quantitative analysis of methionine-enkephalin in human pituitary tissues. *Anal. Chem.* **1990**, *62*: 2395-2400.

[60] Barr, J.R., Maggio, V.L., Patterson, D.G., Jr, Cooper, G.R., Henderson, L.O., Turner, W.E., Smith, S.J., Hannon, W.H., Needham, L.L., Sampson, E.J. Isotope dilution--mass

spectrometric quantification of specific proteins: model application with apolipoprotein A-I. *Clin. Chem.* **1996**, *42*: 1676-1682.

[61] Barnidge, D.R., Dratz, E.A., Martin, T., Bonilla, L.E., Moran, L.B., Lindall, A. Absolute quantification of the G protein-coupled receptor rhodopsin by LC/MS/MS using proteolysis product peptides and synthetic peptide standards. *Anal. Chem.* **2003**, *75*: 445-451.

[62] Gerber, S.A., Rush, J., Stemman, O., Kirschner, M.W., Gygi, S.P. Absolute quantification of proteins and phosphoproteins from cell lysates by tandem MS. *Proc. Natl. Acad. Sci. U.S.A.* **2003**, *100*: 6940-6945.

[63] Kuhn, E., Wu, J., Karl, J., Liao, H., Zolg, W., Guild, B. Quantification of C-reactive protein in the serum of patients with rheumatoid arthritis using multiple reaction monitoring mass spectrometry and ^{13}C-labeled peptide standards. *Proteomics* **2004**, *4*: 1175-1186.

[64] Zhang, F., Bartels, M.J., Stott, W.T. Quantitation of human glutathione S-transferases in complex matrices by liquid chromatography/tandem mass spectrometry with signature peptides. *Rapid Commun. Mass Spectrom.* **2004**, *18*: 491-498.

[65] Hochleitner, E.O., Kastner, B., Fröhlich, T., Schmidt, A., Lührmann, R., Arnold, G., Lottspeich, F. Protein stoichiometry of a multiprotein complex, the human spliceosomal U1 small nuclear ribonucleoprotein: absolute quantification using isotope-coded tags and mass spectrometry. *J. Biol. Chem.* **2005**, *280*: 2536-2542.

[66] Pan, S., Zhang, H., Rush, J., Eng, J., Zhang, N., Patterson, D., Comb, M.J., Aebersold, R. High throughput proteome screening for biomarker detection. *Mol. Cell. Proteomics* **2005**, *4*: 182-190.

[67] Cheng, D., Hoogenraad, C.C., Rush, J., Ramm, E., Schlager, M.A., Duong, D.M., Xu, P., Wijayawardana, S.R., Hanfelt, J., Nakagawa, T., Sheng, M., Peng, J. Relative and absolute quantification of postsynaptic density proteome isolated from rat forebrain and cerebellum. *Mol. Cell. Proteomics* **2006**, *5*: 1158-1170.

[68] Holzmann, J., Pichler, P., Madalinski, M., Kurzbauer, R., Mechtler, K. Stoichiometry determination of the MP1-p14 complex using a novel and cost-efficient method to produce an equimolar mixture of standard peptides. *Anal. Chem.* **2009,** *81*: 10254-10261.

[69] Schmidt, C., Lenz, C., Grote, M., Lührmann, R., Urlaub, H. Determination of protein stoichiometry within protein complexes using absolute quantification and multiple reaction monitoring. *Anal. Chem.* **2010,** *82*: 2784-2796.

[70] Bennett, E.J., Rush, J., Gygi, S.P., Harper, J.W. Dynamics of cullin-RING ubiquitin ligase network revealed by systematic quantitative proteomics. Cell **2010,** *143*: 951-965.

[71] Holzmann, J., Fuchs, J., Pichler, P., Peters, J.M., Mechtler, K. Lesson from the stoichiometry determination of the cohesin complex: a short protease mediated elution increases the recovery from cross-linked antibody-conjugated beads. *J. Proteome Res.* **2011,** *10*: 780-789.

[72] DeSouza, L.V., Taylor, A.M., Li, W., Minkoff, M.S., Romaschin, A.D., Colgan, T.J., Siu, K.W. Multiple reaction monitoring of mTRAQ-labeled peptides enables absolute quantification of endogenous levels of a potential cancer marker in cancerous and normal endometrial tissues. *J. Proteome Res.* **2008,** *7*: 3525-3534.

[73] Li, N., Nemirovskiy, O.V., Zhang, Y., Yuan, H., Mo, J., Ji, C., Zhang, B., Brayman, T.G., Lepsy, C., Heath, T.G., Lai, Y. Absolute quantification of multidrug resistance-associated protein 2 (MRP2/ABCC2) using liquid chromatography tandem mass spectrometry. *Anal. Biochem.* **2008,** *380*: 211-222.

[74] Williams, D.K., Muddiman, D.C. Absolute quantification of C-reactive protein in human plasma derived from patients with epithelial ovarian cancer utilizing protein cleavage isotope dilution mass spectrometry. *J. Proteome Res.* **2009,** *8*: 1085-1090.

[75] DeSouza, L.V., Romaschin, A.D., Colgan, T.J., Siu, K.W. Absolute quantification of potential cancer markers in clinical tissue homogenates using multiple reaction monitoring on a hybrid triple quadrupole/linear ion trap tandem mass spectrometer. *Anal. Chem.* **2009,** *81*: 3462-3470.

[76] Kawakami, H., Ohtsuki, S., Kamiie, J., Suzuki, T., Abe, T., Terasaki, T. Simultaneous absolute quantification of 11 cytochrome P450 isoforms in human liver microsomes by liquid chromatography tandem mass spectrometry with *in silico* target peptide selection. *J. Pharm. Sci.* **2011,** *100*: 341-352.

[77] Stemmann, O., Zou, H., Gerber, S.A., Gygi, S.P., Kirschner, M.W. Dual inhibition of sister chromatid separation at metaphase. *Cell* **2001**, *107*: 715-726.

[78] Mayya, V., Rezual, K., Wu, L., Fong, M.B., Han, D.K. Absolute quantification of multisite phosphorylation by selective reaction monitoring mass spectrometry: determination of inhibitory phosphorylation status of cyclin-dependent kinases. *Mol. Cell. Proteomics* **2006,** *5*: 1146-1157.

[79] Ciccimaro, E., Hanks, S.K., Yu, K.H., Blair, I.A. Absolute quantification of phosphorylation on the kinase activation loop of cellular focal adhesion kinase by stable isotope dilution liquid chromatography/mass spectrometry. *Anal. Chem.* **2009,** *81*: 3304-3313.

[80] Atrihm A., Turnock, D., Sellar, G., Thompson, A., Feuerstein, G., Ferguson, M.A., Huang, J.T. Stoichiometric quantification of Akt phosphorylation using LC-MS/MS. *J. Proteome Res.* **2010,** *9*: 743-751.

[81] Kirkpatrick, D.S., Gerber, S.A., Gygi, S.P. The absolute quantification strategy: a general procedure for the quantification of proteins and post-translational modifications. *Methods* **2005**, *35*: 265-273.

[82] Kirkpatrick, D.S., Hathaway, N.A., Hanna, J., Elsasser, S., Rush, J., Finley, D., King, R.W., Gygi, S.P. Quantitative analysis of *in vitro* ubiquitinated cyclin B1 reveals complex chain topology. *Nat. Cell Biol.* **2006**, *8*: 700-710.

[83] Appels, N.M., Rosing, H., Stephens, T.C., Hughes, A., Schellens, J.H., Beijnen, J.H. Absolute quantification of farnesylated Ras levels in complex samples using liquid chromatography fractionation combined with tryptic digestion and electrospray tandem mass spectrometry. *Anal. Biochem.* **2006**, *352*: 33-40.

[84] Havlis, J., Shevchenko, A. Absolute quantification of proteins in solutions and in polyacrylamide gels by mass spectrometry. *Anal. Chem.* **2004**, *76*: 3029-3036.

[85] Kippen, A.D., Cerini, F., Vadas, L., Stöcklin, R., Vu, L., Offord, R.E., Rose, K. Development of an isotope dilution assay for precise determination of insulin, C-peptide, and proinsulin levels in non-diabetic and type II diabetic individuals with comparison to immunoassay. *J. Biol. Chem.* **1997**, *272*: 12513-12522.

[86] Peng, J., Kim, M.J., Cheng, D., Duong, D.M., Gygi, S.P., Sheng, M. Semiquantitative proteomic analysis of rat forebrain postsynaptic density fractions by mass spectrometry. *J. Biol. Chem.* **2004**, *279*: 21003-21011

[87] Janecki, D.J., Bemis, K.G., Tegeler, T.J., Sanghani, P.C., Zhai, L., Hurley, T.D., Bosron, W.F., Wang, M. A multiple reaction monitoring method for absolute quantification of the human liver alcohol dehydrogenase ADH1C1 isoenzyme. *Anal. Biochem.* **2007**, *369*: 18-26.

[88] Brun, V., Dupuis, A., Adrait, A., Marcellin, M., Thomas, D., Court, M., Vandenesch, F., Garin, J. Isotope-labeled protein standards: toward absolute quantitative proteomics. *Mol. Cell. Proteomics* **2007**, *6*: 2139-2149.

[89] Kaiser, S.E., Riley, B.E., Shaler, T.A., Trevino, R.S., Becker, C.H., Schulman, H., Kopito, R.R. Protein standard absolute quantification (PSAQ) method for the measurement of cellular ubiquitin pools. *Nat. Methods* **2011,** *8*: 691-696.

[90] Singh, S., Springer, M., Steen, J., Kirschner, M.W., Steen, H. FLEXIQuant: a novel tool for the absolute quantification of proteins, and the simultaneous identification and quantification of potentially modified peptides. *J. Proteome Res.* **2009,** *8*: 2201-2210.

[91] Beynon, R.J., Doherty, M.K., Pratt, J.M., Gaskell, S.J. Multiplexed absolute quantification in proteomics using artificial QCAT proteins of concatenated signature peptides. *Nat. Methods* **2005**, *2*: 587-589.

[92] Kito, K., Ota, K., Fujita, T., Ito, T. A synthetic protein approach toward accurate mass spectrometric quantification of component stoichiometry of multiprotein complexes. *J. Proteome Res.* **2007**, *6*: 792-800.

[93] Rivers, J., Simpson, D.M., Robertson, D.H., Gaskell, S.J., Beynon, R.J. Absolute multiplexed quantitative analysis of protein expression during muscle development using QconCAT. *Mol. Cell. Proteomics* **2007**, *6*: 1416-1427.

[94] Anderson, L., Hunter, C.L. Quantitative mass spectrometric multiple reaction monitoring assays for major plasma proteins. *Mol. Cell. Proteomics* **2006**, *5*: 573-588.

[95] Yen, C.Y., Russell, S., Mendoza, A.M., Meyer-Arendt, K., Sun, S., Cios, K.J., Ahn, N.G., Resing, K.A. Improving sensitivity in shotgun proteomics using a peptide-centric database with reduced complexity: protease cleavage and SCX elution rules from data mining of MS/MS spectra. *Anal. Chem.* **2006**, *78*: 1071-1084.

[96] Siepen, J.A., Keevil, E.J., Knight, D., Hubbard, S.J. Prediction of missed cleavage sites in tryptic peptides aids protein identification in proteomics. *J. Proteome Res.* **2007**, *6*: 399-408.

[97] Nanavati, D., Gucek, M., Milne, J.L., Subramaniam, S., Markey, S.P. Stoichiometry and absolute quantification of proteins with mass spectrometry using fluorescent and isotope-labeled concatenated peptide standards. *Mol. Cell. Proteomics* **2008**, *7*: 442-447.

[98] Mirzaei, H., McBee, J., Watts, J., Aebersold, R. Comparative evaluation of current peptide production platforms used in absolute quantification in proteomics. *Mol. Cell. Proteomics* **2007**, in press.

[99] Old, W.M., Meyer-Arendt, K., Aveline-Wolf, L., Pierce, K.G., Mendoza, A., Sevinsky, J.R., Resing, K.A., Ahn, N.G. Comparison of label-free methods for quantifying human proteins by shotgun proteomics. *Mol. Cell. Proteomics* **2005**, *4*: 1487-1502.

[100] Wang, G., Wu, W.W., Zeng, W., Chou, C.L., Shen, R.F. Label-free protein quantification using LC-coupled ion trap or FT mass spectrometry: Reproducibility, linearity, and application with complex proteomes. *J. Proteome Res.* **2006**, *5*: 1214-1223.

[101] Andersen, J.S., Wilkinson, C.J., Mayor, T., Mortensen, P., Nigg, E.A., Mann, M. Proteomic characterization of the human centrosome by protein correlation profiling. *Nature* **2003**, *426*: 570-574.

[102] Rinner, O., Mueller, L.N., Hubálek, M., Müller, M., Gstaiger, M., Aebersold, R. An integrated mass spectrometric and computational framework for the analysis of protein interaction networks. *Nat. Biotechnol.* **2007**, *25*: 345-352.

[103] Griffin, N.M., Yu, J., Long, F., Oh, P., Shore, S., Li, Y., Koziol, J.A., Schnitzer, J.E. Label-free, normalized quantification of complex mass spectrometry data for proteomic analysis. *Nat. Biotechnol.* **2010,** *28*: 83-89.

[104] Colaert, N., Gevaert, K., Martens, L. RIBAR and xRIBAR: Methods for reproducible relative MS/MS-based label-free protein quantification. *J. Proteome Res.* **2011,** *10*: 3183-3189.

[105] Kim, Y.J., Zhan, P., Field, B., Ruben, S.M., He, T. Reproducibility assessment of relative quantitation strategies for LC-MS based proteomics. *Anal. Chem.* **2007**, *79*: 5651-5658.

[106] Bondarenko, P.V., Chelius, D., Shaler, T.A. Identification and relative quantitation of protein mixtures by enzymatic digestion followed by capillary reversed-phase liquid chromatography-tandem mass spectrometry. *Anal. Chem.* **2002**, *74*: 4741-4749.

[107] Wang, W., Zhou, H., Lin, H., Roy, S., Shaler, T.A., Hill, L.R., Norton, S., Kumar, P., Anderle, M., Becker, C.H. Quantification of proteins and metabolites by mass spectrometry without isotopic labeling or spiked standards. *Anal. Chem.* **2003**, *75*: 4818-4826.

[108] Tabata, T., Sato, T., Kuromitsu, J., Oda, Y. Pseudo internal standard approach for label-free quantitative proteomics. *Anal. Chem.* **2007**, *79*: 8440-8445.

[109] Jaitly, N., Monroe, M.E., Petyuk, V.A., Clauss, T.R., Adkins, J.N., Smith, R.D. Robust algorithm for alignment of liquid chromatography-mass spectrometry analyses in an accurate mass and time tag data analysis pipeline. *Anal. Chem.* **2006**, *78*: 7397-7409.

[110] Pang, J.X., Ginanni, N., Dongre, A.R., Hefta, S.A., Opitek, G.J. Biomarker discovery in urine by proteomics. *J. Proteome Res.* **2002**, *1*: 161-169.

[111] Gao, J., Opiteck, G.J., Friedrichs, M.S., Dongre, A.R., Hefta, S.A. Changes in the protein expression of yeast as a function of carbon source. *J. Proteome Res.* **2003**, *2*: 643-649.

[112] Zybailov, B., Coleman, M.K., Florens, L., Washburn, M.P. Correlation of relative abundance ratios derived from peptide ion chromatograms and spectrum counting for quantitative proteomic analysis using stable isotope labeling. *Anal. Chem.* **2005**, *77*: 6218-6224.

[113] Asara, J.M., Christofk, H.R., Freimark, L.M., Cantley, L.C. A label-free quantification method by MS/MS TIC compared to SILAC and spectral counting in a proteomics screen. *Proteomics* **2008**, *8*: 994-999.

[114] Silva, J.C., Gorenstein, M.V., Li, G.Z., Vissers, J.P., Geromanos, S.J. Absolute quantification of proteins by $LCMS^E$: a virtue of parallel MS acquisition. *Mol. Cell. Proteomics* **2006**, *5*: 144-156.

[115] Cheng, F.Y., Blackburn, K., Lin, Y.M., Goshe, M.B., Williamson, J.D. Absolute protein quantification by LC/MS(E) for global analysis of salicylic acid-induced plant protein secretion responses. *J. Proteome Res.* **2009,** *8*: 82-93.

[116] Blondeau, F., Ritter, B., Allaire, P.D., Wasiak, S., Girard, M., Hussain, N.K., Angers, A., Legendre-Guillemin, V., Roy, L., Boismenu, D., Kearney, R.E., Bell, A.W., Bergeron, J.J., McPherson, P.S. Tandem MS analysis of brain clathrin-coated vesicles reveals their critical involvement in synaptic vesicle recycling. *Proc. Natl. Acad. Sci. U.S.A.* **2004**, *101*: 3833-3838.

[117] Powell, D.W., Weaver, C.M., Jennings, J.L., McAfee, K.J., He, Y., Weil, P.A., Link, A.J. Cluster analysis of mass spectrometry data reveals a novel component of SAGA. *Mol. Cell Biol.* **2004**, *24*: 7249-7259.

[118] Zybailov, B., Mosley, A.L., Sardiu, M.E., Coleman, M.K., Florens, L., Washburn, M.P. Statistical analysis of membrane proteome expression changes in *Saccharomyces cerevisiae. J. Proteome Res.* **2006**, *5*: 2339-2347.

[119] Paoletti, A.C., Parmely, T.J., Tomomori-Sato, C., Sato, S., Zhu, D., Conaway, R.C., Conaway, J.W., Florens, L., Washburn, M.P. Quantitative proteomic analysis of distinct mammalian Mediator complexes using normalized spectral abundance factors. *Proc. Natl. Acad. Sci. U.S.A.* **2006**, *103*: 18928-18933.

[120] Rappsilber, J., Ryder, U., Lamond, A.I., Mann, M. Large-scale proteomic analysis of the human spliceosome. *Genome Res.* **2002**, *12*: 1231-1245.

[121] Ishihama, Y., Oda, Y., Tabata, T., Sato, T., Nagasu, T., Rappsilber, J., Mann, M. Exponentially modified protein abundance index (emPAI) for estimation of absolute protein

amount in proteomics by the number of sequenced peptides per protein. *Mol. Cell. Proteomics* **2005**, *4*: 1265-1272.

[122] Schwanhäusser, B., Busse, D., Li, N., Dittmar, G., Schuchhardt, J., Wolf, J., Chen, W., Selbach. M. Global quantification of mammalian gene expression control. *Nature* **2011,** *473*: 337-342.

[123] Tang, H., Arnold, R.J., Alves, P., Xun, Z., Clemmer, D.E., Novotny, M.V., Reilly, J.P., Radivojac, P. A computational approach toward label-free protein quantification using predicted peptide detectability. *Bioinformatics* **2006**, *22*: e481-488.

[124] Lu, P., Vogel, C., Wang, R., Yao, X., Marcotte, E.M. Absolute protein expression profiling estimates the relative contributions of transcriptional and translational regulation. *Nat. Biotechnol.* **2007**, *25*: 117-124.

[125] Vogel, C., Marcotte, E.M. Calculating absolute and relative protein abundance from mass spectrometry-based protein expression data. *Nat. Protoc.* **2008,** *3*: 1444-1451.

[126] Nesvizhskii, A.I., Keller, A., Kolker, E., Aebersold, R. A statistical model for identifying proteins by tandem mass spectrometry. *Anal. Chem.* **2003**, *75*: 4646-4658.

[127] Mallick, P., Schirle, M., Chen, S.S., Flory, M.R., Lee, H., Martin, D., Ranish, J., Raught, B., Schmitt, R., Werner, T., Kuster, B., Aebersold, R. Computational prediction of proteotypic peptides for quantitative proteomics. *Nat. Biotechnol.* **2007**, *25*: 125-131.

[128] Kuster, B., Schirle, M., Mallick, P., Aebersold, R. Scoring proteomes with proteotypic peptide probes. *Nat. Rev. Mol. Cell Biol.* **2005,** *6*: 577-583.

[129] Fusaro, V.A., Mani, D.R., Mesirov, J.P., Carr, S.A. Prediction of high-responding peptides for targeted protein assays by mass spectrometry. *Nat. Biotechnol.* **2009,** *27*: 190-198.

[130] Eyers, C.E., Lawless, C., Wedge, D.C., Lau, K.W., Gaskell, S.J., Hubbard, S.J. CONSeQuence: prediction of reference peptides for absolute quantitative proteomics using consensus machine learning approaches. *Mol. Cell Proteomics* **2011,** *10*: M110.003384.

[131] Craig, R., Cortens, J.P., Beavis, R.C. The use of proteotypic peptide libraries for protein identification. *Rapid Commun. Mass Spectrom.* **2005**, *19*: 1844-1850.

[132] Sanders, W.S., Bridges, S.M., McCarthy, F.M., Nanduri, B., Burgess, S.C. Prediction of peptides observable by mass spectrometry applied at the experimental set level. *BMC Bioinformatics* **2007**, *8*: S23.

[133] Wei, J., Sun, J., Yu, W., Jones, A., Oeller, P., Keller, M., Woodnutt, G., Short, J.M. Global proteome discovery using an online three-dimensional LC-MS/MS. *J. Proteome Res.* **2005**, *4*: 801-808.

[134] de Godoy, L.M., Olsen, J.V., de Souza, G.A., Li, G., Mortensen, P., Mann, M. Status of complete proteome analysis by mass spectrometry: SILAC labeled yeast as a model system. *Genome Biol.* **2006**, *7*: R50.

[135] Hubner, N.C., Ren, S., Mann, M. Peptide separation with immobilized pI strips is an attractive alternative to in-gel protein digestion for proteome analysis. *Proteomics* **2008**, *8*: 4862-4872.

[136] Ghaemmaghami, S., Huh, W.K., Bower, K., Howson, R.W., Belle, A., Dephoure, N., O'Shea, E.K., Weissman, J.S. Global analysis of protein expression in yeast. *Nature* **2003**, *425*: 737-741.

[137] de Godoy, L.M., Olsen, J.V., Cox, J., Nielsen, M.L., Hubner, N.C., Fröhlich, F., Walther, T.C., Mann, M. Comprehensive mass-spectrometry-based proteome quantification of haploid *versus* diploid yeast. *Nature* **2008,** *455*: 1251-1254.

[138] Brunner, E., Ahrens, C.H., Mohanty, S., Baetschmann, H., Loevenich, S., Potthast, F., Deutsch, E.W., Panse, C., de Lichtenberg, U., Rinner, O., Lee, H., Pedrioli, P.G.,

Malmstrom, J., Koehler, K., Schrimpf, S., Krijgsveld, J., Kregenow, F., Heck, A.J., Hafen, E., Schlapbach, R., Aebersold, R. A high-quality catalog of the *Drosophila melanogaster* proteome. *Nat. Biotechnol.* **2007**, *25*: 576-583.

[139] Baerenfaller, K., Grossmann, J., Grobei, M.A., Hull, R., Hirsch-Hoffmann, M., Yalovsky, S., Zimmermann, P., Grossniklaus, U., Gruissem, W., Baginsky, S. Genome-scale proteomics reveals Arabidopsis thaliana gene models and proteome dynamics. *Science* **2008,** *320*: 938-941.

[140] Schrimpf, S.P., Weiss, M., Reiter, L., Ahrens, C.H., Jovanovic, M., Malmström, J., Brunner, E., Mohanty, S., Lercher, M.J., Hunziker, P.E., Aebersold, R., von Mering, C., Hengartner, M.O. Comparative functional analysis of the Caenorhabditis elegans and Drosophila melanogaster proteomes. *PLoS Biol.* **2009,** *7*: e48.

[141] Elias, J.E., Haas, W., Faherty, B.K., Gygi, S.P. Comparative evaluation of mass spectrometry platforms used in large-scale proteomics investigations. *Nat. Methods* **2005**, *2*: 667-675.

[142] Krogan, N.J., Cagney, G., Yu, H., Zhong, G., Guo, X., Ignatchenko, A., Li, J., Pu, S., Datta, N., Tikuisis, A.P., Punna, T., Peregrín-Alvarez, J.M., Shales, M., Zhang, X., Davey, M., Robinson, M.D., Paccanaro, A., Bray, J.E., Sheung, A., Beattie, B., Richards, D.P., Canadien, V., Lalev, A., Mena, F., Wong, P., Starostine, A., Canete, M.M., Vlasblom, J., Wu, S., Orsi, C., Collins, S.R., Chandran, S., Haw, R., Rilstone, J.J., Gandi, K., Thompson, N.J., Musso, G., St Onge, P., Ghanny, S., Lam, M.H., Butland, G., Altaf-Ul, A.M., Kanaya, S., Shilatifard, A., O'Shea, E., Weissman, J.S., Ingles, C.J., Hughes, T.R., Parkinson, J., Gerstein, M., Wodak, S.J., Emili, A., Greenblatt, J.F. Global landscape of protein complexes in the yeast *Saccharomyces cerevisiae*. *Nature* **2006**, *440*: 637-643.

[143] Lange, V., Picotti, P., Domon, B., Aebersold, R. Selected reaction monitoring for quantitative proteomics: a tutorial. *Mol. Syst. Biol.* **2008,** *4*: 222.

[144] Pan, S., Aebersold, R., Chen, R., Rush, J., Goodlett, D.R., McIntosh, M.W., Zhang, J., Brentnall, T.A. Mass spectrometry based targeted protein quantification: methods and applications. *J. Proteome Res.* **2009,** *8*: 787-797.

[145] Keshishian, H., Addona, T., Burgess, M., Kuhn, E., Carr, S.A. Quantitative, multiplexed assays for low abundance proteins in plasma by targeted mass spectrometry and stable isotope dilution. *Mol. Cell. Proteomics* **2007**, *6*: 2212-2229.

[146] Deutsch, E.W., Lam, H., Aebersold, R. PeptideAtlas: a resource for target selection for emerging targeted proteomics workflows. *EMBO Rep.* **2008,** *9*: 429-434.

[147] Picotti, P, Bodenmiller, B., Mueller, L.N., Domon, B., Aebersold, R. Full dynamic range proteome analysis of S. cerevisiae by targeted proteomics. *Cell* **2009,** *138*: 795-806.

[148] Sherwood, C.A., Eastham, A., Lee, L.W., Peterson, A., Eng, J.K., Shteynberg, D., Mendoza, L., Deutsch, E.W., Risler, J., Tasman, N., Aebersold, R., Lam, H., Martin, D.B. MaRiMba: a software application for spectral library-based MRM transition list assembly. *J. Proteome Res.* **2009,** *8*: 4396-4405.

[149] Cham, J.A., Bianco, L., Barton, C., Bessant, C. MRMaid-DB: a repository of published SRM transitions. *J. Proteome Res.* **2010,** *9*: 620-625.

[150] Picotti, P., Rinner, O., Stallmach, R., Dautel, F., Farrah, T., Domon, B., Wenschuh, H., Aebersold, R. High-throughput generation of selected reaction-monitoring assays for proteins and proteomes. *Nat. Methods* **2010,** *7*: 43-46.

[151] Stergachis, A.B., MacLean, B., Lee, K., Stamatoyannopoulos, J.A., MacCoss, M.J. Rapid empirical discovery of optimal peptides for targeted proteomics. *Nat. Methods* **2011,** *8*: 1041-1043.

[152] Mann, M., Kelleher, N.L. Precision proteomics: the case for high resolution and high mass accuracy. *Proc. Natl. Acad. Sci. U.S.A.* **2008,** *105*: 18132-18138.

[153] Seppälä, U., Dauly, C., Robinson, S., Hornshaw, M., Larsen, J.N., Ipsen, H. Absolute quantification of allergens from complex mixtures: a new sensitive tool for standardization of allergen extracts for specific immunotherapy. *J. Proteome Res.* **2011,** *10*: 2113-2122.

[154] Scigelova, M., Makarov, A. Orbitrap mass analyzer - overview and applications in proteomics. *Proteomics* **2006**, *6*: S16-21.

[155] Hanke, S., Besir, H., Oesterhelt, D., Mann, M. Absolute SILAC for accurate quantitation of proteins in complex mixtures down to the attomole level. *J. Proteome Res.* **2008**, *7*: 1118-1130.

[156] Zeiler, M., Straube, W.L., Lundberg, E., Uhlen, M., Mann, M. A Protein Epitope Signature Tag (PrEST) Library Allows SILAC-based Absolute Quantification and Multiplexed Determination of Protein Copy Numbers in Cell Lines. *Mol. Cell Proteomics* **2012,** *11*: O111.009613.

[157] Olinares, P.D., Kim, J., Davis, J.I., van Wijk, K.J. Subunit stoichiometry, evolution, and functional implications of an asymmetric plant plastid ClpP/R protease complex in Arabidopsis. *Plant Cell* **2011,** *23*: 2348-2361.

Send Orders of Reprints at reprints@benthamscince.net
Advances in Genome Science, Vol. 1, 2013, 293-315 293

CHAPTER 12

Regulation of Interactions with Sliding Clamps During DNA Replication and Repair

Francisco J. López de Saro*

Department of Molecular Evolution, Centro de Astrobiología (CSIC-INTA), 28850 Torrejón de Ardoz, Madrid, Spain

Abstract: The molecular machines that replicate the genome consist of many interacting components. Essential to the organization of the replication machinery are ring-shaped proteins, like the prokaryotic β-clamp or eukaryotic PCNA (Proliferating Cell Nuclear Antigen), collectively named sliding clamps. They encircle the DNA molecule and slide on it freely and bidirectionally. Sliding clamps are typically associated to DNA polymerases and provide these enzymes with the processivity required to synthesize large chromosomes. Additionally, they interact with a large array of proteins that perform enzymatic reactions on DNA, targeting and orchestrating their functions. In recent years there have been a large number of studies that have analyzed the structural details of how sliding clamps interact with their ligands. However, much remains to be learned in relation to how these interactions are regulated to occur coordinately and sequentially. Since sliding clamps participate in reactions in which many different enzymes bind and then release from the clamp in an orchestrated way, it is critical to analyze how these changes in affinity take place. In this review I focus the attention on the mechanisms by which various types of enzymes interact with sliding clamps and what is known about the regulation of this binding. Especially I describe emerging paradigms on how enzymes switch places on sliding clamps during DNA replication and repair of prokaryotic and eukaryotic genomes.

Keywords: DNA replication, DNA repair, PCNA, beta clamp, DNA polymerase, MutS.

INTRODUCTION

The sliding clamp proteins, like the prokaryotic β-clamp and the eukaryotic PCNA (Proliferating Cell Nuclear Antigen), are central to the organization of the enzymatic machinery that replicates and repairs DNA in cells [1, 2]. These ring-shaped structures can interact with a large number of enzymes (See [3-5] for

***Address correspondence to Francisco J. López de Saro:** Department of Molecular Evolution, Centro de Astrobiología (CSIC-INTA), 28850 Torrejón de Ardoz, Madrid, SPAIN; Tel.: (+34) 915206436 or 618506337; Fax: (+34) 915201074; E-mail: lopezsfj@cab.inta-csic.es recent updates) and orchestrate their activities during DNA replication and repair. Sliding clamps are uniquely designed to meet two critical needs in DNA synthesis. First, they tether DNA polymerases to the template, thus providing them with high processivity. Second, sliding clamps are recruiting devices that provide a binding surface to target enzymes to active centers of DNA synthesis and repair.

Protein clamps are typically homodimeric or homotrimeric structures, like the β-clamp or PCNA, respectively, with the monomers binding each other in a head-to-tail fashion and completing a circle (See Fig. **1**) [6]. Recently heterotrimeric clamps have been found in archaea associated with replicative polymerases [5, 7], and in higher organisms involved in checkpoint, like the RHR or 911 complex (Rad9-Hus1-Rad1) [8, 9]. Remarkably, across all life domains and despite limited or no sequence between them, sliding clamps are always composed of six domains with a characteristic fold [10]. A continuous layer of pleated sheet structure is the backbone of the ring, and a set of 12 α-helices line the inside of the clamp. Sliding clamps have two distinct faces with very different electrostatic properties [11]. The major site of interaction of clamps with partners is *via* a hydrophobic pocket between the two domains of each monomer (in PCNA) or domains 2 and 3 (in the β-clamp). Contacts have been also observed on the interconnector loop (only in PCNA) or the extreme C-terminal peptide of the clamp.

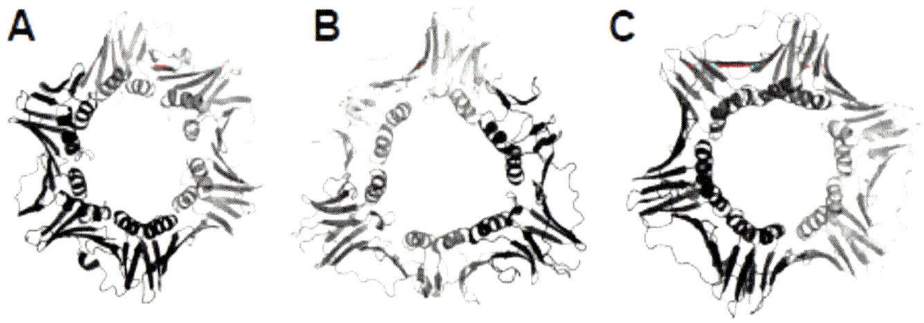

Figure 1: Crystallographic structures of representative sliding clamps. A. The dimeric prokaryotic β-clamp (2POL). B. The homotrimeric clamp from a bacteriophage RB69 (protein gp45) (1B77). C. The homotrimeric eukaryotic clamp PCNA (1SXJ).

Since sliding clamps are co-factors of numerous proteins involved in the metabolism of DNA, interaction with sliding clamps can be a critical point of regulation of the activities of these enzymes. For example, processivity is increased up to 100-times when DNA polymerases are bound to sliding clamps and therefore changes in the affinity of the interaction could be a way by which the activity of the polymerase is regulated. Indeed, as described later, an internal switch within the DNA polymerase III complex in *E. coli* modulates the affinity with the β-clamp and this is the mechanism by which the polymerase regulates its processivity. Alternatively, in eukaryotic replication machineries the modification of either the sliding clamp or the binding surface of the partner enzyme often serves as a regulatory mechanism. Thus, ubiquitylation and sumoylation of PCNA have been described as transient modifications that serve to switch polymerases during the DNA damage response [12]. Phosphorylation of PCNA-binding domains has been observed in numerous proteins involved in DNA replication, such as the cell-cycle regulator p21, Fen1, RFC, Pol δ and DNA ligase and this modification directly affects the interaction. Thus, the process of engagement and disengagement with sliding clamps may be a way of controlling protein function.

Regulation of sliding clamp binding is important in view of the fact that their concentration in the cell is likely to be limiting if compared to the concentration of proteins that bind to them, especially if it is assumed that only DNA-bound clamps are 'active'. It has been estimated that there are only about 300 β-clamps in a *E. coli* cell [13]. In eukaryotes PCNA expression is itself regulated and its levels are cell-cycle dependent. It seems unlikely that each enzyme uses its own sliding clamp exclusively, but rather that individual sliding clamps are trafficked by various proteins, either concurrently or sequentially. Numerous *in vivo* studies using microscopy and fluorescence have shown that PCNA clusters in regions of the chromosome that are being actively replicated, and that it co-localizes with polymerases and DNA repair enzymes [14, 15]. Similarly, in *Bacillus subtilis* the β-clamp seems to cluster in regions of newly replicated DNA, or 'clamp zones', of up to 200 rings [16].

In the last decade we have learned that proteins use similar ways of interacting with sliding clamps but that use distinct mechanisms in the regulation of these interactions. In what follows we will review our knowledge of the structural details of how sliding clamps interact with their ligands, as provided by crystallography and biochemical studies, and then analyze in turn each of the emerging themes of regulatory mechanisms outlined above.

BINDING AND AFFINITY OF PROTEIN LIGANDS TO SLIDING CLAMPS

Most proteins interact with clamps *via* residues in their N- or C-terminal regions [17] although internal binding sites are also critical in many cases. In prokaryotes the β-clamp consensus motif QLxLF has been recognized but is often poorly conserved and seems to be absent in a number of β-binding proteins [18, 19]. On the other hand, the consensus motif QxxLxxFF is present with minor variations in most PCNA-binding proteins [3, 17]. Since the proteins that harbour these motifs belong to many disparate structural families, these short sequences probably reflect convergent evolution for sliding-clamp binding.

The first detailed analysis of the binding surface of a clamp protein and its partner was that of the complex of human PCNA bound to a 22-residue C-terminal peptide of the cell-cycle inhibitor p21 [20, 21]. The p21 peptide uses a extensive array of contacts in binding to PCNA, interacting with the interconnector loop, the hydrophobic pocket and C-terminal residues of the clamp [21]. The C-terminus of p21 is disordered in solution and, in addition to PCNA, binds to a number of other ligands, including calmodulin and the oncoprotein SET [22, 23]. Relative affinities to these various ligands could be regulated by phosphorylation of the C-terminal domain of p21 [24].

The structure of a 11-residue C-terminal peptide of phage RB69 DNA polymerase in complex with its cognate clamp, the gp45 protein, has also been reported [25]. Although the interacting surface is much less extensive than in the case of p21, the peptide targets an analogous hydrophobic crevice on the clamp. Attachment *via* a flexible peptide allows the polymerase to remain tethered to the clamp while alternating between the spatially distant polymerizing and exonuclease modes of DNA binding sites. Another example of a polymerase interacting with its clamp has been recently provided by the structures of the C-terminal, 'little finger', domain of *E. coli* DNA polymerase IV and the β-clamp [26]. As with p21 and RB69, hydrophobic residues at the extended C-terminal peptide of Pol IV (in this case an 'LGL' motif) insert themselves in the surface pocket between domains 2 and 3 of the β-clamp. A distinctive feature observed in the Pol IV-β-clamp structure is a second binding surface that localizes to the interface between the two clamp monomers. A more recent crystal structure of Dpo4, the homologous Pol IV of *Sulfolobus solfataricus*, bound to its cognate clamp has given support to the idea that bound sliding clamp ligands often adopt highly flexible conformations [27].

A number of highly revealing studies of the enzymes involved in Okazaki fragment processing have shown multiple conformations of enzymes bound to PCNA. The interaction of Fen1 to PCNA has been studied using peptides [28] and the whole protein [29], indicating that the binding site is complex and consisting of extensive contacts near the interconnector loop and the C-terminus of PCNA. The binding site contains a hinge region that gives the complex high flexibility, and is highly reminiscent of that found in the case of p21 bound to PCNA [29]. The structure and dynamics of the interaction between *Sulfolobus solfataricus* DNA ligase and the heteromeric PCNA present in organism has been described [30], and that of a peptide of yeast ligase I (Cdc9) and PCNA [31].

Yet another class of structural insights has been provided by sliding clamps complexed to clamp loaders like the γ-complex or RFC. Clamp loaders are pentameric structures that are designed to bind an unloaded sliding clamp (*i.e.*, not topologically bound to DNA) and, in an ATP-dependent reaction, break the ring and place it around DNA at specific sites [10]. The structure of the *Escherichia coli* δ subunit of the γ-complex complexed to a monomeric β-clamp shows that δ binds to β *via* two sites: first, an internal sequence (69-QAMSLF-74) that resembles the β-binding motif found in a number of other proteins interacts with the hydrophobic pocket between domains 2 and 3 of the clamp; the second contact is between α4 of δ and a loop on the β-clamp which as a result becomes restructured [32]. These contacts are important for opening and loading of the ring, and additional contacts of other subunits of the pentameric γ-complex have been demonstrated [32, 33]. On the other hand, biochemical studies have shown that in yeast RFC at least two subunits, RFC1 and RFC3, bind PCNA tightly [34]. These RFC subunits, like δ, bind PCNA *via* an internal sequence and contains conserved hydrophobic residues that have been predicted to interact with the hydrophobic pocket on the ring [34].

Pentameric clamp loaders are highly conserved, reflecting an ancient mechanism of interaction and handling of the sliding clamps [10]. In archaeal organisms the homologous RFC contains only two types of subunits, RFC-L and RFC-S, with a stoichiometry of 1:4 [35]. The crystal structure of a 11-residue C-terminal peptide derived from *Pirococcus furiosus* RFC-L bound to *Pfu*PCNA has been reported, showing an arrangement very similar to peptides bound to other clamps [36]. An extended region of acidic residues adjacent to the consensus binding box is important for binding to the clamp [37].

In general, although interaction with the hydrophobic pocket often represents a strong anchoring site used by clamp loaders and clamp-interacting proteins, other interactions at distinct sites on the rings are usually observed. Detailed studies of the affinity of ligands for the sliding clamps have been performed using peptides of the binding protein as models [19, 22, 38, 39]. However, the *in vivo* and functional relevance of these studies is often difficult to interpret due to the differences in affinity expected depending on whether DNA is present, or if the sliding clamp is topologically bound to DNA or not.

Two main themes emerge from the structural studies of sliding clamp complexes. Firstly, all complexes studied to date (in viral, archaeal, bacteria or eukaryotic systems) interact with the hydrophobic pocket on the same side of PCNA or the β-clamp. This often involves the two main hydrophobic and/or aromatic residues present in the sliding clamp motifs. Secondly, clamp binding is often highly flexible and the various dynamic conformations are dependent on DNA structure, which probably ultimately regulates the affinity of the interaction.

Crystallographic studies have shown us insightful close-ups of the interactions between clamps and their partners. In what follows I will describe the knowledge, mainly biochemical, of the processes and the context in which these dynamic interactions take place.

LOADING OF CLAMPS ON DNA

Sliding clamps are opened and placed on DNA in ATP-dependent reactions by multisubunit complexes named clamp loaders [10]. The biochemistry of these enzymes has been studied in considerable detail for the case of the phage T4 gp44/62, *E. coli*'s γ-complex, and archeal and yeast RFC complexes [40-43]. Despite some distinct peculiarities of each clamp loader, the mechanism of loading of the clamp is very similar in all cases and involves conformational changes in the clamp loader that couple ATP binding to binding to the clamp in solution, and ATP hydrolysis to its loading and release on a DNA primed site [10, 43].

In *E. coli* the γ-complex (consisting of the $\gamma_3\delta\delta'$ subunits) first binds the β-clamp in solution through strong contacts with the δ subunit. However, this only happens when three molecules of ATP have been bound by the γ-complex [32]. The δ subunit binds to the β-clamp *via* a α-helix (α4) which undergoes major conformational changes involving rotation of four hydrophobic residues [32, 33]. In the absence of ATP α4 of δ interacts strongly with the δ' subunit (*via* residues Trp61 and Phe62) and the β-clamp binding residues (Leu73 and Phe74) are sequestered on the surface of δ, unable to bind to the β-clamp. However, upon ATP binding, residues Trp61 and Phe62 loose their interaction to δ' and Leu73 and Phe74 are projected and exposed, binding directly on the hydrophobic pocket of the β-clamp [32, 33]. Upon binding to primed DNA and hydrolysis of ATP, the process is reversed and the clamp is released already topologically bound to DNA.

Therefore the strong interaction of the clamp loader with β-clamp is regulated tightly by conformational changes on the main interacting component, the δ subunit, which depends on ATP binding and hydrolysis and recognition of DNA structure by the clamp loader. Analogous conformational changes are thought to operate in the case of the archaeal [44] or the eukaryotic clamp loaders [34, 40]. Phosphorylation of human RFC by CamKII [45] and cyclin-dependent kinases has been reported [46], in both cases leading to a reduction in its affinity for PCNA. The detailed mechanism and consequences of phosphorylation of RFC are not clear, but they indicate that modulation of the affinity of RFC for PCNA could be used by the cell cycle regulatory machinery to regulate DNA replication and other DNA metabolic processes.

CONTROL OF PROCESSIVITY OF REPLICATIVE DNA POLYMERASES

The high processivity of the DNA polymerases that replicate the chromosome is achieved *via* their binding to sliding clamps. However, in recent years it has become clear that this interaction, rather than being static and permanent, could have a half-life substantially shorter than what could be expected from the measurements of processivity of the polymerases. In a series of elegant kinetic experiments using the phage T4 replication system as a model, it has been shown that each DNA polymerase molecule is replaced by another one at a rate of about every 10 seconds during chromosomal DNA replication (this would result in about 90 exchanges per replicated chromosome *in vivo*)[47]. The replacement of the DNA polymerases is efficient because the sliding clamp allows the simultaneous binding of at least two polymerases, which favors a quick engagement of the second on with the 3'-terminus of the DNA when the first one disengages from it. Thus, the term 'dynamic processivity' has been suggested to reflect the fact that the overall processivity of a reaction may depend on the enzymatic actions of a number of individual DNA polymerase molecules coordinated by the same sliding clamp [47, 48].

If an analogous mechanism of DNA polymerase exchange occurs in bacterial or eukaryotic systems remains to be demonstrated, but a related switch has been observed in the thermophilic euryarchaean *Pyrococcus abissy.* During chromosomal replication in this organism DNA polymerases belonging to different families take up successively the PCNA sliding clamp after it has been loaded on DNA by the clamp loader [49]. Pol D binds to PCNA left by the clamp loader (RCF) and extends the RNA primer. Then, Pol D is replaced by the replicative polymerase Pol B which contains a strong PCNA binding site. The switch is required because Pol B would not be able to extend from a RNA primer efficiently [49].

How do processive replicative DNA polymerases release from the sliding clamp? In *E. coli*, a mechanism for quick disengagement of DNA polymerase III (Pol III) from the clamp has been proposed when the lagging-strand DNA polymerase recycles to a new primer during Okazaki fragment synthesis, a switch that occurs every 2-3 seconds during chromosomal synthesis. How does the polymerase lose its tight grip on its β-clamp at the end of each Okazaki fragment? The alpha catalytic subunit of Pol III can make at least two contacts with the β-clamp, one internal (920-QADMF-924) and another one at the extreme C-terminus of the (1154-QVELEF-1159) [18, 19, 50]. As discussed in the previous section, both sites have been proposed to be important for processivity of the polymerase, although biochemical and crystallographic data suggest that the internal one could be better positioned for interaction with the clamp during elongation [50, 51]. The

C-terminal site of alpha also binds tightly to another subunit of the replisome, the τ subunit of the gamma complex [52, 53], which regulates its processivity [54, 55]. The C-terminal domain of the τ subunit binds to single-stranded and double-stranded DNA [54]. It has been suggested that the τ subunit could compete with the β-clamp for binding to the C-terminal binding site on α. Indeed, in the presence of a primed template the affinity of τ for the C-terminus of α is low, but increases dramatically (>20-fold) in its absence or in the presence of double-stranded DNA [55]. Therefore during processive elongation the α subunit could be engaged with the β-clamp *via* C-terminal residues, which are then sequestered by τ when the complex reaches the double-stranded DNA of the downstream Okazaki fragment and τ-C 'senses' the absence of ssDNA. This DNA-sensing effect can be observed with a minimal system of just τ-C and a synthetic peptide containing the C-terminal 20 amino acids of the α subunit, indicating that these are the minimal players in a 'processivity switch' underlying the dissociation of α from the sliding clamp [55]. Recent crystallographic data obtained by the Steitz laboratory, however, have challenged this view and suggest that it is the C-terminal domain of the α subunit itself, which contains an OB fold which selectively only binds to ssDNA [51], could sense the nature of the DNA substrate and, *via* the τ subunit a conformational change on α, induce the processivity switch [51, 56]. Although speculative given the absence of structural data for the τ subunit (or the highly flexible C-terminal region of the α-subunit of Pol III), these two scenarios hint at possible mechanisms by which a processivity switch could be established. Given the complex and dynamic nature of the interaction of Pol III with the sliding clamp, however, much remains to be discovered about the interactions of Pol III and the sliding clamp [56].

The eukaryotic replication fork has not been explored in sufficient detail to analyze the binding and release of DNA polymerases from the sliding clamp PCNA. A recent study showed that perhaps an intrinsic conformational change on DNA polymerase δ allows it to sense the double stranded DNA and release itself from the DNA to jump to the next primed template during Okazaki fragment synthesis [57]. This 'collision release' model, however, was tested on a highly simplified system and critical effects of other subunits of the replisome on the processivity of the polymerase cannot be ruled out.

It should be noted that, in addition to the moving platform that the sliding clamps provide, additional scaffolds are provided in the replisome in which to which DNA polymerases attach themselves. In *E. coli* this second is provided by the γ/τ-complex which coordinates the actions of the leading and lagging DNA polymerases [58]. On a different system, phage T7 DNA polymerase uses the factor thiorredoxin instead than a sliding clamp to substantially increase its processivity but in addition it binds to a DNA helicase in order to remain associated to the replication fork [59-61]. In this way several T7 DNA polymerases can switch at the fork by simply alternation of those already bound to the DNA helicase [60]. It is remarkable how nature has found analogous solutions to the problem of fast recruitment of polymerases to the active site by binding to multi-attachment site complexes within the replisome, be them in the form of sliding clamps, the γ/τ-complex (clamp loader) or a DNA helicase.

TRANSLESION SYNTHESIS AND THE EXCHANGE OF DNA POLYMERASES

The critical role of sliding clamps in replication and repair has been highlighted in recent years with a large number of studies centered on the switch that occurs when a highly processive replicative DNA polymerase encounters damaged DNA that blocks its progression (reviewed in [62]). Nature's solution to this problem is to allow the transient action of a specialized polymerase that can overcome the damage (trans-lesion synthesis, or TLS). These polymerases have less stringent active sites that allows them to accommodate various types of bulky lesions and therefore are often much less accurate (low-fidelity) than replicative DNA polymerases. TLS polymerases display common features and structural elements, and they all belong to the Y-family of polymerases. The mechanism by which TLS polymerases take over extension of a primer 3'-end from a replicative DNA polymerase is based on their relative affinity for the sliding clamp.

In *E. coli* the Y-family polymerases Pol IV and Pol V act when Pol III stalls at a replication fork. Pol IV binds to the β-clamp at the same position on the clamp as Pol III [19, 63] but since the β-clamp can accommodate two ligands at the same time, both polymerases can simultaneously bind the same sliding clamp [64]. The mechanistic basis of this reaction was evidenced by the determination of the structure of a domain of Pol IV bound to the β-clamp [26]. The complex showed that Pol IV could interact with the sliding clamp *via* a highly flexible hinge that allows the protein to bind to the clamp in two distinct conformations: one extended in which Pol IV does not interact with DNA and would not clash sterically with Pol III bound to the same clamp, and one in which Pol IV would bend over DNA, take over the 3'-end from Pol III to extend the damaged template. Once the damaged DNA block is overcome, Pol III takes over again the 3'-end to resume high-fidelity, processive synthesis [64]. This model has been further supported by recent crystallographic work of the full-size *Sulfolobus solfataricus* Pol IV bound to its cognate clamp [27]. The detailed biochemical requirements also have been extensively studied in the case of *E. coli* Pol V (UmuCD), and reconstituted *in vitro* [65, 66]. Pol V also binds to the β-clamp in its hydrophobic pocket [19, 67] but there is no structural additional data about this interaction.

In eukaryotes a higher level of regulation of binding to sliding clamps is based in the transient covalent modification of PCNA by sumoylation [68-70] or ubiquitination (Fig. **2A**) [12, 68]. Posttranslational modifications have been observed in up to three residues on the surface of PCNA. In yeast PCNA residues K164 and K127 can be sumoylated [68], leading to recruitment of the helicase Srs2 inhibition of recombination ahead of the fork [71]. Ubiquitination of K164 of yeast PCNA by Rad6/Rad18 in response to DNA damaging agents leads to a reduction of affinity for Pol δ or Pol ε and probably of other proteins that bind to PCNA *via* the canonic motif, like the cohesion factor Eco1 or RFC1 [72]. On the other hand, ubiquitination leads to the recruitment of Pol η and Pol ζ [69] and Rev1 [73]. Ubiquitin needs to be removed from PCNA before Pol δ can resume processive DNA synthesis, a function performed by USP1 [39]. Finally, PCNA can be polyubiquitinated at K164 by Ubc13/Mms2/Rad5, but the function of this modification is unknown.

PROCESSING OF OKAZAKI FRAGMENTS

The processing of the RNA/DNA hybrids generated in lagging strand DNA replication, also known as Okazaki fragment maturation, requires the concerted action of one or more nucleases, a DNA polymerase and a DNA ligase [28, 74, 75]. In *E. coli* this process is achieved by DNA polymerase I, which contains a 5'-

3' exonuclease activity within the molecule that digests the RNA primer, and by DNA ligase. Both of these enzymes bind to the β-clamp [76, 77] but it is unknown if their actions are coordinated by it. On the other hand, in the archaeon *Sulfolobus solfataricus* the DNA ligase, the nuclease Fen1 and a DNA polymerase can bind with distinct affinity to each subunit of the heterotrimeric PCNA present in this organism, showing that simultaneous binding to the sliding clamp could be possible and suggesting tight coordination of these enzymes *via* PCNA [7]. In eukaryotes numerous recent studies have highlighted the role of PCNA in coordinating the activities of Pol δ and Fen1 during lagging strand DNA synthesis [28, 78-80].

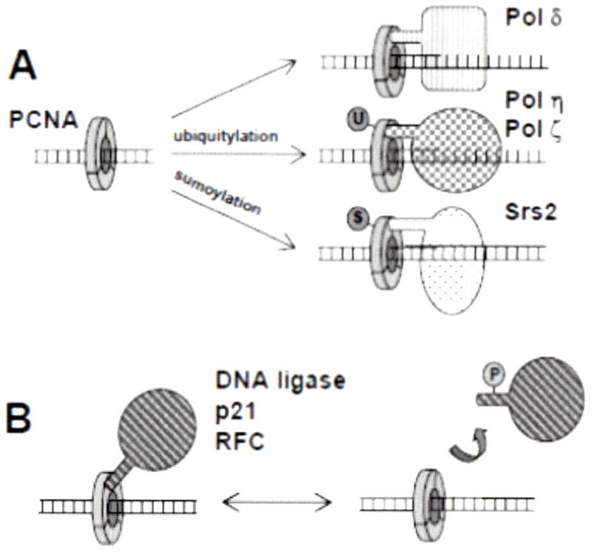

Figure 2: Two modes by which the interactions between eukaryotic sliding clamp PCNA and its partner protein ligands can be regulated. **A**. Ubiquitylation (U) or sumoylation (S) of PCNA leads to changes of affinity for DNA polymerases and, in the case of sumoylation, the recruitment of helicase Srs2. **B**. Phosphorylation of the ligand leads to a reduced affinity for PCNA (see Text). Phosphorylation often occurs in regions close to the binding site to PCNA.

Crystallographic studies of Fen1 bound to PCNA found that the interaction between these proteins is highly flexible [28, 29]. Fen1 binds to PCNA *via* an extended C-terminal peptide which at its base contains a hinge region that can adopt various conformations. Fen1 opens the DNA helix, enforcing a kink that facilitates flap recognition [28]. Multiple conformations of Fen1 could allow other proteins to access the DNA while Fen1 is still bound to the clamp because Fen1 can swivel on the hinge region next to the PCNA binding interface [29]. The effectors of the Fen1 switch from inactive to active conformations are unclear, but a kinked DNA structure is a likely candidate [28, 81].

The interaction of DNA ligase with PCNA and with nicked templates has also been extensively investigated functionally and structurally [82]. DNA ligase is a modular enzyme that can adopt open and closed conformations around DNA, as observed for human Lig1 in complex with DNA [83]. These large conformational changes could coordinate the action of DNA ligase with other PCNA-binding proteins [30], but it is unclear how binding to PCNA stimulates the ligation reaction. It has been suggested that perhaps PCNA somehow rigidifies the enzyme or allows it to distort and kink DNA [83], as observed for Fen1 [28]. In the closed conformation DNA ligase forms with PCNA two stacked rings with extensive contacts, and this perhaps would preclude other proteins from binding to the clamp. Perhaps this is related to the fact that ligation of DNA is typically the last step in Okazaki fragment processing or DNA repair reactions, and no subsequent binding to PCNA by other factors is required [30, 82, 84].

It is unclear how PCNA coordinates the handoff of the enzymes implicated in Okazaki fragment processing [80]. DNA structure could dictate the various conformational switches that must take place for the three enzymes to perform their functions, but the mechanistic details are unknown. Given the extensive contacts made by these enzymes with the sliding clamp [28, 30, 31], it is interesting to consider the question of how the different ligands sever their binding surfaces to the clamp. As in the case of translesion synthesis, protein modification seems to be the answer. Fen1 is phosphorylated by PCNA-bound Cdk1-Cyclin A in late S-phase, leading to its release from PCNA [85]. Likewise, human DNA ligase phosphorylation in multiple sites is cell-cycle dependent and results in its release from replication sites, but the effects of this modification on PCNA affinity are unclear [86-88]. Interestingly, an interaction between human DNA ligase I and the clamp loader RFC has been reported which inhibits ligation, but this effect is reduced by binding of DNA ligase to PCNA [89]. Analysis of the binding surface of yeast PCNA to yeast Ligase I (Cdc9) showed that RFC could compete for the same binding surfaces [31], suggesting a complex interplay between these proteins. One possible explanation for the interaction between DNA ligase I and RFC is that RFC could be required to unload PCNA once DNA has been sealed. Since DNA ligase occludes the face of PCNA that RFC binds, an initial contact between RFC and DNA ligase would become necessary [89].

MODULATION OF DNA REPAIR REACTIONS: MISMATCH REPAIR

The critical involvement of sliding clamps on DNA repair pathways has been well documented, where they act to orchestrate the enzymatic activities required for DNA damage recognition and processing. The mismatch repair pathway corrects errors introduced by the replication machinery and it has received intense scrutiny in the last decade [90]. The sensors of DNA mismatches, the MutS in *E. coli* or the MSH proteins of eukaryotes, can interact with the β-clamp or PCNA, respectively [76, 91]. In the case of MutS two interaction sites have been proposed: a weak one at the N-terminal, mismatch-binding, region of the protein, and one strong at the C-terminal region of the protein [18, 92, 93]. Recent studies of the *Bacillus subtilis* MutS-β-clamp interaction show that that β-clamp promotes the stabilization of MutS at mismatches *via* the C-terminal binding site, suggesting that mismatch recognition and β-clamp binding are tightly coupled to target MutS to active replication sites [93]. MutS could induce the release of DNA polymerase from the replication fork upon mismatch detection, and recent evidence point in to the tight interaction between mismatch repair proteins and the replisome [92, 94]. On the other hand, the binding site of eukaryotic MSH3 or MSH6 to PCNA is located at the extreme N-terminus of these proteins [91, 95-97]. It has been suggested that PCNA could stimulate the preferential binding of MSH2-MSH6 to DNA containing mismatches [95], but recent structural data of the PCNA-binding domain, which shows this domain as highly flexible, seems to argue against this idea [96-98].

E. coli's MutL and its eukaryotic homolog MLH1 also interact with the β-clamp and PCNA, respectively [99-101]. MutL is a multifaceted protein which acts as a 'matchmaker' among the mismatch repair factors and that, as revealed by crystallography, can undergo major conformational changes [102, 103]. The interaction between MutL and the β-clamp seems to require binding of MutL to single-stranded DNA, a natural ligand of MutL [92, 103]. In addition, the interactions of MutL to the β-clamp and MLH1 to PCNA have been tentatively mapped using peptides to an internal loop (Loop 2) that undergoes a conformational change during the ATPase cycle of these protein [92, 103]. This loop is exposed to binding to the sliding clamp in the nucleotide-free MutL, but buried inside the N-terminal domain of MutL when the protein is bound to ATP, suggesting a possible on-off switch for the interaction [103]. The function of these contacts is unknown, but they point towards a highly dynamic interaction in which protein affinity is ultimately modulated by DNA structure and ATP-induced conformational changes. MutL also binds *via* a canonical motif located at the C-terminus of the protein and the interaction could stimulate the endonuclease activity present in MutL [101]. On the other hand, it has been shown that yeast PCNA is essential in orienting the endonuclease activity present in MLH1-PMS1 complexes to discontinuous strands of DNA [104]. While the role of sliding clamps in mismatch repair is undisputed, there are still no clear view as to their function within this pathway. The suggestion has been made that perhaps the mismatch repair enzymes operate in very close proximity or even within the replisome itself, and that their main anchor is the sliding clamp.

FUTURE PROSPECTS

From the point of view of organic chemistry synthesis, the replication of a chromosome is an extraordinarily complex event. Human cells replicate their entire genome ($3x10^9$ base pairs) in just a few hours and with an error rate of 10^{-9} per base pair. The orchestration of the myriad biochemical events that take place during DNA synthesis relies on sequential and finely-tuned protein-protein and protein-DNA interactions of specialized protein modules that perform particular functions in an orderly, thermodynamically favorable, manner. How is it ensured that only the required enzymes coordinate their functions and activities at the replication fork at a given time and place? Part of the answer is given by the observed ability of different enzymes to organize themselves on proteins that accept multiple ligands and act as scaffolds for sequential enzymatic reactions. Within the replisome, at least three such scaffolds have been suggested in the literature: sliding clamps, the sliding clamp loaders [48], and the DNA helicase [60]. Of these, the sliding clamps show the highest versatility and universality. An understanding of what governs attachment or detachment from the sliding clamps is therefore a critical aspect of the study of the multi-enzymatic reactions that drive DNA replication and repair. As reviewed here, cells employ a variety of means to modulate affinity for clamps, but in most systems studied to date the detailed mechanism remains to be analyzed (Table **1**). In prokaryotic systems a combination of conformational changes in the enzyme, ultimately brought about by specific DNA structures and coupled to ATPase activity, seems to be a widespread mechanism of controlling binding to sliding clamps. On the other hand, posttranslational modification of the interacting surfaces is the typical way by which eukaryotic systems modulate access of the various ligands to the PCNA ring. It even has been suggested that these modifications in PCNA could serve as a way to 'reset' the clamp and prevent it from binding any factor [72]. A field that requires more investigation relates to how are clamps removed from DNA, as we don't clearly know what factor or factors are responsible for this process. The clamp loader (γ-complex or RFC) could be responsible or, as it has been proposed, one of the newly identified variants of RFC [1].

Table 1: Documented mechanisms that regulate binding and affinity to sliding clamps

Direct displacement by peptides or other proteins	
- Replication arrest in *Staphylococcus aureus* by phage-derived peptides	[105]
- τ processivity switch in *Escherichia coli*	[50, 51, 54, 55]
Conformational changes on the sliding clamp binding protein	
- γ-complex (clamp loader ATPase cycle and DNA binding)	[32, 33]
- MutL (modulated by ATP and ssDNA binding)	[92]
Phosphorylation of protein	
- p21 (inhibition of binding to PCNA)	[24]
- DNA ligase (effects unknown)	[86, 87]
- Fen1 by PCNA-bound Cdk1-ciclin A (inhibition of binding to PCNA)	[85]
- RFC1 by cell-cycle kinases (inhibition of binding to PCNA)	[45, 46]
Ubiquitination of protein	
- Cdt1 (leads to proteolysis and prevention of re-replication)	[106]
Sumoylation of PCNA	
- on K164 or K127 (recruitment of the helicase Srs2 and inhibition of recombination ahead of the fork)	[68, 71, 72]
Ubiquitination of PCNA	
- on K164 by RAD6 pathway (trans-lesion synthesis by DNA polymerases Pol η and Pol ζ)	[68, 69]
Polyubiquitination of PCNA	

Table 1: cont...

- on K164 by Ubc13/Mms2/Rad5 (trans-lesion synthesis regulation)	[68]
Phosphorylation of PCNA	
- Tyr211 by EGFR (stability on chromatin)	[107]

The diversity of proteins that interact with the β-clamp and PCNA makes sliding clamps one of the most trafficked elements in the cell protein network. Future work is likely to focus on the solving of additional crystal structures of complexes of sliding clamps and their partner proteins, perhaps in combination with relevant DNA structures. The analysis of the architecture of the ternary complexes of sliding clamps bound to active enzymes engaged with their natural substrates should be the goal because it is probably different in the absence of DNA substrate. Detailed analysis of affinities in the case of competition between clamps and other factors for binding to common sites on partner proteins will likely add to our understanding of how sliding clamps are regulated.

ACKNOWLEDGMENTS

I thank Dr. Paula Andrade for her encouragement and support. Funding for this work was provided by the 'Ramón y Cajal' program of the Spanish government (RyC-2005-62) and by a Marie Curie International Reintegration Grant from the European Commission (MC-IRG-21426).

CONFLICT OF INTEREST

The author(s) confirm that this chapter content has no conflict of interest.

DISCLOSURE

This is a revised and updated version of the review article "Regulation of Interactions with Sliding Clamps During DNA Replication and Repair", Current Genomics, 2009, 10: 206-215. Francisco J. López de Saro.

REFERENCES

[1] Majka, J. and Burgers, P.M. The PCNA-RFC families of DNA clamps and clamp loaders. *Prog. Nucleic Acid Res. Mol. Biol.* **2004**, *78*: 227-260.

[2] Johnson, A. and O'Donnell, M. Cellular DNA replicases: components and dynamics at the replication fork. *Annu. Rev. Biochem* **2005**, *74*: 283-315.

[3] Moldovan, G.L., Pfander, B. and Jentsch, S. PCNA, the maestro of the replication fork. *Cell* **2007**, *129*: 665-679.
[4] Naryzhny, S.N. Proliferating cell nuclear antigen: a proteomics view. *Cell Mol. Life Sci.* **2008**, *65*: 3789-3808.
[5] Pan, M., Kelman, L.M. and Kelman, Z. The archaeal PCNA proteins. *Biochem. Soc. Trans.* **2011**, *39*: 20-24.
[6] Kong, X.-P., Onrust, R., O'Donnell, M. and Kuriyan, J. Three dimensional structure of the ß subunit of *Escherichia coli* DNA polymerase III holoenzyme: a sliding clamp. *Cell* **1992**, *69*: 425-437.
[7] Dionne, I., Nookala, R.K., Jackson, S.P., Doherty, A.J. and Bell, S.D. A heterotrimeric PCNA in the hyperthermophilic archaeon *Sulfolobus solfataricus*. *Mol. Cell* **2003**, *11*: 275-282.
[8] Thelen, M.P., Venclovas, C. and Fidelis, K. A sliding clamp model for the Rad1 family of cell cycle checkpoint proteins. *Cell* **1999**, *96*: 769-770.
[9] Jansen, J.G., Fousteri, M.I. and de Wind, N. Send in the clamps: control of DNA translesion synthesis in eukaryotes. *Mol. Cell* **2007**, *28*: 522-529.
[10] Indiani, C. and O'Donnell, M. The replication clamp-loading machine at work in the three domains of life. *Nat. Rev. Mol. Cell. Biol.***2006**, *7*: 751-761.
[11] Krishna, T.S., Kong, X.P., Gary, S., Burgers, P.M. and Kuriyan, J. Crystal structure of the eukaryotic DNA polymerase processivity factor PCNA. *Cell* **1994**, *79*: 1233-1243.
[12] Fox, J.T., Lee, K.Y. and Myung, K. Dynamic regulation of PCNA ubiquitylation/deubiquitylation. *FEBS Lett* **2011**, *585*: 2780-2785.
[13] Burgers, P.M., Kornberg, A. and Sakakibara, Y. The dnaN gene codes for the beta subunit of DNA polymerase III holoenzyme of *Escherichia coli*. *Proc. Natl. Acad. Sci. USA* **1981**, *78*: 5391-5395.
[14] Essers, J., Theil, A.F., Baldeyron, C., van Cappellen, W.A., Houtsmuller, A.B., Kanaar, R. and Vermeulen, W. Nuclear dynamics of PCNA in DNA replication and repair. *Mol. Cell Biol* **2005**, *25*: 9350-9359.
[15] Sporbert, A., Gahl, A., Ankerhold, R., Leonhardt, H. and Cardoso, M.C. DNA polymerase clamp shows little turnover at established replication sites but sequential *de novo* assembly at adjacent origin clusters. *Mol. Cell* **2002**, *10*: 1355-1365.
[16] Su'etsugu, M. and Errington, J. The replicase sliding clamp dynamically accumulates behind progressing replication forks in *Bacillus subtilis* cells. *Mol. Cell* **2011**, *41*: 720-732.
[17] Warbrick, E. The puzzle of PCNA's many partners. *Bioessays* **2000**, *22*: 997-1006.
[18] Dalrymple, B.P., Kongsuwan, K., Wijffels, G., Dixon, N.E. and Jennings, P. A universal protein-protein interaction motif in the eubacterial DNA replication and repair systems. *Proc. Natl. Acad. Sci. USA* **2001**, *98*: 11627-11632.
[19] López de Saro, F.J., Georgescu, R.E., Goodman, M.F. and O'Donnell, M. Competitive processivity-clamp usage by DNA polymerases during DNA replication and repair. *EMBO J.* **2003**, *22*: 6408-6418.
[20] Warbrick, E., Lane, D.P., Glover, D.M. and Cox, L.S. A small peptide inhibitor of DNA replication defines the site of interaction between the cyclin-dependent kinase inhibitor p21WAF1 and proliferating cell nuclear antigen. *Curr. Biol.* **1995**, *5*: 275-282.
[21] Gulbis, J.M., Kelman, Z., Hurwitz, J., O'Donnell, M. and Kuriyan, J. Structure of the C-terminal region of p21(WAF1/CIP1) complexed with human PCNA. *Cell* **1996**, *87*: 297-306.
[22] Esteve, V., Canela, N., Rodriguez-Vilarrupla, A., Aligue, R., Agell, N., Mingarro, I., Bachs, O. and Perez-Paya, E. The structural plasticity of the C terminus of p21Cip1 is a

determinant for target protein recognition. *Chembiochem.* **2003**, *4*: 863-869.

[23] Zheleva, D.I., Zhelev, N.Z., Fischer, P.M., Duff, S.V., Warbrick, E., Blake, D.G. and Lane, D.P. A quantitative study of the *in vitro* binding of the C-terminal domain of p21 to PCNA: affinity, stoichiometry, and thermodynamics. *Biochemistry* **2000**, *39*: 7388-7397.

[24] Scott, M.T., Morrice, N. and Ball, K.L. Reversible phosphorylation at the C-terminal regulatory domain of p21(Waf1/Cip1) modulates proliferating cell nuclear antigen binding. *J. Biol. Chem.* **2000**, *275*: 11529-11537.

[25] Shamoo, Y. and Steitz, T.A. Building a replisome from interacting pieces: sliding clamp complexed to a peptide from DNA polymerase and a polymerase editing complex. *Cell* **1999**, *99*: 155-166.

[26] Bunting, K.A., Roe, S.M. and Pearl, L.H. Structural basis for recruitment of translesion DNA polymerase Pol IV/DinB to the β-clamp. *EMBO J.* **2003**, *22*: 5883-5892.

[27] Xing, G., Kirouac, K., Shin, Y.J., Bell, S.D. and Ling, H. Structural insight into recruitment of translesion DNA polymerase Dpo4 to sliding clamp PCNA. *Mol. Microbiol.* **2009**, *71*: 678-691.

[28] Chapados, B.R., Hosfield, D.J., Han, S., Qiu, J., Yelent, B., Shen, B. and Tainer, J.A. Structural basis for FEN-1 substrate specificity and PCNA-mediated activation in DNA replication and repair. *Cell* **2004**, *116*: 39-50.

[29] Sakurai, S., Kitano, K., Yamaguchi, H., Hamada, K., Okada, K., Fukuda, K., Uchida, M., Ohtsuka, E., Morioka, H. and Hakoshima, T. Structural basis for recruitment of human flap endonuclease 1 to PCNA. *Embo J.* **2005**, *24*: 683-693.

[30] Pascal, J.M., Tsodikov, O.V., Hura, G.L., Song, W., Cotner, E.A., Classen, S., Tomkinson, A.E., Tainer, J.A. and Ellenberger, T. A flexible interface between DNA ligase and PCNA supports conformational switching and efficient ligation of DNA. *Mol. Cell* **2006**, *24*: 279-291.

[31] Vijayakumar, S., Chapados, B.R., Schmidt, K.H., Kolodner, R.D., Tainer, J.A. and Tomkinson, A.E. The C-terminal domain of yeast PCNA is required for physical and functional interactions with Cdc9 DNA ligase. *Nucleic Acids Res.* **2007**, *35*: 1624-1637.

[32] Jeruzalmi, D., Yurieva, O., Zhao, Y., Young, M., Stewart, J., Hingorani, M., O'Donnell, M. and Kuriyan, J. Mechanism of processivity clamp opening by the delta subunit wrench of the clamp loader complex of E. coli DNA polymerase III. *Cell* **2001**, *106*: 417-428.

[33] Jeruzalmi, D., O'Donnell, M. and Kuriyan, J. Crystal structure of the processivity clamp loader gamma (gamma) complex of E. coli DNA polymerase III. *Cell* **2001**, *106*: 429-441.

[34] Bowman, G.D., O'Donnell, M. and Kuriyan, J. Structural analysis of a eukaryotic sliding DNA clamp-clamp loader complex. *Nature* **2004**, *429*: 724-730.

[35] Cann, I.K.O., Ishino, S., Yuasa, M., Daiyasu, H., Hiroyuki, T. and Ishino, Y. Biochemical analysis of replication factor C from the hyperthermophilic archaeon *Pyrococcus furiosus*. *J. Bacteriol.* **2001**, *183*: 2614-2623.

[36] Matsumiya, M., Ishino, S., Ishino, Y. and Morikawa, K. Physical interaction between proliferating cell nuclear antigen and replication factor C from *Pyrococcus furiosus*. *Genes to Cells* **2002**, *7*: 911-922.

[37] Nishida, H., Ishino, S., Miyata, T., Morikawa, K. and Ishino, Y. Identification of the critical region of Replication Factor C from Pyrococcus furiosus for the stable complex formation with Proliferating cell nuclear antigen and DNA. *Genes Genet. Syst.* **2005**, *80*: 83-93.

[38] Bruning, J.B. and Shamoo, Y. Structural and thermodynamic analysis of human PCNA with peptides derived from DNA polymerase- delta p66 subunit and flap endonuclease-1. *Structure* **2004**, *12*: 2209-2219.

[39] Zhuang, Z., Johnson, R.E., Haracska, L., Prakash, L., Prakash, S. and Benkovic, S.J. Regulation of polymerase exchange between Poleta and Poldelta by monoubiquitination of PCNA and the movement of DNA polymerase holoenzyme. *Proc. Natl. Acad. Sci. U S A* **2008**, *105*: 5361-5366.

[40] Yao, N.Y., Johnson, A., Bowman, G.D., Kuriyan, J. and O'Donnell, M. Mechanism of proliferating cell nuclear antigen clamp opening by replication factor C. *J. Biol. Chem.* **2006**, *281*: 17528-17539.

[41] Dionne, I., Brown, N.J., Woodgate, R. and Bell, S.D. On the mechanism of loading the PCNA sliding clamp by RFC. *Mol. Microbiol.* **2008**, *68*: 216-222.

[42] Bowman, G.D., Goedken, E.R., Kazmirski, S.L., O'Donnell, M. and Kuriyan, J. DNA polymerase clamp loaders and DNA recognition. *FEBS Lett.* **2005**, *579*: 863-867.

[43] Kelch, B.A., Makino, D.L., O'Donnell, M. and Kuriyan, J. How a DNA polymerase clamp loader opens a sliding clamp. *Science* **2011**, *334*: 1675-1680.

[44] Seybert, A., Singleton, M.R., Cook, N., Hall, D.R. and Wigley, D.B. Communication between subunits within an archaeal clamp-loader complex. *Embo J.* **2006**, *25*: 2209-2218.

[45] Maga, G., Mossi, R., Fischer, R., Berchtold, M.W. and Hubscher, U. Phosphorylation of the PCNA binding domain of the large subunit of replication factor C by Ca2+/calmodulin-dependent protein kinase II inhibits DNA synthesis. *Biochemistry* **1997**, *36*: 5300-5310.

[46] Salles-Passador, I., Munshi, A., Cannella, D., Pennaneach, V., Koundrioukoff, S., Jaquinod, M., Forest, E., Podust, V., Fotedar, A. and Fotedar, R. Phosphorylation of the PCNA binding domain of the large subunit of replication factor C on Thr506 by cyclin-dependent kinases regulates binding to PCNA. *Nucleic Acids Res.* **2003**, *31*: 5202-5211.

[47] Yang, J., Zhuang, Z., Roccasecca, R.M., Trakselis, M.A. and Benkovic, S.J. The dynamic processivity of the T4 DNA polymerase during replication. *Proc. Natl. Acad. Sci. U S A* **2004**, *101*: 8289-8294.

[48] Lovett, S.T. Polymerase switching in DNA replication. *Mol. Cell* **2007**, *27*: 523-526.

[49] Rouillon, C., Henneke, G., Flament, D., Querellou, J. and Raffin, J.P. DNA polymerase switching on homotrimeric PCNA at the replication fork of the euryarchaea *Pyrococcus abyssi*. *J. Mol. Biol.* **2007**, *369*: 343-355.

[50] Dohrmann, P.R. and McHenry, C.S. A bipartite polymerase-processivity factor interaction: only the internal β binding site of the α subunit is required for processive replication by the DNA polymerase III holoenzyme. *J. Mol. Biol.* **2005**, *350*: 228-239.

[51] Bailey, S., Wing, R.A. and Steitz, T.A. The structure of *T. aquaticus* DNA polymerase III is distinct from eukaryotic replicative DNA polymerases. *Cell* **2006**, *126*: 893-904.

[52] Dallmann, H.G., Kim, S., Pritchard, A.E., Marians, K.J. and McHenry, C.S. Characterization of the unique C terminus of the Escherichia coli tau DnaX protein. Monomeric C-tau binds alpha AND DnaB and can partially replace tau in reconstituted replication forks. *J. Biol. Chem.* **2000**, *275*: 15512-15519.

[53] Gao, D. and McHenry, C.S. tau binds and organizes Escherichia coli replication through distinct domains. Partial proteolysis of terminally tagged tau to determine candidate domains and to assign domain V as the alpha binding domain. *J. Biol. Chem.* **2001**, *276*: 4433-4440.

[54] Leu, F.P., Georgescu, R. and O'Donnell, M. Mechanism of the *E. coli* tau processivity switch during lagging-strand synthesis. *Mol. Cell* **2003**, *11*: 315-327.

[55] Lopez de Saro, F.J., Georgescu, R.E. and O'Donnell, M. A peptide switch regulates DNA polymerase processivity. *Proc. Natl. Acad. Sci. U S A* **2003**, *100*: 14689-14694.

[56] Georgescu, R.E., Kurth, I., Yao, N.Y., Stewart, J., Yurieva, O. and O'Donnell, M. Mechanism of polymerase collision release from sliding clamps on the lagging strand. *Embo J.* **2009**, *28*: 2981-2991.

[57] Langston, L.D. and O'Donnell, M. DNA polymerase delta is highly processive with proliferating cell nuclear antigen and undergoes collision release upon completing DNA. *J. Biol. Chem.* **2008**, *283*: 29522-29531.

[58] McInerney, P., Johnson, A., Katz, F. and O'Donnell, M. Characterization of a triple DNA polymerase replisome. *Mol. Cell* **2007**, *27*: 527-538.

[59] Hamdan, S.M., Marintcheva, B., Cook, T., Lee, S.J., Tabor, S. and Richardson, C.C. A unique loop in T7 DNA polymerase mediates the binding of helicase-primase, DNA binding protein, and processivity factor. *Proc. Natl. Acad. Sci. U S A* **2005**, *102*: 5096-5101.

[60] Hamdan, S.M., Johnson, D.E., Tanner, N.A., Lee, J.B., Qimron, U., Tabor, S., van Oijen, A.M. and Richardson, C.C. Dynamic DNA helicase-DNA polymerase interactions assure processive replication fork movement. *Mol. Cell* **2007**, *27*: 539-549.

[61] Johnson, D.E., Takahashi, M., Hamdan, S.M., Lee, S.J. and Richardson, C.C. Exchange of DNA polymerases at the replication fork of bacteriophage T7. *Proc. Natl. Acad. Sci. U S A* **2007**, *104*: 5312-5317.

[62] Friedberg, E.C., Lehmann, A.R. and Fuchs, R.P. Trading places: how do DNA polymerases switch during translesion DNA synthesis? *Mol. Cell* **2005**, *18*: 499-505.

[63] Burnouf, D.Y., Olieric, V., Wagner, J., Fujii, S., Reinbolt, J., Fuchs, R.P. and Dumas, P. Structural and biochemical analysis of sliding clamp/ligand interactions suggest a competition between replicative and translesion DNA polymerases. *J. Mol. Biol.* **2004**, *335*: 1187-1197.

[64] Indiani, C., McInerney, P., Georgescu, R.E., Goodman, M.F. and O'Donnell, M. A sliding-clamp toolbelt binds high- and low-fidelity DNA polymerases simultaneously. *Mol. Cell* **2005**, *19*: 805-815.

[65] Fujii, S. and Fuchs, R.P. Defining the position of the switches between replicative and bypass DNA polymerases. *Embo J.* **2004**, *23*: 4342-4352.

[66] Schlacher, K., Cox, M.M., Woodgate, R. and Goodman, M.F. RecA acts in trans to allow replication of damaged DNA by DNA polymerase V. *Nature* **2006**, *442*: 883-887.

[67] Becherel, O.J., Fuchs, R.P. and Wagner, J. Pivotal role of the beta-clamp in translesion DNA synthesis and mutagenesis in E. coli cells. *DNA Repair (Amst)* **2002**, *1*: 703-708.

[68] Hoege, C., Pfander, B., Moldovan, G.-L., Pyrowolakis, G. and Jentsch, S. RAD6-dependent DNA repair is linked to modification of PCNA by ubiquitin and SUMO. *Nature* **2002**, *419*: 135-141.

[69] Stelter, P. and Ulrich, H.D. Control of spontaneous and damage-induced mutagenesis by SUMO and ubiquitin conjugation. *Nature* **2003**, *425*: 188-191.

[70] Ulrich, H.D. The SUMO system: an overview. *Methods Mol. Biol.* **2009**, *497*: 3-16.

[71] Pfander, B., Moldovan, G.L., Sacher, M., Hoege, C. and Jentsch, S. SUMO-modified PCNA recruits Srs2 to prevent recombination during S phase. *Nature* **2005**, *436*: 428-433.

[72] Moldovan, G.L., Pfander, B. and Jentsch, S. PCNA controls establishment of sister chromatid cohesion during S phase. *Mol. Cell* **2006**, *23*: 723-732.

[73] Garg, P. and Burgers, P.M. Ubiquitinated proliferating cell nuclear antigen activates translesion DNA polymerases eta and REV1. *Proc. Natl. Acad. Sci. U S A* **2005**, *102*: 18361-18366.

[74] Kao, H.I. and Bambara, R.A. The protein components and mechanism of eukaryotic

Okazaki fragment maturation. *Crit. Rev. Biochem. Mol. Biol.* **2003**, *38*: 433-452.
[75] Liu, Y., Kao, H.I. and Bambara, R.A. Flap endonuclease 1: a central component of DNA metabolism. *Annu. Rev. Biochem.* **2004**, *73*: 589-615.
[76] López de Saro, F.J. and O'Donnell, M. Interaction of the beta sliding clamp with MutS, ligase and DNA polymerase I. *Proc. Natl. Acad. Sci. USA* **2001**, *98*: 8376-8380.
[77] Maul, R.W., Sanders, L.H., Lim, J.B., Benitez, R. and Sutton, M.D. Role of Escherichia coli DNA polymerase I in conferring viability upon the dnaN159 mutant strain. *J. Bacteriol.* **2007**, *189*: 4688-4695.
[78] Garg, P. and Burgers, P.M. DNA polymerases that propagate the eukaryotic DNA replication fork. *Crit. Rev. Biochem. Mol. Biol.* **2005**, *40*: 115-128.
[79] Rossi, M.L. and Bambara, R.A. Reconstituted Okazaki fragment processing indicates two pathways of primer removal. *J. Biol. Chem.* **2006**, *281*: 26051-26061.
[80] Beattie, T.R. and Bell, S.D. The role of the DNA sliding clamp in Okazaki fragment maturation in archaea and eukaryotes. *Biochem. Soc. Trans.* **2011**, *39*: 70-76.
[81] Gomes, X.V. and Burgers, P.M. Two modes of FEN1 binding to PCNA regulated by DNA. *Embo J.* **2000**, *19*: 3811-3821.
[82] Tomkinson, A.E., Vijayakumar, S., Pascal, J.M. and Ellenberger, T. DNA ligases: structure, reaction mechanism, and function. *Chem. Rev.* **2006**, *106*: 687-699.
[83] Pascal, J.M., O'Brien, P.J., Tomkinson, A.E. and Ellenberger, T. Human DNA ligase I completely encircles and partially unwinds nicked DNA. *Nature* **2004**, *432*: 473-478.
[84] Levin, D.S., Bai, W., Yao, N., O'Donnell, M. and Tomkinson, A.E. An interaction between DNA ligase I and proliferating cell nuclear antigen: implications for Okazaki fragment synthesis and joining. *Proc. Natl. Acad. Sci. U S A* **1997**, *94*: 12863-12868.
[85] Henneke, G., Koundrioukoff, S. and Hubscher, U. Phosphorylation of human Fen1 by cyclin-dependent kinase modulates its role in replication fork regulation. *Oncogene* **2003**, *22*: 4301-4313.
[86] Rossi, R., Villa, A., Negri, C., Scovassi, I., Ciarrocchi, G., Biamonti, G. and Montecucco, A. The replication factory targeting sequence/PCNA-binding site is required in G(1) to control the phosphorylation status of DNA ligase I. *Embo J.* **1999**, *18*: 5745-5754.
[87] Koundrioukoff, S., Jonsson, Z.O., Hasan, S., de Jong, R.N., van der Vliet, P.C., Hottiger, M.O. and Hubscher, U. A direct interaction between proliferating cell nuclear antigen (PCNA) and Cdk2 targets PCNA-interacting proteins for phosphorylation. *J. Biol. Chem.* **2000**, *275*: 22882-22887.
[88] Ferrari, G., Rossi, R., Arosio, D., Vindigni, A., Biamonti, G. and Montecucco, A. Cell cycle-dependent phosphorylation of human DNA ligase I at the cyclin-dependent kinase sites. *J. Biol. Chem.* **2003**, *278*: 37761-37767.
[89] Levin, D.S., Vijayakumar, S., Liu, X., Bermudez, V.P., Hurwitz, J. and Tomkinson, A.E. A conserved interaction between the replicative clamp loader and DNA ligase in eukaryotes: implications for Okazaki fragment joining. *J Biol Chem* **2004**, *279*: 55196-55201.
[90] Kunkel, T.A. and Erie, D.A. DNA mismatch repair. *Annu. Rev. Biochem.* **2005**, *74*: 681-710.
[91] Clark, A.B., Valle, F., Drotschmann, K., Gary, R.K. and Kunkel, T.A. Functional interaction of proliferating cell nuclear antigen with MSH2-MSH6 and MSH2-MSH3 complexes. *J. Biol. Chem.* **2000**, *275*: 36498-36501.
[92] Lopez de Saro, F.J., Marinus, M.G., Modrich, P. and O'Donnell, M. The beta sliding clamp binds to multiple sites within MutL and MutS. *J. Biol. Chem.* **2006**, *281*: 14340-14349.
[93] Simmons, L.A., Davies, B.W., Grossman, A.D. and Walker, G.C. Beta clamp directs

localization of mismatch repair in Bacillus subtilis. *Mol. Cell* **2008**, *29*: 291-301.

[94] Klocko, A.D., Schroeder, J.W., Walsh, B.W., Lenhart, J.S., Evans, M.L. and Simmons, L.A. Mismatch repair causes the dynamic release of an essential DNA polymerase from the replication fork. *Mol. Microbiol. 82*: 648-663.

[95] Flores-Rozas, H., Clark, D. and Kolodner, R.D. Proliferating cell nuclear antigen and Msh2p-Msh6p interact to form an active mispair recognition complex. *Nat. Genet.* **2000**, *26*: 375-378.

[96] Shell, S.S., Putnam, C.D. and Kolodner, R.D. The N terminus of *Saccharomyces cerevisiae* Msh6 is an unstructured tether to PCNA. *Mol. Cell* **2007**, *26*: 565-578.

[97] Clark, A.B., Deterding, L., Tomer, K.B. and Kunkel, T.A. Multiple functions for the N-terminal region of Msh6. *Nucleic Acids Res.* **2007**, *35*: 4114-4123.

[98] Iyer, R.R., Pohlhaus, T.J., Chen, S., Hura, G.L., Dzantiev, L., Beese, L.S. and Modrich, P. The MutSalpha-proliferating cell nuclear antigen interaction in human DNA mismatch repair. *J. Biol. Chem.* **2008**, *283*: 13310-13319.

[99] Umar, A., Buermeyer, A.B., Simon, J.A., Thomas, D.C., Clark, A.B., Liskay, R.M. and Kunkel, T.A. Requirement for PCNA in DNA mismatch repair at a step preceding DNA resynthesis. *Cell* **1996**, *87*: 65-73.

[100] Lee, S.D. and Alani, E. Analysis of interactions between mismatch repair initiation factors and the replication processivity factor PCNA. *J. Mol. Biol.* **2006**, *355*: 175-184.

[101] Pillon, M.C., Miller, J.H. and Guarne, A. The endonuclease domain of MutL interacts with the beta sliding clamp. *DNA Repair (Amst)* **2011**, *10*: 87-93.

[102] Ban, C. and Yang, W. Crystal structure and ATPase activity of MutL: implications for DNA repair and mutagenesis. *Cell* **1998**, *95*: 541-552.

[103] Ban, C., Junop, M. and Yang, W. Transformation of MutL by ATP binding and hydrolysis: a switch in DNA mismatch repair. *Cell* **1999**, *97*: 85-97.

[104] Kadyrov, F.A., Holmes, S.F., Arana, M.E., Lukianova, O.A., O'Donnell, M., Kunkel, T.A. and Modrich, P. *Saccharomyces cerevisiae* MutLalpha is a mismatch repair endonuclease. *J. Biol. Chem.* **2007**, *282*: 37181-37190.

[105] Belley, A., Callejo, M., Arhin, F., Dehbi, M., Fadhil, I., Liu, J., McKay, G., Srikumar, R., Bauda, P., Bergeron, D., Ha, N., Dubow, M., Gros, P., Pelletier, J. and Moeck, G. Competition of bacteriophage polypeptides with native replicase proteins for binding to the DNA sliding clamp reveals a novel mechanism for DNA replication arrest in Staphylococcus aureus. *Mol. Microbiol.* **2006**.

[106] Arias, E.E. and Walter, J.C. PCNA functions as a molecular platform to trigger Cdt1 destruction and prevent re-replication. *Nat. Cell. Biol.* **2006**, *8*: 84-90.

[107] Wang, S.C., Nakajima, Y., Yu, Y.L., Xia, W., Chen, C.T., Yang, C.C., McIntush, E.W., Li, L.Y., Hawke, D.H., Kobayashi, R. and Hung, M.C. Tyrosine phosphorylation controls PCNA function through protein stability. *Nat. Cell. Biol.* **2006**, *8*: 1359-1368.

Send Orders of Reprints at reprints@benthamscince.net
Advances in Genome Science, Vol. 1, 2013, 316-351

CHAPTER 13

Revealing Gene Function and Genetic Diversity in Plants and Animals *via* TILLING and EcoTILLING

N.A. Barkley* and M.L. Wang*

USDA-ARS, Plant Genetic Resources Conservation Unit (PGRCU), 1109 Experiment Street, Griffin, GA 30223, USA

Abstract: With the fairly recent advent of inexpensive, rapid sequencing technologies that continues to improve sequencing efficiency and accuracy, many species of animals, plants, and microbes have complete annotated genome information publicly available. The focus on genomics has thus been shifting from the collection of whole sequenced genomes to the study of functional genomics. Reverse genetic approaches have been used for many years to elucidate the gene function and resulting phenotype from DNA sequence content. Many of the currently used approaches (RNAi, gene knockout, site-directed mutagenesis, transposon tagging) rely on the creation of transgenic material, the development of which is not always feasible for many plant or animal species. TILLING is a non-transgenic reverse genetics approach that is applicable to all animal and plant species which can be mutagenized, regardless of its mating / pollinating system, ploidy level, or genome size. This approach, however, does require prior DNA sequence information. TILLING can ultimately provide an allelic series of silent, missense, nonsense, and splice site mutations to examine the effect of various mutations in a gene and their link to a particular phenotype. TILLING has proven to be a practical, efficient, and an effective approach for functional genomic studies in numerous plant and animal species. EcoTILLING, which is a variant of TILLING, examines natural genetic variation in populations and has been successfully utilized in animals and plants to reveal genetic diversity. In this review, TILLING and EcoTILLING techniques are described and reviewed, as well as, their beneficial applications and limitations.

Keywords: Reverse genetics, TILLING (target induced local lesions in genomes), EcoTILLING (Ecotype TILLING), sequencing, SNP (single nucleotide polymorphism), mutant population, functional genomics, germplasm, genetic diversity.

GENOMICS REVOLUTION

The development of what would ultimately become the modern day field of

***Address correspondence to Barkley NA and Wang ML:** USDA-ARS, Plant Genetic Resources Conservation Unit (PGRCU), 1109 Experiment Street, Griffin, GA 30223, USA; Tel: (770)412-4035; E-mail: Elle.Barkley@ars.usda.gov; Tel: (770) 229-3342; Fax: (770) 229-3323; E-mail: Mingli.Wang@ars.usda.gov

Christian Neri (Ed)

genomics began in 1977 with two publications of chemical enzymatic DNA sequencing methods developed by Sanger *et al.* [1], and Maxam and Gilbert [2], as well as, a publication of the first complete DNA sequence of a bacteriophage, phiX174 [3]. Since 1977, the refinement and automation of the Sanger method of dideoxy chain termination sequencing has led to complete sequencing, assembly, and annotation of an extraordinary number of genes and entire genomes. The only sequenced genomes made available from 1977-1995 were either viral or organelles, which have smaller genome sizes [4] than plants and animals. Moreover, in the last decade, due to the progression of sequence automation, computing technology, and bioinformatics, there have been many advances in genomics including a vast number of species (animals, microbes, and plants) with their whole genome sequenced. Rapid sequencing technologies (Sanger dideoxy chain termination automated sequencing [4, 5], pyrosequencing [6], reversible terminator sequencing [6], ligation sequencing [6], nanopore sequencing [5]) that have continually improved in efficiency, accuracy, and a reduction of the overall cost and time required to sequence genomes have revolutionized the ability of scientists to obtain completely assembled genome information on many species of animals, plants, and microbes.

The exponential growth of nucleotide sequence information across eukaryote taxa in the last few years due to improvements and reduction in cost of next generation sequencing, as well as, the *in silico* software capable of analyzing it has allowed reverse genetic approaches to be employed to determine putative gene function. The number of genes deposited into databases in which the function is unknown has soared due to genome sequencing efforts [7]. Since microbes in general have simple and relatively small sized genomes comparatively to eukaryotes, many researchers originally rushed to sequence these less complicated organisms such as *Agrobacterium tumefaciens* (cause of crown gall disease), *Bacillus spp.*, *Campylobacter jejuni* (causes gastroenteritis in humans), *Escherichia coli*, *Fusobacterium nucleatum* (dental plaque), *Helicobacter ssp.*, *Lactobacillus ssp.* (convert lactose to lactic acid), *Mycoplasma ssp* (parasites or saprobe), *Propionibacterium acnes* (acne) *Staphylococcus spp*, and *Streptococcus spp*. Some of the animal sequencing projects that are in progress or are currently assembled include simple to very complex organisms such as *Drosophila melanogaster* (fruit fly),

Caenorhabditis elegans (worm), *Canis lupus familiaris* (dog), *Ciona itestinalis* (sea squirt), *Gallus gallus* (chicken), *Homo sapiens* (human), *Mus musculus* (mouse), *Rattus norvegicus* (rat), *Pan troglodytes* (chimpanzee), and *Takifugu rubripes* (pufferfish). The number of characterized plant genomes including some agriculturally important crops has lagged behind sequencing of microbes and animals due to large genome sizes, high proportion of repetitive DNA, and the complication of various ploidy levels. The first published plant genome available was from the model plant *Arabidopsis thaliana*, but since then others have followed such as *Oryza sativa* (rice), *Populus trichocarpa* (black cottonwood), and *Medicago trunculata*. Since these initial genomes were made publicly available, plant genome sequencing projects have become routine. Some of the crop plants that have genome data include *Cajunus cajon* (pigeon pea), *Carica papaya* (papaya), *Citrus sinensis* (sweet orange), *Cucumis sativus* (cucumber), *Glycine max* (soybean), *Gossypium raimonddi* (cotton), *Linum usiitatissimum* (flax), *Lotus japonicus*, *Malus domestica* (apple), *Manihot esculenta* (cassava), *Solanum lycopersicum* (tomato), *Solanum tuberosum* (potato), *Phaseolus vulgaris* (common bean), *Ricinus communis* (castor bean), *Sorghum bicolor* (sorghum), *Theobroma cacao* (chocolate), and *Zea mays* (maize). (The purpose was not to mention all sequencing projects but to give an overview of some major ones available. Therefore, this is just a partial list of some of the genome sequencing projects completed or ongoing in microbes, animals, and plants).

The international scientific community has cooperated on open access database creation, maintenance, and usage to the extent that these database collections are experiencing exorbitant growth. Due to the exponential growth in sequences deposited to the three major databases (GenBank, EMBL, and DDBJ), there was a need in the scientific community to develop taxa specific databases to manage the ever growing amount of sequence information. Some of these new and expanding databases include: the International Grape Genome Program (IGGP) (http://www.vitaceae.org/index.php/International_Grape_Genome_Program; http://www.genoscope.cns.fr/externe/GenomeBrowser/Vitis/) formed by the genetics international research community with projects funded by the National Science Foundation (NSF) in the US, the cacao genome database (http://www.cacaogenomedb.org/genome-sequence-anon), soybean (http://soybase.org/SequenceIntro.php), rice (http://gramene.org/Oryza_indica/Info/Index), and cucurbits (cumber and watermelon) maintained by the International

Cucurbit Genomics Initiative (www.icugi.org). Many of these databases maintain annotated genome information, genetic maps, QTL's, molecular markers, and other valuable resources for the crop of interest and the scientists studying them. This is not an exhaustive list of all the taxa specific databases that exist on the web, but a sample of some specific to plants.

Many believed that once genomes were sequenced scientists would be able to better understand genome organization including gene function and ultimately manipulate genes in a genome *via* genetic engineering. A main target for genetic engineering would be to alleviate disease problems in plants and animals. One of the main problems for determining the mechanism of human disease such as lupus is that epistasis, or the interaction between genes at different loci, can play a major role in disease susceptibility [8]. This interaction between genes cannot be detected by knowing the sequence content, it must be determined empirically. Further, structural variants and rare alleles can be missed when sequencing read depth is not sufficient or multiple individuals from a species are not targeted to reveal these variants. The 1,000 genome project (www.1000genomes.org) in which a large number of humans are being sequenced at a read depth of 4X and aligned to a reference genome will allow low frequency variants to be revealed. These variants may demonstrate the underlying inheritance of certain traits or diseases. Therefore, pinpointing the mechanism and causes of disease requires more effort than just sequencing a specific genome and identifying a candidate gene.

Another problem in determining gene function for pieces of DNA sequence is the lack of information on the role of many genes in a particular genome that have been deposited into sequence databases. Bioinformatics and computing technology is helping to unravel the coding DNA and predict gene sites in the genome, as well as, gene function based on similarity to other deposited sequences which should provide many insights. However, similarity does not always translate into equivalent function especially when comparing across distant taxa. Furthermore, large portions of sequenced genomes in plants and animals currently have not yet been assigned any function based on homology to known proteins. Therefore, other strategies are necessary to empirically identify unknown genes and elucidate their respective function.

GENETIC APPROACHES FOR FUNCTIONAL GENOMICS

Two main approaches are utilized to link genotype to phenotype known as forward and reverse genetics. Both of these processes aim to determine the function of genes through screening individual mutants' phenotype or genotype and ultimately determine how it is controlled. Traditionally utilized, forward genetics (phenotype to genotype), in which one starts with a particular identified phenotype or biological process and the gene sequence is ultimately deduced through screening large numbers of mutagenized individuals for phenotypic variations is a useful approach. However, forward genetic approaches are not practical for genome wide analysis primarily due to the effort and time involved to identify each gene coding for a particular phenotype [9]. In reverse genetics (genotype to phenotype), the gene sequence is known and mutants are screened to identify individuals with alterations in the gene of interest [10]. This approach is generally less time demanding than forward genetics. Reverse genetic strategies have been successfully used for functional genomics in many animal and plant species and the widespread availability of sequence data allows researchers to rapidly design reverse genetic strategies to determine gene function. Some of the reverse genetic strategies employed in plants and animals include homologous recombination, *Agrobacterium* mediated insertional mutagenesis, transposon tagging, RNAi (RNA interference) or PTGS (post transcriptional gene silencing) and chemical mutagenesis. Even though many of these strategies are effective they can often be organism specific [11]. An overview and the advantages and disadvantages of these strategies will be discussed.

Homologous recombination is a reverse genetic approach that has become routine for some microbes such as *E. coli* and yeast, but has proven to be much less effective or efficient to employ in multi-cellular eukaryotes such as plants [12, 13]. Many attempts have been made in various species of plants to optimize homologous recombination with limited success rates [12, 14]. In mice *Mus musculus*, homologous recombination is a common practice and has mainly been optimized to provide precise mutations in any gene from embryonic stem cells [11, 14]. This procedure is advantageous because it can target specific sites in the genome and disrupt them by reciprocal exchange of DNA. It is thought to occur by double strand breaks, replacement and subsequent repair to ultimately knock out or disrupt the target gene. The disadvantages to this approach include the creation of transgenic material, lack of efficiency, and the limited success rate with many species.

The gram negative bacterium, *Agrobacterium tumefaciens,* which causes crown gall disease (tumors) in many host plants, has been turned into a widely used tool for reverse genetic research in plant systems. This soil borne bacterium naturally infects wounded plants by transferring T-DNA (located on a plasmid known as Ti for tumor inducing plasmid) from the bacteria to the plant, which subsequently integrates into the plant's genome [13]. An engineered strain of *Agrobacterium tumefaciens* has been successfully used to create numerous transgenic plants including many agronomical and horticultural important species [15]. For example, rice plants have been genetically modified to increase the plant's ability to resist disease, tolerate salt and cold stress, and improve nutrition by adding genes encoding for the β-carotene biosynthetic pathway to the endosperm allowing the synthesis of vitamin A [16]. Advantages to this approach includes generation of large mutant populations that can be created and stored, as is currently available for *Arabidopsis* [12], and the potential for improving the nutritional value or disease resistance in crop plants. A disadvantage of this method is inefficient transgene expression [15]. Another potential disadvantage is that in some plants, such as soybean (*Glycine max*), transformation success may be genotype specific [17] and of relatively low efficiency.

Transposons, also known as jumping genes, were first discovered by Barbara McClintock in the 1940's when she analyzed the mosaic pigmentation patterns of maize kernels. Since their discovery, they have been broadly utilized for reverse genetics studies in plants and animals. In 1995 Pioneer Hi-Bred used the Mutator (*Mu*) transposon on a large scale to screen for *Mu* insertions in genes of interest in maize. One of the potential drawbacks of *Mu* is the detection of somatic insertions in genes that are thus not transmitted to the offspring [18]. In animals, transposons known as P-elements have been vastly used to examine gene function in *Drosophilia*. These elements have been genetically engineered for insertional mutagensis and one can order stock that carries an insertion in a gene of interest [19]. Advantages include complete or partial disruption of gene function dependent on the transposon insertion site. Of course, this reverse genetic approach requires the creation of transgenic material which is often times not feasible in certain species.

RNAi has been widely employed and studied in plant and animal systems. RNAi is a gene silencing mechanism, which occurs through the cell's recognition and degradation of double stranded RNA (dsRNA) which ultimately leads to an interruption to a gene's function [20] in diminished protein production. One of the earliest examples of PTGS in plants was found in petunia where genetic constructs were introduced in an attempt to increase expression levels of chalcone synthase (CHS), a key component in flavonoid biosynthesis responsible for plant and flower pigmentation. The experiment unexpectedly resulted in the cosuppression of the endogenous and exogenous CHS gene expression [21]. Research later demonstrated in *Caenorhabditis elegans* [22] that the phenomenon of PTGS was due to dsRNA in a cell eliciting a silencing mechanism. Most eukaryotes have RNAi capability, and thus, can process dsRNA into 22-25 bp pieces, which can proceed to recognize homologous transcripts for degradation or target genes for DNA methylation [12]. RNAi technology is currently being examined to treat human disease. The main hurdles for RNAi therapeutic treatment for disease in humans is delivery and safety [23]. Advantages in certain organisms include that the silencing is heritable and systemic [24]. Disadvantages of this method include low throughput and efficiency of gene silencing can vary and be unpredictable [12].

Since the 1940s, chemical mutagenesis, which was first discovered to produce mutations in *Drosophila* and plants, has long been used as a tool for reverse genetic approaches [25] to obtain mutants for evaluation. A benefit of chemical mutagenesis comparatively to other reverse genetic approaches is that it can provide loss-of-function, as well as, gain-of-function mutations producing null, hypomorphic, neomorphic or hypermorphic phenotypes [7]. Numerous novel varieties of crop plants such as rice, wheat, cotton, and sunflower, to name a few, have been derived through mutation induction [26]. Chemical agents allow researchers to examine a range of various mutations in a gene of interest including silent, missense, nonsense, splice site, and deletions, and furthermore, link a genotype to a particular phenotype. Some of the chemical agents used to create mutants include ethylmethane sulfonate (EMS), N-ethyl-N-nitrosourea (ENU), N-nitroso-N-methylurea (NMU), and ionizing radiation. The advantages of chemical mutagens is that they tend to produce a relatively high density of random mutations [12] throughout the genome including gene knockouts and produce induced point mutations which generate a range of alleles for genetic analysis [27] (Tables **1** and **2**). Additionally, the creation of transgenic material is not required. Furthermore, chemical mutagensis does not require tissue culture and the mutations are heritable in successive generations [17]. Lastly, unlike other reverse genetic approaches chemical mutagenesis is applicable in most taxa that can be mutagenized. The main disadvantage (until recently) of chemically induced mutations was an efficient, practical approach for detection. TILLING, a high throughput mutation detection method, mines induced mutations in a population and results in high mutational density with very low levels of aneuploidy and dominant lethality [28].

Table 1: Overview of the published mutation frequencies, mutagen dose, genome size, and ploidy level reported from TILLING studies in various organisms including plants and animals.

Organism	**Genus Species**	**Mutagen Applied**	**MutagenDose**	**Estimated Genome Size**	**Ploidy**	**Reported Mutation Rate**	**Citation**
Arabidopsis	*Arabidopsis thaliana*	EMS	20-40 mM	125 Mb	2X	1/300 kb	[28] Greene *et al.* 2003 Genetics
Barley	*Hordeum vulgare*	EMS	20-30 mM	5,300 Mb	2X	1/Mb	[51] Caldwell *et al.* 2004 The Plant Journal
Maize	*Zea mays*	EMS	0.0625%	2,500 Mb	2X	0.93/kb B73	[50] Weil and Monde 2007 The Plant Genome
Maize	*Zea mays*	EMS	0.0625%	2,500 Mb	2X	2.10/kb W22	[50] Weil and Monde 2007 The Plant Genome
Maize	*Zea mays*	EMS	1%	2,500 Mb	2X	2/Mb	[49] Till *et al.* 2004 BMC Plant Biology
	Medicago truncatula	EMS	0.15%-0.2%	500 Mb	2X	1/485kb	[44] Le Signor *et al.* 2009 Plant Biotechnology Journal

Table 1: cont…

Melon	*Cucumis melo*	EMS	1.5-2%	450 Mb	2X	1/573 kb	[60] Dahmani-Mardas *et al.* 2010 PLoS One
Pea	*Pisum sativum*	EMS	4 mM	4,300 Mb	2X	1/669 kb	[48] Triques *et al.* 2007 The Plant Journal
Rice	*Oryza sativa*	EMS	1.5%	430 Mb	2X	1/294 kb	[37] Till *et al.* 2006 BMC Plant Biology
Rice	*Oryza sativa*	Az-MNU	1 mM Az-15 mM MNU	430 Mb	2X	1/265 kb	[37] Till *et al.* 2006 BMC Plant Biology
Rice	*Oryza sativa*	EMS	0.8-1%	430 Mb	2X	0.5/Mb	[57] Wu *et al.* 2005 Plant Mol. Bio.
Rice	*Oryza sativa*	EMS	1.6%	430 Mb	2X	1/Mb	[57] Wu *et al.* 2005 Plant Mol. Bio
Soybean	*Glycine max*	NMU	2.5 mM	1,115 Mb	2X	1/140 kb	[17] Cooper 2008 BMC plant bio
Soybean	*Glycine max*	EMS	50 mM	1,115 Mb	2X	1/250 kb	[17] Cooper 2008 BMC plant bio
Soybean	*Glycine max*	EMS	40 mM	1,115 Mb	2X	1/550 kb	[17] Cooper 2008 BMC plant bio
Wheat	*Triticum turgidum* subsp. *durum*	EMS	0.75-1%	12,000 Mb	4X	1/40 kb	[54] Slade *et al.* 2005 Nature biotechnology
Wheat	*Triticum aestivum*	EMS	0.75-1.2%	17,000 Mb	6X	1/24 kb	[54] Slade *et al.* 2005 Nature biotechnology
Fruit fly	*Drosophila*	EMS	50mM	180 Mb	2X	1/156 kb	[65] Winkler *et al.* 2005 Genome Research
Fruit fly	*Drosophila*	EMS	125mM	180 Mb	2X	1/90.5kb	[65] Winkler *et al.* 2005 Genome Research
Fruit fly	*Drosophila melanogaster*	EMS	50mM	180 Mb	2X	1/209kb	[66] Bentley 2000 Genetics
Nematode	*C. elegans*	EMS	0.025 M	100 Mb	2X	1/293 kb	[35] Gilchrist *et al.* 2006 BMC Genomics
Zebrafish	*Danio rerio*	ENU	3.0mM	1,700 Mb	2X	1/235 kb	[34] Wienholds *et al.* 2003 Genome Res

Table 2: A list of the number of target genes and different classes of mutations detected by TILLING from previously published plant and animal studies.

Organism	No. of Genes	No. of Silent Mutations	No. of Missense Mutations	No. of Nonsense Mutations	No. of Splice Site Mutations	Non coding	Total	Citation
Arabidopsis	192	851	946	93	--	--	1,890	[28] Greene *et al.* 2003 Genetics
Barley	2	4	6	0	0	--	10	[51] Caldwell *et al.* 2004 The Plant Journal
Lotus	1	--	6	--	1	--	15	[46] Perry 2003 plant phys
Lotus	4	8	28	2	0	19	57	[47] Horst Plant Phys
Maize	11	7	10	0	0	--	17	[49] Till *et al.* 2004 BMC Plant Biology
Pea	5	19	39	2	0	--	60	[48] Triques *et al.* 2007 The Plant Journal
Rice	10	19	29	1	--	8	57	[37] Till *et al.* 2006 BMC Plant Biology
Soybean	7	42	62	3	0	9	116	[17] Cooper *et al.* 2008 BMC Plant bio
Wheat	3	--	84	3	5	--	246	[54] Slade *et al.* 2005 Nature biotechnology
Fruit fly	3	10	33	1	0	--	44	[65] Winkler *et al.* 2005 Genome Research
Fruit fly	1	6	6	0	0	4	16	[66] Bentley 2000 Genetics
Nematode	10	27	42	2	0	--	71	[35] Gilchrist *et al.* 2006 BMC Genomics
Zebrafish	16	63	119	14	7	52	255	[34] Wienholds *et al.* 2003 Genome Res

HISTORY AND OVERVIEW OF THE TILLING METHOD

TILLING first began in the late 1990's from the effort of a graduate student, Claire McCallum (and collaborators from Fred Hutchinson Cancer Research Center and Howard Hughes Medical Institute), who worked on characterizing the function of two chromomethylase genes in *Arabidopsis* [29]. Claire McCallum utilized reverse genetic approaches such as T-DNA lines and antisense RNA, but, was unable to successfully apply these approaches to characterize CMT2. The approach that was successful turned out to be what is now known as TILLING (Targeting Induced Local Lesions in Genomes). This was accomplished by pooling chemically induced mutagenized plants together, amplifying the region of interest, creating heteroduplexes among the pooled DNA, and performing dHPLC (denaturing high performance liquid chromatography) to detect the mutants by chromatographic alterations [30]. Since the inception of this method, TILLING has been streamlined, automated, and utilized in many plant and animal taxa. Further, since the inception of this technique multiple TILLING facilities both public and private have been established worldwide [31] for researchers to order and characterize mutants in a gene of interest.

TILLING, which is a reverse genetic high throughput approach, aims to identify SNPs (single nucleotide polymorphisms) and / or INDELS (insertions / deletions) in a gene(s) of interest from large populations. Therefore, the first step in TILLING is the creation of a mutagenized population, which is often accomplished by treatment with a chemical mutagen such as EMS (Fig. **1**). Many plant species are well suited for this strategy because they can be self-fertilized and seeds can be saved and stored for long periods of time [29, 32]. Strategies also exist for the creation of mutant populations in animal species [33-35]. In plants, seeds are treated with EMS and grown out to produce M_1 plants, which are subsequently self-fertilized to produce the M_2 generation, of which, DNA is collected for mutational screening [36]. To avoid sampling the same mutation only one M_2 individual from each M_1 is chosen for DNA extraction [37]. The M_2 progeny can be self-fertilized and the resulting M_3 seed can be preserved in long term storage [38]. EMS has been widely used as a chemical mutagen in TILLING in both plant and animal studies to generate mutant populations, although other chemical mutagens can also be effective to generate the populations (Table **1**). EMS has been used frequently and typically produces transition mutations (G/C : A/T) because it alkylates G residues [39]. The alkylated G residue pairs with T instead of the conventional base pairing with C [12]. It is a beneficial strategy for users to try a range of concentrations of the chemical mutagen being applied to evaluate the toxicity and sterility on germinal tissue before preparing large mutant populations [38].

Once the population has been prepared, the genomic DNA targets need to be selected. The web based program CODDLE (http://www.proweb.org/input) allows users to input genomic, cDNA, or protein sequences and evaluate the probable effect of induced or natural polymorphisms on gene function [27]; furthermore, optimal PCR primers can be designed for a functional domain target. The next step is to collect DNA from the population and normalize the DNA concentration (Fig. **1**). It is crucial to ensure that all DNA samples are equivalent so that no biasing of samples occurs. Once samples have been normalized, the DNA should be pooled together. In general, a pool of DNA consisting of two individual samples up to eight individual samples for diploid organisms has been successful for mutation detection. With a larger pool size, the sensitivity of mutation detection will decrease, because the proportion of heteroduplexes compared to homoduplexes in the reaction is reduced [12]. Therefore, depending on ploidy level, heterozygosity, and the amount of naturally occurring SNPs, optimal pooling for a species of interest should be determined empirically.

Once the pooled DNA is arrayed into 96 well microtiter plates, pooled samples are amplified using primers targeting the gene of interest. The forward and reverse primers are differentially 5' end labeled with IRD700 and IRD800 dye labels for fluorescent detection at ~700 nm and ~800 nm, respectively (Fig. **1**). Next, heteroduplexes and homoduplexes are formed from the PCR products of pooled samples (consisting of mutants and the wild type) by heating (denaturing) and cooling (annealing). The endonuclease enzyme CEL I is applied and a short incubation is required for the enzymatic reaction to progress. CEL I, isolated from celery, not only specifically recognizes mismatches in a heteroduplex, it also cleaves the DNA on the 3' side of the mismatch [32]. After the enzyme incubation period, detection of any digested fragments occurs by separation on a denaturing

TILLING and EcoTILLING diagram

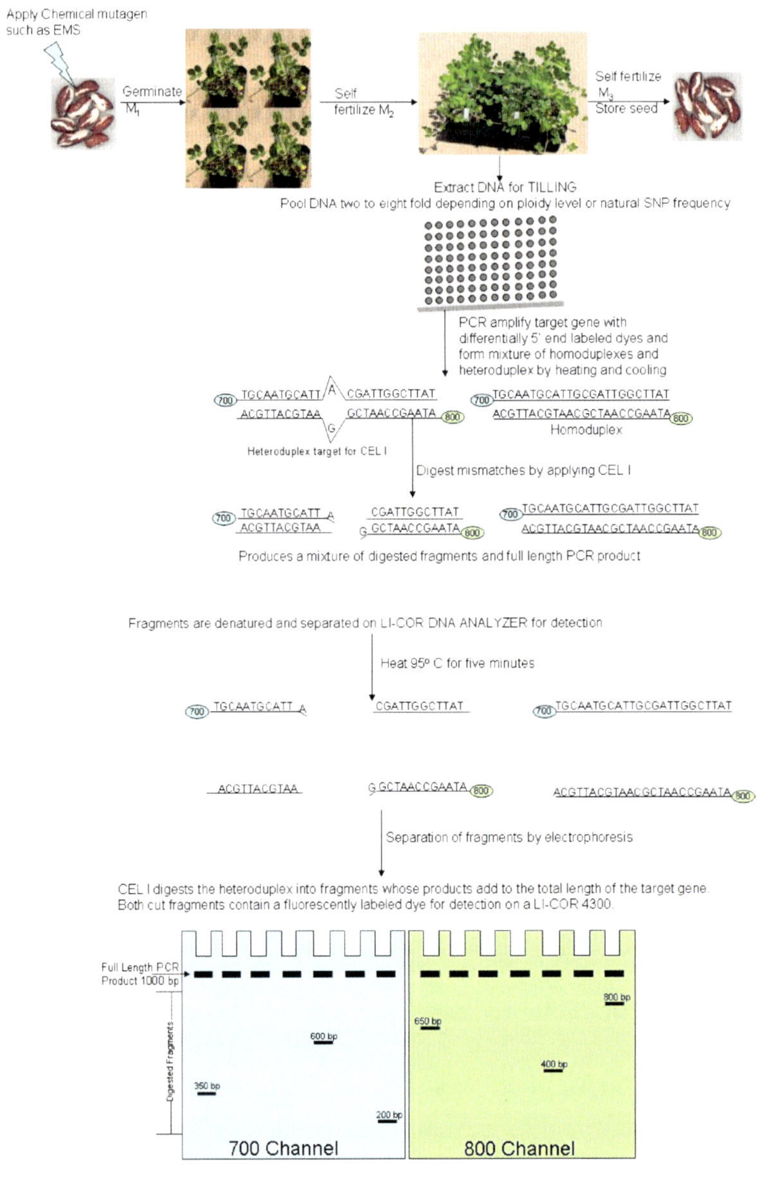

Figure 1: Diagram of the TILLING method in which seeds are mutagenized with a chemical mutagen and germinated to produce M_1 plants. M_1 plants are selfed to produce the M_2 from which DNA is extracted for analysis. The M_2 is allowed to produce seed which can be easily stored for future analysis. Once the DNA is extracted from the mutant population, the DNA is normalized and pooled together. The number of individuals in a pool depends on the ploidy level of the plant and the amount of naturally occurring SNPs, which may require the number of individuals in the pool to be

reduced. The targeted gene is amplified using a forward primer with 700 nm dye label and a reverse primer with an 800 nm dye label attached to the 5' ends. The PCR products are heated and cooled to form heteroduplexes between the accessions in the pool. The resulting pool will contain a mixture of homoduplexes and heteroduplexes. Any mismatches (SNPs or small INDELS) will be detected by a mismatch endonuclease (CEL I) and cleaved into two separate products, which will be detected in the 700 and 800 dye channel of a LI-COR DNA Analyzer. The additive size of the cleaved fragments should equal the total length of the entire product. Once the cleaved fragments and their respective polymorphic site are identified, these individuals are sequenced to verify the induced mutation. EcoTILLING is performed in the same manner except that the seed are not mutagenized; therefore, the process begins by extracting DNA from a reference plant and members of the population and continuing with the remaining steps to determine natural polymorphisms.

polyacrylamide gel attached to a LI-COR 4300 DNA analysis system (Fig. **1**). Pools containing an induced mutation will consist of a mixture of homo- and heteroduplexes; therefore, when fragments are separated a full length product (detected in both 700 and 800 channels) and two cleaved fragments (one IRD700 labeled, one IRD800 labeled) will be visible. The sum of the cleaved fragments should equal the full length PCR product. The size of the cleaved fragments can be estimated by comparison to a size standard, and thus, the approximate location of the mutation will be identified and further confirmed by sequencing. The web based program PARSESNP (http://www.proweb.org/parsesnp/) can be used once mutations are identified to display the locations of the polymorphisms in a gene(s) in a graphical format [40]. (A protocol and materials required for TILLING in *Arabidopsis* using a LI-COR DNA analyzer has been published [41], as well as, protocols for preparing celery juice extract for enzymatic digestion [42] and DNA pooling strategies [43]).

The ever reducing cost of genome sequencing has allowed additional approaches to be employed to evaluate variants in genes. Next generation sequencing has made Sequence Candidate Amplicons in Multiple Parallel Reactions (SCAMPR) a plausible method to detect mutations in a gene at costs comparable to TILLING [7]. In this approach, all samples or mutant lines are sequenced for a gene target by next generation sequencing to identify potential mutations/variants. One of the difficulties, however, is separating real variants from sequencing errors generated by the numerous runs produced [7]. A potential benefit of this method (SCAMPR) would be to save time and resources avoiding the gel separation, nuclease digestion, and interpretation steps by employing massive parallel sequencing to reveal mutant alleles. This approach has only become cost effective in recent years due to the reduction and availability of next generation sequencing.

TILLING IN PLANTS

Since the inception of TILLING, this method has been widely used for the study of functional genomics in plants, especially for the model plant *Arabidopsis thaliana*. In 2003, Greene *et al.* reported that the *Arabidopsis* TILLING Project (ATP), which was set up and introduced as a public service for the *Arabidopsis* community [38], had detected 1,890 mutations in 192 target gene fragments with heterozygote mutations detected at twice the rate of homozygote mutations [28]. Therefore, the mutational density for treatment of *Arabidopsis* with EMS was approximately 1 mutation / 300 kb of DNA screened with these mutations distributed throughout the genome [28]. The numerous mutations in *Arabidopsis thaliana* that have been identified *via* TILLING have provided an allelic series of phenotypes and genotypes to elucidate gene and protein function throughout the genome for *Arabidopsis* researchers.

Medicago truncatula a legume native to the Mediterranean region has been widely used in genetic research. This plant became a model plant to perform genetic and genomic research because it has a small diploid genome of $2n = 2x = 16$, 1C value = 0.48, self fertilizes, has a short generation time, produces a prolific amount of seeds, symbiotic with nitrogen fixing rhizobia, and it can be subjected to genetic transformation. Further, genome sequence data is currently available so researchers can apply reverse genetic approaches to elucidate gene function. TILLING has been employed on *Medicago* [44]. A kill curve analysis demonstrated that plant survival decreased at doses above 0.2% EMS; therefore, dosages of 0.15% and 0.2% were selected to generate two EMS populations. Forty percent of the M_2 plants displayed a visible phenotype. In total, 546 mutants were detected from 56 target genes in the mutaginzed populations *via* TILLING. Overall, a mean of 13 mutations per exon kb examined was revealed of which 67% were missense and 5% were nonsense mutations providing an allelic series of mutations in multiple target genes [44].

Another model plant, *Lotus japonicus*, has also been the focus of elucidating gene function through TILLING. *Lotus japonicus* is a perennial temperate legume that is a model plant for genomic studies because it has a short life cycle, it is a diploid ($2n = 2x = 12$), has a relatively small genome (472 Mb) and can self fertilize [45].

TILLING was used to investigate induced mutations occurring in the SYMRK gene, which is necessary for root symbiosis [46]. The protein kinase domain of the SYMRK locus was the target of this study. Six missense mutations were identified along with a mutation in the splice acceptor site [46]. Nitrogen fixation and the functional role of sucrose synthase was the target of another *Lotus japonicus* TILLING study [47]. Six isoforms of sucrose synthase were identified and several mutations including missense and nonsense were located in four of the six isoforms. Quantitative RT-PCR was performed to examine expression levels in *L. japonicus*, which were determined to have differential expression in various plant organs. Further, EMS null allele mutants were examined and shown to have reduced sucrose synthase activity compared to the wild type; however, mutants still retained the ability for nitrogen fixation [47]. In a separate study of pea (*Pisum sativum*), which also fixes nitrogen and is a member of the legume family, TILLING was applied to identify an allelic series of mutations in five genes with a total of 60 mutants identified [48]. Some of the mutations discovered in the LE gene, which encodes 3β-hydroxylase, were further characterized and determined to affect internode length. Mutants were backcrossed to the wild type and the segregation of the mutations and their respective phenotypes were examined [48].

Due to the success of ATP (now known as Seattle TILLING Project, STP), TILLING has radiated from the model plants such as *Arabidopsis* and *Medicago* with a simple small diploid genomes to other agronomically important crop plants with increasingly complex genomes. In 2004, maize, which is an important staple crop with a large genome, was shown to be conducive to the TILLING method [49]. A total of 11 genes were examined in a population of 750 mutagenized plants and six of these 11 genes had detectable induced mutations. One of the genes examined in this study, DMT102 (chromomethylase gene), has been previously suggested to play a role in non-CpG DNA methylation and gene silencing in *Arabidopsis* [49]. A Maize TILLING Project established in 2005 at Purdue University has already identified 319 mutations in 62 genes, which has in turn greatly assisted the maize community for functional genomics [50]. In a separate study, barley, which is also an important cereal crop with a large genome size of ~5,300 Mb was evaluated for the ability of induced mutations being detectable by TILLING [51]. Two genes (*Hin-a* and *HvFor1*) were examined and 10 variants were identified, six of which were missense mutations. Phenotyping of the M_3 plants revealed novel traits such as leaf characteristics and plant development with 20% of the individuals displaying visible phenotypes [51].

Wheat is an extremely important agronomic staple crop with an estimated production level of 600 million tons per year [52]. A polyploid plant investigation to locate variants in the *waxy* locus (granule-bound starch synthase I, GBSSI) in wheat was implemented [53, 54]. Partial *waxy* wheat cultivars are desirable because production of amylose starch is reduced, which leads to the production of superior flour and noodle products [55] for human consumption. Wheat genetics can be complicated because its genome is more complex than most angiosperms, given it is an allohexaploid and the total genome size is quite large (17,000 Mb). A total of 246 alleles were uncovered in three *waxy* gene homologues (*Wx*-A1, *Wx*-B1, and *Wx*-D1) from allohexaploid and allotetraploid wheat *via* TILLING. This comprehensive allelic series provided 84 missense, three nonsense, and five splice site mutations (Table **2**). Phenotyping of M_3 progeny demonstrated reduction of amylose production. Detecting genetic variants *via* phenotyping in wheat can be difficult because redundant copies of loci in the genome can mask expression. This study identified extensive allelic variation in GBSSI, which was more than any previous report produced in 25 years [53, 54].

Rice, which is also a staple and important economic crop around the world, currently estimated to provide 80% of caloric intake for three billion people [56], has been the focus of a few TILLING studies. The rice genome has been predicted to contain ~50,000 genes [57], of which gene function needs to be determined empirically. In 2005, a report on the generation of a large mutation population (60,000) using multiple chemical mutagens on IR64, a widely grown *indica* rice, demonstrated that TILLING was suitable for reverse genetic studies with mutations detected in two genes [57], albeit the mutational density in the population was fairly low. Furthermore, extensive phenotypic variation was assayed for the various chemical mutagens used to develop the mutant population and albinism was a common phenotype no matter which mutagen was applied [57]. In a separate study, EMS and Az-MNU were used to induce mutations in rice, which produced an elevated mutational density (Table **1**), with 57 polymorphisms identified from 10 target genes by TILLING [37]. Another report on rice TILLING published in 2007, demonstrated the efficacy of TILLING to detect mutations by separation of products on agarose gels and the results were similar to pooling DNA and detecting mutations on a LI-COR DNA Analyzer [58].

Soybean (*Glycine max* L.) contains approximately 35-50% protein and has been shown to be beneficial for human health [59]. Furthermore, it is a widely grown important economic crop and can improve soil quality by fixing nitrogen [17]. Four mutant populations from two genetic backgrounds (Forrest and Williams 82) were created for soybean either by treatment with EMS or NMU and evaluated for induced mutations [17]. Several of the target genes initially tested amplified more than one target; therefore, further work was carried out to produce a single product to employ TILLING so that mutation detection functioned optimally. A total of 116 mutations were identified *via* TILLING from seven target genes. The majority of the mutations uncovered by TILLING were determined to be the expected G/C to A/T transitions. This study demonstrated that soybean is suitable for TILLING studies [17].

Fruit ripening is an important key trait determining food supply, shelf life, and nutritional value. TILLING was applied in melon to delineate mutations in genes controlling fruit ripening [60]. A population of 4,260 M_2 seeds was produced from the EMS treatment. Eleven genes were evaluated with 134 induced mutations revealed by TILLING for a total mutation density rate of 1/573 kb. The *CmAC01* gene which catalyzes the last step of ethylene biosynthesis was targeted in this study to evaluate possible mutants that would have delayed fruit ripening, and thus, a longer shelf life. One mutant G194D was associated with increased fruit firmness, delayed fruit ripening, delayed yellowing, and comparable soluble sugar content to the wild type [60]. In a separate study in tomato, TILLING was applied to reveal mutations that effected carotenoid content in fruit ripening. The target was the phytoene synthase 1 (*Psy1*) gene [61]. A total of eight point mutations were identified in the target. One mutation knocked out function and led to fruit with a yellow flesh; whereas, a separate substitution lead to delay in lycopene and β-carotene accumulation [61].

Even though TILLING was originally designed for and applied mainly in *Arabidopsis* in the early years of its inception, it has been demonstrated to be an extremely versatile reverse genetic approach compared to other reverse genetic approaches previously discussed. This method has proven to be successful to rapidly identify variant genotypes and determine gene function in plants that are diploid and have relatively small genomes such as *Arabidopsis*; but, also can be easily applied to other crop plants with very large genomes that are further complicated by various ploidy levels such as maize and wheat. Moreover, the use of chemical mutagens in diploid and polyploid plants produces a range of various mutations (Table **2**) and a high density of mutations throughout the genome (Table **1**). Obtaining an allelic series for a gene of interest greatly assists in the overall determination of gene function by providing multiple types of phenotypic variants to be analyzed.

TILLING IN ANIMALS

Several model animals such as *Drosophila melanogaster*, zebrafish, rats, and *C. elegans*, which have been historically used to investigate numerous genetic hypotheses, have been subjugated to mutagensis for TILLING experiments to gain insight on functional genomics. Unlike plant TILLING, in which mutant populations (in self pollinated species) can generally be easily created by chemical mutagensis followed by self fertilization and harvesting of seed (Fig. **1**), animals require different tactics to establish and maintain mutant populations. In *C. elegans*, if self-fertilizing hermaphrodites are selected, then the creation of a mutant population is fairly similar to plants, in that the nematode is mutagenized, animals are allowed to self and DNA is extracted for analysis. However, since the majority of animal species cannot self fertilize, outcrossing subsequent to mutagensis may be obligatory. No single strategy is applicable or practical to employ for all animal species. Therefore, multiple strategies exist for the creation, maintenance, preservation and recovery of animal mutant populations and an approach should be chosen based on the life history or availability of tissue preservation / recovery protocols available for the animal of interest.

Since the early 1900's with the work of geneticist Thomas Hunt Morgan, *Drosophila* has been a well studied model organism for scientific research. Currently, complete genomic information is available for 12 separate species classified in the genus *Drosophila*, which allows a critical examination of comparative genomics between species [62] and the availability of genomic information facilitates reverse genetic studies. A commonly used mechanism to evaluate gene function in *Drosophila* is insertional mutagenesis *via* genetic engineering of transposable elements [63]. Homologous recombination is also available to researchers, but is considered a less favorable method [64]. EMS has also been widely used to evaluate numerous genes in *Drosophila* [65]. EMS mutagenesis was used to screen for mutations in the *awd* (abnormal wing discs) gene from *Drosophila* with 16 mutations all G:C to A:T transitions identified by dHPLC [66]. A range of silent, missense and intron based mutations were detected in this study (Table **2**), as well as a noted mutagensis bias for 5'-PuG-3' sites or a middle G base in a stretch of three or more G bases [66]. CEL I based TILLING was applied to pinpoint 44 mutations in three genes in a separate study [65]. One of the targets in this study was determined to be an essential gene with several of the induced mutations causing lethality. Winkler *et al.* (2005) found that increasing the EMS treatment by 1.5 fold augmented the frequency of SNPs detected by 1.5 fold; however, this treatment greatly decreased fly viability [65]. A general TILLING service for the *Drosophila* community was established [39], which helps streamline the effort required for a researcher to determine function in their gene of interest.

Zebrafish is an aquatic vertebrate model organism, particularly valuable for studying embryonic development, because they are easy to maintain and breed, produce a large number of offspring, and females can produce ~15,000-35,000 eggs in their lifetime [67]. Several reverse genetic approaches are available for functional genomics studies in zebrafish. TILLING has been effective in zebrafish and two separate methods of creating and maintaining mutant populations are published [33, 34]. The first TILLING study in zebrafish reported the creation of 4,608 mutants with ENU that were maintained as a live library [34]. TILLING uncovered 255 mutations in 16 genes of which approximately 20% of the mutations were found in noncoding sequences. The mutations detected in coding regions consisted of 119 missense and 14 nonsense [34], all of which should affect the protein function or produce a loss of function, respectively. In a separate study, 1,235 mutagenized zebrafish, which were archived by cryopreservation of dissected male testes, were examined *via* TILLING and resequencing for polymorphism in 54 exons (derived from 17 genes) with targets ranging in size from 74 to 626 bp [33]. Polymorphic sites in the genes examined ranged from zero to 26 per amplicon. Due to a high frequency of SNPs in some amplicons detection of rare induced mutations became somewhat of a challenge [33].

Caenorhabditis elegans (soil nematode) is a model organism that has its complete genome sequenced, which has been publicly available for over ten years, allowing reverse genetic techniques to be applied to gain insight on functional genomics. Available reverse genetic methods include RNAi, trimethylpsoralen and UV radiation (TMP/UV), transposon insertional mutagenesis and homologous recombination, all of which have certain advantages and disadvantages [35]. *C. elegans* is predicted to contain approximately 19,000 genes [68]; however, mutant alleles are only available for 3,400 genes [35]. TILLING can be used as an effective way to gain an allelic series of mutants in the remaining genes to determine gene function. In 2006, a study examined 1,500 EMS mutagenized nematodes in 10 genes with a total of 71 mutations identified, many of which were determined to have an effect on the protein product [35]. Once mutants were identified *via* TILLING, phenotyping analyses were performed to further characterize or substantiate gene function [35].

Another study examined the feasibility of TILLING in rat [69]. Male rats were treated with ENU, which caused a range of sterile to completely fertile animals to be produced. These chemically treated males were crossed with untreated females to produce progeny that were analyzed by TILLING. A total of 768 animals were analyzed and 17 induced mutations were detected of which 70.5% were missense. Rats were further bred to homozygosity and determined to have germ-line mutations due to Mendelian inheritance of these mutations. Hip dysplasia and diphallus were observed in the mutant population [69].

Lastly, the pathogenic bacterium *Photobacterium damselae* subsp. *piscicida*, which is known to cause pseudo-tuberculosis in yellowtail and amberjack fish, has generally been controlled in fish farms with treatment of a quinolone, such as Nalidixic acid [70]. Mutant bacteria selected from cultures containing various concentrations of quinolone which demonstrated varying levels of resistance were pooled with a quinolone susceptible strain for TILLING analysis. The quinolone resistant determining region (QRDR) of the *gyr*A gene was targeted. A single point mutation in all resistant mutants was identified. Although this study was performed on pathogenic bacteria, the results will provide a rapid detection method for resistant bacteria, the avoidance of pointless quinolone treatments to control outbreak of the disease, and possibly prevent economic loss at fish farms [70]. This study not only examined gene function, but also provided data that could be used as a functional marker to rapidly screen pathogenic bacteria and prevent economic loss.

ECOTILLING IN PLANTS AND ANIMALS

EcoTILLING is a molecular technique that is similar to TILLING, except that its objective is to uncover natural genetic variation as opposed to induced mutations (Fig. **1**). Many species are not amenable to chemical mutagenesis; therefore, EcoTILLING can aid in the discovery of natural variants and their putative gene function [27]. This approach allows one to rapidly screen through many samples with a gene of interest to identify naturally occurring SNPs and / or small INDELS. The method has proven to be successful to detect DNA polymorphisms including variations in satellite repeat number [43]. Furthermore, in highly heterozygous outcrossing species, EcoTILLING can be used to determine heterozygosity levels within a gene fragment [27]. EcoTILLIING reduces the time and effort for SNP discovery generally required by weeding out identical haplotypes, and thus, not requiring sequencing of all individuals in a population to identify polymorphisms. It also has the advantage of detecting multiple polymorphisms in a single fragment because CEL I will digest only a small proportion of the heteroduplexes or it can be applied to examine rare DNA polymorphisms [43]. This technique has not been as widely employed as TILLING; however, there are a few published studies on EcoTILLING, which will be further discussed.

The first publication of the EcoTILLING method in which TILLING was modified to mine for natural polymorphisms was in 2004 from work in *Arabidopsis thaliana* [71]. The strategy used was to pool each ecotype with the standard Columbia ecotype (reference) in a 1:1 ratio. A total of 192 accessions were assayed to uncover 55 haplotypes in five different genes that were approximately 1 Kb in length. A large proportion of the variation was detected in the introns. This study demonstrated that CEL I could detect SNPs, INDELS and polymorphisms in microsatellite repeats. Interestingly, a 21 bp deletion was also shown to be detected and cleaved by CEL I enzyme. Overall, EcoTILLING proved to be an efficient method to haplotype individuals without requiring sequencing of all the individuals included in this study [71].

EcoTILLING has also been used to examine DNA variation in natural populations of black cottonwood (*Populus trichocarpa*), which is one of the first large deciduous trees to have its genome completely sequenced [72]. DNA polymorphisms were assessed from 41 trees derived from 41 populations originating from Canada to Oregon. Nine separate mapped loci were evaluated by mixing the members of the population to a reference genotype. Each gene produced some polymorphic sites with the number of SNPs ranging from 1-23 and an overall mean of 6.78. The rate of SNPs for coding and noncoding regions assayed were 1/229 bp and 1/64 bp, respectively. Additionally, this study was the first to successfully report the evaluation of heterozygosity of single individual trees with EcoTILLING. Moreover, nucleotide diversity estimates, selection, and linkage disequilibrium were evaluated [72]. This study demonstrated how useful EcoTILLING can be for the evaluation of genetic diversity in natural populations.

Another potential useful application of EcoTILLING was mining for variation in resistance genes to help speed up the identification process of finding individuals with alleles that could be immune to various diseases. In 2006, allelic variation was examined and identified by using EcoTILLING in *mlo* and *Mla* resistance genes of *Hordeum vulgare* (barley) [73]. These genes are involved in defending the plant from the fungal pathogen that causes powdery mildew. This study demonstrated that compared to classical methods of determining disease resistance alleles, EcoTILLING provided several allelic variants in two resistance genes that can be exploited to breed cultivars with improved resistance [73]. In a similar study, EcoTILLING was employed to screen for natural allelic variation for disease resistance to Melon Necrotic Spot Virus (MNSV) in various *Cucumis species* [74]. High conservation of eIF4E, a translation initiation factor, was found among 113 accessions evaluated. Six polymorphisms were identified; however, only one site produced an amino acid change that correlated with disease resistance [74]. In general, EcoTILLING shows great promise of speeding up the process of identifying natural disease resistance alleles, which upon identification, can be used to breed improved cultivars.

EcoTILLING was employed to identify polymorphisms in a germplasm collection of mung bean (*Vigna radiata* (L.) R. Wilczek), which previously showed limited diversity [75]. *Vigna radiata*, which is classified in the family Fabaceae, is an important economic crop and a dietary staple in many developing countries [76]. The species *radiata* can be further subdivided into two botanical varieties *radiata* and *sublobata*, of which *sublobata* is currently acknowledged as the putative progenitor of *radiata* [77, 78]. Polymorphic sites were abundant when comparing *V. radiata* var. *sublobata* to *V. radiata* var. *radiata*; however, when accessions of *V. radiata* var. *radiata* were pooled together relatively few polymorphisms were identified suggesting that accessions classified as *V. radiata* var. *radiata* could have a narrow genetic base. Morphological data taken from accessions of *V. radiata* var. *radiata* also demonstrated limited diversity in the flowers and pod descriptors. The majority of polymorphisms detected, however, between *sublobata* and *radiata* were found in putative introns. The banding patterns varied from simple to complex as the number of DNA polymorphisms between two pooled samples increased. Overall, this modification of TILLING can be effectively used as an efficient, rapid technique to identify DNA polymorphisms in populations with high genetic identity and to mine for SNPs in collections of plant germplasm [75].

In a pioneering animal study, genotypic variation from humans was analyzed by EcoTILLING to detect rare SNPs [79]. Current resequencing efforts for the detection of SNPs in humans tend to identify common polymorphisms not the rare variation that exists in the genome. Detecting rare SNPs can be difficult and quite expensive using standard sequencing methods. In total, 384 human samples were evaluated in five target genes using the EcoTILLING method. This effort produced the detection of 28 rare SNPs, some of which were predicted to have a damaging effect to the protein. This study also examined the precision of EcoTILLING for the detection of SNPs by comparing the SNP dataset detected through public resequencing efforts to those identified by EcoTILLING. Of the 25 SNPs previously identified through resequencing, EcoTILLING uncovered 24 of the 25, and furthermore, uncovered seven new alleles [79], which demonstrates the precision and efficiency of this technology for the detection of rare polymorphisms.

ADVANTAGES AND DISADVANTAGES OF TILLING AND ECOTILLING

TILLING is a non-transgenic, high throughput reverse genetic approach to rapidly identify induced mutants in a population for gene(s) of interest in order to provide an allelic series of mutations to examine the gene(s) function *via* the resulting mutant genotype(s) and phenotype(s). This technique unlike other SNP detection methods, provides the approximate location within a few base pairs of the induced mutation [29, 36], which allows targeted sequencing in the area of the induced mutation opposed to the entire fragment. Since chemical mutagensis produces a range of various mutations throughout the genome such as nonsense, splice site, and missense, all of which potentially can affect the protein structure and the resulting phenotype, it has been used for decades to obtain mutants for genetic studies. Therefore, through mutagensis one can obtain partial loss or complete loss of function and new novel functions, which can provide valuable insight into the true role of a gene [11] in a species of interest. As discussed previously using chemical mutagensis and TILLING to pinpoint these mutations has been highly effective in the elucidation of gene function in plants and animals without the production of transgenic material. TILLING has been demonstrated to be sensitive enough to detect induced mutations and naturally occurring SNPs [67], as well as, the detection of heterozygotes. EcoTILLING, which has been less frequently employed in the current literature, can also be a valuable tool for mining for SNPs in germplasm, assessing heterozygosity, uncovering variants for disease resistance, or to ascertain the function of a gene or regulatory element by detecting natural variants. EcoTILLING can be a good technique to employ especially when working with a well-established population with thoroughly characterized morphological data.

One of the main advantages of TILLING for a scientist is the amount of time and money this method can potentially save by not requiring resequencing of all individuals in a population to mine for frequent or rare SNPs. As a general rule for a diploid organism, TILLING is performed by pooling eight individuals of a population at a time and assessing differences by endonuclease digestion of mismatches in a heteroduplex. Ordinarily, the majority of the samples screened in TILLING have the same haplotype with very few samples in the population having an induced mutation in the gene of interest due to a relatively low frequency of induced mutants by utilizing chemical mutagenesis (Table **1**). Furthermore, TILLING is sensitive enough to detect homozygous mutations as well as heterozygous mutations in an 8-fold pool, which represent 1 of the 16 genomes in pools from diploid species [36, 49, 53]. This method allows one to weed out the identical individuals and only focus resources on individuals with rare chemically induced DNA polymorphisms. EcoTILLING also shares this same advantage as a technique except that it focuses on naturally occurring variation as opposed to induced variation as in TILLING. Most EcoTILLING experiments have used a twofold-pool strategy [71, 72, 74]; however, one study in which rare human nucleotide differences were examined an eight-fold pool strategy was employed [79]. One of the potential disadvantages of EcoTILLING is that when the number of polymorphic sites is high for a gene or PCR fragment of interest across all samples in the population, then an eight-fold pooling strategy requires much more labor and time to identify and analyze SNPs.

False negatives and false positives can be a potential disadvantage to any reverse genetic application including TILLING and EcoTILLING. This issue has not been extensively examined in the current literature. A study of EcoTILLING of human SNPs did report a fairly low false negative rate of 5% and false positive rate of 4% [79]. False negatives and false positives often can be due to human error in scoring gel images. Generally, this becomes less of a problem as researchers become more experienced in scoring these images. One strategy is to sequence a small percentage of pools determined to be negative to validate if any false negatives are occurring. An advantage for TILLING and EcoTILLING is that the method helps guard against false positives by double-end labeling the target so that a cleaved product should produce a fragment in both the 700 nm and 800 nm fluorescent channel [28] and the sizes of the cleaved products should total the full length product. Furthermore, two dimensional arraying/pooling previously described [43, 79], in which samples are mixed in duplicate both by column and by row, also helps reduce false positives because cleaved fragments produced from an induced mutant are replicated, and thus, will appear in two separate lanes in a gel. This pooling strategy when used in TILLING experiments or EcoTILLING with very low SNP frequencies will also allow one to identify the positive sample carrying the mutation without having to remix the positive pool [43].

One of the initially largest expenses to TILLING and EcoTILLING experiments are the use of robotic equipment and the purchase of automated sequencers such as the LI-COR DNA Analyzer commonly used for cleaved fragment detection. However, these products are not essential and experiments can be carried out without the use of robotics. Capillary electrophoresis or other automated sequencers such as ABI 377 have been demonstrated in previous studies to be effective in the separation and detection of digested fragments [46, 73, 80, 81] in lieu of a LI-COR DNA Analyzer. Moreover, PCR products and cleaved fragments can be separated by more economical methods such as agarose gels as reported previously for rice [58] or possibly detection *via* silver staining of polyacrylamide gels for fragment separation. The only disadvantage of using agarose gels would be lower resolution between fragments that are similar in size in comparison to polyacrylamide gels. An advantage to using agarose or silver stained gels is that fluorescently labeled primers would not be required, which would remove an additional expense of the technique. If automated sequencers are already purchased or available, one can reduce the experimental costs by purchasing labeled primers using a universal primer strategy [34, 65, 79], in which all reactions are fluorescently labeled with the same primer such as M13 or T7. Lastly, one could also reduce the expense in these experiments by purifying CEL I for endonuclease mismatch detection from celery [42, 43, 73]. Therefore, these techniques can be adapted to less expensive common laboratory equipment, and thus, possible for all researchers to employ.

TECHNICAL CHALLENGES FOR APPLICATION OF TILLING AND ECOTILLING

There are some technical challenges in employing TILLING or EcoTILLING experiments; however, most of these challenges are not insurmountable. The creation of a mutant population can be somewhat of a challenge and sufficient time needs to be allocated for the development of a high-quality population. One to two years can be expected to be devoted to population development [51, 53]. The first step to create a population should include varying the concentration of the chemical mutagen being applied to assess lethality and to find an optimal concentration to generate a high density of mutants with few lethal embryos [12, 38] before too much time is invested in the creation of the population. This can be a little bit of a challenge because species and varieties of a species can respond differently to chemical mutagenesis so that a dose of mutagen applied in barley may be highly lethal, but the same dose in *Arabidopsis* will not produce a highly lethal effect [37]. An ideal population would maximize mutational load, but still allow the majority of the population to remain fertile [50] and thus viable. Creating mutant populations for plants that propagate vegetatively or have long generation times [53] could also slow down progress of generating a mutant population. Working with species that are highly heterozygous may limit mutation detection producing lots of cleaved fragments due to natural polymorphisms, which ultimately may inhibit detection of rare induced mutations [43]. Variation can be minimized by choosing a single parent to produce thousands of progeny in a single or multiple generations [39].

In most plants, production and maintenance of seeds to preserve mutants for future analysis is fairly straightforward; however, the collection and maintenance of gametes in animals can be somewhat problematic [11]. Therefore, multiple strategies exist for the creation and maintenance of animal mutant populations [11, 33-35], and one must carefully examine the species of interest and decide on a suitable strategy. In animal studies, maintaining live animals as a strategy to preserve the library of mutants as a living resource for analysis [34] can require a large amount of space and some labor for the caretaking of thc animals. A living mutant resource may be the only option if prior effective cryopreservation methods have not been established for the species of interest such as reported in *Drosophila* and rat [11, 69]. In zebrafish, caudal fins regenerate so that DNA can be extracted from live fish, which can be maintained for a limited time for breeding when mutations are identified [11]. A limitation to a living library is that the library must be screened within the generation time of the animal [34], which may not be an issue in animals that have fairly long life spans, but can be a limitation in animals with short life spans especially if a large number of genes are targeted in a study.

Once a suitable mutant population, whether in plants or animals, has been established, high quality DNA needs to be extracted and normalized. It is important that all DNA extracts be equivalent so that they are all equally represented in the pools being analyzed. Otherwise, unique induced mutations may fail to be identified because as the amount of individual DNA decreases in comparison to others in a pool the sensitivity of mutation detection decreases [49]. Another challenge especially in regards to plants can be choosing target genes that only exist as a single copy within the genome. This becomes more of a problem when working with polyploid plants. Overcoming this challenge can be accomplished by designing primers that are specific to a single homoeologous gene, which may require some additional effort [82, 83]. Another strategy is to sequence the target genes to reveal any restriction site differences between the targets, so that DNA can be digested prior to TILLING, which would cleave the unwanted target leaving the desired gene intact [17] for analysis. TILLING in polyploid plants also generally requires a reduction of the number of samples mixed in a pool [49, 54], so throughput may be somewhat decreased in comparison to diploids. This is because recognition of mismatches becomes more difficult with an increasing number of genomes. Detection of a heterozygous mutation in a tetraploid or hexaploid individual in an 8-fold pool could be overwhelmed unless the pool size is decreased because detection would require recognition of a mutation in 1 of 32 genomes or 1 of 48, respectively as opposed to 1 of 16 in a heterozygous diploid pool.

Other issues to consider carefully when designing a TILLING or EcoTILLING experiment pertain to the target gene being examined. In general, maximizing the size of the target amplicon (~1 Kb – 1.5 Kb) gives an increased probability of detecting cleaved fragments, especially when working with highly conserved exons or a species that contains a narrow genetic base. Several research papers have previously reported the difficulty of tracking SNPs that are on the ends ~100-200 bp of the target sequence [28, 37, 58, 71, 72, 79]. SNPs detected in our experience in TILLING or EcoTILLING experiments also proved to be difficult to detect when on the ends of the target. One way to avoid this problem if possible is to design primers to target an area slightly larger than the exon so that the ends of the fragment include ~100 bp. of the intron. Alternatively, one could design multiple primer sets for a particular exon of interest such that these amplicons overlap by at least 100 bases [58]. Additionally, another potential difficulty in EcoTILLING is that as the number of SNPs detected per fragment increased, the scoring and tracking of cleaved fragments becomes more difficult. Single SNPs detected in a heteroduplex produce a higher concentrated product compared to multiple mismatch sites [58, 71].

Another point for consideration when designing an EcoTILLING or TILLING experiment is the choice of the nuclease used to digest the mismatches in the heteroduplexes. CEL I, which is derived from celery, is an endonuclease that recognizes and cleaves mismatches in a heteroduplex and also contains 5' to 3' exonucleolytic activity [12, 81]. Therefore, CEL I not only digest mismatches in a heteroduplex but also will digest the full length PCR product starting with the 5' fluorescent label. Therefore, care needs to be taken not to over digest DNA samples, and thus, lose the fluorescent signal of the PCR products. In our TILLING and EcoTILLING work, the mismatch enzyme CEL I was purchased from Transgenomic, Inc. (Omaha, NE) instead of purifying it in the laboratory. This enzyme has an optimal DNA concentration range suggested for efficient digestion; therefore, in our studies amplicons were quantified and an aliquot of the reaction fitting the target range was used to ensure efficient digestions. If one decides to purify CEL I or other endonucleases, which can reduce some of the cost for the experiment, then optimum digestion conditions and appropriate enzyme concentrations for enzymatic cleavage should be determined empirically [43, 84]. Other nucleases that are either available for purchase or that can be purified in the laboratory and have previously been tested for efficient mismatch digestion ability include mung bean nuclease, S1, and ENDO1 [48, 84].

The last challenge to consider is assigning a particular phenotype to a genotype and inferring the putative function of a gene. Chemical mutagensis is known to introduce background mutations, which can at times make phenotyping analysis difficult [32]. This may require several generations of outcrossing or backcrossing [12, 32] to ascertain. Generally, outcrossing M_3 plant lines may be unnecessary to attribute a phenotype to a mutant genotype because one can evaluate the M_3 lines generated from a heterozygous selfed M_2 plant for typical Mendelian segregation ratios, which should allow a correlation between genotype and phenotype. Background mutations, if unlinked to the gene of interest, can be distinguished because they will segregate in the M_3 generation [12]. Another strategy is to cross two independent TILLED lines to produce heteroallelic individuals so that background mutations will independently assort [12, 32]. Of course, if there is any epistasis or pleiotropic effects from the background mutations or other wild type alleles then assigning a function will be more problematic. Another concern for polyploid plants is that double or triple mutants may be necessary to assign a phenotype to genotype [50] because redundant copies of loci in the genome can mask expression [53].

PERSPECTIVE

TILLING and EcoTILLING have been proven to be highly effective reverse genetic tools for functional genomic studies in plants and animals. These methods are now well established for many model plant and animal systems regardless of their mating system, genome size, or ploidy level. TILLING is one of the few reverse genetic applications that has not been proven to be applicable in a species-specific manner unlike other approaches (*i.e.*-RNAi or homologous recombination), which potentially makes this application available for all species. The main limitation for TILLING is that the species is capable of being mutagenized. Since the inception of these techniques, many researchers have gained indispensable insight on gene function in order to identify natural and induced variants. As time goes on, more genomic information will become readily available for other plant and animal species, and thus, reverse genetic approaches will be necessary to assign putative gene function. Although recently high-throughput and low-cost sequencing techniques are available, TILLING and EcoTILLING are still easy approaches to gain valuable insight on linking a specific mutation to a phenotype of interest.

ACKNOWLEDGEMENTS

The authors would like to thank Dr. Peggy Ozias-Akins and Dr. Gary Pederson for their suggestions and corrections to improve the quality of this manuscript.

CONFLICT OF INTEREST

The author(s) confirm that this chapter content has no conflict of interest.

DISCLOSURE

This is an updated report of the earlier review article By N.A. Barkley and M.L. Wang "Application of TILLING and EcoTILLING as reverse genetic approaches to elucidate the function of genes in plants and animals". Current Genomics, 2008 9(4):212-216.

REFERENCES

[1] Sanger, F., Nicklen, S,, Coulson, A.R. DNA sequencing with chain-terminating inhibitors. *Proc. Natl. Acad. Sci. U S A* **1977**, *74*(12):5463-5467.

[2] Maxam, A.M., Gilbert, W. A new method for sequencing DNA. *Proc. Natl. Acad. Sci. U S A* **1977**, 74(2):560-564.

[3] Sanger, F., Air, G.M., Barrell, B.G,, Brown, N.L., Coulson, A.R, Fiddes, C.A., Hutchison, C.A,, Slocombe, P.M., Smith, M. Nucleotide sequence of bacteriophage phi X174 DNA. *Nature* **1977**, 265(5596):687-695.

[4] Hutchison, C.A. 3rd: DNA sequencing: bench to bedside and beyond. *Nucleic Acids Res* **2007**, *35*(18):6227-6237.

[5] Bonetta ,L.Genome sequencing in the fast lane. *Natu.Methods* **2006**, 3(2):141-147.

[6] Rusk, N., Kiermer, V. Primer: Sequencing--the next generation. *Nat. Methods* **2008**, *5*(1):15.

[7] Gilchrist, E., Haughn, G. Reverse genetic techniques: engineering loss and gain of gene function in plants. *Briefings Functi. Genomics* **2010**, *9*(2):103-110.

[8] Nagel, R.L. Epistasis and the genetics of human diseases. *C R Biol* **2005**, *328*(7):606-615.

[9] Alonso, J.M., Ecker, J.R. Moving forward in reverse: genetic technologies to enable genome-wide phenomic screens in Arabidopsis. *Nat.Rev. Genetics* **2006**, *7*(7):524-536.

[10] Nagy, A., Perrimon, N., Sandmeyer. S., Plasterk. R. Tailoring the genome: the power of genetic approaches. *Nat.Genetics* **2003**, *33*:276-284.

[11] Stemple, D.L. TILLING--a high-throughput harvest for functional genomics. *Nat Rev Genet* **2004**, *5*(2):145-150.

[12] Henikoff, S., Comai, L. Single-nucleotide mutations for plant functional genomics. *Annu Rev Plant Biol* **2003**, *54*:375-401.

[13] Tissier, A., Bourgeois, P. Reverse Genetics in Plants. *Curr.Genomics* **2001**, *2*:269-284.

[14] Iida, S., Terada, R. A tale of two integrations, transgene and T-DNA: gene targeting by homologous recombination in rice. *Curr .Opin .Biotechnol.* **2004**, *15*(2):132-138.

[15] Gelvin, S.B. Agrobacterium-mediated plant transformation: the biology behind the "gene-jockeying" tool. *Microbiol Mol Biol Rev* **2003**, *67*(1):16-37, table of contents.

[16] Malabika, R., Rajinder, K.J., Jai, R.S., Wu ,.: Production of agronomically superior transgenic rice plants using *Agrobacterium* transformation methods: Present status and future perspectives. *Curr. Sci.* **2000**, *79*(7):954-960.

[17] Cooper, J.L., Till, B.J., Laport, R.G., Darlow, M.C., Kleffner, J.M., Jamai, A., El-Mellouki, T., Liu, S., Ritchie, R., Nielsen, N., *et al*: TILLING to detect induced mutations in soybean. *BMC Plant Biol* **2008**, *8*:9.

[18] Brutnell ,T.P. Transposon tagging in maize. *Funct Integr Genomics* **2002**, *2*(1-2):4-12.

[19] Adams, M.D., Sekelsky, J.J. From sequence to phenotype: reverse genetics in Drosophila melanogaster. *Nat Rev Genet* **2002**, *3*(3):189-198.

[20] Moffat, J., Sabatini, D.M. Building mammalian signalling pathways with RNAi screens. *Nat Rev Mol Cell Biol* **2006**, *7*(3):177-187.

[21] Napoli, C., Lemieux, C., Jorgensen, R. Introduction of a Chimeric Chalcone Synthase Gene into Petunia Results in Reversible Co-Suppression of Homologous Genes in trans. *Plant Cell* **1990**, *2*(4):279-289.

[22] Fire, A., Xu, S.Q., Montgomery, M.K., Kostas, S.A., Driver, S.E., Mello, C.C. Potent and specific genetic interference by double-stranded RNA in Caenorhabditis elegans. *Nature* **1998**, *391*(6669):806-811.

[23] Grimm, D., Kay, M.A. ***Therapeutic application of RNAi: is mRNA targeting finally ready for prime time?*** *J Clin Invest* **2007**, *117*(12):3633-3641.

[24] Mello, C.C. Return to the RNAi world: rethinking gene expression and evolution. *Cell Death Differ* **2007**, *14*(12):2013-2020.

[25] Natarajan, A.T. Chemical mutagenesis: From plants to human. *Curr.Sci.* **2005**, *89*(2):312-317.

[26] Ahloowalia, B.S., Maluszynski, M., Nichterlein, K. Global impact of mutation-derived varieties. *Euphytica* **2004**, *135*(2):187-204.

[27] Gilchrist, E.J., Haughn, G.W. TILLING without a plough: a new method with applications for reverse genetics. *Curr Opin Plant Biol* **2005**, *8*(2):211-215.

[28] Greene, E.A., Codomo, C.A., Taylor, N.E., Henikoff, J.G., Till, B.J., Reynolds, S.H., Enns, L.C., Burtner, C., Johnson, J.E., Odden. A,R. *et al*: Spectrum of chemically induced mutations from a large-scale reverse-genetic screen in Arabidopsis. *Genetics* **2003**, *164*(2):731-740.

[29] Henikoff, S., Till ,B.J, Comai, L. TILLING. Traditional mutagenesis meets functional genomics. *Plant Physiol* **2004**, *135*(2):630-636.

[30] McCallum, C.M., Comai, L., Greene, E.A., Henikoff, S. Targeted screening for induced mutations. *Nat Biotechnol* **2000**, *18*(4):455-457.

[31] Weil, C.F. TILLING in grass species. *Plant Physiol* **2009**, *149*(1):158-164.

[32] McCallum, C.M., Comai, L., Greene, E.A., Henikoff, S. Targeting induced local lesions IN genomes (TILLING) for plant functional genomics. *Plant Physiol* **2000**, *123*(2):439-442.

[33] Sood, R., English, M.A., Jones, M., Mullikin, J., Wang, D.M., Anderson, M., Wu, D,, Chandrasekharappa, S.C., Yu, J., Zhang, J., *et al*: Methods for reverse genetic screening in zebrafish by resequencing and TILLING. *Methods* **2006**, *39*(3):220-227.

[34] Wienholds, E., van Eeden, F., Kosters, M., Mudde, J., Plasterk, R.H., Cuppen, E. Efficient target-selected mutagenesis in zebrafish. *Genome Res* **2003**, *13*(12):2700-2707.

[35] Gilchrist, E.J., O'Neil, N.J., Rose, A.M., Zetka, M.C., Haughn, G.W. TILLING is an effective reverse genetics technique for Caenorhabditis elegans. *BMC Genomics* **2006**, *7*:262.

[36] Colbert, T., Till, B.J., Tompa, R., Reynolds, S., Steine, M.N., Yeung, A.T., McCallum, C.M., Comai, L., Henikoff, S. High-throughput screening for induced point mutations. *Plant Physiol* **2001**, *126*(2):480-484.

[37] Till, B.J., Cooper, J., Tai, T.H., Colowit, P., Greene, E.A., Henikoff, S., Comai, L. Discovery of chemically induced mutations in rice by TILLING. *BMC Plant Biol* **2007**, *7*:19.

[38] Till, B.J., Reynolds, S.H., Greene, E.A., Codomo, C.A., Enns, L.C., Johnson, J.E., Burtner, C., Odden, A.R., Young, K., Taylor, N.E., *et al*: Large-scale discovery of induced point mutations with high-throughput TILLING. *Genome Res* **2003**, *13*(3):524-530.

[39] Comai, L., Henikoff, S. TILLING: practical single-nucleotide mutation discovery. *Plant J* **2006**, *45*(4):684-694.

[40] Taylor, N.E., Greene, E.A. PARSESNP: a tool for the analysis of nucleotide polymorphisms. *Nucleic Acids Research* **2003**, *31*(13):3808-3811.

[41] Till, B.J., Colbert, T., Codomo, C., Enns, L., Johnson, J., Reynolds, S.H., Henikoff, J.G., Greene, E.A., Steine, M,N., Comai, L., *et al*: High-throughput TILLING for Arabidopsis. *Methods Mol Biol* **2006**, *323*:127-135.

[42] Till, B.J., Colbert, T., Tompa, R., Enns, L.C., Codomo, C.A., Johnson, J.E., Reynolds, S.H., Henikoff, J.G., Greene, E.A., Steine, M.N., *et al*: High-throughput TILLING for functional genomics. *Methods Mol Biol* **2003**, *236*:205-220.

[43] Till ,B.J., Zerr, T., Comai, L., Henikoff, S. A protocol for TILLING and Ecotilling in plants and animals. *Nat Protoc* **2006**, *1*(5):2465-2477.

[44] Le Signor, C., Savois, V., Aubert, G., Verdier, J., Nicolas, M., Pagny, G., Moussy, F., Sanchez, M., Baker, D., Clarke, J *et al*: Optimizing TILLING populations for reverse genetics in Medicago truncatula. *Plant Biotechnol J* **2009**, *7*(5):430-441.

[45] Sato, S., Tabata, S. Lotus japonicus as a platform for legume research. *Curr Opin Plant Biol* **2006**, *9*(2):128-132.

[46] Perry, J.A., Wang, T.L., Welham, T.J., Gardner, S., Pike, J.M,, Yoshida, S., Parniske, M. A TILLING reverse genetics tool and a web-accessible collection of mutants of the legume Lotus japonicus. *Plant Physiol* **2003**, *131*(3):866-871.

[47] Horst, I., Welham, T., Kelly, S., Kaneko, T., Sato, S., Tabata, S., Parniske, M., Wang, T.L. TILLING mutants of Lotus japonicus reveal that nitrogen assimilation and fixation can occur in the absence of nodule-enhanced sucrose synthase. *Plant Physiol* **2007**, *144*(2):806-820.

[48] Triques, K., Sturbois, B., Gallais, S., Dalmais, M., Chauvin, S., Clepet, C., Aubourg, S., Rameau, C., Caboche, M., Bendahmane, A. Characterization of Arabidopsis thaliana mismatch specific endonucleases: application to mutation discovery by TILLING in pea. *Plant J* **2007**, *51*(6):1116-1125.

[49] Till, B.J., Reynolds, S.H., Weil, C., Springer, N., Burtner, C., Young, K., Bowers, E., Codomo, C.A., Enns, L.C., Odden, A.R., *et al*: Discovery of induced point mutations in maize genes by TILLING. *BMC Plant Biol* **2004**, *4*:12.

[50] Weil, C.F., Monde, R.A. Getting the point - Mutations in maize. *Crop Sci.,* **2007**, *47*(1):S60-S67.

[51] Caldwell, D.G., McCallum, N., Shaw, P., Muehlbauer, G.J., Marshall, D.F., Waugh, R. A structured mutant population for forward and reverse genetics in barley (*Hordeum vulgare* L.). *Plant J.***2004**, *40*:143-150.

[52] Bagge, M., Xia, X., Lubberstedt, T. Functional markers in wheat. *Curr Opin Plant Biol* **2007**, *10*(2):211-216.

[53] Slade, A.J., Knauf, V.C. TILLING moves beyond functional genomics into crop improvement. *Transgenic Res* **2005**, *14*(2):109-115.

[54] Slade, A.J., Fuerstenberg, S.I., Loeffler, D., Steine, M.N., Facciotti, D. A reverse genetic, nontransgenic approach to wheat crop improvement by TILLING. *Nat Biotechnol* **2005**, *23*(1):75-81.

[55] Graybosch, R.A. Waxy wheats: origin, properties and prospects. *Trends Food Sci Technol* **1998**, *9*:135-142.

[56] Storozhenko, S., De Brouwer, V., Volckaert, M., Navarrete, O., Blancquaert, D., Zhang, G.F., Lambert, W., Van Der Straeten, D. Folate fortification of rice by metabolic engineering. *Nat Biotechnol* **2007**, *25*(11):1277-1279.

[57] Wu, J.L., Wu, C., Lei, C., Baraoidan, M., Bordeos, A., Madamba, M.R., Ramos-Pamplona, M., Mauleon, R., Portugal, A., Ulat, V.J., *et al*: Chemical- and irradiation-induced mutants of indica rice IR64 for forward and reverse genetics. *Plant Mol Biol* **2005**, *59*(1):85-97.

[58] Raghavan, C., Naredo, M.E.B., Wang, H.H., Atienza, G., Liu, B., Qiu, F.L., McNally, K.L., Leung, H. Rapid method for detecting SNPs on agarose gels and its application in candidate gene mapping. *Molecular Breeding* **2007**, *19*(2):87-101.

[59] Krishnan, H.B. Engineering soybean for enhanced sulfur amino acid content. *Crop Sci.* **2005**, *45*(2):454-461.

[60] Dahmani-Mardas, F., Troadec, C., Boualem, A., Leveque, S., Alsadon, A.A,, Aldoss, A.A., Dogimont, C., Bendahmane, A:. Engineering Melon Plants with Improved Fruit Shelf Life Using the TILLING Approach. *Plos One* **2010**, *5*(12).

[61] Gady, A.L., Vriezen, W.H., Van de Wal, M.H., Huang, P., Bovy, A.G., Visser, R.G., Bachem, C.W. Induced point mutations in the phytoene synthase 1 gene cause differences in carotenoid content during tomato fruit ripening. *Mol Breed* **2012**, *29*(3):801-812.

[62] Markow, T.A., O'Grady, P.M. Drosophila biology in the genomic age. *Genetics* **2007**, *177*(3):1269-1276.

[63] Bellen, H.J., Levis, R.W., Liao, G., He, Y., Carlson, J.W., Tsang, G., Evans-Holm, M., Hiesinger, P.R., Schulze, K.L., Rubin, G.M., *et al*: The BDGP gene disruption project: single transposon insertions associated with 40% of Drosophila genes. *Genetics* **2004**, *167*(2):761-781.

[64] O'Keefe, L.V., Smibert, P., Colella, A., Chataway, T.K., Saint, R., Richards, R.I. Know thy fly. *Trends Genet* **2007**, *23*(5):238-242.

[65] Winkler, S., Schwabedissen, A., Backasch, D., Bokel, C., Seidel, C., Bonisch, S., Furthauer, M., Kuhrs, A., Cobreros, L., Brand, M., *et al*: Target-selected mutant screen by TILLING in Drosophila. *Genome Res* **2005**, *15*(5):718-723.

[66] Bentley, A., MacLennan, B., Calvo, J., Dearolf, C.R. Targeted recovery of mutations in Drosophila. *Genetics* **2000**, *156*(3):1169-1173.

[67] Dahm, R,, Geisler, R. Learning from small fry: the zebrafish as a genetic model organism for aquaculture fish species. *Mar Biotechnol (NY)* **2006**, *8*(4):329-345.

[68] Stein, L.D., Bao, Z., Blasiar, D., Blumenthal, T., Brent, M.R., Chen, N., Chinwalla, A., Clarke, L., Clee, C, Coghlan A *et al*: The genome sequence of Caenorhabditis briggsae: a platform for comparative genomics. *PLoS Biol* **2003**, *1*(2):E45.

[69] Smits, B.M.G., Mudde, J., Plasterk, R.H.A., Cuppen, E: Target-selected mutagenesis of the rat. *Genomics* **2003**, *83*:332-334.

[70] Kim, M.J., Hirono, I., Aoki, T: Detection of quinolone-resistance genes in Photobacterium damselae subsp. piscicida strains by targeting-induced local lesions in genomes. *J Fish Dis* **2005**, *28*(8):463-471.

[71] Comai, L., Young, K., Till, B.J., Reynolds, S.H., Greene, E.A., Codomo. C.A., Enns, L.C., Johnson, J.E., Burtner, C., Odden, A.R. *et al*: Efficient discovery of DNA polymorphisms in natural populations by Ecotilling. *Plant J* **2004**, *37*(5):778-786.

[72] Gilchrist, E.J., Haughn, G.W., Ying, C.C., Otto, S.P., Zhuang, J., Cheung, D., Hamberger, B., Aboutorabi, F., Kalynyak, T., Johnson, L. *et al*: Use of Ecotilling as an efficient SNP discovery tool to survey genetic variation in wild populations of Populus trichocarpa. *Mol Ecol* **2006**, *15*(5):1367-1378.

[73] Mejlhede, N., Kyjovska, Z., Backes, G., Burhenne, K., Rasmussen, S.K., Jahoor, A: EcoTILLING for the identification of allelic variation in the powdery mildew resistance genes mlo and Mla of barley. *Plant Breeding* **2006**, *125*(5):461-467.

[74] Nieto, C., Piron, F., Dalmais, M., Marco, C.F., Moriones, E., Gomez-Guillamon, M.L., Truniger, V., Gomez P, Garcia-Mas J, Aranda MA *et al*: EcoTILLING for the identification of allelic variants of melon eIF4E, a factor that controls virus susceptibility. *BMC Plant Biol* **2007**, *7*(1):34.

[75] Barkley, N.A., Wang, M.L., Gillaspie, A.G., Dean, R.E., Pederson, G.A., Jenkins, T.M: Discovering and verifying DNA polymorphisms in a mung bean [V. radiata (L.) R. Wilczek] collection by EcoTILLING and sequencing. *BMC Research Notes* **2008**, *1*:28.

[76] Fery, R.L: **New opportunities in *Vigna***. In: *Trends in new crops and new uses.* Edited by Janick J, Whipkey A. Alexandria, VA.: American Society of Horticultural Sciences Press; **2002**: 424-428.

[77] Goel, S., Raina, S.N., Ogihara, Y: Molecular evolution and phylogenetic implications of internal transcribed spacer sequences of nuclear ribosomal DNA in the Phaseolus-Vigna complex. *Mol Phylogenet Evol* **2002**, *22*(1):1-19.

[78] Bisht, I.S., Bhat, K.V., Lakhanpaul, S., Latha, M., Jayan, P.K., Biswas, B.K., Singh. A.K: Diversity and genetic resources of wild Vigna species in India. *Genetic Resources and Crop Evolution* **2005**, *52*(1):53-68.

[79] Till ,B.J., Zerr, T., Bowers, E., Greene, E.A., Comai, L., Henikoff, S: High-throughput discovery of rare human nucleotide polymorphisms by Ecotilling. *Nucleic Acids Res* **2006**, *34*(13):e99.

[80] Cordeiro, G., Eliott, F.G., Henry, R.J: An optimized ecotilling protocol for polyploids or pooled samples using a capillary electrophoresis system. *Anal Biochem* **2006**, *355*(1):145-147.

[81] Yeung, A.T., Hattangadi, D., Blakesley, L, Nicolas, E: Enzymatic mutation detection technologies. *BioTechniques* **2005**, *38*:749-758.

[82] Ramos, M.L, Fleming, G., Chu, Y,, Akiyama, Y., Gallo, M., Ozias-Akins, P: Chromosomal and phylogenetic context for conglutin genes in Arachis based on genomic sequence. *Mol Genet Genomics* **2006**, *275*(6):578-592.

[83] Ozias-Akins, P., Ramos, M.L., Chu, Y: **Food Allergy**. Washington, D. C.: ASM Press; 2006.

[84] Till, B.J., Burtner, C., Comai, L., Henikoff, S: Mismatch cleavage by single-strand specific nucleases. *Nucleic Acids Res* **2004**, *32*(8):2632-2641.

Send Orders of Reprints at reprints@benthamscince.net
352 *Advances in Genome Science*, **Vol. 1, 2013, 352-376**

CHAPTER 14

Variations Around the Arrestin Fold

Laurence Aubry, Dorian Guetta and Gérard Klein*

[1]CEA, DSV, iRTSV, Laboratoire Biologie à Grande Echelle, F-38054 Grenoble, France; [2]INSERM, U1038, F-38054 Grenoble, France and [3]Université Joseph Fourier, Grenoble 1, F-38000, France

Abstract: Endocytosis of ligand-activated plasma membrane receptors has been shown to contribute to the regulation of their downstream signaling. β-arrestins interact with the phosphorylated tail of activated receptors and act as scaffolds for the recruitment of adaptor proteins and clathrin, that constitute the machinery used for receptor endocytosis. Visual- and β-arrestins have a two-lobe, immunoglobulin-like, β-strand sandwich structure. The recent resolution of the crystal structure of VPS26, one of the retromer subunits, unexpectedly evidenced an arrestin fold in this protein otherwise unrelated to arrestins. From a functional point of view, VPS26 is involved in the retrograde transport of the mannose 6-P receptor from the endosomes to the trans-Golgi network. In addition to the group of true arrestins and Vps26, mammalian cells harbor a vast repertoire of proteins that are related to arrestins on the basis of their PFAM Nter and Cter arrestin-domains, which are named Arrestin Domain-Containing proteins (ARRDCs). The biological role of ARRDC proteins is still poorly understood. The three subfamilies have been merged into an arrestin-related protein clan.

This paper provides an overall analysis of arrestin clan proteins. The structures and functions of members of the subfamilies are reviewed in mammals and model organisms such as *Drosophila*, *Caenorhabditis*, *Saccharomyces* and *Dictyostelium*.

Keywords: Arrestins; Vps26; GPCR; Retromer; Trafficking; Endocytosis.

TUNING OF RECEPTOR-DEPENDENT SIGNALING BY ARRESTINS

Visual arrestin, also known as arrestin 1 or arrestin S, was discovered in the mid eighties in bovine retinal rods. Arrestin arrests the signaling downstream of light-activated rhodopsin [1, 2]. Four true arrestins have now been characterized in mammals. Two of them are specific of visual cells with arrestin 1/arrestin S found in rods and arrestin 4/arrestin C in cones. The other two isoforms, β-arrestin 1 also

***Address correspondence to Gérard Klein:** BGE/OdyCell, Institut de Recherches en Technologies et Sciences pour le Vivant, CEA-Grenoble, 17 avenue des Martyrs, 38054 Grenoble Cedex 9, France; Tel: +33 438 784 661; Email: gklein@cea.fr described as arrestin 2 and β-arrestin 2 aka arrestin 3 have an ubiquitous expression (for reviews, see [3-6]). True arrestins interact with agonist activated G-protein coupled receptors (GPCRs) and play an important role in their desensitization. The consequences of the GPCR-arrestin interaction are multiple: 1) a conformational change of the arrestins that unfolds their C-terminal tail and reveals binding sites for clathrin and adaptin; 2) an arrest of the signaling downstream of the receptors through a physical interference of arrestins with heterotrimeric G-proteins; 3) the internalization of GPCRs in coated vesicles nucleated by arrestin-bound clathrin and adaptor protein AP-2. Internalized GPCRs are either rapidly recycled to the plasma membrane leading to a transient desensitization or routed for degradation in the lysosomes leading to a long term desensitization. More recently, β-arrestins have been described as G-protein independent signaling platforms through interactions with MAP kinases [3, 4, 7-12]. β-arrestins are able to translocate to the nucleus, associate with transcription cofactors and thus regulate transcription [10].

Given the size of the GPCR family (over 1000 different GPCRs in humans responding to a variety of stimuli including hormonal signals, neurotransmitters, chemokines, chemoattractants and sensory stimuli such as light, odor or taste), the way in which GPCRs interact with arrestins, and the novel roles of arrestins in cytoplasmic signaling and transcription, it seems that we are just scratching the surface of the potential involvement of the arrestin family in a rapidly expanding list of signaling pathways.

ARRESTINS AND THEIR PARTNERS AS DISCLOSED FROM STRUCTURAL DATA

The crystal structure of visual arrestins and β-arrestins has been solved [13-20]. Arrestin proteins are organized as two modules, the N- and C-domains (PFAM signatures PF00339 and PF02752, respectively), related by a pseudo two-fold rotation axis and connected by a flexible hinge (Fig. **1A**). Each module consists of a curved sandwich of two antiparallel β-sheets. Interestingly, this structure is found without any sequence conservation in fibronectin and IgGs [21].

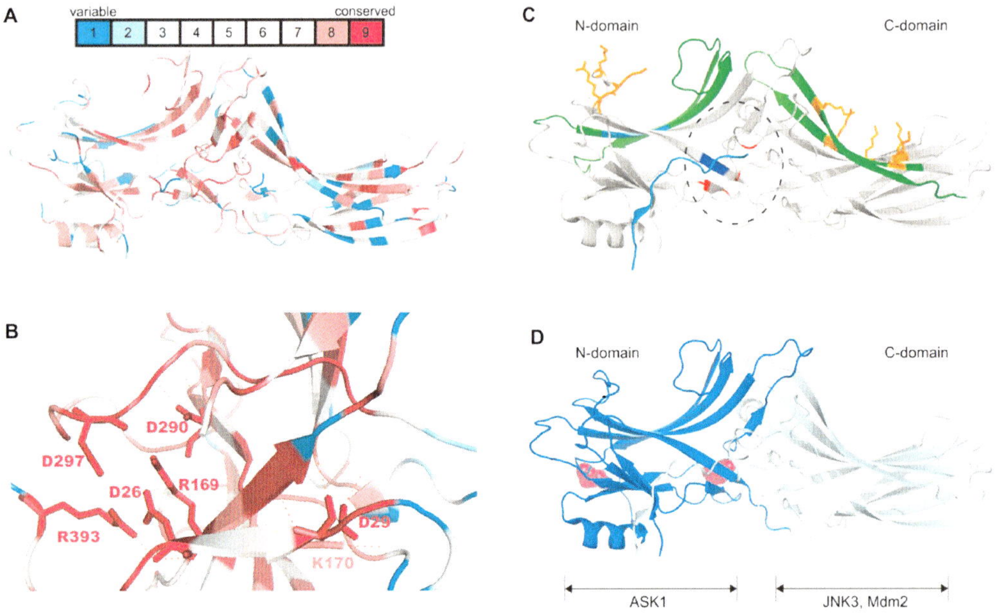

Figure 1: Residue conservation among members of the arrestin family. Localization of binding sites on β-arrestins. A. Amino acid conservation in human rod and cone arrestins and β-arrestins 1 and 2 was calculated with the online Consurf software (http://consurf.tau.ac.il) and represented with PyMol using Consurf coloring (see scale) on the structure of β-arrestin 1 (1ZSH). B. Blowup of the polar core showing the high conservation of the residues important for the polar core highlighted as sticks. Electrostatic bonds between core residues are indicated as dotted lines. C. The structure of β-arrestin 1 (PDB 1ZSH) is represented with the PyMol software. The localization of the polar core is indicated by a dashed ellipse with positive residues in blue and negative residues in red. Receptor binding β-strands are indicated in green. The adaptin AP-2 binding site is indicated in cyan. Low- and high-affinity binding sites for $InsP_6$ in the N- and C-domains, respectively are indicated as orange sticks. D. The N- and C-domains of β-arrestin 2 (PDB 1JSY) are indicated in dark and light blue, respectively. Prolines involved in Src binding appear as magenta spheres. The binding domains of ASK1, JNK3 and Mdm2 are indicated.

The mechanism of interaction of β-arrestins with their cognate receptors involves a electrostatic interaction domain with two strategic lysine residues ($K^{10}K^{11}$ in β-arrestin 1) and a polar core generated by charged residues of the N-domain, a lariat loop of the C-domain and the C-tail (Fig. **1B**). The binding of the negatively-charged phosphorylated C-terminus of activated GPCRs disrupts the electrostatic interactions within the polar core and triggers a conformational rearrangement of arrestin and the subsequent unmasking of the C-tail carrying a clathrin-binding site (L[L/I][D/E/N][L/F][D/E], [22, 23]) and an adaptin-binding site (IVFxxFxRxR, [24, 25]). These sites drive the recruitment of the endocytic machinery and the internalization of the β-arrestin-bound receptors in clathrin-coated vesicles. Other proteins involved in endocytosis are also scaffolded by β-arrestins, the monomeric G protein Arf6 and its GEF (ARNO) and GAP (GIT) [26-29]. Arrestins Arr1 and Arr2 from the fruit fly promote clathrin-dependent endocytosis of rhodopsin in physiological or hyperstimulation conditions [30, 31]. Visual arrestins lack the classical clathrin and AP-2 binding sites in their C-terminal tail, but could possibly use for this endocytosis function a novel clathrin box similar to that discovered in an alternatively spliced form of β-arrestin 1 [32].

β-arrestins are regulated by post-translational modifications: phosphorylation, ubiquitination, SUMOylation and nitrosylation [11]. Inactive cytosolic β-arrestins 1 and 2 are phosphorylated by ERK or casein kinase II on sites located on their C-terminal tail, close to the clathrin-adaptin binding sites. Upon their recruitment to an activated GPCR, β-arrestins are rapidly dephosphorylated by an unknown phosphatase, a step required for the binding of clathrin and adaptin. Soon after the receptor internalization step, β-arrestins are rephosphorylated [33-37].

At the membrane, the stability of the association of β-arrestin 2 with GPCRs is dependent on its ubiquitination by the ubiquitin E3 ligase Mdm2 [38]. The 2 ubiquitinated lysines are the same that serve as phosphate-sensors when they bind activated and phosphorylated GPCRs. Using the same set of lysines to recognize an activated receptor and to stabilize the GPCR-β-arrestin interaction may serve as a sequential switch to temporally order interaction events. A stable receptor-arrestin complex on the endosome is essential for the arrestin-dependent signaling platform [39]. The reciprocal activities on β-arrestins of the ubiquitin E3 ligase Mdm2 and the deubiquitinating protease USP33 are essential for the regulation of GPCR trafficking and downstream signaling [38].

Bovine β-arrestins 1 and 2 are also modified by SUMO-1 upon activation of β_2-adrenergic receptor and this modification regulates GPCR internalization [40].

It has been demonstrated that β-arrestin 2 serves as a scaffold for endothelial NO synthase that sequentially S-nitrosylates β-arrestin 2 and β_2-adrenergic receptor. This modification potentiates the interaction between β-arrestin, clathrin and the adaptor protein AP-2, and the receptor internalization [41].

Visual and β-arrestins display apparent selectivity towards their respective physiological receptor(s). However, the existence of a one-to-one correspondence between β-arrestins and receptors cannot be envisioned, as the mammalian genome harbors close to a thousand GPCRs and only a few arrestin proteins. The current hypothesis is that the flexibility of the β-arrestin N-domain helps to accommodate the diverse GPCRs, whereas a rigid visual arrestin N-domain best fits its specialized partner, rhodopsin [14]. However, class B GPCRs (see definition below) are able to interact with visual arrestin [42], and cone arrestin can bind non-visual receptors in heterologous expression situations [18]. GPCRs have been divided into two classes on the basis of their affinity for β-arrestins and the duration of this interaction. Class A receptors (β_2-adrenergic receptor, μ-opioid receptor, dopamine receptor) interact preferentially with β-arrestin 2. Once internalized, they rapidly dissociate from β-arrestin and are rapid recycled and resensitized. Class B receptors (V2-vasopressin receptor, angiotensin receptor) bind β-arrestin 1 and β-arrestin 2 equally well, maintain a stable association with β-arrestins on the endosomes and recycle slowly. The class A or B information is contained in the cytosolic tail of GPCRs. Swapping the C-terminus of a class A receptor with that of a class B receptor, or vice-versa, converts the GPCR class [43]. The formation of a stable receptor-β-arrestin complex is dependent upon the presence on the receptor of phospho-serine and phospho-threonine clusters [44]. The detection of the phosphorylated state of GPCRs by arrestins is mediated by their polar core, and an arrestin mutated by charge inversion in this core binds non-phosphorylated and phosphorylated receptors equally well [45, 46]. The conformational changes induced in arrestins by the sensing of the phosphate moiety allow receptor-binding on surfaces located on the concave side of both the N-domain (β-strands 5 and 6) and the C-domain (β-strands 15 and 16) for both visual arrestin and β-arrestin 1 [47] (Fig. **1C**).

Phosphoinositides and phosphoinositols such as phytic acid (inositol hexakisphosphate, $InsP_6$) regulate the oligomerization state of the arrestins, their subcellular location and their interaction with receptors [16, 48, 49]. The concentration of $InsP_6$ in cell lines and tissues is in the 10-100 μM range [50]. As β-arrestin has an affinity for $InsP_6$ between 50 nM and 1 μM, the cytosolic basal-state arrestin is mostly stored in an oligomeric $InsP_6$-bound form. Non-visual arrestins bind $InsP_6$ on two low- and high -affinity sites located in the N- and C-domains of the arrestins (Fig. **1C**). Both $InsP_6$ binding sites are defined by positive residues patches [16, 17]. The high-affinity $InsP_6$ binding residues (K^{233}, R^{237}, K^{251} in human β-arrestin 2) are located on the concave surface of the C-domain, also interacting with the phosphorylated GPCR [17]. The possibility that the arrestin-GPCR interaction is modulated by $InsP_6$ can thus be envisioned. The C-domain site is also able to bind membrane-embedded phosphoinositides, with the highest affinities for PtdIns(4,5)P_2 and PtdIns(3,4,5)P_3. Mutations in this binding site impair β_2-adrenergic receptor endocytosis [51]. Phosphoinositides may act as co-receptors for coincidence detection, increasing the binding affinity/selectivity of activated arrestins for a phosphorylated GPCR.

EMERGING AREAS IN β-ARRESTIN SIGNALING

Extending the List of Arrestin-Regulated Cargoes

Recent data demonstrate that the role of arrestins is not restricted to serpentine receptors [52, 53], but includes single transmembrane-domain receptors with or without tyrosine kinase activity (type III TGF-β receptor, insulin-like growth factor I receptor, Notch and inhibitory killer-cell immunoglobulin-like KIR receptor), 12-membrane spanning transporters (Na^+/H^+ exchanger NHE5) or voltage-dependent cation channels (calcium channel $Ca_v1.2$) [54-60]. Three elements in the concave N- and C-domains of β-arrestin 2 interact with NHE5 [56]. Interestingly, the first binding site (residues 52-78) corresponds to the conserved β-strand 5 required for GPCR binding [47]. Whether phosphorylation of NHE5 is required for β-arrestin binding has not been elucidated. In natural killer cells, β-arrestin 2 interacts with KIR engaged with ligands. Interaction necessitates the tyrosine phosphorylation of consensus immunoreceptor tyrosine-based inhibition motifs (ITIM) in KIR's cytoplasmic tail. This interaction facilitates the recruitment of SHP-1 and -2 phosphatases into a ternary complex that transduces an inhibitory signal abrogating the signal arising from activating receptors [60, 61].

Extending the List of Arrestin Sites of Action

Recent literature indicates that arrestins are not solely used to squelch receptor-dependent signaling at the plasma membrane. Their role is increasingly comparable to that of heterotrimeric G-proteins, as they act as a secondary signaling platform on the endosomes. They bind c-Src, various MAPKs, PI3K and Akt as well as non-kinase partners such as the cytoskeletal component RhoA (Fig. **1D**) [17]. In the cytosol, β-arrestins also interact with Mdm2 and IκBα, and thereby indirectly regulate p53- and NFκB-dependent transcription [10]. The nuclear functions of arrestins have also been documented recently [3]. Both β-arrestins translocate to the nucleus and interact with transcription cofactors p300 and cAMP-response element-binding protein (CREB) [10]. Furthermore, β-arrestin 1 negatively regulates the transcription factor STAT1 (signal transducer and activator of transcription 1). When the antiviral cytokine interferon–γ binds its receptor, it activates the phosphorylation of STAT1, inducing its dimerization, translocation to the nucleus and transcriptional activity. β-arrestin 1 promotes the dephosphorylation of STAT1 by recruiting the tyrosine phosphatase TC45 in a ternary complex with STAT1, thereby turning STAT1 off and reducing interferon–γ-induced gene transcription. It should be noted that this effect is not observed with β-arrestin 2, though it can bind STAT1, possibly because of a difference in its nucleo-cytoplasmic shuttling [62, 63]. Indeed, the presence of a functional NES in the C-domain of β-arrestin 2 (^{388}FEDFARLRL^{396} sequence), which is absent in β-arrestin 1, prevents its nuclear accumulation. This nucleo-cytoplasmic shuttling leads to the cytosolic location of c-Jun N-terminal kinase 3 (JNK3) when the kinase is docked on the ^{196}RRSLHL^{201} sequence in the C-terminus of β-arrestin 2 [64].

β-arrestin 2, and not β-arrestin 1, is located at the centrosome in both interphasic and mitotic cells, but is not involved in the classical functions of the centrosome, such as nucleation and anchoring of microtubules. Lack of β-arrestin 2 correlates with a reduction in primary cilium formation and cell cycle deregulation [65]. Upon primary cilium formation in fibroblasts, β-arrestin 2 is translocated to the axoneme [65]. Its role there is connected with the Smoothened serpentine protein (Smo) in the Hedgehog (Hh) pathway. Smo is inhibited by Patched (Ptc), a 12 transmembrane-spanning protein. Binding of Hh to Ptc relieves the inhibition of Smo, and this event constitutes the initiation step of the developmental processes. Phosphorylation of Smo by GRK2 induces the recruitment of β-arrestin 2 and the endocytosis of Smo in clathrin-coated vesicles [66]. Once activated, Smo regulates the expression of Hh target genes by controlling the activity of the Glioma-associated oncogene homologue (Gli) [67]. It has now been demonstrated that β-arrestins 1 and -2 promote the interaction of activated Smo with the kinesin motor protein Kif3A, the subsequent movement of Smo into the primary cilium mediated by components of the intracellular transport complex, the production of the activator form of the Gli transcription factor and gene expression [68, 69].

THE ARRESTIN FOLD, A WIDELY USED MOLECULAR ARCHITECTURE

Vps26

The arrestin-fold is also shared by an apparently unrelated protein, Vps26, which is present in human as two isoforms, Vps26A and Vps26B [70, 71] (Figs. **2**, **3**). Vps26 is a subunit of a large multimeric complex, termed a retromer, which is located on early/recycling endosomes in mammals and which mediates the retrograde transport of the cation-independent mannose 6-phosphate receptor (CI-MPR) and its yeast homologue Vps10 from the endosomes to the Golgi apparatus. The retromer comprises two sub-complexes. The Vps26/Vps29/Vps35 subcomplex is involved in cargo-recognition and loading. The second subcomplex is made up of sorting nexins (SNX) that participate, *via* their PX and BAR domains, in the location of the retromer on PtdIns(3)P-rich highly curved membranes of early and recycling endosomes [72-74]. Vps26 interacts with both Vps35 and SNX1.

At present, it is not known whether the structural similarities between Vps26 and the arrestin families correlate with the existence of a functional kinship beyond their role in trafficking processes. In both families, polar and electrostatic contacts play a key role in interdomain interactions. However, when compared to arrestins, the polar core of Vps26 involves different side chains and is generated by distinct secondary structure elements. Mutations of polar core residues in Vps26 do not interfere with either its endosomal location in human cells or with CPY sorting in yeast, thus calling into question its role in the activation step and/or the function of Vps26 [71]. In the retromer complex, the position of Vps26 relative to the neighboring subunits may allow access to the concave surface by target cargoes. Vps35 is the retromer subunit generally used for the binding of the transported cargoes *via* a [FW]L[MV] sequence [75]. VPS26 can bind the VPS10 receptor family SorlA that transports the amyloid precursor protein to the Golgi through it's sorting motif FANSHY [76]. Both recognition motifs can be merged into a consensus "aromatic aa-x5-aromatic aa" signature.

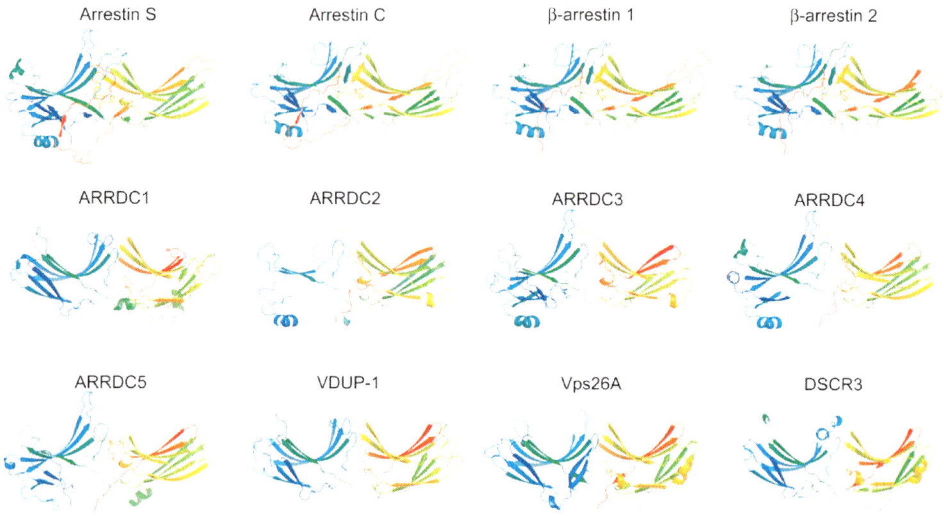

Figure 2: 3D-structure of human arrestin clan proteins. The structure of human arrestin clan proteins was modeled with the Phyre software (http://www.sbg.bio.ic.ac.uk/~phyre/) and visualized with the Pymol software (http://www.pymol.org). The PDB templates used for the modeling were 1CF1 for arrestin S, ARRDC2, ARRDC3, ARRDC4 and ARRDC5, 1JSY for β-arrestin 1, β-arrestin 2 and arrestin C, 2FAU for ARRDC1, VDUP, Vps26A and DSCR3.

DSCR3

The human genome harbors the DSCR3 gene encoding a protein belonging to the arrestin clan (see definition below) (Fig. **3**). Interestingly, it has been predicted that human DSCR3 contains overlapping PFAM PF00339 (Arrestin, N-terminal domain, E-value 1.9×10^{-3}) and PFAM PF03643 (Vps26, E-value 1.1×10^{-155}) domains. Thus, DSCR3 shares sequence kinship with both arrestins and Vps26 proteins and represents an intermediate link between these two families. The structure of full-length DSCR3 can be modeled by homology (Fig. **2**) using the crystal structure of Vps26 (PDB 2fauA) as a template. DSCR3 belongs to the subset of about 20 genes located within locus 21q22, called the **D**own **S**yndrome **C**ritical **R**egion [77], which is involved in the partial or full trisomy of chromosome 21 that is responsible for Down syndrome. Up to now, DSCR3 attracted only little biochemical interest. Because DSCR3 protein is found in the nucleus, a transcriptional regulation function has been proposed for it [78]. A possible role for DSCR3 and 3 other chromosome 21 trisomy proteins from the DSCR region in a mitogen-activated protein kinase pathway has been predicted on the basis of a computational approach [79].

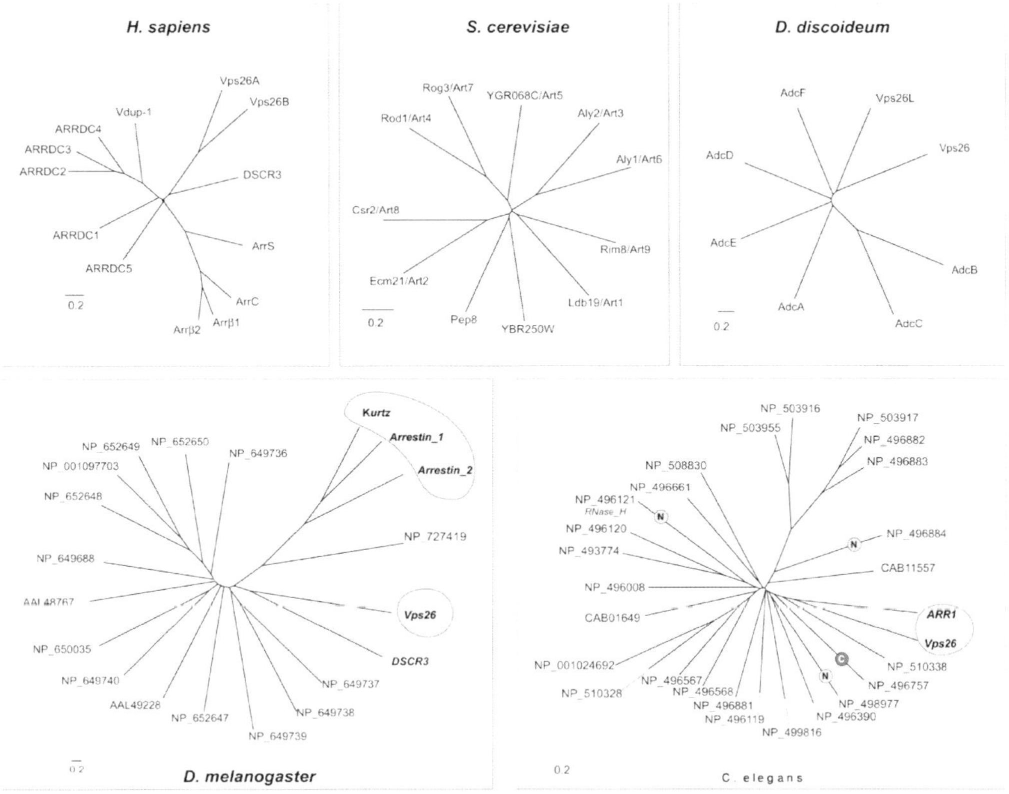

Figure 3: Diversity of arrestin clan proteins in human, fruit fly, nematode, yeast and amoeba. A) Sequences of human arrestin clan proteins were aligned with ClustalW software [128]. An unrooted phylogenetic tree was drawn with TreeDyn for each species (http://www.phylogeny.fr) [129, 130]. Labels (N, C) in the nematode tree indicate genes in which only the arrestin N- or C-domain was found. A grey cloud highlights proteins in nematode and fruit fly for which literature is available.

Mammalian Arrestin-Domain Containing Proteins (ARRDC)

Mammalian genomes harbor genes coding for arrestin-domain containing proteins (ARRDCs + TXNIP/VDUP-1) besides true arrestins. The encoded proteins harbor the PFAM signatures of the N- and C-arrestin domains and their secondary structure as disclosed by PsiPred is organized as β-strands. Their identity scores with arrestins

(11-15%) are however very low and ARRDCs cluster away from arrestins in a phylogenic tree (Fig. **3**). Surprisingly, the structure of all ARRDCs can be modeled by homology using the structure of crystalized arrestins as a template (Fig. **2**). The interface between the modeled N- and C-domains does not seem to involve a polar core with matched positively and negatively charged amino acids, but this point awaits biochemical confirmation. The classical clathrin and adaptin sites are missing. Data pertaining to the cellular function of these ARRDCs are scarce. The mRNA of TXNIP/VDUP-1 increases 4-5 fold upon treatment of HL60 cells with 1,25-dihydroxyvitamin D_3 [80]. In a yeast two-hybrid screen, TXNIP/VDUP-1 has been shown to interact with thioredoxin [81, 82] and hence to regulate the redox state of the cell [83, 84]. Suppression of thioredoxin activity by TXNIP/VDUP-1 has antiproliferative and apoptotic effects [85]. The expression of ARRDC2 has been shown to respond to lysergic acid diethylamine in rat and the gene was therefore named *ILAD-1* (induced by lysergic acid diethylamine) [86].

ARRDC3 binds the β_2-adrenergic receptor and β-4 integrin and controls their endocytosis, ubiquitination and final degradation [87, 88]. ARRDC3 and ARRDC4 are associated with the plasma membrane, endosomes and lysosomes whereas TXNIP/VDUP-1 and ARRDC2 are essentially nuclear [89].

ARRDC1-4 and TXNIP/VDUP-1 proteins show interactions with subunits of the ESCRT machinery, namely Alix and Tsg101 (known as Vps23 in yeast). They all possess PPxY motifs and they interact indeed with several E3 ubiquitin ligases : WWP1, WWP2, Nedd4 and Itch. ARRDC1 and possibly the other ARRDCs are ubiquitinated by WWP1. The current hypothesis is that ARRDC1 is involved at the PM in PPxY-dependent viral budding by bridging HECT ubiquitin ligases and ESCRT-III components [90]. Interestingly, cells are able to shed in a virus-independent process microvesicles (ARMMs, ARRDC1-mediated microvesicles) that differ from exosomes. This budding is driven by the specific interaction between a PSAP sequence of TSG101/VPS23 and ARRDC1 [91].

Variations in Other Models

Alternative models were studied soon after the discovery of vertebrate arrestins and important roles established in flies or chordates. Data now extend to unicellular eukaryotes.

Drosophila melanogaster

Visual arrestins 1 and 2 were described in *Drosophila* rapidly after the discovery of arrestin in bovine retinas [92-94]. Arrestins 1 and 2 are expressed in fruit fly photoreceptor cells. Binding of arrestin 2 to rhodopsin is not dependent on its phosphorylation, whereas arrestin 1 is recruited on phosphorylated rhodopsin. This latter interaction promotes light-induced rhodopsin endocytosis. Arrestin 1 is also required for light-independent photoreceptor survival [31].

More recently, a nonvisual arrestin, Kurtz, has been shown to be an essential neural gene in *Drosophila* [95]. In an arrestin-translocation assay, Kurtz binds amine and various peptide receptors in response to ligand stimulation. Kurtz null mutants are hypersensitive to osmotic stress, involving a still unknown GPCR desensitization in the stress response [96]. Kurtz downregulates the Notch cell surface receptor, a single transmembrane protein carrying EGF-like repeats that, together with its ligand Delta, determines the ectodermal or neuronal fate of cells. Kurtz is able to complex Deltex, a E3 ubiquitin ligase. Deltex, in a ternary complex with Notch and Kurtz, will ubiquitinate the intracellular domain of Notch. The outcome of this ubiquitination is the degradation of Notch, possibly by the proteasome [59, 97].

In addition to these 3 arrestins, the fruit fly repertoire of arrestin clan proteins includes Vps26 and DSCR3 proteins, 14 ARRDCs and an uncharacterized branch, NP_727419. This gene encodes a long isoform (804 aa) branching off before the divergence of arrestins 1 and 2 and Kurtz (Fig. **3**).

Work on *Drosophila* and on the nematode *Caenorhabditis elegans* (see below) has revealed the essential role of the retromer machinery and Vps26 during development. This role is conserved during mouse embryogenesis [98]. Important clues for the understanding of the regulation mechanism of Wg/Wnt secretion by the retromer have now been produced in fruit fly experiments in which Vps26 and Vps35, two subunits of the retromer, are eliminated by RNA-interference. The Wingless (Wg) morphogen is normally carried from the Golgi apparatus to the cell surface by Wntless (Wls), a multi-pass transmembrane protein that acts as a Wg cargo receptor. Once unloaded, Wls is recaptured by endocytosis and recycled to the trans-Golgi network by the retromer complex. In the absence of a functional retromer, Wls is targeted to the lysosome and degraded, thereby preventing Wls to direct Wg/Wnt secretion [99-101].

Caenorhabditis elegans

The repertoire of nematode arrestin clan proteins comprises a single arrestin, a Vps26 protein and a crown of ARRDCs (Fig. **3**). The presence of some minor branches may be due to not yet stable entries in the databases. ARR-1, the sole genuine arrestin encoded by the nematode genome, is highly similar to fruit fly Kurtz. While the disruption of the G protein-coupled receptor kinase 2 (grk-2) gene strongly affects chemotaxis, disruption of ARR-1 does not perturb chemosensation, which is a paradox, since GRKs and arrestins cooperate in the regulation of receptor-dependent signaling [102]. ARR-1 is dispensable for development, but has been shown to regulate olfactory adaptation and recovery. The C-terminal domain harboring a conserved clathrin-binding box (LIQLH) and a β_2-adaptin binding box (FxxxR), is essential to promote internalization of the β_2-adrenergic receptor when nematode ARR-1 is expressed in HEK293 cells [103].

The first evidence that the retromer is involved in the trafficking of Wntless was provided by studies on *C. elegans*. The retromer functions in the formation of a gradient of EGL-20, the nematode ortholog of Wnt, as disclosed by RNA-mediated interference against *vps35*, *vps29* and *vps26,* and regulates neuronal polarity [104, 105].

The NP_496121 gene encodes a fusion protein of the N-domain of arrestin and RNAseH. No data are available for the ARRDCs.

Saccharomyces cerevisiae

In accordance with the claim that arrestins are restricted to metazoan, genes coding for β-arrestins are absent in yeast despite the presence of a few GPCRs. Instead, the *Saccharomyces* genome encodes 10 arrestin-related proteins that have been coined ARTs (arrestin-related trafficking adaptors) [106]. ARTs have a central arrestin core with extensions on either side. They have only about 10% identity with visual arrestin. Their secondary structures are essentially organized as β-strands.

ARTs are responsible for the **R**esistance of yeast to **o**-**d**initrobenzene (ART4/Rod1p and ART7/Rog3p), involved in the response to alkaline ambient pH (ART9/Rim8/PalF) or cadmium (ART2/Ecm21 and ART8/Csr2) or interacting with the autophagic machinery (ART6/Aly1p and ART3/Aly2p) (Fig. **3**) [107-114]. PalF in *A. niger* and Ecm21 in *S. cerevisiae* promote the internalization of the pH sensor PalH or the yeast manganese transporter Smf1, respectively [109, 112]. Three genes loosely related to arrestins complete the repertoire of yeast arrestin domain containing proteins: ART5/YGR068C, ART1/Ldb19 (low dye binding) and ART10/YLR392C.

ARTs (and human ARRDCs) have ([L,P]PxY motifs that bind WW sequences present on E3 HECT-type ubiquitin ligases. These "PY" sites have been used by Alvarez to discriminate between ancient α-arrestins (with a PY motif) and visual/β-arrestins (without a PY motif). In addition, visual/β-arrestins possess a short α-helix in their N domain that is missing in α-arrestins [115]. Most ART proteins mediate the ubiquitination of their plasma membrane cargoes and participate in their endocytosis [106, 112, 116]. ART9 (previously known as Rim8) binds both Rim21, a 7TM pH sensor and VPS23, a subunit of the ESCRT-I subcomplex. Binding of VPS23 recruits *in fine* the whole ESCRT-I subcomplex on Rim8. This interaction is essential for Rim signaling [117].

Phosphorylation is another important regulation step of the interaction of arrestins with GPCRs in mammals. ART4/Rod1 is phosphorylated by Snf1-kinase and the kinase activity is important for drug resistance [113]. ART3/Aly2p is phosphorylated by Cdk1 [118].

Pep8 is the genuine yeast Vps26. It has been demonstrated that it forms the retromer complex with Vps35, Vps29, Vps17 and Vps5 [119] and bridges the cargo-binding protein Vps35 and the membrane deformation subcomplex Vps5/Vps17 [120]. Pep8 is essential for the correct sorting of carboxypeptidase Y to the vacuole [121]. The retromer cooperates in endosomal cargo retrieval back to the Golgi apparatus with BTN2, the yeast homolog of mammalian Hook1, through the direct interaction of BTN2 with Vps26 [122]. YBR250W/Spo23, a VPS26-like protein, is a member of thc yeast sporulation pathway [123]. The null mutant is viable and has no sporulation defect [123, 124].

Dictyostelium discoideum

The unveiling of PalF and more recently ART2/Ecm21, two fungal proteins with arrestin domains and functions, prompted us to screen for whether the genome of other unicellular eukaryotes, in particular the amoeba *Dictyostelium discoideum,* codes for proteins of the arrestin clan. Eight candidates were found in *D. discoideum* (Fig. **3**). Most obviously, hits for Vps26 and DSCR3 were identified on the basis of their strong conservation with human orthologs ($1e^{-109}$ expect value and 55% identity for DDB_G0269168/Vps26; $7e^{-94}$ expect value and 56% identity for DDB_G0292212/DSCR3/Vps26L) (Fig. **3**). As for the human protein, amoebal DSCR3/Vps26L has a Vps26 PFAM domain that overlaps in its first half with an arrestin C-domain.

Besides these 2 proteins, 6 candidates harbor either both arrestin N- and C domains (AdcA, -B, -C, -D, -F) or the C-domain only (AdcE). Interestingly, most arrestin-related proteins from *Dictyostelium* exhibit new pairings of known domains. They are extended on both sides of the arrestin core, with consensus protein-protein interaction domains (SAM, MIT, LIM) or protein-phospholipid interaction domains (FYVE and C2) (Fig. **4**). Such a domain organization is also found in *Entamoeba histolytica* arrestins and seems to be limited to Amoebozoae. AdcA and AdcD have been found in the proteome of the macropinocytic pathway [125]. AdcA's targeting to endosomes is dependent on its FYVE domain [126]. AdcA is also present in the protein inventory of the centrosome [127]. The presence of C2 and FYVE domains may indicate that ancient arrestins have targeted functions at the plasma membrane and the endosome, respectively.

AdcB, AdcC and AdcD harbor "PY" sequences outside their arrestin core (AdcB: PPVY; AdcC: LPRY) or within the arrestin N-domain (AdcD: LPKY) as do mammalian α-arrestins and yeast ARTs, which may serve to recruit a HECT domain-containing ubiquitin ligase. However, there is no evidence for the presence in the amoebal genome of a gene coding a Nedd4/Rsp5 paralog. Whether these sequences are not used or are used to recruit another type of ubiquitin E3 ligase is not known. The structure of whole AdcA, AdcB and Vps26 has been modeled by homology on the structure of crystalized arrestin S or Vps26A [126]. It shows the general conserved structure of arrestins with an α-helix in the N-domain of AdcA as do visual and non-visual arrestins. The modeled structure of AdcA does not show a polar core with opposite-charged residues able to act as the sensor for a phosphorylated receptor, possibly suggesting a different mode of activation, but this awaits biochemical demonstration.

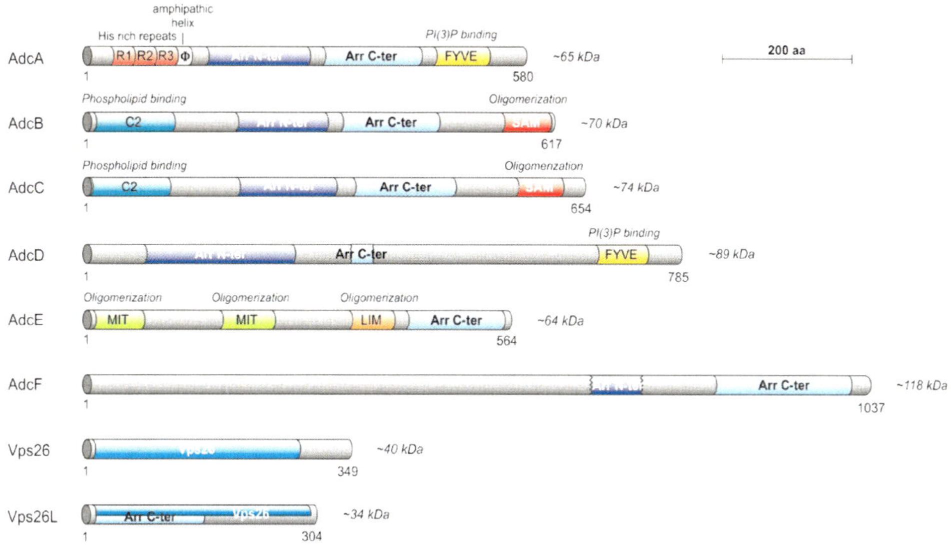

Figure 4: Domain organization of *Dictyostelium* arrestin clan proteins. *Dictyostelium* arrestin-domain proteins (ADCs), Vps26 and Vps26L/DSCR3 are represented with their colored domains as indicated. Number of amino acids and molecular masses of each protein are indicated. The putative functions of each domain are indicated on top.

CONCLUSION

The arrestin clan comprises visual and non-visual true arrestins, arrestin-domain containing proteins (ARRDCs), the retromer subunit Vps26 and DSCR3, all proteins which share an common arrestin-fold. In invertebrates, members of all subfamilies can be found with at least one β-arrestin, a visual arrestin when the animal has eyes or ocelli (*Drosophila*, *Ciona*), Vps26 and DSCR3, and a broad representation of ARRDCs. Arrestin-related proteins have also been found in unicellular eukaryotes (yeast and *Dictyostelium*). In most cases where studies have been performed, arrestin clan proteins have been linked to intracellular trafficking, though they belong to different machineries in different cellular locations. This points towards the existence of different functions for proteins built on a conserved structure, in which the 3D-structure is essential for scaffolding the various affiliates.

ACKNOWLEDGMENTS

This work was supported in part by the Commissariat à l'Energie Atomique, the Centre National de la Recherche Scientifique, the Université Joseph Fourier Grenoble, the Ministère pour la Recherche et la Technologie (ACI Biologie du Développement et Physiologie Intégrative) and by grants from the ANR (Agence Nationale de la Recherche).

CONFLICT OF INTEREST

The author(s) confirm that this chapter content has no conflict of interest.

DISCLOSURE

This is an updated report of the earlier review article by A. di Masi and A. Antoccia " NBS1 heterozygosity and cancer risk ". Curr. Genomics, 2008, 9: 275-281.

REFERENCES

[1] Zuckerman, R., Buzdygon, B., Philp, N., Liebman, P., Sitaramayya, A. Arrestin: an ATP/ADP exchange protein that regulates cGMP phosphodiesterase activity in retinal rod disk membranes (RDM). *Biophys. J.* **1985**, *47*: 37a.

[2] Zuckerman, R., Cheasty, J.E. A 48 kDa protein arrests cGMP phosphodiesterase activation in retinal rod disk membranes. *FEBS Lett.* **1986**, *207*: 35-41.

[3] DeWire, S.M., Ahn, S., Lefkowitz, R.J., Shenoy, S.K. Beta-arrestins and cell signaling. *Annu. Rev. Physiol.* **2007**, *69*: 483-510.

[4] Gurevich, E.V., Gurevich, V.V. Arrestins: ubiquitous regulators of cellular signaling pathways. *Genome Biol.* **2006**, *7*: 236.

[5] Moore, C.A., Milano, S.K., Benovic, J.L. Regulation of receptor trafficking by GRKs and arrestins. *Annu. Rev. Physiol.* **2007**, *69*: 451-482.

[6] Premont, R.T., Gainetdinov, R.R. Physiological roles of G protein-coupled receptor kinases and arrestins. *Annu. Rev. Physiol.* **2007**, *69*: 511-534.

[7] Barki-Harrington, L., Rockman, H.A. Beta-arrestins: multifunctional cellular mediators. *Physiology* **2008**, *23*: 17-22.

[8] DeFea, K.A. Beta-arrestins as regulators of signal termination and transduction: how do they determine what to scaffold? *Cell. Signal.* **2011**, *23*: 621-629.

[9] Luttrell, L.M., Gesty-Palmer, D. Beyond desensitization: physiological relevance of arrestin-dependent signaling. *Pharmacol. Rev.* **2010**, *62*: 305-330.

[10] Ma, L., Pei, G. Beta-arrestin signaling and regulation of transcription. *J. Cell Sci.* **2007**, *120*: 213-218.

[11] Shenoy, S.K., Lefkowitz, R.J. beta-Arrestin-mediated receptor trafficking and signal transduction. *Trends Pharmacol. Sci.* **2011**, *32*: 521-533.

[12] Shukla, A.K., Xiao, K., Lefkowitz, R.J. Emerging paradigms of beta-arrestin-dependent seven transmembrane receptor signaling. *Trends Biochem. Sci.* **2011**, *36*: 457-469.

[13] Granzin, J., Wilden, U., Choe, H.W., Labahn, J., Krafft, B., Buldt, G. X-ray crystal structure of arrestin from bovine rod outer segments. *Nature* **1998**, *391*: 918-921.

[14] Han, M., Gurevich, V.V., Vishnivetskiy, S.A., Sigler, P.B., Schubert, C. Crystal structure of beta-arrestin at 1.9 Å: possible mechanism of receptor binding and membrane Translocation. *Structure* **2001**, *9*: 869-880.

[15] Hirsch, J.A., Schubert, C., Gurevich, V.V., Sigler, P.B. The 2.8 Å crystal structure of visual arrestin: a model for arrestin's regulation. *Cell* **1999**, *97*: 257-269.

[16] Milano, S.K., Kim, Y.M., Stefano, F.P., Benovic, J.L., Brenner, C. Nonvisual arrestin oligomerization and cellular localization are regulated by inositol hexakisphosphate binding. *J. Biol. Chem.* **2006**, *281*: 9812-9823.

[17] Milano, S.K., Pace, H.C., Kim, Y.M., Brenner, C., Benovic, J.L. Scaffolding functions of arrestin-2 revealed by crystal structure and mutagenesis. *Biochemistry* **2002**, *41*: 3321-3328.

[18] Sutton, R.B., Vishnivetskiy, S.A., Robert, J., Hanson, S.M., Raman, D., Knox, B.E., Kono, M., Navarro, J., Gurevich, V.V. Crystal structure of cone arrestin at 2.3 Å: evolution of receptor specificity. *J. Mol. Biol.* **2005**, *354*: 1069-1080.

[19] Granzin, J., Cousin, A., Weirauch, M., Schlesinger, R., Buldt, G., Batra-Safferling, R. Crystal structure of p44, a constitutively active splice variant of visual arrestin. *J. Mol. Biol.* **2012**, *416*: 611-618.

[20] Zhan, X., Gimenez, L.E., Gurevich, V.V., Spiller, B.W. Crystal structure of arrestin-3 reveals the basis of the difference in receptor binding between two non-visual subtypes. *J. Mol. Biol.* **2011**, *406*: 467-478.

[21] Campbell, I.D., Spitzfaden, C. Building proteins with fibronectin type III modules. *Structure* **1994**, *2*: 333-337.

[22] Dell'Angelica, E.C. Clathrin-binding proteins: got a motif? Join the network! *Trends Cell Biol.* **2001**, *11*: 315-318.

[23] Krupnick, J.G., Goodman, O.B., Jr., Keen, J.H., Benovic, J.L. Arrestin/clathrin interaction. Localization of the clathrin binding domain of nonvisual arrestins to the carboxy terminus. *J. Biol. Chem.* **1997**, *272*: 15011-15006.

[24] Burtey, A., Schmid, E.M., Ford, M.G., Rappoport, J.Z., Scott, M.G., Marullo, S., Simon, S.M., McMahon, H.T., Benmerah, A. The conserved isoleucine-valine-phenylalanine motif couples activation state and endocytic functions of beta-arrestins. *Traffic* **2007**, *8*: 914-931.

[25] Laporte, S.A., Oakley, R.H., Holt, J.A., Barak, L.S., Caron, M.G. The interaction of beta-arrestin with the AP-2 adaptor is required for the clustering of beta 2-adrenergic receptor into clathrin-coated pits. *J. Biol. Chem.* **2000**, *275*: 23120-23126.

[26] Bouschet, T., Martin, S., Kanamarlapudi, V., Mundell, S., Henley, J.M. The calcium-sensing receptor changes cell shape *via* a beta-arrestin-1 ARNO ARF6 ELMO protein network. *J. Cell Sci.* **2007**, *120*: 2489-2497.

[27] Claing, A., Chen, W., Miller, W.E., Vitale, N., Moss, J., Premont, R.T., Lefkowitz, R.J. beta-Arrestin-mediated ADP-ribosylation factor 6 activation and beta 2-adrenergic receptor endocytosis. *J. Biol. Chem.* **2001**, *276*: 42509-42513.

[28] Hunzicker-Dunn, M., Gurevich, V.V., Casanova, J.E., Mukherjee, S. ARF6: a newly appreciated player in G protein-coupled receptor desensitization. *FEBS Lett.* **2002**, *521*: 3-8.

[29] Shenoy, S.K., Lefkowitz, R.J. Multifaceted roles of beta-arrestins in the regulation of seven-membrane-spanning receptor trafficking and signalling. *Biochem. J.* **2003**, *375*: 503-515.

[30] Kiselev, A., Socolich, M., Vinos, J., Hardy, R.W., Zuker, C.S., Ranganathan, R. A molecular pathway for light-dependent photoreceptor apoptosis in *Drosophila*. *Neuron* **2000**, *28*: 139-152.

[31] Satoh, A.K., Ready, D.F. Arrestin1 mediates light-dependent rhodopsin endocytosis and cell survival. *Curr. Biol.* **2005**, *15*: 1722-1733.

[32] Kang, D.S., Kern, R.C., Puthenveedu, M.A., von Zastrow, M., Williams, J.C., Benovic, J.L. Structure of an arrestin2-clathrin complex reveals a novel clathrin binding domain that modulates receptor trafficking. *J. Biol. Chem.* **2009**, *284*: 29860-29872.

[33] Barthet, G., Carrat, G., Cassier, E., Barker, B., Gaven, F., Pillot, M., Framery, B., Pellissier, L.P., Augier, J., Kang, D.S., Claeysen, S., Reiter, E., Baneres, J.L., Benovic, J.L., Marin, P., Bockaert, J., Dumuis, A. Beta-arrestin1 phosphorylation by GRK5 regulates G protein-independent 5-HT4 receptor signalling. *EMBO J.* **2009**, *28*: 2706-2718.

[34] Kim, Y.M., Barak, L.S., Caron, M.G., Benovic, J.L. Regulation of arrestin-3 phosphorylation by casein kinase II. *J. Biol. Chem.* **2002**, *277*: 16837-16846.

[35] Lin, F.T., Chen, W., Shenoy, S., Cong, M., Exum, S.T., Lefkowitz, R.J. Phosphorylation of beta-arrestin2 regulates its function in internalization of beta(2)-adrenergic receptors. *Biochemistry* **2002**, *41*: 10692-10699.

[36] Lin, F.T., Krueger, K.M., Kendall, H.E., Daaka, Y., Fredericks, Z.L., Pitcher, J.A., Lefkowitz, R.J. Clathrin-mediated endocytosis of the beta-adrenergic receptor is regulated by phosphorylation/dephosphorylation of beta-arrestin1. *J. Biol. Chem.* **1997**, *272*: 31051-31057.

[37] Lin, F.T., Miller, W.E., Luttrell, L.M., Lefkowitz, R.J. Feedback regulation of beta-arrestin1 function by extracellular signal-regulated kinases. *J. Biol. Chem.* **1999**, *274*: 15971-15974.

[38] Shenoy, S.K., Modi, A.S., Shukla, A.K., Xiao, K., Berthouze, M., Ahn, S., Wilkinson, K.D., Miller, W.E., Lefkowitz, R.J. Beta-arrestin-dependent signaling and trafficking of 7-transmembrane receptors is reciprocally regulated by the deubiquitinase USP33 and the E3 ligase Mdm2. *Proc. Natl. Acad. Sci. USA* **2009**, *106*: 6650-6655.

[39] Shenoy, S.K., Lefkowitz, R.J. Receptor-specific ubiquitination of beta-arrestin directs assembly and targeting of seven-transmembrane receptor signalosomes. *J. Biol. Chem.* **2005**, *280*: 15315-15324.

[40] Wyatt, D., Malik, R., Vesecky, A.C., Marchese, A. Small ubiquitin-like modifier modification of arrestin-3 regulates receptor trafficking. *J. Biol. Chem.* **2011**, *286*: 3884-3893.

[41] Ozawa, K., Whalen, E.J., Nelson, C.D., Mu, Y., Hess, D.T., Lefkowitz, R.J., Stamler, J.S. S-nitrosylation of beta-arrestin regulates beta-adrenergic receptor trafficking. *Mol. Cell* **2008**, *31*: 395-405.

[42] Oakley, R.H., Laporte, S.A., Holt, J.A., Caron, M.G., Barak, L.S. Differential affinities of visual arrestin, beta arrestin1, and beta arrestin2 for G protein-coupled receptors delineate two major classes of receptors. *J. Biol. Chem.* **2000**, *275*: 17201-17210.

[43] Oakley, R.H., Laporte, S.A., Holt, J.A., Barak, L.S., Caron, M.G. Association of beta-arrestin with G protein-coupled receptors during clathrin-mediated endocytosis dictates the profile of receptor resensitization. *J. Biol. Chem.* **1999**, *274*: 32248-32257.

[44] Oakley, R.H., Laporte, S.A., Holt, J.A., Barak, L.S., Caron, M.G. Molecular determinants underlying the formation of stable intracellular G protein-coupled receptor-beta-arrestin complexes after receptor endocytosis. *J. Biol. Chem.* **2001**, *276*: 19452-19460.

[45] Gurevich, V.V., Benovic, J.L. Mechanism of phosphorylation-recognition by visual arrestin and the transition of arrestin into a high affinity binding state. *Mol. Pharmacol.* **1997**, *51*: 161-169.

[46] Vishnivetskiy, S.A., Paz, C.L., Schubert, C., Hirsch, J.A., Sigler, P.B., Gurevich, V.V. How does arrestin respond to the phosphorylated state of rhodopsin? *J. Biol. Chem.* **1999**, *274*: 11451-11454.

[47] Vishnivetskiy, S.A., Hosey, M.M., Benovic, J.L., Gurevich, V.V. Mapping the arrestin-receptor interface. Structural elements responsible for receptor specificity of arrestin proteins. *J. Biol. Chem.* **2004**, *279*: 1262-1268.

[48] Palczewski, K., Pulvermuller, A., Buczylko, J., Gutmann, C., Hofmann, K.P. Binding of inositol phosphates to arrestin. *FEBS Lett.* **1991**, *295*: 195-199.

[49] Hanson, S.M., Vishnivetskiy, S.A., Hubbell, W.L., Gurevich, V.V. Opposing effects of inositol hexakisphosphate on rod arrestin and arrestin2 self-association. *Biochemistry* **2008**, *47*: 1070-1075.

[50] Pittet, D., Schlegel, W., Lew, D.P., Monod, A., Mayr, G.W. Mass changes in inositol tetrakis- and pentakisphosphate isomers induced by chemotactic peptide stimulation in HL-60 cells. *J. Biol. Chem.* **1989**, *264*: 18489-18493.

[51] Gaidarov, I., Krupnick, J.G., Falck, J.R., Benovic, J.L., Keen, J.H. Arrestin function in G protein-coupled receptor endocytosis requires phosphoinositide binding. *EMBO J.* **1999**, *18*: 871-881.

[52] Hupfeld, C.J., Olefsky, J.M. Regulation of receptor tyrosine kinase signaling by GRKs and beta-arrestins. *Annu. Rev. Physiol.* **2007**, *69*: 561-577.

[53] Lefkowitz, R.J., Rajagopal, K., Whalen, E.J. New roles for beta-arrestins in cell signaling: not just for seven-transmembrane receptors. *Mol. Cell* **2006**, *24*: 643-652.

[54] Chen, W., Kirkbride, K.C., How, T., Nelson, C.D., Mo, J., Frederick, J.P., Wang, X.F., Lefkowitz, R.J., Blobe, G.C. Beta-arrestin 2 mediates endocytosis of type III TGF-beta receptor and down-regulation of its signaling. *Science* **2003**, *301*: 1394-1397.

[55] Lin, F.T., Daaka, Y., Lefkowitz, R.J. beta-arrestins regulate mitogenic signaling and clathrin-mediated endocytosis of the insulin-like growth factor I receptor. *J. Biol. Chem.* **1998**, *273*: 31640-31643.

[56] Szabo, E.Z., Numata, M., Lukashova, V., Iannuzzi, P., Orlowski, J. beta-Arrestins bind and decrease cell-surface abundance of the Na^+/H^+ exchanger NHE5 isoform. *Proc. Natl. Acad. Sci. USA* **2005**, *102*: 2790-2795.

[57] Wu, J.H., Peppel, K., Nelson, C.D., Lin, F.T., Kohout, T.A., Miller, W.E., Exum, S.T., Freedman, N.J. The adaptor protein beta-arrestin2 enhances endocytosis of the low density lipoprotein receptor. *J. Biol. Chem.* **2003**, *278*: 44238-44245.

[58] Lipsky, R., Potts, E.M., Tarzami, S.T., Puckerin, A.A., Stocks, J., Schecter, A.D., Sobie, E.A., Akar, F.G., Diverse-Pierluissi, M.A. beta-Adrenergic receptor activation induces internalization of cardiac Cav1.2 channel complexes through a beta-Arrestin 1-mediated pathway. *J. Biol. Chem.* **2008**, *283*: 17221-17226.

[59] Mukherjee, A., Veraksa, A., Bauer, A., Rosse, C., Camonis, J., Artavanis-Tsakonas, S. Regulation of Notch signalling by non-visual beta-arrestin. *Nat. Cell Biol.* **2005**, *7*: 1191-1201.

[60] Yu, M.C., Su, L.L., Zou, L., Liu, Y., Wu, N., Kong, L., Zhuang, Z.H., Sun, L., Liu, H.P., Hu, J.H., Li, D., Strominger, J.L., Zang, J.W., Pei, G., Ge, B.X. An essential function for beta-arrestin 2 in the inhibitory signaling of natural killer cells. *Nat. Immunol.* **2008**, *9*: 898-907.

[61] Bryceson, Y.T., Ljunggren, H.G. Arrestin NK cell cytotoxicity. *Nat. Immunol.* **2008**, *9*: 835-836.

[62] Lohse, M.J., Klenk, C. Blocking them all: beta-arrestins inhibit cellular signaling. *Mol. Cell* **2008**, *31*: 619-621.

[63] Mo, W., Zhang, L., Yang, G., Zhai, J., Hu, Z., Chen, Y., Chen, X., Hui, L., Huang, R., Hu, G. Nuclear beta-arrestin1 functions as a scaffold for the dephosphorylation of STAT1 and moderates the antiviral activity of IFN-gamma. *Mol. Cell* **2008**, *31*: 695-707.

[64] Scott, M.G., Le Rouzic, E., Perianin, A., Pierotti, V., Enslen, H., Benichou, S., Marullo, S., Benmerah, A. Differential nucleocytoplasmic shuttling of beta-arrestins. Characterization of a leucine-rich nuclear export signal in beta-arrestin2. *J. Biol. Chem.* **2002**, *277*: 37693-37701.

[65] Molla-Herman, A., Boularan, C., Ghossoub, R., Scott, M.G.H., Burtey, A., Zarka, M., Saunier, S., Concordet, J.P., Marullo, S., Benmerah, A. Targeting of ß-arrestin 2 to the centrosome and primary cilium: role in cell proliferation control. *PLoS One* **2008**, *3*: e3728.

[66] Chen, W., Ren, X.R., Nelson, C.D., Barak, L.S., Chen, J.K., Beachy, P.A., de Sauvage, F., Lefkowitz, R.J. Activity-dependent internalization of smoothened mediated by beta-arrestin 2 and GRK2. *Science* **2004**, *306*: 2257-2260.

[67] Ingham, P.W., Placzek, M. Orchestrating ontogenesis: variations on a theme by sonic hedgehog. *Nat. Rev. Genet.* **2006**, *7*: 841-850.

[68] Kovacs, J.J., Whalen, E.J., Liu, R., Xiao, K., Kim, J., Chen, M., Wang, J., Chen, W., Lefkowitz, R.J. Beta-arrestin-mediated localization of smoothened to the primary cilium. *Science* **2008**, *320*: 1777-1781.

[69] Rohatgi, R., Scott, M.P. Arrestin' movement in cilia. *Science* **2008**, *320*: 1726-1727.

[70] Collins, B.M., Norwood, S.J., Kerr, M.C., Mahony, D., Seaman, M.N., Teasdale, R.D., Owen, D.J. Structure of Vps26B and mapping of its interaction with the retromer protein complex. *Traffic* **2008**, *9*: 366-379.

[71] Shi, H., Rojas, R., Bonifacino, J.S., Hurley, J.H. The retromer subunit Vps26 has an arrestin fold and binds Vps35 through its C-terminal domain. *Nat. Struct. Mol. Biol.* **2006**, *13*: 540-548.

[72] Bonifacino, J.S., Hurley, J.H. Retromer. *Curr. Opin. Cell Biol.* **2008**, *20*: 1-10.

[73] Collins, B.M. The structure and function of the retromer protein complex. *Traffic* **2008**, *9*: 1811-1822.

[74] Hierro, A., Rojas, A.L., Rojas, R., Murthy, N., Effantin, G., Kajava, A.V., Steven, A.C., Bonifacino, J.S., Hurley, J.H. Functional architecture of the retromer cargo-recognition complex. *Nature* **2007**, *449*: 1063-1067.

[75] Seaman, M.N. Identification of a novel conserved sorting motif required for retromer-mediated endosome-to-TGN retrieval. *J. Cell Sci.* **2007**, *120*: 2378-2389.

[76] Fjorback, A.W., Seaman, M., Gustafsen, C., Mehmedbasic, A., Gokool, S., Wu, C., Militz, D., Schmidt, V., Madsen, P., Nyengaard, J.R., Willnow, T.E., Christensen, E.I., Mobley, W.B., Nykjaer, A., Andersen, O.M. Retromer binds the FANSHY sorting motif in SorLA

to regulate amyloid precursor protein sorting and processing. *J. Neurosci.* **2012**, *32*: 1467-1480.

[77] Toyoda, A., Noguchi, H., Taylor, T.D., Ito, T., Pletcher, M.T., Sakaki, Y., Reeves, R.H., Hattori, M. Comparative genomic sequence analysis of the human chromosome 21 Down syndrome critical region. *Genome Res.* **2002**, *12*: 1323-1332.

[78] Hu, Y.H., Warnatz, H.J., Vanhecke, D., Wagner, F., Fiebitz, A., Thamm, S., Kahlem, P., Lehrach, H., Yaspo, M.L., Janitz, M. Cell array-based intracellular localization screening reveals novel functional features of human chromosome 21 proteins. *BMC Genom.* **2006**, *7*: 155.

[79] Pellegrini-Calace, M., Tramontano, A. Identification of a novel putative mitogen-activated kinase cascade on human chromosome 21 by computational approaches. *Bioinformatics* **2006**, *22*: 775-778.

[80] Chen, K.S., DeLuca, H.F. Isolation and characterization of a novel cDNA from HL-60 cells treated with 1,25-dihydroxyvitamin D-3. *Biochim. Biophys. Acta* **1994**, *1219*: 26-32.

[81] Nishiyama, A., Matsui, M., Iwata, S., Hirota, K., Masutani, H., Nakamura, H., Takagi, Y., Sono, H., Gon, Y., Yodoi, J. Identification of thioredoxin-binding protein-2/vitamin D(3) up-regulated protein 1 as a negative regulator of thioredoxin function and expression. *J. Biol. Chem.* **1999**, *274*: 21645-21650.

[82] Yamanaka, H., Maehira, F., Oshiro, M., Asato, T., Yanagawa, Y., Takei, H., Nakashima, Y. A possible interaction of thioredoxin with VDUP1 in HeLa cells detected in a yeast two-hybrid system. *Biochem. Biophys. Res. Commun.* **2000**, *271*: 796-800.

[83] Junn, E., Han, S.H., Im, J.Y., Yang, Y., Cho, E.W., Um, H.D., Kim, D.K., Lee, K.W., Han, P.L., Rhee, S.G., Choi, I. Vitamin D3 up-regulated protein 1 mediates oxidative stress *via* suppressing the thioredoxin function. *J. Immunol.* **2000**, *164*: 6287-6295.

[84] Schulze, P.C., De Keulenaer, G.W., Yoshioka, J., Kassik, K.A., Lee, R.T. Vitamin D3-upregulated protein-1 (VDUP-1) regulates redox-dependent vascular smooth muscle cell proliferation through interaction with thioredoxin. *Circ. Res.* **2002**, *91*: 689-695.

[85] Billiet, L., Furman, C., Larigauderie, G., Copin, C., Page, S., Fruchart, J.C., Brand, K., Rouis, M. Enhanced VDUP-1 gene expression by PPARgamma agonist induces apoptosis in human macrophage. *J. Cell Physiol.* **2008**, *214*: 183-191.

[86] Nichols, C.D., Sanders-Bush, E. Molecular genetic responses to lysergic acid diethylamide include transcriptional activation of MAP kinase phosphatase-1, C/EBP-beta and ILAD-1, a novel gene with homology to arrestins. *J. Neurochem.* **2004**, *90*: 576-584.

[87] Draheim, K.M., Chen, H.B., Tao, Q., Moore, N., Roche, M., Lyle, S. ARRDC3 suppresses breast cancer progression by negatively regulating integrin beta4. *Oncogene* **2010**, *29*: 5032-5047.

[88] Nabhan, J.F., Pan, H., Lu, Q. Arrestin domain-containing protein 3 recruits the NEDD4 E3 ligase to mediate ubiquitination of the beta2-adrenergic receptor. *EMBO Rep.* **2010**, *11*: 605-611.

[89] Oka, S., Masutani, H., Liu, W., Horita, H., Wang, D., Kizaka-Kondoh, S., Yodoi, J. Thioredoxin-binding protein-2-like inducible membrane protein is a novel vitamin D3 and peroxisome proliferator-activated receptor (PPAR)gamma ligand target protein that regulates PPARgamma signaling. *Endocrinology* **2006**, *147*: 733-743.

[90] Rauch, S., Martin-Serrano, J. Multiple interactions between the ESCRT machinery and arrestin-related proteins: implications for PPXY-dependent budding. *J. Virol.* **2011**, *85*: 3546-3556.

[91] Nabhan, J.F., Hu, R., Oh, R.S., Cohen, S.N., Lu, Q. Formation and release of arrestin domain-containing protein 1-mediated microvesicles (ARMMs) at plasma membrane by recruitment of TSG101 protein. *Proc. Natl. Acad. Sci. USA* **2012**, *109*: 4146-4151.

[92] Hyde, D.R., Mecklenburg, K.L., Pollock, J.A., Vihtelic, T.S., Benzer, S. Twenty *Drosophila* visual system cDNA clones: one is a homolog of human arrestin. *Proc. Natl. Acad. Sci. USA* **1990**, *87*: 1008-1012.

[93] LeVine, H., 3rd, Smith, D.P., Whitney, M., Malicki, D.M., Dolph, P.J., Smith, G.F., Burkhart, W., Zuker, C.S. Isolation of a novel visual-system-specific arrestin: an *in vivo* substrate for light-dependent phosphorylation. *Mech. Dev.* **1990**, *33*: 19-25.

[94] Smith, D.P., Shieh, B.H., Zuker, C.S. Isolation and structure of an arrestin gene from *Drosophila*. *Proc. Natl. Acad. Sci. USA* **1990**, *87*: 1003-1007.

[95] Roman, G., He, J., Davis, R.L. kurtz, a novel nonvisual arrestin, is an essential neural gene in *Drosophila*. *Genetics* **2000**, *155*: 1281-1295.

[96] Johnson, E.C., Tift, F.W., McCauley, A., Liu, L., Roman, G. Functional characterization of kurtz, a *Drosophila* non-visual arrestin, reveals conservation of GPCR desensitization mechanisms. *Insect Biochem. Mol. Biol.* **2008**, *38*: 1016-1022.

[97] Shenoy, S.K., Lefkowitz, R.J. Receptor regulation: beta-arrestin moves up a notch. *Nat. Cell Biol.* **2005**, *7*: 1159-1161.

[98] Verges, M. Retromer and sorting nexins in development. *Front. Biosci.* **2007**, *12*: 3825-3851.

[99] Belenkaya, T.Y., Wu, Y., Tang, X., Zhou, B., Cheng, L., Sharma, Y.V., Yan, D., Selva, E.M., Lin, X. The retromer complex influences Wnt secretion by recycling wntless from endosomes to the trans-Golgi network. *Dev. Cell* **2008**, *14*: 120-131.

[100] Port, F., Kuster, M., Herr, P., Furger, E., Banziger, C., Hausmann, G., Basler, K. Wingless secretion promotes and requires retromer-dependent cycling of Wntless. *Nat. Cell Biol.* **2008**, *10*: 178-185.

[101] Franch-Marro, X., Wendler, F., Guidato, S., Griffith, J., Baena-Lopez, A., Itasaki, N., Maurice, M.M., Vincent, J.P. Wingless secretion requires endosome-to-Golgi retrieval of Wntless/Evi/Sprinter by the retromer complex. *Nat. Cell Biol.* **2008**, *10*: 170-177.

[102] Fukuto, H.S., Ferkey, D.M., Apicella, A.J., Lans, H., Sharmeen, T., Chen, W., Lefkowitz, R.J., Jansen, G., Schafer, W.R., Hart, A.C. G protein-coupled receptor kinase function is essential for chemosensation in C. elegans. *Neuron* **2004**, *42*: 581-593.

[103] Palmitessa, A., Hess, H.A., Bany, I.A., Kim, Y.M., Koelle, M.R., Benovic, J.L. *Caenorhabditis elegans* arrestin regulates neural G protein signaling and olfactory adaptation and recovery. *J. Biol. Chem.* **2005**, *280*: 24649-24662.

[104] Coudreuse, D.Y., Roel, G., Betist, M.C., Destree, O., Korswagen, H.C. Wnt gradient formation requires retromer function in Wnt-producing cells. *Science* **2006**, *312*: 921-924.

[105] Prasad, B.C., Clark, S.G. Wnt signaling establishes anteroposterior neuronal polarity and requires retromer in *C. elegans*. *Development* **2006**, *133*: 1757-1766.

[106] Lin, C.H., Macgurn, J.A., Chu, T., Stefan, C.J., Emr, S.D. Arrestin-Related Ubiquitin-Ligase Adaptors Regulate Endocytosis and Protein Turnover at the Cell Surface. *Cell* **2008**, *135*: 714-725.

[107] Andoh, T., Hirata, Y., Kikuchi, A. PY motifs of Rod1 are required for binding to Rsp5 and for drug resistance. *FEBS Lett.* **2002**, *525*: 131-134.

[108] Gupta, R., Kus, B., Fladd, C., Wasmuth, J., Tonikian, R., Sidhu, S., Krogan, N.J., Parkinson, J., Rotin, D. Ubiquitination screen using protein microarrays for comprehensive identification of Rsp5 substrates in yeast. *Mol. Syst. Biol.* **2007**, *3*: 116.

[109] Herranz, S., Rodriguez, J.M., Bussink, H.J., Sanchez-Ferrero, J.C., Arst, H.N., Jr., Penalva, M.A., Vincent, O. Arrestin-related proteins mediate pH signaling in fungi. *Proc. Natl. Acad. Sci. USA* **2005**, *102*: 12141-12146.
[110] Huh, W.K., Falvo, J.V., Gerke, L.C., Carroll, A.S., Howson, R.W., Weissman, J.S., O'Shea, E.K. Global analysis of protein localization in budding yeast. *Nature* **2003**, *425*: 686-691.
[111] Kumar, A., Agarwal, S., Heyman, J.A., Matson, S., Heidtman, M., Piccirillo, S., Umansky, L., Drawid, A., Jansen, R., Liu, Y., Cheung, K.H., Miller, P., Gerstein, M., Roeder, G.S., Snyder, M. Subcellular localization of the yeast proteome. *Genes Dev.* **2002**, *16*: 707-719.
[112] Nikko, E., Sullivan, J.A., Pelham, H.R. Arrestin-like proteins mediate ubiquitination and endocytosis of the yeast metal transporter Smf1. *EMBO Rep.* **2008**, *9*: 1216-1221.
[113] Shinoda, J., Kikuchi, Y. Rod1, an arrestin-related protein, is phosphorylated by Snf1-kinase in *Saccharomyces cerevisiae*. *Biochem. Biophys. Res. Commun.* **2007**, *364*: 258-263.
[114] Wu, A.L., Hallstrom, T.C., Moye-Rowley, W.S. ROD1, a novel gene conferring multiple resistance phenotypes in *Saccharomyces cerevisiae*. *J. Biol. Chem.* **1996**, *271*: 2914-2920.
[115] Alvarez, C.E. On the origins of arrestin and rhodopsin. *BMC Evol.Biol.* **2008**, *8*: 222.
[116] Becuwe, M., Vieira, N., Lara, D., Gomes-Rezende, J., Soares-Cunha, C., Casal, M., Haguenauer-Tsapis, R., Vincent, O., Paiva, S., Leon, S. A molecular switch on an arrestin-like protein relays glucose signaling to transporter endocytosis. *J. Cell Biol.* **2012**, *196*: 247-259.
[117] Herrador, A., Herranz, S., Lara, D., Vincent, O. Recruitment of the ESCRT machinery to a putative seven-transmembrane-domain receptor is mediated by an arrestin-related protein. *Mol. Cell. Biol.* **2010**, *30*: 897-907.
[118] Ubersax, J.A., Woodbury, E.L., Quang, P.N., Paraz, M., Blethrow, J.D., Shah, K., Shokat, K.M., Morgan, D.O. Targets of the cyclin-dependent kinase Cdk1. *Nature* **2003**, *425*: 859-864.
[119] Seaman, M.N., McCaffery, J.M., Emr, S.D. A membrane coat complex essential for endosome-to-Golgi retrograde transport in yeast. *J. Cell Biol.* **1998**, *142*: 665-681.
[120] Reddy, J.V., Seaman, M.N. Vps26p, a component of retromer, directs the interactions of Vps35p in endosome-to-Golgi retrieval. *Mol. Biol. Cell* **2001**, *12*: 3242-3256.
[121] Bachhawat, A.K., Suhan, J., Jones, E.W. The yeast homolog of Hß58, a mouse gene essential for embryogenesis, performs a role in the delivery of proteins to the vacuole. *Genes Dev.* **1994**, *8*: 1379-1387.
[122] Kama, R., Robinson, M., Gerst, J.E. Btn2, a Hook1 ortholog and potential Batten disease-related protein, mediates late endosome-Golgi protein sorting in yeast. *Mol. Cell Biol.* **2007**, *27*: 605-621.
[123] Tevzadze, G.G., Pierce, J.V., Esposito, R.E. Genetic evidence for a SPO1-dependent signaling pathway controlling meiotic progression in yeast. *Genetics* **2007**, *175*: 1213-1227.
[124] Giaever, G., Chu, A.M., Ni, L., Connelly, C., Riles, L., Veronneau, S., Dow, S., Lucau-Danila, A., Anderson, K., Andre, B., Arkin, A.P., Astromoff, A., El-Bakkoury, M., Bangham, R., Benito, R., Brachat, S., Campanaro, S., Curtiss, M., Davis, K., Deutschbauer, A., Entian, K.D., Flaherty, P., Foury, F., Garfinkel, D.J., Gerstein, M., Gotte, D., Guldener, U., Hegemann, J.H., Hempel, S., Herman, Z., Jaramillo, D.F., Kelly, D.E., Kelly, S.L., Kotter, P., LaBonte, D., Lamb, D.C., Lan, N., Liang, H., Liao, H., Liu, L., Luo, C., Lussier, M., Mao, R., Menard, P., Ooi, S.L., Revuelta, J.L., Roberts, C.J., Rose, M., Ross-Macdonald, P., Scherens, B., Schimmack, G., Shafer, B., Shoemaker, D.D., Sookhai-Mahadeo, S., Storms, R.K., Strathern, J.N., Valle, G., Voet, M., Volckaert, G., Wang, C.Y., Ward, T.R., Wilhelmy, J., Winzeler, E.A., Yang, Y., Yen, G., Youngman, E., Yu, K.,

Bussey, H., Boeke, J.D., Snyder, M., Philippsen, P., Davis, R.W., Johnston, M. Functional profiling of the *Saccharomyces cerevisiae* genome. *Nature* **2002**, *418*: 387-391.

[125] Journet, A., Klein, G., Brugière, S., Vandenbrouck, Y., Chapel, A., Kieffer, S., Bruley, C., Masselon, C., Aubry, L. Investigating the macropinocytic proteome of *Dictyostelium* amoebae by high-resolution mass spectrometry *Proteomics* **2012**, *12*: 241-245.

[126] Guetta, D., Langou, K., Grunwald, D., Klein, G., Aubry, L. FYVE-dependent endosomal targeting of an arrestin-related protein in amoeba. *PLoS One* **2010**, *5*: e15249.

[127] Reinders, Y., Schulz, I., Graf, R., Sickmann, A. Identification of novel centrosomal proteins in *Dictyostelium discoideum* by comparative proteomic approaches. *J. Proteome Res.* **2006**, *5*: 589-598.

[128] Chenna, R., Sugawara, H., Koike, T., Lopez, R., Gibson, T.J., Higgins, D.G., Thompson, J.D. Multiple sequence alignment with the Clustal series of programs. *Nucleic Acids Res.* **2003**, *31*: 3497-3500.

[129] Dereeper, A., Guignon, V., Blanc, G., Audic, S., Buffet, S., Chevenet, F., Dufayard, J.F., Guindon, S., Lefort, V., Lescot, M., Claverie, J.M., Gascuel, O. Phylogeny.fr: robust phylogenetic analysis for the non-specialist. *Nucleic Acids Res.* **2008**, *36*: W465-W469.

[130] Chevenet, F., Brun, C., Banuls, A.L., Jacq, B., Christen, R. TreeDyn: towards dynamic graphics and annotations for analyses of trees. *BMC Bioinformatics* **2006**, *7*: 439.

Send Orders of Reprints at reprints@benthamscince.net
Advances in Genome Science, Vol. 1, 2013, 377-425

CHAPTER 15

Role of Calcium Regulated Kinases: Expressional and Functional Analysis in Abiotic Stress Signaling

Ritika Das[1], Amita Pandey[2] and Girdhar K. Pandey[2,*]

[1]Department of Biology, University of Rochester, Rochester, New York 14627, USA and [2]Department of Plant Molecular Biology, University of Delhi South Campus, Benito Juarez Road, Dhaula Kuan, New Delhi-110021, India

Abstract: Perception of stimuli and activation of a signaling cascade is an intrinsic characteristic feature of all living organisms. Till date, several signaling pathways have been elucidated that are involved in multiple facets of growth and development of an organism. Exposure to unfavorable stimuli or stress condition activates different signaling cascades in both plants and animal. Being sessile, plants cannot move away from an unfavorable condition, and hence activate the molecular machinery to cope up or adjust against that particular stress condition. In plants, role of calcium as second messenger has been studied in detail in both abiotic and biotic stress signaling. Several calcium sensor proteins such as calmodulin (CaM), calcium dependent protein kinases (CDPK) and calcinuerin B-like (CBL) protein were discovered to play a crucial role in abiotic stress signaling in plants. Unlike CDPK, CBL and CaM are calcium-binding proteins, which do not have any protein kinase enzyme activity and interact with a target protein kinase termed as CBL-interacting protein kinase (CIPK) and CaM kinases respectively. Genome sequence analysis of Arabidopsis and rice has led to the identification of multigene families of these calcium signaling protein kinases. Individual and global gene expression analysis of these protein kinase family members has been analyzed under several developmental and different abiotic stress conditions. Here, we are making an attempt to overview and emphasize the expressional analysis of calcium signaling protein kinases under different abiotic stress and developmental stages, and linking the expression to possible function for these kinases.

Keywords: Calcium, abiotic stress, kinase, CBL, CIPK, CDPK, CaM, signal transduction.

INTRODUCTION

Plants are constantly challenged in an environment by several stimuli and factors, which guide in regulating physiological and developmental programs. Plants need

[*]**Address correspondence to Girdhar K. Pandey:** Department of Plant Molecular Biology, University of Delhi South Campus, Benito Juarez Road, Dhaula Kuan, New Delhi-110021, India; Tel: +91-11-24113106; Ext. 387; Fax: +91-11-24111208; E-mail: gkpandey@south.du.ac.in

Christian Neri (Ed)

to sense and process the environmental stimuli such as light, temperature, water status, and nutrient availability, in order to grow and develop normally. But under adverse environmental condition, several signals restrict the normal growth and development. There are two major classes of stresses, namely, abiotic and biotic, which can hinder and threaten plant viability.

Agronomically, environmental stresses are the major cause of loss of crop productivity worldwide and pose a basic threat to the plant ecosystem as a whole. Therefore, it is of immense importance to understand the molecular and biochemical basis of how plants perceive and combat environmental stresses. In response to these adverse signals, plants activate their cellular machinery comprising of a plethora of proteins, which are involved in perception, signal transduction and respond to these stimuli that help plants to adjust or adapt to these adverse conditions, and which ultimately leads to tolerance or survival of the plants. Signal transduction pathway comprises of multiple arrays of proteins, especially protein kinases, which are involved in the phosphorylation-mediated pathway, and play a crucial role for plants to tackle the environmental stresses they encounter.

Calcium serves as ubiquitous and a central hub in a large number of signaling pathways. Multiple extracellular signals such as light, hormones, biotic and abiotic stimuli elicit changes in calcium levels in the cell [1-4]. Biologists were fascinated at how several different stimuli or stress signal leads to changes in the calcium concentration, and how specificity in signaling network is being maintained. A number of studies in animal and plants have already proven the fact that calcium signal is not only represented as changes in intracellular concentration but also as spatial and temporal information that is also encoded along with [5-7]. In addition, a terminology called "Calcium Signature" has been coined for these spatial and temporal changes in calcium concentration, which generates specificity in calcium-mediated signaling pathway [5-7]. These specific calcium changes include calcium oscillations or spiking with features such as frequency, amplitude and shape that are specified by nature of the stimulus [8]. Multiple organelles such as endoplasmic reticulum (ER), vacuole, apoplast, mitochondria and peroxisome have been reported to be not only regulated by calcium but also themselves controlling calcium signaling in the cell [9].

Alterations in the level of calcium is an event that triggers a whole range of signal transduction pathways *via* different calcium sensors that are either induced or are already present in the cell. As stated above, calcium has been implicated in vast number of physiological processes such as symbiotic interaction, stomatal closure, pollen tube and root hair growth involving various sensor and responder proteins [8]. All the eukaryotic cells have a multiple number of calcium sensors, which are responsible for detecting the changes in calcium concentration. Therefore, a very precise level of regulation is being mediated by these calcium binding proteins, which act as sensors of calcium [10-14]. In plants, calcium sensor proteins are categorized as calcium sensor responder and sensor relay [12]. The calcium relay protein bind calcium and affect their target protein since they themselves do not have enzymatic activity, and typical example of these are calmodulin (CaM) and calcineurin B-like (CBL) protein [9]. In contrast, the sensor responder proteins bind calcium and a change in conformation takes place, and hence modulates their own activity by intra-molecular interaction [15]. Calcium dependent protein kinases (CDPKs) are the best characterized protein family, which has calmodulin-like calcium binding domain and a Ser/Thr protein kinase domain in a single protein. Upon binding calcium, they are directly activated and transduce the signal *via* phosphorylation cascades and regulation of gene expression [1, 11, 16-18]. CaM and calmodulin-like protein have been best characterized as a calcium sensor relay, which do not have enzymatic activity, and hence activate or deactivate their interacting proteins [11]. Calcium levels have also been shown to regulate gene expression *via* CaM and CaM like proteins (CML) [19]. Protein chip studies identified a host of CaM binding proteins, many of which were transcription factors belonging to the CAMTA (Calmodulin binding transcription activators) family that regulate gene expression in response to both abiotic and biotic stresses. Direct activation of light inducible genes by CaM7 provides an example where calcium directly triggers changes in gene expression *via* a calcium sensor that binds to DNA elements [20]. A new family of calcium sensors from Arabidopsis has been identified, which is similar to calcineurin B-subunit and neuronal calcium sensor from animals [11, 21, 22]. These plant calcium sensors were referred as calcineurin B-like (CBL) proteins [22]. Like calmodulin, CBL proteins are calcium sensor relay, which upon binding calcium undergo changes in conformation and activate their target proteins. CBL protein interacts with a novel SNF1-like protein kinase family called CIPK [11,14, 23-26]. Besides these well characterized kinases involved in signal transduction pathways, there are several other calcium regulated kinases such as Ca^{2+}/CaM kinase (CCaMK) and CDPK-related protein kinase (CRK) that are responsive to stress, and also exhibit development specific expression pattern [27]. However, their exact nature and role in stress signaling requires detailed investigation.

Besides their well-characterized role in diverse abiotic stress response, calcium signaling and activated protein kinases have also been found to be important in innate immunity. Increasing evidence suggests the role of CDPK, CBL and CaM in defense response against microbial pathogens and herbivore attack by promoting the synthesis of enzymes and molecules necessary for biotic stress resistance [28].

Here, we are attempting to focus on expressional analysis of these calcium-signaling kinases emphasizing on the functional regulation by expression profiling, combined with functional analysis of these different calcium-signaling kinases under diverse stress conditions, development and stage specific studies will be overviewed. Moreover, the expression data together with cellular localization and protein-protein interaction will also be touched upon for some of these kinases in activating signaling cascades upon sensing environmental stress.

CALCIUM DEPENDENT PROTEIN KINASES (CDPKS) AND ABIOTIC STRESS

Calcium dependent protein kinases or CDPKs that are the most extensively studied calcium signaling kinases act in a manner which is independent of upstream sensors since they directly bind and sense calcium ions to activate the downstream signaling in response to different stress and development cues [16,17, 29]. CDPKs have been found in a wide range of plant species and in some of the protists [17]. Other eukaryotic genomes such as that of yeast and other animals have not shown the presence of CDPKs [16]. CDPKs are predicted to be having molecular weight between 54.3 kD to 72.2 kD with the differences in variable domains mainly contributing to this size difference [16]. N-terminal domain possesses potential myristoylation (Glycine residues) and palmitoylation sites

(Cysteine residues), which might be important for membrane association as demonstrated by various studies [30, 31]. Such membrane associations are important for sensing the changes in calcium fluxes, which occur mostly at these membrane sites. An adjacent junction domain follows the kinase domain, which is near to the N-terminal, is inhibitory in function. Next to this domain lies the calmodulin-like, calcium-binding domain which comprises of four EF-hands responsible for binding to the calcium ions. The kinase activity of the CDPKs has been confirmed not only by the identification of physiological substrates that are shown to be phosphorylated by CDPKs [32], but also using cell free systems where synthetic peptides used as substrates were shown to be phosphorylated in presence of calcium by CDPKs translated from cloned CDPKs using wheat germ extract [33]. Of the twenty six Arabidopsis CDPKs studied, seven were found to be showing both autophosphorylation as well as phosphorylation of the biotinylated peptide at the threonine residue. In addition to calcium ions, phosphorylation, membrane phospholipids and 14-3-3 proteins are also found to be positively regulating the activity of different CDPK proteins [34]. Besides showing up regulation under different stress conditions, significant numbers of reports have also indicated the role of CDPKs in hormonal response, regulation by light and pathogen attack [35-37]. With respect to tissue specific expression, it was seen that most CDPKs are expressed in almost all tissues [29]. However, some CDPK isoforms such as OsCDPK2 and OsCDPK11 have been found to be showing differential expression pattern under specific stimulus and stage of plant development [38]. Various enzymes and proteins have been identified as putative CDPK substrates utilizing *in vitro* phosphorylation experiments. This includes a broad range of substrates, some of which include enzymes involved in carbon and nitrogen metabolism, ion channels and transporters, ethylene biosynthesis pathway and component of phospholipid metabolism [34]. Some of these substrates have been directly linked to the abiotic stress response-signaling pathway mediated by the CDPKs [39].

CDPKs in Arabidopsis

Genome wide analysis of CDPKs in Arabidopsis has revealed the presence of 34 CDPKs. Many of these CDPKs are found to be showing differential pattern of expression under various developmental and abiotic stress conditions. Some of the

CDPKs have been shown to be important in Abscisic acid (ABA) signaling which indicates their function in bringing about ABA-mediated abiotic stress response. AtCPK10 and AtCPK30 are known to activate ABA inducible and stress related promoter [40]. HVA1 promoter (a barley stress promoter) used in the study was fused to green fluorescent protein (GFP) and the effect of constitutively active CDPKs was monitored in the transfected maize protoplasts. GFP expression indicating activation of this stress promoter even in the absence of stress signals provided evidence for the important role of CDPK activation in abiotic stress signaling cascade. Further, a deletion in the kinase domain in one of the two CDPKs, abolished GFP expression showing that the protein kinase domain is essential for stress promoter activation. *AtCPK10* and *AtCPK11* reported to be induced under drought and salt stress conditions indicating their possible role in osmotic stress signaling [41]. ABA is a stress specific phytohormone that is formed when plants encounter various environmental stress conditions and mediates signaling pathway, which helps generate adaptive responses to encounter such adverse conditions. Significant evidence has accumulated, which shows the involvement of multiple CDPKs as positive regulators of ABA mediated signal transduction pathway under abiotic stress conditions. AtCPK32 regulates ABA mediated seed germination [42] while AtCPK3 and AtCPK6 were reported to be controlling ABA mediated closure of stomata [43]. *CPK3* and *CPK6* double mutant, *cpk3ckp6* in Arabidopsis, was found to be impaired in activity of S (slow)-type anion channel in guard cells, which results only in partial closure of stomata in response to ABA [43]. The partial closure was due to partly overlapping functions of S-type and R-type of anion channels in the guard cell where R or rapid ion channel functions in a Ca^{2+}-independent fashion.

ABF4, which is a transcriptional regulator of ABA responsive gene was phosphorylated at Ser^{110} by AtCPK32, AtCPK10 and AtCPK30 (CDPK subgroup III). ABF4 is a basic leucine zipper (bZIP) type transcription factor that interacts with the kinase domain of AtCPK32, and this domain although necessary for interaction with ABF4, is not sufficient alone since this domain as well as the EF-hand domain is important for complete interaction [32]. Using deletion constructs and yeast two-hybrid studies, a conserved domain within the ABF4, namely, C2-C3 domain was determined to be the region responsible for interaction with AtCPK32. This again points towards an example where a CDPK might bring about the abiotic stress response by positively modulating its substrate (in this case a transcription factor). Overexpression of *AtCPK32* confers hypersensitive response to ABA and salt stress conditions. Also, the expression of certain ABA responsive genes such as *RAB18*, *RD29A* and *RD29B* was up regulated in the overexpression lines, indicating the function of AtCPK32 as a positive regulator of ABA mediated stress-signaling pathway. Two more CDPKs from Arabidopsis, that include, CPK4 and CPK11, have also been shown to be positively regulating ABA signal transduction pathway as concluded from both mutant as well as overexpression studies [43]. *cpk4* and *cpk11* loss of function mutants displayed an ABA insensitive phenotype with respect to seed germination, seedling growth and root growth. Also, these mutants showed reduced sensitivity to ABA in terms of stomatal closure and had bigger stomatal aperture upon ABA treatment in comparison to the wild type. An opposite phenotype could be seen in the overexpression lines that showed increased sensitivity to ABA in terms of stomatal closure, seed germination, seedling and root growth. A similar contrasting pattern could be observed upon performing real time PCR to determine the expression of various ABA responsive genes in both mutant as well as the overexpression lines. Expression of several ABA responsive genes such as ABF1, ABF2, ABF4, ABA insensitive 4 (ABI4), and ABI5 was significantly down regulated in the mutant lines while considerable up regulation could be seen for these genes in the overexpression transgenic lines. *In gel* phosphorylation assays demonstrated ABF1 and ABF4 to be phosphorylated by immunoprecipitated CPK4 and CPK11, which was reduced to a great extent in the *cpk4* and *cpk11* mutant lines. The molecular genetic evidence combined with biochemical approach employed in this study lends strong support to the role of CPK4 and CPK11 in ABA mediated signal transduction. As observed by the authors, AtCPK4 and AtCPK11 are localized in both cytoplasm and the nucleus, they could be participating in initial or early as well as delayed phase of stress response. Sensing of calcium fluxes brought about by stress stimuli, and phosphorylation of targets already present in the cell controlling responses such as stomatal closure could form an early response while phosphorylation of transcription factors at the nucleus affecting gene expression would form a later or delayed part of the response. In contrast to all these studies, which show the role of CDPKs as positive regulators of stress signaling mediated by ABA, one of the

CDPK gene, namely, *AtCPK23* is believed to be serving as a negative regulator of abiotic stress signaling as observed from the *cpk23* mutant, which has improved tolerance to drought and salt stress [44]. However, whether this negative regulation of abiotic stress response pathway is *via* negative control of *AtCPK23* over ABA signaling pathway is not known.

Besides their role in ABA mediated signal transduction under stress conditions, the CDPKs could also function to relay ABA independent stress signaling as observed for AtCPK11 interacting with a zinc finger protein, namely, AtDi19-1 that was identified in a yeast two-hybrid screen. *AtDi19-1* (drought induced) belongs to a novel stress regulated gene family (*AtDi19*) and encodes a zinc finger protein with nuclear localization signal/nuclear export signal (NLS/NES) sequences [45]. AtDi19 gene family members are believed to be participating in an ABA independent pathway during drought, salt and light signaling pathways as none of the members were induced upon ABA treatment. NLS/NES domain of AtDi19-1 is sufficient for both interactions with AtCPK11 as well as phosphorylation by AtCIPK11. Both NLS/NES and the C2H2 zinc finger domains are present in the central region of AtDi19. The NLS/NES domain could be important for shuttling both proteins between the nucleus and the cytoplasm during stress signaling.

Control over calcium levels in the cell forms an important means of regulation of calcium mediated signal transduction pathways, and a report on AtCPK1 shows how calcium sensing CDPK could possibly play an important role in controlling calcium ions during signaling cascades [46]. CPK1 isoform in Arabidopsis was found to be inhibiting type IIB Ca^{2+}-ATPase channel, which is a calmodulin-stimulated calcium pump (ACA2) (involved in calcium efflux from cytoplasm) present in the endoplasmic reticulum by phosphorylation at Ser^{45}. This inhibition not only blocks the basal activity of this channel but also prevents its activation by calmodulin [46]. Here, crosstalk between two opposing pathways (by CaM and CDPK) was perhaps regulating the calcium levels and oscillations in the cell that might be crucial in regulating some specific responses to environmental stress stimuli.

Sub-cellular targeting of different CDPKs has provided insight into the possible functions specific to different isoforms of CDPK. CDPKs were localized at multiple cellular sites as demonstrated in different studies. GFP fusion constructs of nine AtCPK isoforms targeted in Arabidopsis roots has shown AtCPK3 and AtCPK4 to be having both cytoplasmic and nuclear distribution while AtCPK1, 7, 8, 9, 16, 21, and 28 were found to be associated with the membrane [47]. Of these, AtCPK1 was localized to the peroxisome where it was believed to be perhaps involved in oxidative stress and lipid metabolism. Besides AtCPK1, the other membrane anchored CDPK isoforms were found to be present in the plasma membrane.

AtCPK2 that has high similarity to AtCPK1, was also shown to be localized in the endoplasmic reticulum (ER), and is believed to be having a similar potential substrate as AtCPK1, namely, ACA2, which is also localized at ER [48]. AtCPK21 has been found to be activating SLAC1 (Slow type anion channel 1), which control stomatal closure in response to CO_2, ABA, calcium, ozone, light and humidity [49]. The mutant of *slac1* displayed defects in slow type anion current and stomatal closure upon ABA treatment. Split-YFP assay using SLAC1 as a bait identified CPK21 as one of the interacting partners and this CPK21 mediated activation of SLAC1 was found to be calcium sensitive suggesting that ABA activation of calcium dependent pathway promote closure of stomata *via* CPK21-SLAC1.

CDPKs in Rice

A large multigene family has also been identified in rice comprising of 31 CDPK genes in rice genome annotation and transcriptomic analysis [50, 51]. Studies aiming to uncover the expression profile of these multiple genes at both global as well as single gene level highlight the relevance of CDPKs in abiotic stress and development. RT-PCR (Reverse transcriptase-PCR) and RNA gel blot studies [50] have indicated the presence of 17 OsCPK transcripts in response to cold, drought, salt and heat stress. These include *OsCPK1*, *OsCPK4*, *OsCPK 6*, *OsCPK7*, *OsCPK8*, *OsCPK9*, *OsCPK10*, *OsCPK 13*, *OsCPK14*, *OsCPK15*, *OsCPK16*, *OsCPK17*, *OsCPK19*, *OsCPK23*, *OsCPK24*, *OsCPK25* and *OsCPK29*. Of these, four genes show differential response to these stresses. *OsCPK17* was down regulated by cold, drought and salt stress. *OsCPK6* was up regulated by drought and *OsCPK25* was showing up regulation in response to heat stress. *OsCPK13* (encoding OsCDPK7) was up regulated by salt, drought and cold stress. Microarray based expression studies of rice seedling grown under cold, drought, and salinity stress for 3 hours has revealed the induction of six CDPK genes (*OsCPK4, 10, 12, 13, 15, 21*) and down regulation of one gene *OsCPK1* [4, 51]. Interestingly, *OsCPK13* was induced under all stress conditions. *OsCPK4* was induced under cold stress while the rest of the CDPK genes, *OsCPK 10, 12, 15* and *21* were up regulated specifically under desiccation stress. Least up regulation was observed under salt stress where none of these genes showed more than 2 fold change in expression. In fact, *OsCPK1* transcript levels were found to be decreasing by 2 folds under salinity stress. The lack of correlation in the expression profile for some of the rice CPKs such as *OsCPK6* studied from RNA gel blots and microarray could possibly be due to difference in the rice variety and experimental conditions employed in these studies. Additional support for the involvement of CDPKs under abiotic stress conditions has come from the analysis of *cis*-elements of the CDPK genes [50]. *OsCPK21* was recently studied in greater detail by Asano *et al.* [52] who used FOX based system [53] to generate 250 independent transgenic rice lines overexpressing each of the rice CDPK and screened for salt tolerance phenotype. OsCPK21-FOX lines also exhibited greater inhibition of seedling growth by ABA and induction of ABA as well as salinity inducible genes suggesting that it positively regulate both ABA as well as salt mediated responses. *OsCPK12* has also been studied in greater detail and was shown to be negatively regulating ABA induced seedling germination and growth [54]. This negative role of OsCPK12 was shown by biochemical assays to be mediated by its direct interaction and phosphorylation of type 2C phosphatase ABI2 as well as two other ABA responsive transcription factors ABF1 and ABF4.

Of the different elements studied, the rice CDPK genes were found to be having one or more number of such different 1kb elements upstream of the gene. Most of the rice CDPK genes were found to be having an EBOX *cis*-element (EBOXBNNAPA), which is having an ABA responsive element (ABRE) motif characterized by the sequence CANNTG. Presence of such *cis*-elements upstream of the CDPK genes points towards the ability of such sequences to bind with stress specific transcription factors such as dehydration responsive element/C-repeat (DREBs/CBFs) that can then control the expression of stress responsive CDPK genes under various dehydration stress conditions such as low temperature, salt and drought.

Previous studies at the level of single genes have conclusively shown the direct role played by some of the CDPKs in abiotic stress signaling cascades. OsCDPK13 and OsCDPK7 are rice CDPKs for which expression and functional analyses has been carried out in detail [55, 56]. OsCDPK13 gene is induced by cold and gibberellin in sheath as well as in callus as determined by northern blot analyses [55]. However, the mRNA levels for this gene were suppressed by drought and salt stress along with brassinolide and ABA treatment. Consistent with this observation, a similar pattern of CDPK13 protein expression was seen from immunoblot experiments. Additional studies have implicated OsCDPK13 (encoded by *OsCPK7*) as a mediator of both gibberellic acid (GA) and cold signaling [57]. Both GA and cold stress are known to be inducing changes in cytosolic Ca^{2+} levels [1], and thus said to be involved in calcium-mediated signaling. Some other proteins such as calreticulin, are also perhaps involved in the cold tolerance pathway mediated by OsCDPK13 and calreticulin interacting protein1 (CRTintP1). This was concluded from 2D-polyacrylamide gel electrophoresis (PAGE) profiles of transgenic lines overexpressing OsCDPK13 and CRTintP1, both of which, showed higher levels of calreticulin [58]. Also, cold tolerant varieties had higher levels of both of these proteins. Proteomic analysis of these overexpression lines revealed the up regulation of fructokinase, cytoplasmic malate dehydrogenase and tubulin. Such proteins might have a protective role in cold stress response by repairing damage caused under such conditions. Subcellular localization study using immuno-localization experiments showed OsCDPK13 to be a cytosolic protein [55]. Another CDPK (encoded by *OsCDPK14*) was also found to be localized in the cytoplasm and is believed to be involved in mediating calcium signaling in the cytosol [59].

Overexpression of OsCDPK7 confers the rice plants with enhanced tolerance to both cold as well as salt stress [60]. In the same study, both *in situ* hybridization experiments to detect mRNA levels and immuno-localization study to detect protein levels showed that it mainly expressed in vascular tissues of crowns and roots, vascular bundles and central cylinder, respectively. These are the tissues where water stress occurs most strongly, and overexpression of OsCDPK7 is seen to be causing higher transcript accumulation of not only OsCDPK7, but also increased accumulation of *RAB16A*, which is a putative target gene of OsCDPK7.

OsCDPK7 has been proposed to be serving as a branch point of cold and salt/drought signaling pathway acting upstream of these target genes such as *RAB16A*. Overexpression of OsCDPK7 in sorghum was found to be not improving salt or cold tolerance of the transgenic lines [61]. However, other unexpected phenotypes at the transcript level such as up-regulation of pathogen related proteins and photosynthesis genes were observed. This possibly hints towards the fact that overexpression of calcium kinases could have off target effects, as they are involved in multiple cellular pathways and functions.

Most of the CDPKs also showed a tissue specific expression as determined from various studies in rice, Arabidopsis, and wheat. For example, in rice, expression of 19 CDPK genes was detected using northern blot and RT-PCR based studies in roots, stems, leaves and panicles [50]. A comprehensive expression profiling of 31 CDPK genes in rice employing microarray has been performed [51] in three vegetative stages, six stages of panicle and five stages of seed development. Following this, the genes have been grouped into eleven categories with genes showing maximum expression levels being placed in the first category. Twenty-three transcripts were found to be showing a differential pattern of expression in the reproductive development stages. *OsCPK23* showed maximum up regulation of up to 87-folds in panicle stages and 1,724 folds in the seed development stages relative to its expression in mature leaf. Some genes like *OsCPK5* and *16* showed up regulation during panicle development and were down regulated at the seed stage. In contrast, *OsCPK8* transcript levels were increased in the seed developmental stage and decreased in the panicle stages. Genes such as *OsCPK1, 4*, 13, 15, *19*, *20*, and *24* were specifically down regulated in late stages of seed development. *OsCPK24* was found to be specifically up regulated in S1 stage of seed development, and has earlier been shown to be up regulated during wounding [50]. As proposed by the authors, early seed development involving pollen tube growth is known to be similar to wounding response or invasion by fungal hyphae, and that both elicit similar responses [62]. *OsCPK9* had highest expression in the mature leaf with undetectable transcript levels in the panicle stages. Opposite to this, *OsCPK21* was observed to be having almost no expression in the vegetative stage, but higher transcript level was observed in seed and panicle stages correlated with the presence of a *cis*-regulatory element motif required for endosperm expression in *OsCPK21*. All the genes that were placed in the eighth category expressed in the P6 stage of panicle development, these include *OsCPK6, 14, 22, 25*, *26*, and *29* suggesting their expression in mature pollen. Also, *OsCPK25* and *26* had high similarity to a maize CDPK gene, which also expressed in the late panicle stages, and whose inhibition by antisense mechanism resulted in impaired germination and pollen tube growth [63]. Thus, *OsCPK25* and *26* together with other genes of this category could be specifically involved in pollen maturation. Still others that include *OsCPK11*, *27* and *31* were found to be showing almost no expression, which could possibly be due to some other specific stimuli, or stage that may be responsible for inducing the expression of these genes.

OsCPK2 was detected at higher levels only at the P6 stage in the microarray analysis, which was in agreement with RNA gel blot studies that also showed the presence of this transcript specifically at the panicle stage [50, 51].

Microarray based expression analysis of some CDPK genes under stress conditions has also been documented in RED (Rice Expression Database; http://red.dna.affrc.go.jp/RED/). Both *OsCPK 7* and *15* are not regulated by salt stress in the 7 days old leaf stage as indicated by their relative expression values close to 1. However, at 24 days old leaf stage, under salt stress conditions, the expression of *OsCPK7* was seen to fall below 1 while that of *OsCPK15* was increased by more than 1 [50]. This indicates that some isoforms of CDPKs are possibly devoted to abiotic stress tolerance in specific tissues or developmental stages.

All such detailed expression profiling of the CDPK gene transcripts showing their differential regulation underlines the functional role and importance of this large family of calcium sensing kinases in both abiotic stress and different development stages in rice, which is an important model crop plant, and provides an avenue to carry out further functional characterization of these genes using other molecular tools and techniques.

CDPKs in Wheat and Other Plants

In addition to rice, genome wide analysis and expression studies of CDPKs has been performed in wheat where 20 CDPK genes have been identified based on hidden Markov model (HMM) generated using rice CDPK proteins followed by full length cDNA isolation for some of these genes [64]. Based on such analysis, detailed features of the 20 CDPK genes identified have been elucidated with respect to details such as the number of EF-hands, protein length, subgroup classification and N-terminal amino acids along with information regarding the presence of myristoylation motif and subcellular localization. Like Arabidopsis and rice, the wheat CDPKs also showed a tissue specific expression pattern, and 13 of the CDPKs studied exhibited variable pattern of expression in the leaf, root, stem, young spikes and immature seeds as tested by semi quantitative RT-PCR. *TaCPK1*, *2*, *5*, and *16* were detected in all the tissues. *TaCPK3*, *6*, *9*, *12*, *14*, and *15* were found to be expressing mainly in the root, stem, leaf and young spikes. *TaCPK10* was expressed significantly in all the tested tissues except stem [64]. A more limited expression pattern was observed for *TaCPK8* (predominant expression in leaves and immature seeds) and *TaCPK13* (mainly expressed only in young spikes). This type of expression pattern during defined developmental stages points towards the role of CDPKs in plant development besides their crucial role in abiotic stress response. An elaborate study to determine the expression of the wheat CDPK genes under abiotic and biotic stress was also undertaken. Seedling tissues exposed to abiotic stresses such as cold, drought, salt, hydrogen peroxide (oxidative stress), and hormones such as ABA and GA were observed for specific expression of the identified CDPK genes. Multiple CDPKs were found to be expressing under a given treatment and different stress treatments were also seen to be inducing the expression of some common CDPKs, implying the possible cross talk or common point, which might exist between such different stress signaling pathways. Maximum numbers of the identified genes were responding to hydrogen peroxide as an abiotic stress as determined by the differential expression of *TaCPK1*, *2*, *5*, *7*, *9*, *10*, *12*, and *18* using RT-PCR. Cold stress was seen to be affecting the expression of seven *TaCPK* genes, namely, *TaCPK3*, *4*, *5*, *6*, *7*, *12*, and *15*. Five genes, *TaCPK 4, 9, 10, 18, 19* responded to salt stress while only four genes, *TaCPK1*, *6*, *9*, and *18* were shown to be affected by drought [64]. ABA commonly associated with stress responses was seen to be causing the induction of *TaCPK3*, *4*, *5*, *6*, *7*, *9* and *10*. Expression of one of the CDPKs, *TaCPK4* was common to all the tested hormones and abiotic stress conditions except drought. In fact, *TaCPK4* expression was also observed under the biotic stress condition tested that involved infection with the fungal pathogen *Blumeria graminis tritici* (causal agent of powdery mildew). TaCPK2 and OsCPK13 share high identity at the level of protein sequence and both of these were detected in leaf, stem and root tissues. However, comparison of the expression of *TaCPK2* [64] with *OsCPK13* (a previously well studied CDPK gene in rice identified as a putative *TaCPK2* ortholog) revealed that both respond to pathogen challenge in their respective host, but unlike *OsCPK13*, *TaCPK2* appears to have lost its ability to respond to cold, drought and salt stress conditions.

To correlate the *in silico* structural features of the wheat CDPKs like the predicted myristoylation motif with their respective *in vivo* function such as membrane association, subcellular localization studies utilizing onion epidermal cells were performed [64]. Out of the 12 TaCDPKs tested for their localization, seven TaCDPK-GFP constructs were found to be expressing and yielding the fluorescence in the cell. TaCPK2 and 5 that had previously been identified to be containing a myristoylation motif were found to be membrane associated. TaCPK3 and 15, which did not harbor such myristoylation motif were also observed to be localizing to plasma membrane [64]. This showed that other membrane association motifs might be present that were not detected by the *in silico* methodology. Both TaCPK3 and TaCPK15 were determined to be consisting of a bipartite NLS or nuclear localization in their junction domain, which according to the authors indicated the possibility of migration of these CDPKs from the plasma membrane to the nucleus during signaling [64]. Similarly, TaCPK1, 4, and 9 that were not found to be having any NLS sequence or myristoylation motif [64] were seen to be both plasma membrane as well as nuclear localized, indicating the presence of uncharacterized motifs, which are responsible for the observed localization pattern.

Various studies in other plant species have also established the role of CDPKs under environmental stress conditions. *PaCDPK1* from Orchid (*Phalaenopsis amabilis*) was induced by cold, wounding and pathogen attack as shown by promoter analysis of this gene utilizing β-glucuronidase (GUS) expression study as well as RNA blot experiments to detect the transcript levels. Northern blot hybridization showed high levels in labellum and peloric flowers [65]. *VfCPK1* from *Vicia faba* or broad bean showed increased accumulation upon drought stress and ABA treatment. Analysis in different parts of this plant showed

VfCPK1 expression to be particularly high in leaves especially the epidermal peels [66]. NtCDPK1 from *Nicotiana tabacum* is detected largely in rapidly proliferating tissues such as roots, stems, and flowers. It was also induced by multiple stress agents such as salt, wounding, calcium, ABA, GA, cytokinin, jasmonate, fungal elicitors and chitosin reflecting its possible role in a wide variety of stress and hormone signaling pathways [35]. Immuno-precipitation studies found NtCDPK1 to be associated with the membrane fractions. *NtCPK4* is another CDPK from tobacco that has an important function in development as well as abiotic stress response as concluded from the transcript levels for this CDPK that showed a time and space dependent expression in the vegetative and reproductive tissue [67]. Higher expression was detected in zone of cell division and vascular bundle. A varied expression was seen in different parts of the flower where large amount was detected in the tapetum and anther wall. During the pollen development, level of *NtCPK4* gradually declined. Similarly during ovule development, it was present in higher level in placenta, ovule primordia and ovule. After fertilization, it was seen to be accumulating in the embryo where a peak in the expression was detected at the pro-embryo stage. Increased *NtCPK4* mRNA accumulation was also seen after salt and GA treatment. *LeCDPK1* from tomato was observed to be induced upon mechanical wounding [68], and was also shown to be conferring the plants with salt tolerance [69]. This was a type of cross-tolerance where resistance to one type of stress provides additional resistance to another form of stress. *LeCDPK1* transcripts were also seen to be increasing upon both salt and ABA treatment. Jasmonic acid (JA) treatment enhanced the expression of *LeCDPK1* mRNA which indicates that perhaps during wounding, it was JA signaling pathway that up regulates *LeCDPK1* expression whereas during abiotic stress such as salinity, it was ABA that activated its expression. *Mesembryanthemum crystallinum* calcium dependent protein kinase 1 (*McCPK1*) transcripts were induced in response to salt and dehydration stress [70]. McCPK1 also exhibited an interesting localization pattern where in response to salt and dehydration stress, it was seen to be moving from the plasma membrane to the nucleus. McCPK1 interacts with McCAP1 and McCAP2 (McCPK1 Adaptor Protein 1 and 2) as determined from yeast two-hybrid studies. McCAP1 is a cytoskeletal protein whereas McCAP2 is observed to be associated with vesicle structure. Under low humidity conditions, both McCPK1 and

McCAP1 co-localize to the nucleus and cytoplasmic strands suggesting possible role of McCAP1 in tethering McCPK1 under stress conditions. McCAP2 on the other hand remains on the vesicular structures under high humidity conditions while McCPK1 moves to the plasma membrane. Another CDPK from mungbean, *VrCDPK* was expressing under mechanical stress [71] while *MsCPK3* (in alfalfa) was induced under heat stress [72]. EST library of *Ulva compressa* generated from plants grown with Cu^{2+} showed the presence of CDPK gene suggesting the possible involvement of CDPKs in heavy metal tolerance [73]. Several CDPK genes from herb plant *Panax ginseng* such as *PgCDPK1c*, *PgCDPK2c* and *PgCDPK4a* showed increased accumulation upon salt treatment [74]. *ACPK1*, a CDPK gene from grape berries has been found to be a positive modulator of ABA signaling as concluded from ABA hypersensitive phenotype in transgenic Arabidopsis that showed increased seedling growth and stomatal movement [75].

Several potential CDPK substrates have been identified, some of which provide a clue towards the downstream targets of CDPKs in abiotic stress signaling. Many ion transporters and channels have been found to be serving as substrates for CDPKs. A 59 kDa CDPK from *Vicia faba* guard cell has been reported to be phosphorylating and inhibiting the K^+-inward channel, KAT1 [39]. This kind of targeting of the potassium channel might point towards the role of CDPKs in mediating stomatal closure during ABA signaling that occurs during stress conditions. Similarly, another CDPK from the guard cells of *Vicia faba* has been shown to be activating an influx chloride channel on the vacuole. However, in the latter case it was still not clear if the CDPK acts directly to activate the channel or if it first phosphorylates an intermediate regulatory protein, which then activated the channel. Here, ABA biosynthesis was up regulated under stress conditions such as drought and has been shown to be involved in mediating signal transduction pathway for stomatal closure [39]. Calcium plays an important role in the direct activation of CDPK kinase activity as demonstrated *in vitro* in a few studies [35, 67]. In the case of potato, a bZIP transcription factor StABF1 was found as the target of StCDPK2. StABF1 was up regulated under drought, salt, ABA and cold and was found to be phosphorylated by StCDPK2 under salt and ABA treatment [76].

Overall, expression pattern of several CDPKs have been found to be regulated under multiple abiotic stress condition in several plants species and Table **1** briefly represent the different CDPK isoforms regulated by multiple stimuli. Functional role of CDPKs in plants is summarized in Fig. **1** where CDPKs act as the critical component of calcium mediated signaling network to regulate several downstream targets in the plant cell.

Table 1: Summary of CDPK genes known to be induced under different abiotic stress conditions from genome-wide and single gene studies in Arabidopsis and rice. Microarray, northern blots and quantitative PCR have revealed up regulation of specific CDPK genes under environmental stress condition. Many of the genes such as *CPK3, 4, 6, 10, 11, 12, 13, 21* show induction by multiple stimuli such as ABA, salt, drought and cold that suggests their involvement in wide range of stress signaling cascades and abiotic stress tolerance.

Stress/stimuli	*CPKs* induced
ABA	***CPK3, 4, 6, 10, 11, 12, 21, 30, 32***
Salt	***CPK4, 11***
Drought	***CPK 6, 10, 12, 13, 15, 21***
Cold	***CPK 13***

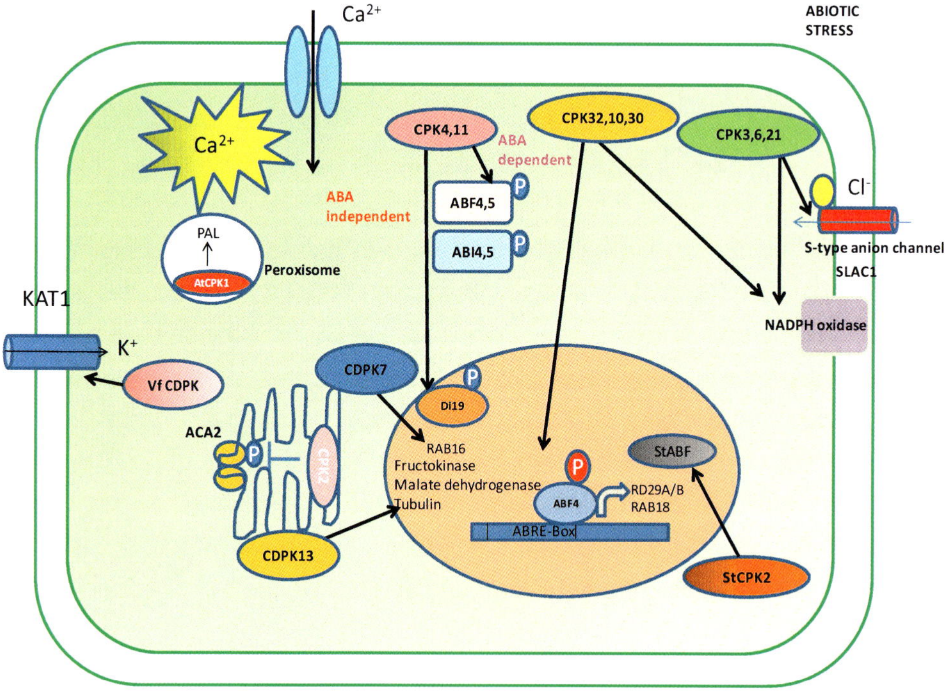

Figure 1: CDPK and some of its known targets in the cell involved in abiotic stress tolerance. The model shows different CDPKs activated under various abiotic stress conditions for

which known targets have been identified based on genetic and biochemical studies. Abiotic stress triggers calcium changes in the cell, which turns on specific subset of CDPKs that then activate various targets at both transcriptional as well as post-translational level. Transcriptional control is achieved *via* activation of transcription factor whereas specific cellular proteins such as ion transporters and enzymes are positively regulated by phosphorylation. Abbreviations used: PAL-Phenylalanine lyase, ABRE – ABA response element, ABF- ABRE binding factor, ABI-ABA insensitive, ACA2- Calmodulin stimulated calcium ATPase, SLAC1 – Slow anion channel 1, KAT1- K^+ channel

CALCINEURIN B-LIKE INTERACTING PROTEIN KINASE (CIPK) AND ABIOTIC STRESS

CIPKs are a class of novel serine-threonine protein kinases in plants involved specifically in interacting with calcium sensing protein calcineurin B-like protein (CBL) and transducing the various abiotic stresses, nutrient and hormonal signals further downstream in the signaling pathway [11, 14, 26, 77, 78]. Recently, significant evidence has come, which suggests their role in developmental processes [79, 80]. CIPKs belong to the SnRK3 subgroup of protein kinases and show similarity at the structural level to SNF1 (sucrose non-fermenting) kinase from yeast and AMPK (AMP activated protein kinases) from animals. CIPKs have a catalytic N-terminal kinase domain and a regulatory C-terminal domain. Another domain, which is highly conserved, is NAF/FISL motif, which has a stretch of 24 hydrophobic amino acid residues, and is responsible for mediating interaction with CBLs. Also, the NAF domain serves to inhibit the kinase activity of CIPKs, which is relieved by binding of CBLs to this domain [11, 14, 26, 77]. Since CIPKs do not have any localization motif of their own, it is the myristoylation/palmitoylation motifs present within CBLs that help anchor CIPKs to the membranes [81]. The kinase activity of CIPKs was confirmed using commonly employed or artificial substrates for kinase assay such as myelin basic protein (MBP) and histone IIIS that were found to be phosphorylated by OsCK1, which is a stress responsive CIPK member from rice [82]. Besides, many other physiological substrates for CIPKs such as potassium channel (AKT1) and Na^+/H^+-antiporter (SOS1) were phosphorylated by CIPKs confirming the *in vivo* kinase activity of CIPKs [83, 84].

CIPKs in Arabidopsis

Genome analysis in Arabidopsis has revealed the presence of 26 CIPKs in Arabidopsis [85]. RT-PCR based cDNA cloning of all the AtCIPKs confirmed their expression and also validated theoretically predicted gene structure for some while leading to the correction for others. Since CIPKs do not sense calcium directly, and require CBLs as calcium sensors to relieve the inhibition of their kinase activity, protein interaction analysis using yeast two-hybrid approach has been utilized to identify potential CBL-CIPK interacting partners to gain a clue about interactions, which could actually be taking place *in vivo* in response to abiotic stress. Yeast two-hybrid analysis to study the interaction of Arabidopsis CBL1 and CBL9 with all the CIPKs have shown that CBLs that are closely related in sequence may not necessarily interacts with similar subset of CIPKs [85]. For instance, only four AtCIPKs (AtCIPK1, 8, 18, 23 and 24) exhibited strong interaction with both AtCBL1 and 9. Others such as AtCIPK21 interacted more significantly or strongly with AtCBL9 while AtCIPK7 and 17 preferentially interacted with AtCBL1. Some of the CIPKs such as AtCIPK5, 11, 12 and 23 interacted weakly with AtCBL1. Of these, AtCIPK5, 11 and 12 exhibited a weak interaction even with AtCBL9 along with other weakly interacting partners, which include AtCIPK6, 10 and 16 [85]. Another study [86] has found AtCBL3 to be interacting specifically with AtCIPK11 in a calcium dependent manner. This shows that each CIPK could be interacting with a specific subset of CBLs at varying strengths, and that phylogenetic relationship was not indicative of CBL-CIPK interacting partners as concluded from the observed interaction of evolutionarily separated intron harboring (AtCIPK1, AtCIPK8, AtCIPK17 and AtCIPK24) and intron free (AtCIPK7 and AtCIPK18) with AtCBL1.

CIPKs are found to be differentially regulated by a variety of conditions as demonstrated by their expression pattern under cold, drought, salt, ABA, low K^+ ion conditions and various developmental stages [25, 79, 87, 88]. Studies based on mutant analysis under different stress conditions involving stress marker gene expression pattern, seed germination assays on stress media combined with biochemical approach have clearly implicated different CBL-CIPK partners to be involved in sensing specific or multiple stress conditions by modulating different downstream targets such as potassium channels (AKT1), sodium efflux pumps (Na^+/H^+-antiporter) and other proteins with as of yet unknown function such as ECT1 [83, 84,89]. Recently, the role of CIPK in development has been clearly defined from studies based on specific CIPK members such as *AtCIPK6* from Arabidopsis [79].

In CIPK gene family in Arabidopsis, *CIPK3* transcript was strongly inducible by cold, drought, salinity, wounding and ABA [87]. One of the important report by Hu *et al.* 2008 [78], showed the involvement of CIPK8 in early nitrate signaling. CIPK8 expression profiling in Arabidopsis plants by quantitative PCR showed that CIPK8 was induced very rapidly by NO^{-3}. The other CIPKs, which were responsive to abiotic stress include, *CIPK9* (responsive to low potassium levels) [25], *CIPK11 /PKS5,* which is responsive to ABA, drought and salinity [90].

Beside *CIPK9*, which was strongly up regulated under potassium deprivation stress condition [25], the transcript of another member of this family, *CIPK23*, was also slightly up regulated under potassium deficient condition [91]. Moreover, some of these CIPKs were also regulated spatially and temporally and at different developmental stages of plants [11, 14].

Numerous studies at the level of single genes have elucidated the role of CIPK in abiotic stress conditions. Both CBL1 as well as CBL9 were found to be interacting with CIPK1 as studied by yeast two-hybrid method and this interaction was seen to be targeting CIPK1 to the plasma membrane which was determined from GFP localization studies in *Nicotiana benthamiana* protoplasts [88]. Further, *cipk1* mutants like *cbl9* mutants exhibited hypersensitive response to ABA while *cbl1* mutant was not affected in ABA response. This showed that CIPK1 upon interaction with CBL9 serves as the negative regulator of ABA response under osmotic stress conditions. Such experimental data indicated the presence of a plausible pathway where alternate complex formation of CIPK1 with either CBL1 or CBL9 decides the activation of an ABA dependent or ABA independent pathway under abiotic stress.

In the CIPK gene family, *CIPK21* is strongly up regulated under multiple abiotic stresses such as salt and osmotic stress conditions shown to be positively regulating salinity and osmotic stress responses (Pandey *et al.*, unpublished). By reverse genetic mutant analysis, *cipk21* mutants were showing hypersensitive growth pattern in high salt and osmotic stress conditions. The authors have suggested the role of CIPK21 in regulating nuclear gene expression as well as preferential complex formation of CBL2/CBL3-CIPK21 at the vacuolar membrane in regulating unknown transporter/channels at the tonoplast (Pandey *et al.* unpublished).

Under phytohormone ABA and abiotic stresses, CIPK3 was implicated as nodal point of calcium mediated signaling in Arabidopsis [87] and shown to be interacting with CBL9 [92]. *cipk3* mutant seed germination was found to be inhibited more by ABA and salt than the wild type plants, which suggested its role as a negative regulator of ABA response [92]. Like CIPK1, CIPK3 was also shown to be interacting with CBL9 in bringing about this response as concluded from ABA hypersensitive phenotype observed in *cbl9* mutants, which was rescued by the overexpression of constitutively active CIPK3 but not by that of the wild type CIPK3 indicating that CIPK3 functions downstream of CBL9, which serves to activate CIPK3 in response to ABA during seed germination [92]. Expression analysis by RNA gel blot of *CIPK3* under ABA and other abiotic stress treatment indicated that it was strongly induced under ABA, salt, drought, cold as well as wounding [87]. GUS expression analysis and RT-PCR of different developmental stages showed that the expression of *CIPK3* was highest in the seedling stage in contrast to the other adult stages such as mature leaf, stem, root, flower and silique, which had lower expression of this gene. This expression pattern of CIPK3 reflects its role and importance under stress conditions in the seedling stage as determined from mutant studies, and might also indicate its unknown function in development. To test the effect of *CIPK3* mutation at the downstream level of stress signaling pathway, several of the stress marker genes were studied. On comparison between the *cipk3* mutant and wild type plants, it was concluded that for some of the genes such as *KIN1*, *RD29A/B* under cold and salt treatment, the early period of induction was being controlled by *CIPK3* [87]. *CIPK3* perhaps regulates ABA independent salt stress pathway since no effect on stress gene expression was observed in case of *cipk3* mutant plants that were given drought stress, and because salt and drought have a common ABA dependent pathway, so it can be concluded that *CIPK3* brings about salt stress response in a manner, which is independent of ABA. More importantly, CIPK3 is believed to be acting as a cross talk node for ABA, salt and cold stress signaling since in *cipk3* mutants, expression pattern of the stress marker genes was simultaneously altered in all these conditions [87]. Interestingly as noted by the authors, CIPK3 is possibly a cross talk node for cold and ABA signal transduction that were previously thought to be independent of each other. Based on this study, it can be postulated that ABA together with cold has a synergistic effect by inducing greater changes in calcium levels, which would then activate higher levels of CIPK3 node and bring about cold induced gene expression, which would otherwise be impaired or reduced in ABA deficient mutants.

In a recent study, CBL1-CIPK7 interaction was found to be important for cold stress response and adaptation [93]. Pull down assays and affinity chromatography were used to confirm the interaction, and real time PCR revealed an up regulation of both CIPK7 and CBL1 under cold treatment. Intriguingly, CIPK7 induction under cold stress was found to be dependent on CBL1 as concluded from compromised induction levels of *cipk7* in *cbl1* mutants [93].

Another report also establishes the function of CIPK15/PKS3, which interacts with CBL1 as a negative regulator of ABA signaling pathway [94]. In this study, silencing of both *CBL1* and *CIPK15* resulted in ABA hypersensitive phenotype. Also, CIPK15 was found to be interacting with PP2C phosphatases, ABI1 and ABI2, which were speculated to be interacting with other downstream targets or proteins modulating ABA regulated gene expression. Thus, CIPKs together with CBLs could be having an opposite effect on ABA signaling as compared to CDPKs, which promote or serve as a positive modulator of ABA signaling.

Plants have evolved to utilize potassium as an important mineral nutrient while sodium ion is toxic even at low concentration [95]. The saline soil environment in which plants typically grow has low K^+ and high Na^+. To accumulate potassium against concentration gradient and expel sodium out of the cell, plants have a host of transporter proteins that are regulated by calcium sensing kinases. Some of the identified transporters include NHX family of Na^+/H^+ antiporters such as SOS1 and symporters such as HKT1 (High affinity K^+ transporters) identified in wheat that are critical to maintain ion homeostasis within the cell.

Potassium deficiency is a form of nutrient stress in which involvement of CIPK9 has been elucidated [25]. *CIPK9* expression was studied under various abiotic stress conditions using semi-quantitative RT-PCR, and it was found to be induced under osmotic stress, salt, cold, wounding and low potassium conditions (low-K^+ stress was provided by eliminating KCl from the MS medium). Similar analyses for *CIPK9* expression in different tissues were conducted using RT-PCR and GUS expression analysis. Anthers, stigma, petals, sepals and siliques in particular, showed a strong expression of this gene. Increase in *CIPK9* transcript accumulation was seen till after 5 days of low potassium stress. Such a strong induction of a CIPK gene under low potassium stress was perhaps unique to *CIPK9*, which might be specific to this form of nutrient stress. This was supported by the fact that T-DNA mutant lines of *cipk9* were specifically altered in their response to low potassium stress and exhibited a hypersensitive response under low-K^+ conditions.

Other studies have also shown CBL-CIPK members to be involved in low potassium signaling. For instance, CIPK23 has been implicated in potassium signaling where it was proposed to be acting by interacting with upstream calcium sensors such as CBL1/CBL9 as shown by yeast two-hybrid studies that was confirmed using bimolecular fluorescence complementation assays (BiFC). Such interaction brings about phosphorylation of K^+ transporter channel, AKT1 [83, 95, 96]. However, CIPK9 has a different mode of action in this respect since no change in K^+-uptake was seen in *cipk9* mutants, and it can thus be said that CIPK9 does not directly interact with the K^+-transporters involved in potassium uptake. An effort has been made to determine the interacting partners of various CIPKs using yeast two-hybrid studies. In comprehensive yeast two-hybrid study, two more CIPKs, CIPK6 and CIPK16 were also found to be interacting with AKT1 in addition to CIPK23. Similarly, two more CBLs, CBL2 and 3 in addition to CBL1 and CBL9 already reported in earlier studies [96, 97], were found to interacting with all three CIPKs, namely, CIPK6, CIPK16 and CIPK23 [83]. Additional layer of regulation of AKT1 by CBL/CIPK/PP2C was uncovered using yeast two hybrid studies that identified specific PP2C members to be interacting with CIPK6, 16 and 23 [98]. The PP2C mediated inhibition of CIPK was observed to be depended on its interaction with the kinase domain. Further, CBLs were shown to be directly binding to PP2C and preventing AKT1 inhibition. *In vitro* phosphatase assays showed CBL1 and 2 to be inhibiting PP2C function in a dose dependent manner. Taken together the study established the role of PP2C and CIPK as a phosphorylation dephosphorylation pair controlling AKT1 activity.

The mechanistic interplay of regulation of CBL-CIPK network has identified the phosphorylation of calcium sensor, CBL by CIPK as an important step in regulation of downstream target or substrate. This detail has been investigated recently by Hashimoto *et al.* [99] which actually showed that full activation of CIPK23 target AKT1 *in vivo* depends on phosphorylation of its interacting partners CBL 1 and 9. Using *in vitro* biochemical assays, the authors dissected the role of serine phosphorylation in the C-terminus of CBL1 leads to activation of the potassium channel by CBL1/CBL9-CIPK23 complex [99]. The same study also characterized the NAF kinase-inhibiting domain to be necessary for this CBL phosphorylation suggesting that NAF-CBL interaction is essential for this modification and subsequent activation of the CIPK target protein. The authors further tested and found this to be true for other CBLs *i.e.* CBL4 and 10 and their interacting partner CIPK24 supporting the conclusion that this is a general mechanism of CIPK activation by their respective CBLs.

Like CIPK1, CIPK24 or salt overly sensitive2 (SOS2) is another calcium signaling kinase that mediates salt signaling pathway by associating with CBL10 in the tonoplast [100] or CBL4 in the plasma membrane [101]. The regulation of ion homeostasis by CBL10 follows a unique mechanism as revealed by the *cbl10* mutants that show lower Na^+ ions under both normal as well as high salt conditions unlike other salt sensitive mutants [100]. Yeast two-hybrid experiments identified CIPK24 as a strong candidate for interacting with CBL10, which was further confirmed utilizing GST fusion constructs of CIPK24 that co purified CBL10 in the pull down assay. BiFC proved the localization of CBL10-CIPK24 at the vacuolar membrane. Such low Na^+ ion content may be due to sequestration of these excess Na^+ ions in the vacuole, a mechanism most suitable for shoot tissue where CBL10 is primarily expressed. In contrast, CBL4, which localizes to the cell membrane, and was strongly expressed in roots, might be mediating the export of excess sodium ions outside into the soil *via* interaction with CIPK24 to activate Na^+/H^+-antiporter [101].

CIPKs have been found to be playing a critical role in cold, drought, salinity and ABA responses, and target proteins for some of the CBL-CIPK complexes have been identified. The Na^+/H^+-antiporter or SOS1 is a well-established target identified for the SOS3 (CBL4) and SOS2 (CIPK24) complex that is activated, and functions under salt stress [24,102]. Response of *sos1, sos2* and *sos3* mutants under salinity stress indicated the involvement of these three proteins in a common signaling pathway regulating salt toxicity. *sos3* mutant did not show up regulation of SOS1 mRNA under salt stress in either root or shoot whereas *sos2* mutant exhibited this loss in up regulation of SOS1 in roots but not in shoots, which can be explained by the fact that CBL4 that is plasma membrane localized, functions mainly in the salt tolerance pathway actively in the root tissue [103]. *sos1* mutant displayed greater sensitivity towards Na^+ and Li^+ stress in comparison with *sos2* and *sos3* plants [104]. SOS1 was identified to be a 127 kDa Na^+/H^+-antiporter that was related to plasma membrane Na^+/H^+-antiporters found in bacterial species such as *E. coli* and *Pseudomonas aeruginosa*. SOS2 and SOS3 are believed to be activating the Na^+/H^+-antiporter in response to NaCl and subsequent rise in calcium levels caused by the stress. SOS1 functions to pump sodium ions out of the cytosol into the environment, and hence protects the cell from salt stress. A rice homolog of SOS1, namely, OsSOS1 demonstrated to be phosphorylated by AtSOS2 and AtSOS3 complex [84]. Yeast two-hybrid approach has identified yet another transporter H^+-ATPase, which is present in the vacuole, to be interacting with SOS2 [105], and regulating plant responses to salt stress since such interaction was found to be enhanced under salt stress conditions.

CIPK11 interacting with CBL2 was found to be regulating a plasma membrane H^+-ATPase referred to as AHA2. In the study [90] using reverse genetics approach, it was found that AHA2 was negatively regulated by PKS5 or CIPK11, which phosphorylated the pump at Ser^{931}. This prevented the activation of AHA2 by 14-3-3 proteins that binds to phosphorylated Thr^{937} of this proton pump. In *pks5* mutants utilized for this study, the PM (plasma membrane) H^+-ATPase pump was de-repressed in mutant, and enhanced the survival of plants under high pH. This occurs primarily due to the ability of AHA2 to pump protons outside into the extracellular space that helps in acidification of the local rhizosphere. Under normal conditions, repression by CIPK11 might be useful for membrane depolarization and cytoplasm acidification in response to external stimuli [90].

A more recently identified class of novel CIPK interacting proteins are ECT1 and ECT2, which are named so due to the presence of a conserved 180 amino acids long C-terminal domain. This class of protein was found in eukaryotic organisms such as monocot plants, yeast and humans, and around 11 such ECT proteins were found in the Arabidopsis genome. These two proteins have been reported as the target for CIPK1 [89], based on yeast two-hybrid screening and pull down assays. Using deletion constructs, it was found that the C-terminal domain of CIPK1 alone was sufficient for interaction with ECT proteins. Similarly, the C-terminal conserved residues in ECT1 and ECT2 were enough for this interaction. Also, it was found that only ECT1 and ECT2 specifically interact with CIPK1 since other tested ECTs (ECT3, ECT6, ECT10 and ECT11) were not found to be showing interaction with CIPK1. In addition, only ECT11 was interacting with CIPK3. Further, it was found that CBL1 and CIPK1 co-expressed with ECT1 are localized into the nucleus. A nuclear localization signal was subsequently found in the C-terminal domain (rich in positively charged amino acids) of ECT proteins. Detailed molecular analysis of these proteins by genetic and molecular approach will shed more light in their role in calcium mediated abiotic stress signaling pathway, which need more experimental input.

Very recently, Tripathi *et al.* [79] have identified the role of Arabidopsis CIPK6 in plant growth and development. A mutation in *AtCIPK6* significantly reduced shoot-to-root and root basipetal auxin transport, in conjunction with reduced expression of a number of genes involved in auxin transport and abiotic stress response. Also, the plants exhibited developmental defects such as fused cotyledons, swollen hypocotyls and compromised lateral root formation. The Arabidopsis mutant was more sensitive to salt stress as compared to wild type, and overexpression of a constitutively active mutant of CIPK6 promoted salt tolerance in transgenic tobacco. A recent report [106] also showed a novel regulator of CIPK6, namely, calreticulin, which is an ER localized calcium binding protein. The authors found that overexpression of calcium binding peptide or CBP (C-terminus of calreticulin) mediated enhanced salt tolerance was depended on AtCIPK6. CBP overexpression transgenic lines showed up regulated levels of CIPK6 and other downstream stress response genes such as DREB1A and RD29A and this activation depended on calcium levels. Null mutation of CIPK6 completely blocked this salt tolerant phenotype. This study demonstrated the sufficiency of calcium levels alone in modulating gene expression and hints towards the possibility of engineering stress tolerant plant varieties by altering calcium-signaling component.

Additional role of CIPK6 was recently uncovered by Held *et al.* [107] where they have done comprehensive studies to demonstrate direct role of CBL4-CIPK6 in targeting K^+ channel AKT2 from the ER to plasma membrane in a calcium dependent manner [107]. Loss of *cbl4* or *cipk6* showed similar phenotype as loss of *akt2* including delayed flowering and reduced leaf number. *In vitro* phosphorylation assays showed this AKT2 activation mediated by CBL4-CIPK6 is phosphorylation independent. CBL4 lacking N- terminal palmotylation and myristoylation residues failed to target CIPK6-AKT2 to the membrane suggesting that CBL4 is essential for the potassium channel translocation from ER to the membrane.

In an interesting study by Hu *et al.* [78], the possible function of another CIPK family member, AtCIPK8 was determined, which was seen to be up regulated by nitrate mineral nutrient, and expression of nitrate-regulated gene transcripts of several nitrate transporters in *cipk8* mutant plant were affected. Induction of *CIPK8* transcripts by nitrate was predicted to be positively regulating nitrate signaling. This study provided the first evidence for the involvement of a CIPK in nutrient sensing besides their function in abiotic stress and potassium deficiency.

Similar to regulation of potassium high affinity transporter, AKT1 by CBL1/9-CIPK23, this complex was also implicated in regulating nitrate sensing and uptake for a wide range of nitrate concentration by phosphorylation of nitrate transporter, CHL1 [135]. Based on the genetic and biochemical analysis, authors of this study reported the primary nitrate responses could be changed to low and high affinity based on the phosphorylation and dephosphorylating state of CHL1. *In vitro* and *in vivo* studies showed that, low-level primary response can be mediated by phosphorylation of CHL1 by CIPK23, and CHL1 uses dual-affinity binding and a phosphorylation switch to sense a wide range of nitrate concentrations in the soil, thereby functioning as an ion sensor in higher plants [135].

CIPKs in Rice

Comparison of Arabidopsis CIPKs with ESTs of two rice subspecies (*Oryza sativa* subspecies *japonica* and *Oryza sativa* subspecies *indica*) led to the identification of 30 CIPKs in both these rice varieties. In addition to the homology, other structural features such as the occurrence of NAF domain were taken into account for uncovering CIPK genes in rice [85]. *OsCIPK1* isolated by a differential cDNA screening approach was strongly induced by cold, light, salt, sugar and cytokinins [82]. Of the various tissues tested, significant expression could be detected only in shoots. The expression profiling of *OsCIPK1-30* under ABA, cold, drought, polyethylene glycol (PEG) and salinity has shown most of them to be inducible by at least one type of these stress or stimuli. RNA gel blot as well as real time PCR and RT-PCR based experiments were conducted to detect the rise in *OsCIPK* transcripts if any under these stress conditions [108]. It was found that of the 30 CIPKs, 27 *OsCIPK*s (except *OsCIPK13*, *14* and *27*) were detected either in the RNA blot /PCR or both. Out of 27 *OsCIPK*s, 20 were induced under various stress stimuli. 15 genes (*OsCIPK1*, *2*, *5*, *9*, *11*, *12*, *15*, *17*, *20*, *21*, *22*, *23*, *24*, *29*, and *30*) responded to drought, 12 genes (*OsCIPK7*, *8*, *9*, *10*, *11*, *15*, *16*, *17*, *21*, *22*, *23*, *24* and *29*) were salt inducible, 12 of these genes (*OsCIPK1*, *3*, *9*, *12*, *15*, *16*, *17*, *21*, *22*, *23*, *24* and *29*) were induced by PEG, 16 of the *OsCIPK*s (*OsCIPK 1*, *2*, *3*, *5*, *7*, *9*, *11*, *12*, *15*, *16*, *17*, *20*, *22*, *24*, *29* and *30*) were ABA inducible, and only 3 genes (*OsCIPK1*, *3* and *9*) showed response to cold. As the data indicates, some genes like *OsCIPK1, 3, 9, 15, 16, 17, 21, 22, 29* and *30* were induced by at least three of the tested stress conditions while others such as *OsCIPK3* show significant induction under cold stress with small rise on ABA and PEG treatment [108]. In addition, most of the genes regulated by salt stress were also induced by drought while those *OsCIPK* genes that were induced by either ABA or PEG, also showed response to both drought and salt stress. This observation as noted by the authors lends support to the concept of cross-talk or interaction between different abiotic stress signaling pathways where most of the genes mediating response to drought or salt also bring about the response to ABA, which is a stress hormone. Since many of the abiotic stress conditions cause dehydration and osmotic stress, therefore, common CIPKs might be induced in that response. At the same time, many of the *OsCIPK*s that were induced by salt or drought do not respond to ABA [108]. This again hints toward the possible existence of ABA and ABA-independent pathway in plants for sensing the abiotic stress conditions.

For *OsCIPK03*, which shows prominent expression under cold stress conditions alone, transgenic lines of *japonica* rice var Zhongua11 overexpressing this gene were generated and these plants were found to be exhibiting an improved tolerance to cold [108]. The transgenic lines were found to be having a better survival rate in comparison to the wild type rice plants. Similarly, overexpression of *OsCIPK12* and *OsCIPK15* rendered the transgenic rice plants with enhanced survival rates under drought and salt stress conditions respectively. Further, to examine the reason for such resistance that is brought about by these genes, the level of proline, which is an osmolyte, as well as that of soluble sugars was tested [108]. In both *OsCIPK03* and *OsCIPK12* overexpressing transgenic lines, higher accumulation of both of these substances was found, which correlated with higher expression levels for two proline biosynthesis genes as well as for two proline transporter genes [108]. Such improvement in tolerance toward the abiotic stresses achieved by higher expression of these CIPK genes establishes the role and importance of the CBL-CIPK signaling pathway as a tool for creating transgenic plants tolerant to environmental stresses. *OsCIPK15* has also been studied in the context of submergence/ flooding stress, which creates low oxygen conditions [109]. Under this condition, *OsCIPK15* has been found to be activating SnRK1 (energy sensor kinase) by phosphorylation, which further activate the transcription of α amylase through MYBS1 transcriptional activator. This promotes the conversion of starch to glucose for increased ATP production. Consistent with the role of CIPK15, *cipk15* loss-of-function mutant seeds do not germinate under water unless provided with sucrose. A recent study has shown the overlapping role of a rice *CIPK* gene namely, *OsCIPK23*, in both development as well as abiotic stress response [80]. High expression of this gene was observed in anther and pistil, which was up regulated after pollination. Also, *OsCIPK23* was induced in response to abiotic stress conditions such as cold, drought and salt. Induction of this gene was also observed after giberrellin and ABA treatment. Both RNAi as well as overexpression approach showed *OsCIPK23* to be essential for seed development and drought tolerance. Subcellular localization using GFP constructs showed this kinase to be mainly a nuclear protein with weak expression in cytosol [80].

Another kinase from rice, *OsCIPK31* has also been studied in greater detail in seedling stage and was found to be induced by salt or mannitol treatment [110]. The mutant lines showed hypersensitivity towards ABA, salt and glucose. Expression of several stress responsive genes was elevated in these lines at seedling or germination stage. However, no change in the levels of ABA or ABA regulated genes was found.

Recently, role of *OsCIPK14* and 15 has been uncovered in innate immunity in rice cultured cells where MAMP (microbe associated molecular patterns) induced cytosolic calcium changes were observed to be regulating phytoalexin biosynthesis and defense gene expression *via* these CIPK members [111]. *OsCIPK14/15* RNAi transgenic lines exhibited compromised defense response such as mitochondrial dysfunction and hypersensitive cell death suggesting their importance in biotic stress signaling.

CIPKs in Other Plants

Study aiming to determine CIPK genes in rice also identified the CBL-CIPK gene network in other crop plant such as wheat (11 CBLs and 29 CIPKs), barley (9 CBLs and 14 CIPKs) and soybean (7 CBLs and 13 CIPKs) [85]. Besides these, other plants in which these genes were found to be present includes *Medicago truncula ta* (9 CBLs and 11 CIPKs), Gymnosperm sp. *Pinus* sp. (2 CBLs and 7 CIPKs) and moss *Physcomirtella patens* (4 CBLs and 3 CIPKs) [85]. Genome-wide analysis of Poplar (*Populus trichocarpa*), which is a woody plant with known genome sequence, has revealed the existence of 27 CIPKs in this tree species [112]. Characterization of CIPK24 in a related species (*Populus euphratica*), namely PeCIPK24 (homolog of AtCIPK23) identified this kinase member to be interacting with two shaker-like potassium channels (PeKC1 and PeKC2) that help in potassium absorption [113]. Further, the report found that overexpression of these two potassium channels could rescue the low K^+ stress phenotype of *akt-1* mutants showing that these potassium channels can complement the function of *akt-1*by acting possibly through the CBL-CIPK network.

Studies on other plant species such as maize have also provided support for the existence of CBL-CIPK signaling pathway in these plants where expression of different CIPK members under stress conditions combined with other molecular genetic evidence and localization studies clearly establishes the importance of this plant specific CBL-CIPK signaling pathway in abiotic stress adaptation. Bioinformatics analysis identified 43 putative CIPKs in maize that showed high evolutionary relationship to rice CIPKs. Microarray and RT-PCR showed these genes to be responsive to abiotic stresses. *ZmCIPK 24, 31, 20* and *19* showed up regulation in response to salt, drought, heat and cold stress respectively. Further analysis revealed differences in CIPK gene expression pattern between cold tolerant and cold sensitive verities [114]. ZmCIPK16, which shares high level of amino acid identity with OsCIPK16 (79%) and AtCIPK16 (51%) was found to be showing up regulation at the transcript level under drought, ABA, PEG, cold, salt and heat treatment [115]. RNA blots have indicated significant rise in the level of *ZmCIPK16* in PEG treated root and shoot. Maize seedlings were found to be accumulating highest level of *ZmCIPK16* transcripts after 12h of salt stress, 24h of heat stress, and after 6h in case of both ABA and drought treatment. However, no such rise was observed in the level of *ZmCIPK16* after cold treatment. Expression of *ZmCIPK16* in *sos2* mutants of Arabidopsis partially restored the salt sensitive phenotype in these plants providing strong support for the involvement of *ZmCIPK16* in tolerating salt stress. Also, expression of *sos1* was enhanced in the transgenic plants overexpressing the *ZmCIPK16*, which showed that ZmCIPK16 was acting by directly activating the *AtSOS1* expression. As an effort to map the interacting partners of *Zm*CIPK16 from maize, yeast two-hybrid studies were also performed, and ZmCIPK16 was found to be interacting with all the tested ZmCBLs, namely, ZmCBL3, 4, 5 and 8 at varying strength. Strongest interaction was observed between ZmCBL4 and ZmCIPK16, which perhaps indicate the role of both these interacting partners in the salt stress signaling as previous study has shown *ZmCBL4* expression in Arabidopsis to be improving salt tolerance [116]. ZmCIPK16 was localized to the nucleus, cell membrane and to a lesser extent in the cytosol. The ZmCBL3, 4 and 5 were found to be restricted to the plasma membrane, and the complex of these maize CBLs with ZmCIPK16 in the BiFC assay were found to be plasma membrane localized. Of these, ZmCBL4 and 5 were determined to be having the N-myristoylation motif and ZmCBL3, which lacks this membrane-anchoring motif, is believed to be having several palmitoylation sites, which might aid in its localization to the membrane. Similar studies in rice have shown that OsCBL2 and OsCBL3, which do not have any such predicted membrane localization domains, are anchored to the tonoplast while OsCBL4 with a myristoylation motif is located at the plasma membrane [117]. All such localization signals anchoring the CIPK in specific cell compartments by CBLs are reflective of their site of function where the CBL-CIPK complex might be acting on other target proteins.

In a very recently published study [118] in apple (*Mallus domestica*), MdCIPK6L was shown to be interacting with AtSOS3 suggesting that it might function in salt tolerance pathway like AtSOS2. Overexpression of both MdCIPK6L and MdCIPK6 T175D (mutant version with threonine site mutated to aspartic acid) enhanced salt, drought and cold tolerance in apple, Arabidopsis and tomato. The fact that MdCIPK6T175D could also improve abiotic stress tolerance across the species implied the similarity of function performed by calcium signaling components in performing biological function. Another study in apple characterized the role of MdSOS2 and found it to be complementing the Arabidopsis *sos2* mutant in addition to conferring salt tolerance phenotype in transgenic Arabidopsis [119]. Yeast two-hybrid analyses showed N- terminal of MdSOS2 to be interacting with MdSOS3 and AtSOS3 demonstrating the existence and conservation of SOS pathway in apple tree.

Recent report (Cuellar *et al.*) [120] in grapevine identified shaker channel VvK 1.1 as the Arbidopsis AKT1 ortholog in controlling potassium influx across the plasma membrane by CBL/CIPK network [120]. Other distinctive features of this channel included activation by a more negative membrane potential and strong induction by drought stress and ABA. More than 10 fold up regulation at the transcript level was observed upon drought stress suggesting that it plays an important role in potassium loading in the berry tissues under desiccation conditions.

An interesting study in *Vicia* guard cells showed Vf CIPK1 to be localized on the outer mitochondrial membrane and interacting with VfCBL1 [120]. This interaction was weakened by increasing calcium concentration and showed maximum activation in the absence of calcium. This is a first report showing negative role of calcium in CIPK activity mediated by CBL.

Study in Pea (*Pisum sativum*), which is a legume plant, has also shown the existence of CBL-CIPK signaling network, and its role in both abiotic and biotic stress conditions [121]. PsCIPK and PsCBL were induced under salt, cold and wounding stress, but no difference in expression was found upon dehydration or

ABA treatment. Transcript accumulation for both PsCBL as well as PsCIPK was observed upon salicylic acid and calcium treatment. This might be due to the involvement of this CBL-CIPK combination in biotic stresses, as salicylic acid is commonly produced when plants face a pathogenic attack causing biotic stress. Immunolocalization experiments found PsCIPK to be both cytosolic as well as membrane associated. Also, PsCIPK and PsCBL were found to be interacting in yeast two-hybrid analysis, which might reflect their actual interaction *in vivo* that helps localize PsCIPK to the membrane. Two CBL and one CIPK genes were also identified in cotton (*Gossypium hirsutum*) that showed overlapping expression in the cotton fibers indicating their possible function in fiber elongation [122]. Selective interaction could be seen of GhCIPK1 with both GhCBL2 and GhCBL3 during yeast two-hybrid assay [122].

In summary, CIPKs in several plants have been discovered to be present as multi-gene family and many members are differentially regulated at the transcript levels and summarized briefly in the Table **2**. In Arabidopsis, several members of CIPKs are shown to be regulated by multiple stimuli and for few of them the physiological substrates or targets have been assigned. Fig. **2** summarizes a few of the CIPKs from Arabidopsis involved to regulate different subcellular targets upon activation of calcium signaling mediated by different stress or stimuli.

Table 2: Summary of CIPK gene expression induced under different abiotic stress conditions as determined from quantitative PCR, microarray and northern blot studies in Arabidopsis and rice. Various abiotic stress stimuli are known to activate CIPK genes at the transcription level, which then regulate a host of cellular targets to promote stress tolerance. Few of the CIPK members such as *CIPK10, 21, 23* are induced by salt and drought but not by ABA suggesting possible existence of ABA independent pathways that control stress tolerance distinctly from ABA.

Stress or Stimuli	CIPK genes induced
ABA	***CIPK 1, 2, 3, 5, 7, 9,11, 12, 15, 16, 17, 20, 22, 24, 29 and 30***
Cold	***CIPK1, 3 and 9***
Drought	***CIPK1, 2, 5, 9, 11, 12, 15, 17, 20, 21, 22, 23,24, 29, and 30***
Salt	***CIPK7, 8, 9, 10,11, 15, 16, 17, 21, 22, 23, 24 and 29***
Nutrient	***CIPK6, 9, 23*** **(potassium)** ***CIPK8*** **(nitrate)**

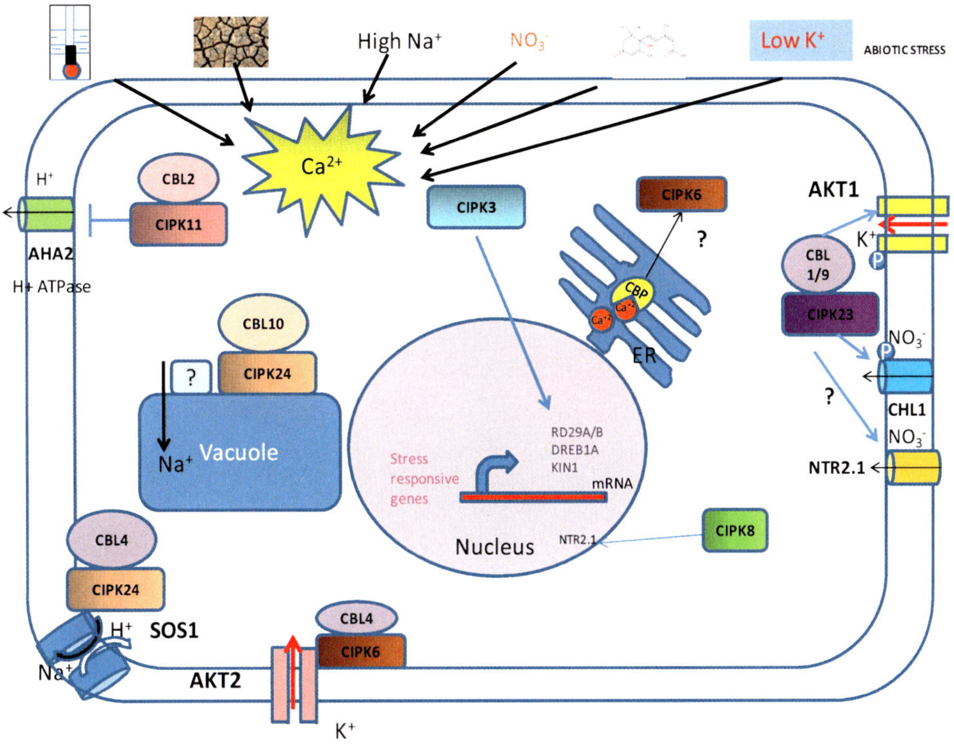

Figure 2: A general model depicting cellular functions and known targets of CBL-CIPK signaling identified in different plant species. Different abiotic stress stimuli such as cold, drought, salinity, ABA and nutrient stress triggers defined changes in calcium levels in the cell, which then activates specific CBL-CIPK members within the cell. These CBL-CIPK components target a host of transporters on the plasma membrane and vacuole to regulate ion homeostasis and stress tolerance of the plant. In addition to this, some CIPK members such as CIPK3 mediate the activation of stress responsive genes such as RD29A/B and DREB1A at the transcriptional level to promote abiotic stress tolerance. Abbreviations used: AHA2- H^+ ATPase, AKT-1 – K^+ Transporter, CBP- calcium binding peptide, NTR2.1- Nitrate transporter, SOS-1 – Salt overly sensitive

OTHER CALCIUM REGULATED PROTEIN KINASES AND ABIOTIC STRESS

Besides CDPKs and CIPKs, there are other classes of calcium regulated kinases, which are yet to be explored in detail, but are postulated to be playing an important role in abiotic stress signaling based on several reports. CDPK-related protein kinase or (CRK) and Ca^{2+}/CaM kinase or (CCaMK) are two such classes of CDPK-related kinases that have been studied in different plant varieties. CRKs like CDPKs possess a N-terminal myristoylation motif, which might help in plasma membrane associations. However, CRKs have degenerate EF-hands that are incapable of binding calcium ions [27]. CDPK related protein kinase or CRKs have been studied in plant species such as maize and Arabidopsis. Almost all of the CRKs studied till date is found to be showing kinase activity that is both Ca^{2+} and CaM-independent. AtCRK3 from Arabidopsis like ZmCRK from maize exhibits autophosphorylation and substrate phosphorylation that is independent of both Ca^{2+} and CaM [123,124]. However, kinase activity of AtCRK3 was inhibited by 1mM of Ca^{2+}. Although this is much higher than physiological concentration of calcium, it indicates a possible mechanism of calcium regulation of AtCRK3. *AtCRK3* expression was induced by ABA as analyzed by northern hybridization. Since ABA is commonly produced under stress conditions, and induces cytosolic changes in Ca^{2+}, it was proposed by the authors that AtCRK3 could possibly be participating in ABA stress signaling by phosphorylating specific substrates. Abundant transcript levels could be detected in vascular bundles of stem and leaf as well as in tapetum layer of pollen mother cells [123]. A characteristic differential expression pattern was observed during pollen development. The *AtCRK3* transcripts gradually increased and showed highest expression at the tetrad stage, and then, declined below detection level at the mature pollen stage. Tapetum, which is responsible for providing nutrition to the growing pollen, also had a similar expression profile of AtCRK3 indicating that this kinase might be important for function and metabolism of the tapetal layer.

AtGLN1 (cytosolic glutamine synthetase involved in nitrogen assimilation) has been identified as a potential substrate for AtCRK3 using yeast two-hybrid approach [125]. This interaction was also confirmed using co-immunoprecipitation assays. Both AtGLN1 and AtCRK3 were induced under early senescence stage, and AtCRK3 specifically phosphorylated AtGLN1. Such interaction is believed to be important for nitrogen mobilization during leaf senescence.

Identification and biochemical characterization of another CRK from Arabidopsis, AtCRK1, has shown this kinase to be binding to CaM isoforms in a calcium dependent manner [126]. Binding of the different CaM-isoforms, namely, AtCaM2, AtCaM4, AtCaM7 and AtCaM8 stimulated the kinase activity of AtCRK1 to similar extent. However, the autophosphorylation and substrate phosphorylation activities for this kinase were found to be calcium independent. Thus, different CRKs sharing high sequence identity (as in case of AtCRK1 and AtCRK3) could be regulated by diverse mechanisms specific to the protein kinase.

Calmodulin binding kinase (CaMK), and calcium and calmodulin regulated kinase (CCaMK), are both calmodulin binding protein kinases (CBKs), which possess a CaM binding domain. CaMKs form a large family of enzymes in animals [16]. In contrast, very few members of this type of kinase have been reported in plants such as apple and tobacco [127,128]. CCaMKs are structurally similar to CDPKs; however, these have three EF-hands, which are similar to the C-terminal domain of brain protein visinin [27]. Autophosphorylation of CCaMK is promoted by binding of calcium, which further stimulates its association with Ca^{2+}/CaM complex. This Ca^{2+}/CaM complex is found to enhance substrate phosphorylation by CCaMK. Several reports have provided insight into the biochemical as well as expression based characteristics of CCaMK in different plant species such as carrot, apple, maize and lily [129]. However, no representative CaMK or CCaMK has been reported from Arabidopsis [16].

NtCBK2 is a CaM binding protein kinase, which is stimulated specifically by GA and salt stress [130]. Of the different stages, developing anther, pistil and embryo showed high expression of this gene. AtCBK3 is also a Ca^{2+}/CaM kinase that has been shown to be important in heat stress signaling. Yeast two-hybrid study found AtCBK3 to be interacting with AtHSF1a [131]. MCK1 and MCK2 are CRK and Ca^{2+}/CaM kinase, respectively from maize that showed a temporal and spatial variation in the pattern of expression during development [132]. MCK1 was identified as a CCaMK [133] and later classified as a CRK based on sequence analysis [27].

PsCCaMK from pea was seen to be involved in stress signaling [134] while *NtCaMK1* is up regulated upon salt stress treatment [128]. Increased accumulation of *PsCCaMK* was observed in root tissue upon low temperature and salinity stress [134]. Tissue specific expression studies have been conducted for CCaMK from lily and tobacco where lily CCaMK was restricted to pollen mother and tapetal cells [128]. Tobacco CCaMK1 and CCaMK2 show mRNA and protein expression specific to the meiosis stage of the anther development [128]. Despite several expression analyses studies of these calcium regulated kinases, and biochemical characterization for a few of them, not much has been elucidated regarding their detail *in planta* function, and future studies are required to understand the exact molecular function performed by these proteins in signaling pathways.

Although the physiological targets or substrates of CRK have not been investigated in detail yet, the CRK family of kinase are also thought to be regulating several physiological processes by targeting metabolic enzymes involved in nitrogen metabolism, heat stress and other abiotic stress by affecting the physiological activity or regulating gene expression as summarized in the Table **3** and Fig. **3**.

Table 3: Summary of CRK and CCaMK genes known to be induced under different abiotic stress conditions from studies in different plant species. Several studies have shown the induction/ activation of different CRK/CCaMK under abiotic stress conditions. AtCBK3, which is a calcium binding kinase is found to be interacting with HSF1 to possibly control heat stress tolerance while studies in other species such as pea and tobacco, mRNA accumulation of a few CCaMK/CRK has been observed under cold, salt and drought stress.

Stress or stimuli	Gene/Protein
Cold	***PsCCaMK***
Salt	***NtCaMK1, PsCCaMK***
ABA	***AtCRK3***
Heat	**AtCBK3**

CONCLUSION AND FUTURE PERSPECTIVE

Calcium plays a pivotal role in multiple biological processes including signaling events mediated in abiotic stress in plants. Calcium sensors are the proteins, which sense the "calcium signature" and transduce the signal downstream through a number of protein components. In plants, there are three major classes of calcium signaling kinases, which exist as multigene families. Gene expression profiling at single and whole genome level have provided a crucial biological clue for deciphering the function of these genes at different developmental and environmental conditions. Expression analysis for these calcium signaling kinases at both transcript (transcriptomic) and protein (proteomic) level has caught

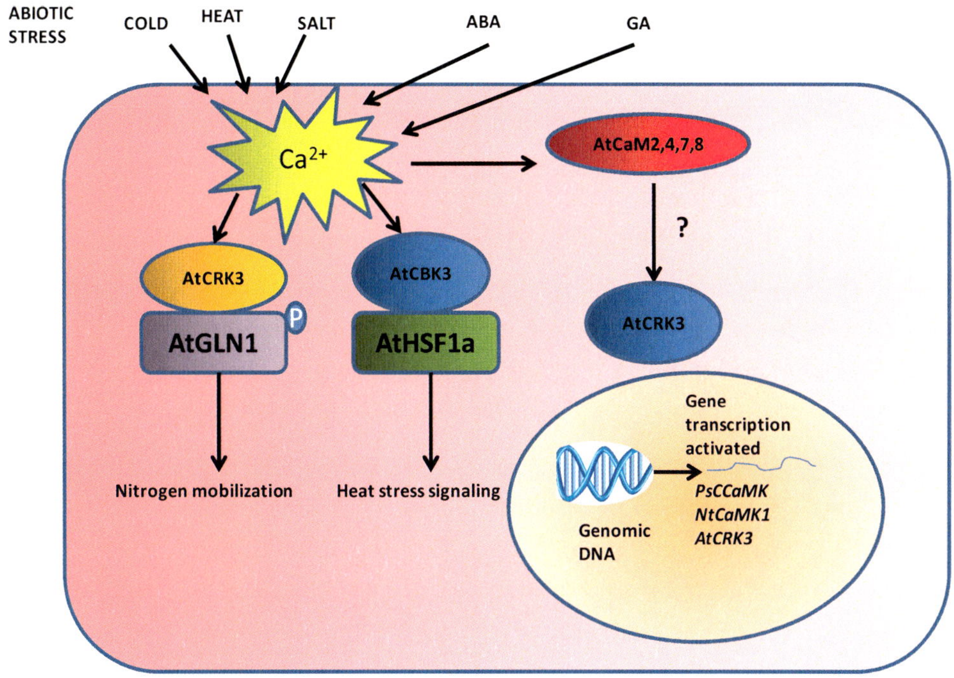

Figure 3: Model showing some of the CRK and CCaMK involved in abiotic stress tolerance and their known cellular targets. Various abiotic stress stimuli trigger calcium oscillations that then promote activation of specific metabolic and stress signaling pathways *via* CRKs or CaM proteins, a few of which have been depicted here. AtCRK3 directly phosphorylates glutamine synthetase, which is an enzyme involved in nitorgen metabolism while AtCBK3 directly interacts with heat shock transcription factor (HSF1a) to promote heat stress signaling and tolerance under thermal stress. Specific members of the CRK and CCaMK in different species are known to be induced at the transcript level by different stress stimuli. Abbreviations used: CRK- CDPK related kinase, CCaMK- Ca/CaM actiavted kinase, CBK Calcium binding kinase, HSF1- Heat shock transcription factor, CaM- calmodulin, GLN1- Glutamine synthetase 1.

attention of several leading research groups who are now attempting to determine, and integrate this important information for investigating the functional analysis of these genes or proteins families.

Multiple approaches are required to understand the functions of these proteins that include site-directed mutagenesis or deletion of key residues or domains, reverse genetics (knockouts) to identify null mutations, overexpression studies, protein interaction screens to identify potential substrates, biochemical analyses to characterize kinetic properties, and integration of expression and localization studies to clarify where and when various family members are expressed.

Existence of multigene family for these calcium-signaling kinases that are redundant in function poses a major challenge for functionally characterizing these by reverse genetic, overexpression studies, and other molecular and biochemical approaches. Therefore, a detailed expression profiling of these families under different developmental and stress conditions will provide crucial biological information to plan an in-depth molecular dissection of individual and multiple genes in plants. Moreover, a comprehensive identification of downstream targets of these kinases is required in order to understand the mechanistic details of a given signaling pathway. In addition, a major challenge for the future research is the elucidation of the interconnections and synergistic functions of these gene families in the diverse signaling network in plants. The cross talk of this calcium signaling kinases with other signaling pathways such as receptor like-kinase (RLKs), two-component systems, and mitogen associated protein kinases (MAPK) cascades also require further investigation. Ultimately, utilizing the tools of transgenic technology and molecular breeding to enhance abiotic stress tolerance by employing key components of these calcium-signaling kinases in agriculturally important crop needs to be undertaken.

ACKNOWLEDGEMENT

We are thankful to Delhi University, Department of Biotechnology (DBT) and Department of Science and Technology (DST), India for supporting the research work in GKP's lab. Because of space constraint we could not discuss several related research papers.

CONFLICT OF INTEREST

The author(s) confirm that this chapter content has no conflict of interest.

DISCLOSURE

"This is an updated report of the earlier review article by Ritika Das and Girdhar K. Pandey" Expressional Analysis and Role of Calcium Regulated Kinases in Abiotic Stress Signaling". Current Genomics, 2010, *11,* 2-13".

REFERENCES

[1] Sanders, D; Brownlee C;Harper, J.F. Communicating with calcium. *Plant Cell.* **1999**, *11*, 691–706.

[2] Rudd, J.J.; Franklin-Tong, V.E. Unraveling response-specificity in Ca^{2+} signaling pathways in plant cells. *New Phytol.* **2001**, *151*, 7–33.

[3] Harper J.F. Dissecting calcium oscillators in plant cells. *Trends Plant Sci.* **2001**, *6*, 395–397.

[4] Knight, H; Knight, M.R. Abiotic stress signaling pathways: Specificity and cross-talk. *Trends Plant Sci.* **2001**, *6*, 262–267.

[5] McAinsh, M.R.; Hetherington, A.M. Encoding specificity in Ca^{2+} signaling systems. *Trends Plant Sci.* **1998**, *3*, 32–36.

[6] Allen, G.J.; Chu, S.P.; Harrington, C.L.; Schumacher, K; Hoffmann, T;Tang, Y.Y.;Grill, E; Schroeder, J.I. A defined range of guard cell calcium oscillation parameters encodes stomatal movements. *Nature.* **2001**, *411*, 1053–1057.

[7] Hetherington, A.M.;Brownlee, C. The generation of Ca^{2+} signals in plants. *Annu. Rev. Plant Physiol. Plant Mol. Biol.* 2004, *55*, 401–427.

[8] Dodd, AN, Kudla, J, Sanders, D. The Language of Calcium Signaling. Annu Rev Plant Biol. 2010, *61*, 593-620.

[9] Stael, S; Wurzinger, B; Mair, A; Mehlmer, N;Vothknecht, UC; Teige, M. Plant organellar calcium signalling: an emerging field.*J Exp Bot.* **2012**, *63*(4), 1525-42.

[10] Snedden, W; Fromm, H. Calmodulin, calmodulin-related proteins and plant responses to the environment. *Trends Plant Sci.* **1998**, *3*, 299–304.

[11] Luan S, Kudla J, Rodriguez-Concepcion M, Yalovsky S, Gruissem W. Calmodulins and calcineurin B-like proteins: calcium sensors for specific signal response coupling in plants. *Plant Cell.* **2002**, *14*, 389–400.

[12] Sanders, D;Pelloux, J;Brownlee, C; Harper, J.F. Calcium at the crossroads of signaling. *Plant Cell.* **2002**, *14*, 401–417.

[13] Reddy, A.S.N. Calcium: silver bullet in signaling. *Plant Sci.* **2001**, *160*, 381–404.

[14] Batistic, O;Kudla, J. Integration and channeling of calcium signaling through the CBL calcium sensor/CIPK protein kinase network. *Planta.* **2004**, *219*, 915–924.

[15] Harmon, A.C.;Gribskov, M;Gubrium, E; Harper, J.F. The CDPK superfamily of protein kinases. *New Phytol.* **2001**, *151*, 175–183.

[16] Hrabak, E.M.; Chan, C.W.; Gribskov, M; Harper, J.F.; Choi, J.H.; Halford, N; Kudla, J;Luan S; Nimmo, H.G.; Sussman, M.R.; Thomas, M; Walker-Simmons, K; Zhu, J.K.; Harmon, A.C. The Arabidopsis CDPK-SnRK superfamily of protein kinases. *Plant Physiol.* **2003**, *132*, 666–80.

[17] Cheng, S.H.; Willmann, M.R.; Chen, H.C.; Sheen, J. Calcium signaling through protein kinases. The Arabidopsis calcium-dependent protein kinase gene family. *Plant Physiol.* **2002**, *129*, 469–85.

[18] Harper, J.F.;Breton, G; Harmon, A. Decoding Ca (2+) signals through plant protein kinases. *Annu. Rev. Plant Biol.* **2004**, *55*, 263–288.

[19] Reddy, AS; Ali, GS; Celesnik, H; Day, IS.Coping with stresses: roles of calcium- and calcium/calmodulin-regulated gene expression.*Plant Cell.* **2011** *23*, 6, 2010-32

[20] Hashimoto, K; Kudla, J Calcium decoding mechanisms in plants. Biochimie. **2011** 93, *12*, 2054-9.

[21] Liu, J; Zhu, J.K. A calcium sensor homolog required for plant salt tolerance. *Science*. **1998**, *280*, 1943–1945.

[22] Kudla, J; Xu, Q;Harter, K; Gruissem, W; Luan, S. Genes for calcineurin B-like proteins in Arabidopsis are differentially regulated by stress signals. *Proc. Natl. Acad. Sci. USA*. **1999**, *96*, 4718–4723.

[23] Shi, J; Kim, K.N.;Ritz, O; Albrecht, V; Gupta, R; Harter, K; Luan, S; Kudla, J. Novel protein kinases associated with calcineurin B-like calcium sensors in Arabidopsis. *Plant Cell*. **1999**, *11*, 2393–2405.

[24] Halfter U, Ishitani M, Zhu J.K. The Arabidopsis SOS2 protein kinase physically interacts with and is activated by the calcium-binding protein SOS3. Proc. Natl. Acad. Sci. USA. **2000**, *97*, 3735–3740.

[25] Pandey, G.K.; Cheong, Y.H.; Kim, B.G.; Grant, J.J.; Li, L; Luan, S. CIPK9: a calcium sensor-interacting protein kinase required for low-potassium tolerance in Arabidopsis. *Cell Res*. **2007**, *17*, 411–42.

[26] Luan, S. The CBL-CIPK network in plant calcium signaling. *Trends Plant Sci*. **2009**;14, 37–42.

[27] Harmon, A.C. Calcium-regulated protein kinases of plants. *Gravity. Space Biol. Bull*. **2003**, *16*, 83–90.

[28] Tena, G; Boudsocq, M; Sheen, J. Protein kinase signaling networks in plant innate immunity.*Curr Opin Plant Biol*. **2011**, *14*, 5, 519-29.

[29] Ludwig, A.A.; Romeis, T, Jones J.D. CDPK-mediated signaling pathways: Specificity and cross-talk. *J. Exp. Bot*. **2004**, *395*, 181–188.

[30] Martín, M.L.; Busconi, L. Membrane localization of a rice calcium-dependent protein kinase (CDPK) is mediated by myristoylation and palmitoylation. *Plant J*. **2000**, *24*, 429–435.

[31] Milligan, G; Parenti, M; Magee, A.I. The Dynamic role of palmitoylation in signal transduction. *Trends Biol. Sci*. **1995**, *20*, 181–186.

[32] Choi, H.I.; Park, H.J.; Park, J.H.;Kim, S; Im, M.Y.; Seo, H.H.; Kim, Y.W.; Hwang, I; Kim, S.Y. Arabidopsis calcium-dependent protein kinase AtCPK32 interacts with ABF4, a transcriptional regulator of abscisic acid-responsive gene expression, and modulates its activity. *Plant Physiol*. **2005**, *139*, 1750–1761.

[33] Sawasaki, T; Hasegawa, Y; Morishita, R; Seki, M;Shinozaki, K; Endo, Y. Genome-scale, biochemical annotation method based on the wheat germ cell-free protein synthesis system. *Phytochemistry*. **2004**, *65*, 1549–1555.

[34] Klimecka, M; Muszynska, G. Structure and functions of plant calcium-dependent protein kinases. *Acta Biochim. Pol*. **2007**, *5*, 219–33.

[35] Yoon, G.M.; Cho, H.S.; Ha, H.J.; Liu, J.R.; Lee, H.S. Characterization of NtCDPK1, a calcium-dependent protein kinase gene in *Nicotiana tabacum*, and the activity of its encoded protein. *Plant Mol. Biol*. **1999**, *39*, 991–1001.

[36] Saijo, Y;Hata, S; Sheen, J; Izui, K. cDNA cloning and prokaryotic expression of maize calcium-dependent protein kinases. *Biochim. Biophys. Acta*. **1997**, *1350*, 109–114.

[37] Murillo, I; Jaeck, E; Cordero, M.J.; San Segundo, B. Transcriptional activation of a maize calcium-dependent protein kinase gene in response to fungal elicitors and infection. *Plant Mol. Biol*. **2001**, *45*, 145–158.

[38] Frattini, M; Morello, L; Breviario, D. Rice calcium-dependent protein kinase isoforms OsCDPK2 and OsCDPK11 show different responses to light and different expression patterns during seed development. *Plant Mol. Biol*. **1999**, *41*, 753–764.

[39] Köhler, B;Blatt, M.R. Protein phosphorylation activates the guard cell Ca^{2+} channel and is a prerequisite for gating by abscisic acid. *Plant J.* **2002**, *32*, 185–194.

[40] Sheen J. Ca^{2+}-dependent protein kinases and stress signal transduction in plants. *Science.* **1996**, *274*, 1900–1902.

[41] Urao, T; Katagiri, T; Mizoguchi, T; Yamaguchi-Shinozaki, K; Hayashida, N; Shinozaki, K. Two genes that encode Ca(2+)-dependent protein kinases are induced by drought and high-salt stresses in Arabidopsis thaliana. *Mol. Gen. Genet.* **1994**, *244*, 331–340.

[42] Mori, I.C.; Y, Yang; Y, Munemasa, S; Wang, Y.F.;Andreoli, S;Tiriac, H; Alonso, J.M.; Harper, J.F.; Ecker, J.R.; Kwak, J.M.; Schroeder, J.I. CDPKs CPK6 and CPK3 function in ABA regulation of guard cell S-type anion- and Ca (2+)-permeable channels and stomatal closure. *PLoS Biol.* **2006**, *4*, 1749–1762.

[43] Zhu, S.Y.; Yu, X.C.; Wang, X.J.; Zhao, R;Li, Y;Fan, R.C.; Shang, Y; Du, S.Y.; Wang, X.F.; Wu, F.Q.; Xu, Y.H.; Zhang, X.Y.; Zhang, D.P. Two calcium dependent protein kinases, CPK4 and CPK11, regulate abscisic acid signal transduction in Arabidopsis. *Plant Cell.* **2007**, *19*, 3019–3036.

[44] Ma, S;Wu, W. AtCPK23 functions in Arabidopsis responses to drought and salt stresses. *Plant Mol. Biol.* **2007**, *65*, 511–518.

[45] Rodriguez, Milla M.A.;Uno, Y; Chang, I.F.; Townsend, J; Maher, E.A.; Quilici, D; Cushman, J.C. A novel yeast two-hybrid approach to identify CDPK substrates: characterization of the interaction between AtCPK11 and AtDi19, a nuclear zinc- finger protein. *FEBS Lett.* **2006**, *580*, 904–911.

[46] Hwang, I; Sze, H; Harper, J.F. A calcium-dependent protein kinase can inhibit a Calmodulin-stimulated Ca2+ pump (ACA2) located in the endoplasmic reticulum of Arabidopsis. *Proc. Natl. Acad. Sci. USA.* **2000**, 6224–6229.

[47] Dammann, C;Ichida, A; Hong, B; Romanowsky, S.M.; Hrabak, E.M.; Harmon, A.C.; Pickard, B.G.; Harper, J.F. Subcellular targeting of nine calcium-dependent protein kinase isoforms from Arabidopsis. *Plant Physiol.* **2003**, *132*, 1840–1848.

[48] Lu, S.X.; Hrabak, E.M. An Arabidopsis calcium-dependent protein kinase is associated with the endoplasmic reticulum. *Plant Physiol.* **2002**, *128*, 1008–1021.

[49] Geiger, D; Scherzer, S; Mumm, P;Marten, I;Ache, P; Matschi, S;Liese, A, ;Wellmann, C;Al-Rasheid, KA; Grill, E; Romeis, T; Hedrich, R Guard cell anion channel SLAC1 is regulated by CDPK protein kinases with distinct Ca2+ affinities .*Proc Natl Acad Sci U S A.* **2010**, *107*, 17, 8023-8.

[50] Wan, B;Lin, Y; Mou, T. Expression of rice Ca (2+)-dependent protein kinases (CDPKs) genes under different environmental stresses. *FEBS Lett.* **2007**, *581*, 1179–1189.

[51] Ray, S; Agarwal, P; Arora, R; Kapoor, S; Tyagi, A.K. Expression analysis of calcium-dependent protein kinase gene family during reproductive development and abiotic stress conditions in rice (Oryza sativa L. ssp. indica) *Mol. Genet. Genom.* **2007**, *278*, 493–505.

[52] Asano, T;Hakata, M; Nakamura, H; Aoki, N; Komatsu, S; Ichikawa, H; Hirochika, H; R. Functional characterisation of OsCPK21, a calcium-dependent protein kinase that confers salt tolerance in rice *Plant Mol Biol.* **2011**, *75*(1-2), 179-91

[53] Nakamura, H; Hakata, M; Amano, K; Miyao, A; Toki, N; Kajikawa, M; Pang, J; Higashi, N;Ando, S; Toki, S; Fujita, M; Enju, A; Seki, M;Nakazawa, M; Ichikawa, T; Shinozaki, K; Matsui, M; Nagamura, Y;Hirochika, H;Ichikawa, H.A genome-wide gain-of function analysis of rice genes using the FOX-hunting system. *Plant Mol Biol.* **2007**, *65*(4), 357-71

[54] Zhao, R; Sun, HL; Mei, C; Wang, XJ; Yan, L; Liu, R; Zhang, XF; Wang, XF;Zhang, DP The *Arabidopsis* Ca(2+)-dependent protein kinase CPK12 negatively regulates Abscisic acid signaling in seed germination and post-germination growth. *New Phytologist* **2011**, *192*(1), 61-73

[55] Abbasi, F; Onodera, H; Toki, S; Tanaka, H; Komatsu, S. OsCDPK13, a calcium-dependent protein kinase gene from rice, is induced by cold and gibberellin in rice leaf sheath. *Plant Mol. Biol.* **2004**, *55*, 541–55.

[56] Saijo, Y; Hata, S; Kyozuka, J;Shimamoto, K; Izui, K. Over-expression of a single Ca2+-dependent protein kinase confers both cold and salt/drought tolerance on rice plants. *Plant J.* **2000**, *23*, 319–327.

[57] Yang, G; Shen, S; Yang, S; Komatsu, S. OsCDPK13, a calcium-dependent protein kinase gene from rice, is induced in response to cold and gibberellin. *Plant Physiol. Biochem.* **2003**, *41*, 369–374.

[58] Komatsu, S; Yang, G; Khan, M; Onodera, H; Toki, S; Yamaguchi, M. Over- expression of calcium-dependent protein kinase 13 and calreticulin interacting protein 1 confers cold tolerance on rice plants. *Mol. Genet. Genom.* **2007**, *277*, 713–723.

[59] Zhang, T; Wang, Q; Chen, X; Tian, C; Wang, X; Xing, T; Li, Y; Wang, Y. Cloning and biochemical properties of CDPK gene OsCDPK14 from rice. *J. Plant Physiol.* **2005**, *162*, 1149–1159.

[60] Saijo, Y; Kinoshita, N; Ishiyama, K; Hata, S; Kyozuka, J; Hayakawa, T; Nakamura, T; Shimamoto, K; Yamaya, T; Izui, K. A Ca(2+)-dependent protein kinase that endows rice plants with cold- and salt-stress tolerance functions in vascular bundles. *Plant Cell Physiol.* **2001**, *42*, 1228–1233.

[61] Mall, TK; Dweikat, I; Sato, SJ; Neresian, N;Xu, K; Ge, Z; Wang, D; Elthon, T; Clemente, T Expression of the rice *CDPK-7* in sorghum: molecular and phenotypic analyses. *Plant Molecular Biology* **2011**, *75* (4-5), 467-479.

[62] Lan, L; Li, M;Lai, Y;Xu, W;Kong, Z; Ying, K; Han, B;Xue, Y. Microarray analysis reveals similarities and variations in genetic programs controlling pollination/fertilization and stress responses in rice (*Oryza sativa* L.) *Plant Mol. Biol.* **2005**, *59*, 151–164.

[63] Estruch, J.J.; Kadwell, S; Merlin, E; Crossland, L. Cloning and characterization of a maize pollen-specific calcium-dependent calmodulin-independent protein kinase. *Proc. Natl. Acad. Sci. USA.* **1994**, *91*, 8837–8841.

[64] Li, A.L.; Zhu, Y.F.; Tan, X.M.; Wang, X; Wei, B; Guo, H.Z.; Zhang, Z.L.; Chen, X.B.; Zhao, G.Y.; Kong, X.Y.; Jia, J.Z.; Mao, L. Evolutionary and functional study of the CDPK gene family in wheat (*Triticum aestivum* L.) *Plant Mol. Biol.* **2008**, *66*, 429–443.

[65] Tsai, T; Chen, Y.R.; Kao, T.W.; Tsay, W.S.; Wu, C.P.; Huang, D.D.; Chen, W.H.; Chang C.C.; Huang H.J. *PaCDPK1*, a gene encoding calcium-dependent protein kinase from orchid, *Phalaenopsis amabilis*, is induced by cold, wounding, and pathogen challenge. *Plant Cell Rep.* **2007**, *26*, 1899–908.

[66] Liu, G; Chen, J; Wang, X. *VfCPK1*, a gene encoding calcium-dependent protein kinase from *Vicia faba*, is induced by drought and abscisic acid. *Plant Cell Environ.* **2006**, *29*, 2091–2099.

[67] Zhang, M; Liang, S; Lu, Y.T. Cloning and functional characterization of NtCPK4, a new tobacco calcium-dependent protein kinase. *Biochim. Biophys. Acta.* **2005**, *1729*, 174–185.

[68] Chico, J.M.; Raíces, M;Téllez-Iñón, M.T.; Ulloa, R.M. A calcium- dependent protein kinase is systemically induced upon wounding in tomato plants. *Plant Physiol.* **2002**, *128*, 256–270.

[69] Capiati, D.A.;País, S.M.; Téllez-Iñón, M.T. Wounding increases salt tolerance in tomato plants: evidence on the participation of calmodulin-like activities in cross-tolerance signalling. *J. Exp. Bot.* **2006**, *57*, 2391–2400.

[70] Patharkar, O. R.; Cushman, J. C. A novel coiled-coil protein co-localizes and interacts with a calcium-dependent protein kinase in the common ice plant during low-humidity stress. *Planta*. **2006**, *225*, 57–73.

[71] Botella, J.R.; Arteca, J.M.; Somodevilla, M; Arteca, R.N. Calcium-dependent protein kinase gene expression in response to physical and chemical stimuli in mungbean (*Vigna radiata*) *Plant Mol. Biol.* **1996**, *30*, 1129–1137.

[72] Davletova, S; Mészáros, T; Miskolczi, P; Oberschall, A; Török, K; Magyar, Z; Dudits, D; Deák, M. Auxin and heat shock activation of a novel member of the calmodulin like domain protein kinase gene family in cultured alfalfa cells. *J. Exp. Bot.* **2001**, *52*, 215–221.

[73] Contreras-Porcia, L;Dennett, G;González, A; Vergara, E; Medina, C; Correa, J.A; Moenne, A Identification of copper-induced genes in the marine alga *Ulva compressa* (Chlorophyta). *Marine Biotechnology* **2011**, *13*(3), 544-56.

[74] Kiselev, KV; Grishchenko OV, Zhuravlev YN CDPK gene expression in salt tolerant *rolB* and *rolC* transformed cell cultures of *Panax ginseng*. *Biologia Plantarum* **2009**, *54*(4), 621-630.

[75] Yu, X.C; Li, MJ;Gao, G.F; Feng, H.Z; Geng, X.Q; Peng, C.C; Zhu, S.Y;Wang, X.L; Shen, Y.Y, ;Zhang, D.P Abscisic acid stimulates a calcium-dependent protein kinase in grape berry. *Plant Physiology* **2006**, *140*, 558–579.

[76] Muñiz García, M; Giammaria, V; Grandellis, C; Téllez-Iñón, MT; Ulloa, R.M; Capiati, D.A. Characterization of StABF1, a stress-responsive bZIP transcription factor from *Solanum tuberosum* L. that is phosphorylated by StCDPK2 *in vitro*. *Planta* **2011**, DOI: 10.1007/s00425-011-1540-7.

[77] Pandey G.K. Emergence of a Novel Calcium Signaling Pathway in Plants: CBL- CIPK signaling network. *Physiol. Mol. Biol. Plants*. **2008**, *14*, 51–68.

[78] Hu, H.C.; Wang, Y.Y.; Tsay, Y.F. AtCIPK8, a CBL-interacting protein kinase, regulates the low-affinity phase of the primary nitrate response. *Plant J.* **2009**, *57*, 264–278.

[79] Tripathi, V;Parasuraman, B; Laxmi, A; Chattopadhyay, D. CIPK6, a CBL-interacting protein kinase is required for development and salt tolerance in plant. *Plant J.* **2009** DOI, 10.1111/j.1365-313X.2009.03812.

[80] Yang, W; Kong, Z; Omo-Ikerodah, E; Xu, W;Li, Q; Xue, Y. Calcineurin B-like interacting protein kinase OsCIPK23 functions in pollination and drought stress responses in rice (*Oryza sativa L.*) *J. Genet. Genom.* **2008**, *35*, 531–543.

[81] Batistic, O; Kudla, J. Plant calcineurin B-like proteins and their interacting protein Kinases. *Biochim. Biophys. Acta*. **2009**, *1793*, 985–92.

[82] Kim, K.N.; Lee, J.S.; Han, H; Choi, S.A.; Go, S.J.; Yoon, I.S. Isolation and characterization of a novel rice Ca2+-regulated protein kinase gene involved in responses to diverse signals including cold, light, cytokinins, sugars and salts. *Plant Mol. Biol.* **2003**, *52*, 1191–1202.

[83] Lee, S.C.; Lan, W.Z.; Kim, B.G.; Li, L; Cheong, Y.H.; Pandey, G.K.; Lu, G, Buchanan, B.B.; Luan, S. A protein phosphorylation/dephosphorylation network regulates a plant potassium channel. *Proc. Natl. Acad. Sci. USA*. **2007**, *104*, 15959–15964.

[84] Martínez-Atienza, J; Jiang, X; Garciadeblas, B; Mendoza, I; Zhu, J.K.; Pardo, J. M.; Quintero F. J. Conservation of the SOS salt tolerance pathway in rice. *Plant Physiol.* **2007**, *143*, 1001–1012. [

[85] Kolukisaoglu, U; Weinl, S; Blazevic, D; Batistic, O; Kudla, J. Calcium sensors and their interacting protein kinases: genomics of the Arabidopsis and rice CBL-CIPK signaling networks. *Plant Physiol.* **2004**, *134*, 43–58.

[86] Jeong, H. J.; Jwa, N.S.; Kim, K.N. Identification and characterization of protein kinases that interact with the CBL3 calcium sensor in Arabidopsis. *Plant Sci.* **2005**, *169*, 1125–1135.

[87] Kim, K.N.; Cheong, Y.H.; Grant, J.J.; Pandey, G.K.; Luan, S. CIPK3, a calcium sensor-associated protein kinase that regulates abscisic acid and cold signal transduction in Arabidopsis. **Plant Cell**. *2003*, *15*, 411–423.

[88] D'Angelo, C; Weinl, S; Batistic, O; Pandey, G.K.; Cheong, Y.H.; Schültke, S; Albrecht, V; Ehlert, B; Schulz, B; Harter, K; Luan, S; Bock, R; Kudla J. Alternative complex formation of the Ca-regulated protein kinase CIPK1 controls abscisic acid-dependent and independent stress responses in Arabidopsis. *Plant J.* **2006**, *48*, 857–872.

[89] Ok, S.H.; Jeong, H.J.; Bae, J.M.;Shin, J.S.; Luan, S; Kim, K.N. Novel CIPK1-associated proteins in Arabidopsis contain an evolutionarily conserved C-terminal region that mediates nuclear localization. *Plant Physiol.* **2005**, *139*, 138–150.

[90] Fuglsang, A.T.; Guo, Y; Cuin, T.A.; Qiu, Q; Song, C; Kristiansen, K.A.; Bych, K; Schulz, A; Shabala, S; Schumaker, K.S.; Palmgren, M.G.; Zhu, J.K. Arabidopsis protein kinase PKS5 inhibits the plasma membrane H+ -ATPase by preventing interaction with 14-3-3 protein. *Plant Cell.* **2007**, *19*, 1617–1634.

[91] Cheong, Y.H.; Pandey, G.K.; Grant, J.J.; Batistic, O; Li, L; Kim, B.G.; Lee, S.C.; Kudla, J;Luan, S. Two calcineurin B-like calcium sensors, interacting with protein kinase CIPK23, regulate leaf transpiration and root potassium uptake in Arabidopsis. *Plant J.* **2007**, *52*, 223–239.

[92] Pandey, G.K.; Grant, J.J.; Cheong, Y.H.; Kim, B.G.; Li, L; Luan, S. Calcineurin-B-like protein CBL9 interacts with target kinase CIPK3 in the regulation of ABA response in seed germination. *Mol. Plants.* **2008**, *1*, 238–248.

[93] Huang, C; Ding, S; Zhang, H; Du, H; An, L.CIPK7 is involved in cold response by interacting with CBL1 in Arabidopsis thaliana. *Plant Sci.* **2011**, *181*(1), 57-64.

[94] Guo, Y; Xiong, L; Song, C.P.; Gong, D; Halfter, U; Zhu, J.K. A calcium sensor and its interacting protein kinase are global regulators of abscisic acid signaling in Arabidopsis. *Dev. Cell.* **2002**, *3*, 233–44.

[95] Luan, S; Lan, W; Chul Lee, S. Potassium nutrition, sodium toxicity, and calcium signaling: connections through the CBL-CIPK network.*Curr Opin Plant Biol.* **2009**, 12(3), 339-46.

[96] Li, L; Kim, B.G.;Cheong, Y.H.; Pandey, G.K.; Luan, S. A Ca(2)+ signaling pathway regulates a K(+) channel for low-K response in Arabidopsis. *Proc. Natl. Acad. Sci. USA.* **2006**, *103*, 12625–126530.

[97] Xu, J; Li, H.D.; Chen, L.Q.; Wang, Y; Liu, L.L.; He, L; Wu, W.H. A protein kinase, interacting with two calcineurin B-like proteins, regulates K(+) transporter AKT1 in Arabidopsis. *Cell.* **2006**, *125*, 1347–1360.

[98] Lan, W.Z.; Lee, S.C.;Che, Y.F.; Jiang, Y.Q.; Luan, S. Mechanistic analysis of AKT1 regulation by the CBL-CIPK-PP2CA interactions.*Mol Plant.* **2011**, 4 (3), 527-36.

[99] Hashimoto, K; Eckert, C; Anschütz, U; Scholz, M; Held, K; Waadt, R; Reyer, A; Hippler, M; Becker, D; Kudla, J. Phosphorylation of Calcineurin B-like (CBL) Calcium Sensor Proteins by their CBL-interacting Protein Kinases (CIPKs) Is Required for Full Activity of CBL-CIPK Complexes toward Their Target Proteins.*J Biol Chem.* **2012,** *9*, 287(11), 7956-68.

[100] Kim, B.G.; Waadt, R; Cheong, Y.H.; Pandey, G.K.; Dominguez-Solis, J.R.; Schültke, S; Lee S.C.; Kudla, J;Luan, S. The calcium sensor CBL10 mediates salt tolerance by regulating ion homeostasis in Arabidopsis. *Plant J.* **2007**, *52*, 473–84.

[101] Zhu J.K. Regulation of ion homeostasis under salt stress. *Curr. Opin. Plant Biol.* **2003**, *6*, 441–445.

[102]Mahajan, S; Pandey, G.K.; Tuteja, N. Calcium- and salt-stress signaling in plants: shedding light on SOS pathway. *Arch. Biochem. Biophys.* **2008**, *471*, 146–158.

[103] Shi, H; Ishitani, M; Kim, C; Zhu, J.K. The *Arabidopsis thaliana* salt tolerance gene *SOS1* encodes a putative Na^+/H^+ antiporter. *Proc. Natl. Acad. Sci. USA.* **2000**, *97*, 6896–6901.

[104] Zhu, J.K.; Liu, J; Xiong, L. Genetic analysis of salt tolerance in Arabidopsis: Evidence for a critical role of potassium nutrition. *Plant Cell.* **1998**, *10*, 1181–1191.

[105] Batelli, G; Verslues, P.E.; Agius, F; Qiu, Q; Fujii, H; Pan, S; Schumaker, K.S.; Grillo, S;Zhu, J.K. SOS2 promotes salt tolerance in part by interacting with the vacuolar H+-ATPase and upregulating its transport activity. *Mol. Cell Biol.* **2007**, *27*, 7781–7790.

[106] Tsou, P.L.; Lee, S.Y.; Allen, N.S; Winter-Sederoff, H; Robertson, D. An ER-targeted calcium-binding peptide confers salt and drought tolerance mediated by CIPK6 in Arabidopsis.*Planta.* **2012,** *235*(3), 539-52.

[107] Held, K; Pascaud, F; Eckert, C;Gajdanowicz, P; Hashimoto, K; Corratgé-Faillie, ; Offenborn, JN;Lacombe, B; Dreyer, I; Thibaud, JB;Kudla, J.Calcium-dependent modulation and plasma membrane targeting of the AKT2 potassium channel by the CBL4/CIPK6 calcium sensor/protein kinase complex.*Cell Res.* **2011**, *21*(7), 1116-30. doi: 10.1038/cr.2011.50

[108] Xiang, Y; Huang, Y; Xiong, L. Characterization of stress-responsive CIPK genes in rice for stress tolerance improvement. *Plant Physiol.* **2007**, *144*, 1416–1428.

[109] Bailey-Serres, J, ;Voesenek, L.A. Life in the balance: a signaling network controlling survival of flooding.*Curr Opin Plant Biol.* **2010**, *13*(5), 489-94.

[110] Piao, H.L.; Xuan, Y.H.; Park, S.H.; Je, B.I.; Park, S.J.; Park, S.H.; Kim, C.M.;Huang, J.;Wang, G.K.;Kim, M.J.; Kang, S.M.;Lee, I.J.; Kwon, T.R.; Kim, Y.H.; Yeo, U.S.;Yi, G.; Kim, M.J. OsCIPK31, a CBL-interacting protein kinase is involved in germination and seedling growth under abiotic stress conditions in rice plants. *Molecules and Cells* **2010**, *30*(1), 19-27, DOI. 10.1007/s10059-010-0084-1.

[111] Kurusu, T; Hamada, J; Hamada, H, Hanamata, S; Kuchitsu, K.Roles of calcineurin B-like protein-interacting protein kinases in innate immunity in rice.*Plant Signal Behav.* **2010** *5*(8), 1045-7.

[112] Yu, Y; Xia, X; Yin, W; Zhang, H. Comparative genomic analysis of CIPK gene family in Arabidopsis and populus. *Plant Growth Regul.* **2007**, *52*, 101–110.

[113] Zhang, H; Yin, W; Xia, X. Shaker-like potassium channels in Populus, regulated by the CBL-CIPK signal transduction pathway, increase tolerance to low-K+ stress. *Plant Cell Rep.* **2010**, *29*(9), 1007-12.

[114] Chen, X; Gu, Z; Xin, D; Hao, L; Liu, C; Huang, J; Ma, B; Zhang, H.Identification and characterization of putative CIPK genes in maize. *J Genet Genomics.* **2011**, *38*(2), 77-87

[115] Zhao, J; Sun, Z; Zheng, J; Guo, X; Dong, Z; Huai, J; Gou, M; He, J; Jin, Y;Wang, J; Wang, G. Cloning and characterization of a novel CBL- interacting protein kinase from maize. *Plant Mol. Biol.* **2009**, *69*, 661–674.

[116] Wang, M; Gu, D; Liu, T; Wang, Z; Guo, X; Hou, W; Bai, Y; Chen, X; Wang, G. Overexpression of a putative maize calcineurin B-like protein in Arabidopsis confers salt tolerance. *Plant Mol. Biol.* **2007**, *65*, 733–746.

[117] Hwang, Y.S.; Bethke, P.C.; Cheong, Y.H.; Chang, H.S.; Zhu, T; Jones, R.L. A gibberellin-regulated calcineurin B in rice localizes to the tonoplast and is implicated in vacuole function. *Plant Physiol.* **2005**, *138*, 1347–1358.

[118] Wang, RK; Li, LL; Cao, ZH; Zhao, Q; Li, M; Zhang, LY; Hao, YJ Molecular cloning and functional characterization of a novel apple MdCIPK6L gene reveals its involvement in multiple abiotic stress tolerance in transgenic plants. *Plant Mol Biol.* 2012 Mar 1.

[119] Hu, D.G.; Li, M; Luo, H; Dong, Q.L.;Yao, Y.X.; You, C.X.; Hao, Y.J. Molecular cloning and functional characterization of MdSOS2 reveals its involvement in salt tolerance in apple callus and Arabidopsis. *Plant Cell Rep.* **2012**, *31*(4), 713-22

[120] Cuéllar, T;Pascaud, F; Verdeil, JL; Torregrosa, L;Adam-Blondon, A.F.; Thibaud, J.B.;Sentenac, H;Gaillard, I. A grapevine Shaker inward K(+) channel activated by the calcineurin B-like calcium sensor 1-protein kinase CIPK23 network is expressed in grape berries under drought stress conditions. *Plant J.* **2010**, *61*(1), 58-69.

[121]] Mahajan, S;Sopory, S.K.; Tuteja, N. Cloning and characterization of CBL- CIPK signalling components from a legume (Pisum sativum) *FEBS J.* **2006**, *273*, 907–925.

[122] Gao, P; Zhao, P.M.; Wang, J; Wang, H.Y.; Du, X.M.; Wang, G.L.; Xia, G.X. Co-expression and preferential interaction between two calcineurin B-like proteins and a CBL-interacting protein kinase from cotton. *Plant Physiol. Biochem.* **2008**, *46*, 935–940.

[123] Du, W; Wang, Y; Liang, S; Lu, Y.T. Biochemical and expression analysis of an Arabidopsis calcium-dependent protein kinase- related kinase. *Plant Sci.* **2004**, *168*, 1181–1192.

[124] Furumoto, T;Ogawa, N; Hata, S; Izui, K. Plant calcium-dependent protein kinase-related kinases (CRKs) do not require calcium for their activities. *FEBS Lett.* **1996**, *396*, 147–151.

[125] Li, R.J.; Hua, W; Lu, Y.T. Arabidopsis cytosolic glutamine synthetase AtGLN1; 1 is a potential substrate of AtCRK3 involved in leaf senescence. *Biochem. Biophys. Res. Commun.* **2006**, *342*, 119–126.

[126] Wang, Y; Liang, S; Xie, Q.G.; Lu, Y.T. Characterization of a calmodulin-regulated Ca2+-dependent-protein-kinase related protein kinase, AtCRK1, from Arabidopsis. *Biochem. J.* **2004**, *383*, 73–81.

]127] Watillon, B; Kettmann, R;Boxus, P; Burny, A. Structure of a calmodulin-binding protein kinase gene from apple. *Plant Physiol.* **1995**, *108*, 847–848.

[128] Zhang, L; Lu, Y.T. Calmodulin-binding protein kinases in plants. *Trends Plant Sci.* **2003**, *8*, 123–127.

[129] Pandey, G.K.; Reddy, M.K.;Sopory, S.K.;Pareek, S.L.S. Calcium Homeostasis in Plants: Role of calcium binding proteins in abiotic stress tolerance *Indian J. Biotech.* **2002**, *1*, 135–137.

[130] Hua, W; Zhang, L; Liang, S; Jones, R.L.; Lu, Y.T. A tobacco calcium/calmodulin-binding protein kinase functions as a negative regulator of flowering. *J. Biol.Chem.* **2004**;*279*, 31483–31494.

[131] Liu, H.T.; Gao, F; Li, G.L.; Han, J.L.; Liu, D.L.; Sun, D.Y.; Zhou, R.G. The calmodulin-binding protein kinase 3 is part of heat-shock signal transduction in Arabidopsis thaliana. Plant J. **2008**, *55*, 760–773.

[132] Wang, L; Liang, S; Lu, Y.T. Characterization, physical location and expression of the genes encoding calcium/calmodulin-dependent protein kinases in maize (Zea mays L.) *Planta.* **2001**, *213*, 556–564.

[133] Lu, Y.T.; Hidaka, H; Feldman, L.J. Characterization of a calcium/calmodulin-dependent protein kinase homolog from maize roots showing light regulated gravitropism. *Planta.* **1996**, *199*, 18–24.

[134] Pandey, S; Tiwari, S.B.;Tyagi, W; Reddy, M.K.; Upadhyaya, K.C.; Sopory, S.K. A Ca2+/CaM-dependent kinase from pea is stress regulated and *in vitro* phosphorylates a protein that binds to AtCaM5 promoter. *Eur. J. Biochem.* **2002**, *269*, 3193–3204.

[135] Ho, C.H.; Lin, S.H.; Hu, H.C.; Tsay, Y.F. CHL1 functions as a nitrate sensor in plants. Cell. **2009**, *138* (6), 1184-94.

Send Orders of Reprints at reprints@benthamscince.net

Advances in Genome Science, Vol. 1, 2013, 426-451

CHAPTER 16

Multiple Hsp70 Isoforms in the Eukaryotic Cytosol: Evidence for Both Redundant and Specialized Functions

Mehdi Kabani* **and Céline N. Martineau**

Laboratoire d'Enzymologie et Biochimie Structurales (LEBS), CNRS, Bât. 34, Avenue de la Terrasse, 91198 Gif-sur-Yvette, France

Abstract: Hsp70 molecular chaperones play a variety of functions in every organism, cell type and organelle, and their activities have been implicated in a number of human pathologies, ranging from cancer to neurodegenerative diseases. The functions, regulations and structure of Hsp70s were intensively studied for about three decades, yet much still remains to be learned about these essential folding enzymes. Genome sequencing efforts revealed that most genomes contain multiple members of the Hsp70 family, some of which co-exist in the same cellular compartment. For example, the human cytosol and nucleus contain six highly homologous Hsp70 proteins while the yeast *Saccharomyces cerevisiae* contains four canonical Hsp70s and three fungal-specific ribosome-associated and specialized Hsp70s. The reasons and significance of the requirement for multiple Hsp70s is still a subject of debate. It has been postulated for a long time that these Hsp70 isoforms are functionally redundant and differ only by their spatio-temporal expression patterns. However, several studies in yeast and higher eukaryotic organisms challenged this widely accepted idea by demonstrating functional specificity among Hsp70 isoforms. Another element of complexity is brought about by specific cofactors, such as Hsp40s or nucleotide exchange factors that modulate the activity of Hsp70s and their binding to client proteins. Hence, a dynamic network of chaperone/co-chaperone interactions has evolved in each organism to efficiently take advantage of the multiple cellular roles Hsp70s can play. We summarize here our current knowledge of the functions and regulations of these molecular chaperones, and shed light on the known functional specificities among isoforms.

Keywords: Hsp70, Ssa1, chaperone network, functional specificity.

INTRODUCTION

The 70-kDa heat-shock proteins (Hsp70s) are a ubiquitous family of molecular chaperones found in all organisms and sub-cellular compartments where they play

***Address correspondence to Mehdi Kabani:** Laboratoire d'Enzymologie et Biochimie Structurales (LEBS), CNRS, Bât. 34, Avenue de la Terrasse, 91198 Gif-sur-Yvette, France; Tel: + 33169823491, Fax +33169823129; E-mail: Mehdi.Kabani@lebs.cnrs-gif.fr

Christian Neri (Ed)

essential housekeeping functions in protein folding, synthesis, assembly, transport across biological membranes and degradation. They are also involved in quality control processes, such as protein refolding after a stress injury, and control the activity of regulatory proteins in signal transduction pathways [1]. Hsp70s constitute one of the most conserved protein families in evolution (Fig. **1**), and members of this family may be constitutively expressed or stress-inducible (hence their classification as heat-shock proteins) [2-6].

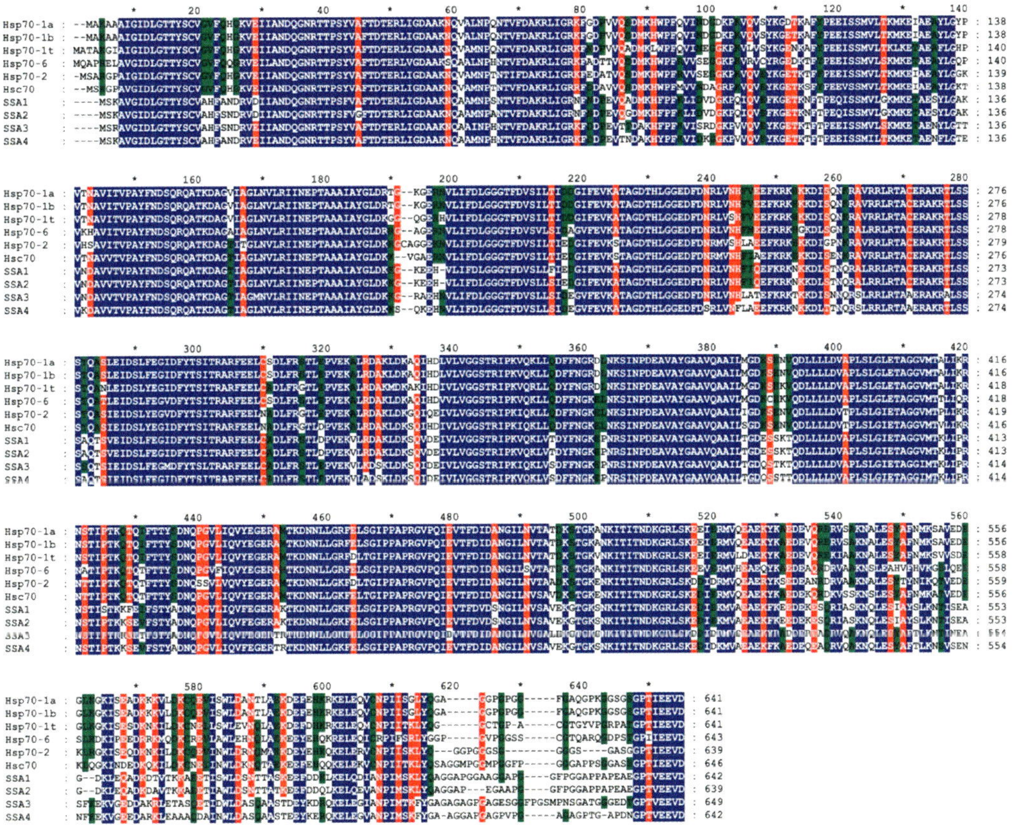

Figure 1: The alignment of yeast and human cytosolic Hsp70s shows their remarkable evolutionary conservation and highlights the variability of the C-terminal lid domain (this figure was made using clustalX 1.83 and GeneDoc v2.6).

The functional pleiotropy of Hsp70s is achieved through the evolutionary amplification and diversification of *HSP70* genes, cofactors that recruit and regulate Hsp70s for specific cellular functions, and cooperation of Hsp70s with other chaperone systems such as TRiC/CCT or Hsp90 [1]. All of these cellular activities depend on the ability of Hsp70s to interact with hydrophobic stretches of proteins in an ATP-dependent manner. Hsp70s are highly conserved and are composed of an N-terminal 44-kDa ATPase domain (also named adenine nucleotide-binding domain or NBD), an 18-kDa peptide-binding domain (PBD) and a C-terminal 10-kDa "lid" domain. The NBD has a bilobal structure and each globular lobe (I and II) is further conventionally divided into two subdomains (A and B) [7]. The interdomain linker connecting the NBD to the PBD is highly conserved and plays a critical role in the allosteric regulation of Hsp70s [8, 9]. Cytosolic Hsp70s also contain a G/P-rich C-terminal region containing an EEVD-motif that mediates their binding to Tetra-Trico-Peptide repeat (TPR)-domain containing co-chaperones such as the C-terminus-Hsp70-Interacting Protein CHIP. The EEVD motif is absent from specialized Hsp70s such as the ribosome-bound Ssb1/2p in yeast (see below).

Hsp70s cycle between two stable conformations with different affinities for substrates: in the ATP-bound state, Hsp70s display fast on-and-off rates of peptide binding, whereas in the ADP-bound state these constants are slowed, resulting in tighter association of the substrate within the PBD [10, 11]. The modulation of the affinity for polypeptide substrates is triggered by a conformational change in the lid that, upon ATP hydrolysis, closes on the substrate that is then trapped within the PBD [12]. The lid may also play a more complex and previously unsuspected role in substrate selection and binding by Hsp70 molecular chaperones [13]. It appears that ADP-ATP exchange is critical for substrate release and for the recycling of the molecular chaperone. This allosteric regulation of Hsp70s activity is tightly controlled by specific co-chaperones (Fig. **1**).

The well characterized Hsp40/DnaJ family members share a common ~70-amino-acid J-domain that mediates binding to Hsp70s and activation of their weak intrinsic ATPase activity (Fig. **2**). These cofactors, by virtue of their cellular localization and/or association to specific complexes, recruit Hsp70s to specific sites of the cell and therefore greatly contribute to the wide range of functions assigned to Hsp70s (reviewed in [14-18]).

Whereas ATP hydrolysis is stimulated by members of a common Hsp40/DnaJ family, ADP-ATP exchange is catalyzed by several evolutionary unrelated classes of nucleotide exchange factors (NEFs) (Fig. **2**); these include GrpE homologs in prokaryotes and the Bag1, HspBP1/Fes1p and Hsp110 families in eukaryotes (reviewed in [1, 19, 20]). It should be noted that Hsp110s are molecular chaperones homologous in sequence and structure to Hsp70s and together these proteins form the super-family of Hsp70s [21, 22]. The domain architecture of Hsp110 is therefore very similar to that of Hsp70s; the main noticeable difference lies in large (~100-140 amino acids) acidic insertions between the PBD and the lid and also at the C-terminus [22]. Hsp110s are not protein folding chaperones, yet they act as efficient "holdases" by binding to and preventing aggregation of denatured proteins [23, 24]. Mammalian Hsp110 and yeast Sse1p were shown to functionally and physically interact with counterpart Hsp70s [25-28], and to be potent nucleotide-activated NEFs [29-32].

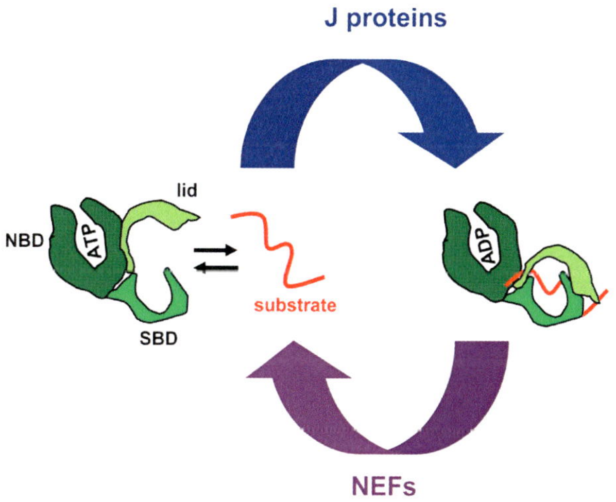

Figure 2: The Hsp70 ATPase cycle. In the ATP-bound state, Hsp70 have low affinity for substrates. Upon J-protein-stimulated ATP hydrolysis, a conformational switch of the lid tightly locks the substrate within the substrate binding domain. NEF catalyzed-ADP release and subsequent ATP re-binding trigger substrate release.

THE EUKARYOTIC CYTOSOL CONTAINS MULTIPLE MEMBERS OF THE HSP70 SUPER-FAMILY

Genome sequencing efforts confirmed early observations that eukaryotes have several Hsp70 encoding genes [4]. Evidently, some of these Hsp70s are compartment specific and fulfil unique and essential functions in the endoplasmic reticulum, mitochondria or plastids. These organelle-specific Hsp70s are generally encoded by a single gene in most organisms. Strikingly, multiple Hsp70 isoforms very often co-exist in the same cytosol and are encoded by different homologous genes. In Table **1** we describe the cytosolic Hsp70 machinery of various fully sequenced eukaryotic organisms, ranging from yeast to human. For most of these selected species, the Hsp70 system has been recently described in details at the genomic and/or functional levels [6, 33-41].

With the exception of ribosome-associated specialized Hsp70s (that do not contain the C-terminal EEVD motif), canonical Hsp70s are generally believed to be functionally redundant, the main differences between isoforms lying in their spatio-temporal expression. This is exemplified by studies in yeast that showed that only one member of the Ssa group of Hsp70s can support yeast growth if expressed at sufficient levels [41]. Additionally, heterologously expressed Hsp70s were shown to protect mammalian cells and transgenic animals from various stresses, demonstrating that the protective role of these molecular chaperones is highly conserved [42-45].

However, many studies also raised the possibility of functional specificity among Hsp70 isoforms, meaning that a particular Hsp70 will for instance preferentially bind to a subset of client proteins and/or co-chaperones to perform a unique task. Such a specialization among Hsp70s may explain the need for several Hsp70-encoding genes and their different tissue-expression patterns in complex multicellular organisms. In this review, we will discuss examples of functional specificities among Hsp70s in yeast and humans, hypothesize on the molecular basis that may govern this specificity, and suggest directions for future research in this field.

REDUNDANT AND SPECIALIZED FUNCTIONS AMONG HSP70S IN YEAST

The Cytosolic Hsp70 System in Yeast

The yeast *Saccharomyces cerevisiae* contains seven cytosolic Hsp70s that fall into two major groups: 1) the canonical Ssa1, Ssa2p, Ssa3p and Ssa4p proteins and 2) the ribosome-associated Ssb1p, Ssb2p and Ssz1p proteins (Table **1**) [4].

Table 1: The cytosolic Hsp70s of various eukaryotes are shown together with some important co-chaperones (because of their high number, J-proteins are not shown). UniProt accession numbers are provided in most cases, along with the alternative acronyms by which these proteins were described when appropriate.

	Yeasts				Nematode
	S. cerevisiae	*S. pombe*	*C. albicans*	*Y. lipolytica*	*C. elegans*
Canonical Hsp70's	Ssa1p (P10591) Ssa2p (P10592) Ssa3p (P09435) Ssa4p (P22202)	*Sp.* Ssa1p (Q10265) *Sp.* Ssa2p (O59855)	*Ca.* Ssa1p (P41797) *Ca.* Ssa2p (P46587)	Ssa5p (Q6C0E9) Ssa6p (Q6C3G5) Ssa7p (Q6C9V0) Ssa8p (Q6C864)	Hsp70-1 (P09446) Hsp70-7 (Q9XTL8) Hsp70-8 (Q9XTL8) Hsp70-9 (O45246)
Other Hsp70's	Ssb1p (P11484) Ssb2p (P40150)	*Sp.* Ssb1p (Sks2) (Q10284)	*Ca.* Ssb1p (P87222)	*Yl.* Ssb1p (Q6CIA7)	Hsp70-10 (Q9TW52)
	Ssz1p (P38788)	*Sp.* Ssz1p (P87142)	*Ca.* Ssz1p (Q5A678)	*Yl.* Ssz1p (Q6CEW0)	-
Hsp110's	Sse1p (P32589) Sse2p (P32590)	*Sp.* Sse1p (Pss1) (O59838)	*Ca.* Sse1p (Q96VB9)	*Yl.* Sse1p (Q6C618)	Hsp110-1 (Q05036)
HspBP1	Fes1p (P38260)	*Sp.* Fes1p (O43030)	*Ca.* Fes1p (Q59NN8)	*Yl.* Fes1p (Q6C239)	-
Bag proteins	Snl1p (P40548)	Bag-1A (O59739) Bag-1B (O59739)	*Ca.* Snl1p (Q59NB3)	*Yl.* Bag1 (Q6C9J3)	*Ce.* Bag-1 (044739) Unc-23a (O61980) Unc-23b (Q86S24) Unc-23c (Q5TKA9)
CHIP	-	-	-	*Yl.* Chn1p (Q6C1F4)	Chn-1 (Q9BMU2)

	green algae	ascidian	Plant	Fruit fly	Human
	C. reinhardtii	*C. intestinalis*	*A. thaliana*	*D. melanogaster*	*H. sapiens*
Canonical Hsp70's	*Cr.* Hsp70-3 (P25840)	*Ci.* HSPA1/6/7-like (P91874) *Ci.* HSPA2/8	*At.* Hsp70-1 (P22953) *At.* Hsp70-2 (P22954) *At.* Hsp70-3 (O65719) *At.* Hsp70-4 (Q9LHA8) *At.* Hsp70-5 (Q959N1)	Hsp70Aa (P82910) Hsp70Ab (P02825) Hsp70Ba (Q8INI8) Hsp70Bbb (Q9VG58) Hsp70Bb (Q9BIS2) Hsp70Bc (Q9BIR7) Hsc70-1 (P29843) Hsc70-2 (P11146) Hsc70-4 (P11147) Hsp68 (097125)	HSPA1A (Hsp70-1a, Hsp70, Hsp72, Hsp70-1) (P08107) HSPA1B (Hsp70-1b, Hsp70, Hsp72, Hsp70-1) (P08107) HSPA1L (Hsp70-1t, Hsp70-hom) (P34931) HSPA2 (Hsp70-2, Hsp70-3) (P54652) HSPA6 (Hsp70-6, Hsp70B') (P17066) HSPA8 (Hsc70, Hsp70-8, Hsp73) (P11142)
Other Hsp70's	-	-	-	-	-
	-	-	-	-	HSPA14 (Hsp70-14, Hsp70L1) (Q0VDF9)
Hsp110's	-	*Ci.* HSPA4/4L/HSPH1	*At.* Hsp70-14 (Q957C0) *At.* Hsp70-15 (Q9CA95) *At.* Hsp70-16 (A8MRM9)	Hsc70Cb (Q9XZT5)	HSPH2 (Hsp110, Hsp70-4) (P34932) HSPH3 (Hsp110, Hsp70-4L) (O95757) HSPH1 (Hsp105) (Q92598)
HspBP1	-	-	*At.* HspBP1-1 (Q9M346) *At.* HspBP1-2 (Q84J81) *At.* HspBP1-3 (Q9LZL7)	?	HspBP1 (Q9NZL4)
Bag proteins	-	*Ci.* BAG1 *Ci.* BAG3 *Ci.* BAT3	*At.* Bag-1 (Q0WUQ1) *At.* Bag-2 (Q8LEP8) *At.* Bag-3 (Q9LYP4) *At.* Bag-4 (O65021) *At.* Bag-5 (O65373) *At.* Bag-6 (O82345) *At.* Bag-7 (Q9LVA0)	starvin, isoform A (Q9VU81) starvin, isoform B (Q9VU83) starvin, isoform C (Q9VU82)	Bag-1L, Bag-1M, Bag-1S (Q99933) Bag-2 (O95816) Bag-3 (O95817) Bag-4 (O95429) Bag-5 (Q9UL15) Bag-6, BAT3 (P46379)
CHIP	*Cr.* CHIP (A8J756)	-	*At.* CHIP (Q9SRS9)	*Dm.* CHIP (Q9XYW6)	CHIP (Q9UNE7)

The highly homologous Ssa proteins differ by their expression pattern: the *SSA2* gene is constitutively expressed at high levels; the *SSA1* gene is constitutively expressed, yet at lower levels than *SSA2,* and is also induced by stress; the *SSA3* and *SSA4* genes are not expressed during normal vegetative growth, but their expression is dramatically induced upon stress; additionally, Ssa3p levels are increased upon entry into stationary phase [41, 46-49]. The simultaneous deletion of these four genes is lethal, but can be complemented by the over-expression of either one of them, suggesting redundant functions for these molecular chaperones [41]. It should be noted that a Δ*ssa1*Δ*ssa2* mutant is viable and constitutively thermotolerant because of the up-regulation of *SSA3*, *SSA4* and other components of the general stress response [41, 46, 47]. However, the Δ*ssa1*Δ*ssa2* mutant forms small colonies at 23°C and is thermosensitive for growth at 37°C, suggesting that the functions carried by Ssa1p and/or Ssa2p cannot be fully complemented by the induction of *SSA3* and *SSA4* [41, 46, 47].

The nearly identical Ssb1p and Ssb2p proteins interact with nascent chains at the yeast ribosome and provide a dispensable function in the process of translation and in early folding events [50, 51]. These Hsp70s are unable to complement the lack of Ssa proteins, indicating that these two groups of Hsp70s clearly have distinct non overlapping functions [4]. Ssb1p and Ssb2p are recruited to the ribosome through their interaction with the stable heterodimeric Ribosome-Associated Complex (RAC) that is composed of the J-domain protein Zuo1p (or zuotin) and the atypical Hsp70 Ssz1p [52, 53]. The simultaneous deletion of *SSB1* and *SSB2* results in slow growth, cold sensitivity and hypersensitivity to translation-inhibiting drugs such as hygromycin B or paromomycin. Identical phenotypes are observed upon deletion of *SSZ1* or *ZUO1*, suggesting that Ssb1/2p and RAC form a functional triad *in vivo* [52-55]. The Ssb1/2p proteins are not stimulated by other J-proteins than Zuo1p and require their PBD for function [56, 57]. The Ssz1p protein diverges markedly from other Hsp70s: it contains an unusually short PBD and lacks the C-terminal lid domain; in fact, truncated Ssz1p lacking nearly the entire PBD is fully functional *in vivo* [58]; Ssz1p is not an ATPase *in vitro* and its nucleotide binding ability is not strictly required *in vivo* [59]. Ssz1p has been proposed to serve as a structural scaffold between Ssb1/2p and Zuo1p [60], although this may not be its only function [59]. While Ssb proteins are generally considered fungal-specific, we have identified an uncharacterized Ssb homolog in *C. elegans* (Table **1**). Moreover, a mammalian RAC complex has been recently described, and is comprised of the Zuo1 homolog MPP11 and the Ssz1 homolog Hsp70L1 (or Hsp70-14; Table **1**) [61, 62].

In *S. cerevisiae*, the Ssa proteins are recruited and activated by 11 cytosolic J-proteins [63], and both Ssa and Ssb proteins can be regulated by the Fes1p and Sse1/2p NEFs [26, 27, 30, 31, 64, 65]. Whether the different Ssa isoforms have preferences and/or different affinities for distinct J-proteins or NEFs has not been systematically tested, but remains an attractive possibility that could partly explain the functional specificities observed among Hsp70 isoforms. Indeed, a functional distinction between cytosolic, ER or mitochondrial Hsp70s has been greatly documented and was attributed to specific interactions between these Hsp70s and their partner J-proteins. In particular, cytosolic and ER lumenal Hsp70s could not replace one another in *in vitro* reconstitution assays of protein translocation across the ER membrane [66, 67]. Moreover, mitochondrial Hsp70 was unable to replace the ER lumenal Hsp70 in similar assays [68]. In both cases, this has been attributed to impaired interactions of the "extraneous" Hsp70 with the "endogenous" compartment-specific J-proteins, suggesting convergent evolution of chaperone and co-chaperones for proper interaction [17, 66, 68].

As shown in Table **1**, Ssa, Ssb and Ssz homologues are present in most yeasts that mainly differ with respect to the number of Ssa-encoding genes: only two in *Schizosaccharomyces pombe* and *Candida albicans*, and four in *Y. lipolytica*. A comparison of the Hsp70 systems in distantly related yeasts [37, 39] will provide useful information on the plasticity and dynamics of Hsp70-containing chaperone networks.

Examples of Functional Specificity Among the Ssa Group of Hsp70s

Early evidence for functional specificity among Ssa proteins was brought by a study from Eisenberg and collaborators that investigated the ability of yeast cytoplasmic Hsp70s to dissociate clathrin from bovine brain coated vesicles [69]. Using purified Hsp70 preparations from wild-type yeast, they first showed that the *in vitro* clathrin uncoating activity is associated with Ssa but not Ssb proteins. They then showed that higher uncoating activities could be achieved using Hsp70 preparations from a Δ*ssa1* mutant than from a Δ*ssa2* mutant. Because Ssa3p and Ssa4p are nearly undetectable in these preparations, they concluded that Ssa2p has a markedly higher uncoating activity than Ssa1p. Moreover, they showed that Ssa1p inhibits the uncoating activity of Ssa2p, probably by blocking Hsp70-binding sites on clathrin baskets [69]. The molecular basis of this functional difference in clathrin uncoating between these highly homologous Hsp70s remains to be directly addressed. One possible explanation is that Ssa2p and Ssa1p could have markedly different affinities for clathrin and/or the J-protein auxilin (Swa2p) which is specifically required to activate Hsc70 during endocytosis [70]. In humans and fruit fly, clathrin uncoating was similarly shown to depend on Hsc70 and Hsc70-4, respectively, which are the most abundant and constitutive Hsp70s in these organisms [71, 72]. This is not surprising given that Hsp70 acts stoichiometrically in clathrin uncoating, and therefore fair amounts of the molecular chaperone are required to fulfil this vital cellular process [73].

Another functional distinction between Ssa1p and Ssa2p has been reported with respect to their roles in prion propagation. The yeast epigenetic factors [*PSI*$^+$], [*URE3*] and [*PIN*$^+$] are the prion forms of the Sup35p, Ure2p and Rnq1p proteins, respectively [74-76]. These infectious proteins self-propagate as amyloids [74, 75, 77-80] and therefore constitute a powerful and safe model to investigate mammalian prion-related diseases such as sheep scrapie or transmissible spongiform encephalopathies [81]. Molecular chaperones of the Hsp104, Hsp70 and Hsp40 families play critical roles in prion formation and propagation [82, 83]. The fungal specific Hsp104 molecular chaperone promotes the ATP-dependent solubilization of aggregated misfolded proteins with the help of Hsp70 and Hsp40 [84]. Hsp104 is strictly required for yeast prion propagation, as the deletion of the *HSP104* gene or the inactivation of Hsp104 by growth in the presence of 5 mM guanidium chloride cure cells from [*PSI*$^+$], [*URE3*] and [*PIN*$^+$]. In contrast, the overexpression of Hsp104 cures [*PSI*$^+$] but not [*URE3*] [75, 85-88]. Hsp104 was proposed to mediate protein-only inheritance by remodelling large prion aggregates into new self-replicating particles (or seeds) [77, 89-91]. The effects of Hsp70s and Hsp40s on prion propagation are complex and prion-dependent [83]. Mutation of the cytosolic Hsp70 Ssa1p impairs [*PSI*$^+$] propagation while its overexpression was shown to inhibit [*PSI*$^+$] curing by overexpressed Hsp104 [92, 93]. The overproduction of

Ssa1p, but not Ssa2p, cures [*URE3*] [94], whereas mutations in *SSA2,* but not *SSA1,* impair [*URE3*] propagation [95]. The overexpression of Ydj1p, the most abundant yeast Hsp40, cures cells from [*URE3*] [86], whereas mutations in Sis1p cure [PIN^+] [96]. Similarly, NEFs have been shown to influence prion propagation. Daniel Masison and collaborators have isolated an *SSA1-21* mutation that weakens [PSI^+] propagation [92]. They then showed that in the *SSA1-21* background the depletion of Fes1p further destabilized [PSI^+], whereas its overexpression restored almost normal [PSI+] propagation [97]. However, contrasting results have been reported for [PSI^+] propagation when *FES1* is deleted in an otherwise wild-type background [97, 98]. The deletion of *SSE1* and *FES1* completely blocked [*URE3*] propagation, but only the overexpression of Sse1p cured it [98-100]. In addition, the deletion of *SSE1* severely inhibited [PSI^+] formation while the overexpression of Sse1p improved it [100]. The Hsp70 machinery appears as an important modulator of amyloid formation and protein-only inheritance, and the complex phenotypes obtained upon mutations in its components indicates that a dynamic network of specific chaperone/co-chaperone/substrate interactions occur *in vivo*. A better understanding of the role of the Hsp70 system in the process of amyloidogenesis will benefit from *in vitro* approaches where the effects of purified chaperones and co-chaperones on the assembly of amyloid fibrils can be precisely tested [101, 102].

In a recent study, we provided the first side-by-side intra- and inter- species comparison of Ssa proteins in a yeast model [39]. We compared the ability of Ssa1p-Ssa4p from *S. cerevisiae* and Ssa5p-Ssa8p from the evolutionary distant yeast *Y. lipolytica* to perform several Hsp70-dependent tasks when expressed as the sole Hsp70 for *S. cerevisiae in vivo* [39]. We found that all these Hsp70 proteins supported yeast viability yet with markedly different growth rates, in agreement with other studies that showed that primate and human Hsp70 proteins were not all able to support yeast cell viability [61, 103]. We found that Ssa1-8p varied widely in their ability to propagate the [PSI^+] and [*URE3*] prions, differently influenced [*URE3*] curing by Ydj1p overexpression, differed in their ability to refold a model substrate, but did not significantly affect the proteasomal degradation rate of CFTR [39]. Our study clearly demonstrates that despite high homology and overlapping functions, Hsp70 proteins have evolved to possess distinct activities that are required to cope with different types of substrates or stress situations [39].

We have also showed that mutations in components of the cytosolic Hsp70 machinery (*e.g. FES1, SSE1, YDJ1, SSA1, SWA2)* dramatically impair biofilm (or 'mat') formation in yeast [104, 105]. We showed that the overexpression of *FES1* rescues the defect in biofilm formation of a Δ*sse1* strain, but the opposite is not true: the Δ*fes1* mutant is unable to form 'mats' on semi-solid medium with or without *SSE1/2* overexpression [104]. Thus, Fes1p and Sse1/2p play both overlapping and distinct functions as suggested by earlier studies [31, 106]. Remarkably, we showed that a Δ*ssa1* mutant, in an otherwise wild-type background for the other SSA genes, is severely affected in biofilm formation [104]. The deletion of *SSA2* had more subtle effects suggesting that Ssa1p is specifically required for 'mat' formation [104]. Moreover, we showed that the additional deletion of *SSA3* and *SSA4* aggravated the phenotypes of the Δ*ssa1* and Δ*ssa2* mutants, indicating a possible functional cooperation between the constitutive and inducible Hsp70s [104].

THE HUMAN HSP70 SYTEM

The Cytosolic Hsp70 System in Humans

Several reviews and studies attempted to provide an accurate description of the human Hsp70 family, a task that has been complicated by confusing and conflicting nomenclature in the literature [6, 33, 107, 108]. However, extensive literature and database search, facilitated by the availability of the human genome, recently allowed Alberto Macario and collaborators to provide a detailed description of human Hsp70 encoding genes [33]. They identified up to 17 Hsp70-encoding genes that fall into evolutionary and functionally distinct groups, as well as 30 pseudogenes [33]. For each of these genes, an impressively high number of mRNA variants and isoforms were identified through EST analysis, although their functional relevance is still unknown [33]. In Table (**1**), we chose to present only the *bona fide* human cytosolic Hsp70s for which expression has been proved and that correspond to Ssa or Sse-like proteins [6, 33, 108].

The human cytosolic Hsp70s are comprised of six canonical members, named HSPA1A (Hsp70-1), HSPA1B (Hsp70-2), HSPA1L, HSPA2, HSPA6 and HSPA8 (Hsc70), and of an Ssz-like protein named HSPA14 [6, 33, 107, 108] (Table **1**). These proteins were also described with different acronyms that are indicated in

Table 1. While the expression of HSPA1A, HSPA1B and HSPA6 is strongly induced by stress [6, 33], all these cytosolic Hsp70s differ considerably with respect to their expression patterns in the different tissues and developmental stages, as predicted by assessing the relative number of ESTs per tissue type [33].

The heat-shock cognate HSPA8 (Hsc70) is by far the most expressed Hsp70 in all tissues with particularly high levels in the vascular tissue [33], and is believed to play essential house-keeping functions in protein folding, transport across biological membranes, prevention of protein aggregation and uncoating of clathrin vesicles [6, 109]. The intron-less genes encoding HSPA1A, HSPA1B and HSPA1L are all closely linked and located within the same MHC class III region on chromosome 6 [110, 111], yet their expression patterns differ [6, 33, 112]. HSPA1A and HSPA1B are expressed in most tissues, and in most cases HSPA1A levels are much higher than HSPA1B levels [6, 33]. Moreover, very high expression levels were observed for HSPA1A in the spleen [33]. On the other hand, HSPA1L is expressed at almost undetectable levels in most tissues, except in testis, adipose tissue and pituitary glands where its levels are much higher [6, 33, 112]. In contrast to HSPA1A and HSPA1B, the levels of HSPA1L are not induced by heat stress [112]. HSPA2 is expressed at low to undetectable levels in most tissues, and at high levels in testis, brain and umbilical cord [6, 33]. The expression of HSPA6 is only induced after severe stress insults [113, 114], although moderate levels of this protein have been reported in some blood cells as well as in the bladder, trachea and esophagus [6, 33].

Thus, all cytosolic Hsp70s are constitutively expressed at varying levels making each tissue and developmental stage unique in its Hsp70 content [6, 33]. Important efforts will be needed to define the specific roles of each Hsp70 in each cell and tissue type, taking into account the possibility that these Hsp70s may functionally interact with one another, either positively or negatively, depending on their respective abundance. Whether the Hsp70 co-chaperones content also varies among cell and tissue types has yet to be determined, but could also be a major determinant of the functional specialization of Hsp70s.

Examples of Functional Specificity Among the Human Cytosolic Hsp70s

Understanding the need for up to six Hsp70 family members in human cells and sorting out their redundant from their specialized functions have proven difficult and technically challenging. Most of the published data on human Hsp70s were obtained by studying the constitutive Hsc70 (HSPA8) or the major heat-inducible Hsp70 (which in fact refers to both HSPA1A and HSPA1B) that were often considered as equivalent and functionally interchangeable. [6, 115]. However, a clear distinction between constitutive and inducible Hsp70 isoforms emerged from several studies, including the characterization of mouse knock-out models [6]. It appeared very logically that constitutive Hsp70s mostly play important housekeeping functions while inducible Hsp70s are required to cope with stress situations [6].

Hence, Hsc70 was shown to be essential for viability in vertebrates, including fruit fly [116], mouse [117] and human cell lines [118, 119], whereas mice deficient for the homologues of human HSPA1A and HSPA1B (Hsp70.1 and Hsp70.3) develop normally in standard conditions but show significantly decreased resistance to stress situations such as heat shock, cardiac ischemia or radiation [6, 120-128].

Additionally, Hsp70 and Hsc70 were shown to have differential and antagonistic effects with regard to the intracellular trafficking of ENaC, an epithelial sodium chloride channel: Hsp70 promoted the maturation and functional cell surface expression of ENaC in *Xenopus* oocytes, whereas Hsc70 disfavoured it [115]. Similarly, both the overexpression of Hsp70 and the decrease of Hsc70 levels *via* the use of sodium 4-phenylbutyrate were shown to promote the intracellular trafficking of the ΔF508 mutant of the cystic fibrosis transmembrane conductance regulator (CFTR) [129-131].

The mammalian RAC complex was recently showed to preferentially recruit Hsp70, but not Hsc70, to nascent chains on translating ribosomes, further demonstrating a functional distinction between constitutive and inducible Hsp70 proteins [61].

Despite very high sequence conservation, members of the HSPA family of Hsp70 molecular chaperones were recently showed by Kampinga and colleagues to display remarkably distinct chaperone-like activities [35], akin to their yeast Ssa counterparts [39]. The authors of this study assessed the ability of Hsp70, Hsp110 and Hsp40 family members to refold heat-inactivated luciferase and to prevent the aggregation of expanded polyglutamine (polyQ)-containing untingtin fragments when overexpressed in mammalian cells [35]. They found that some Hsp70 proteins can stimulate luciferase refolding (*e.g.* HSPA1A, HSPA8) but cannot suppress polyQ aggregation [35]. Conversely, other Hsp70 proteins were able to suppress polyQ aggregation (*e.g.* HSPA2) but were poorly effective towards luciferase [35]. The strictly heat-inducible HSPA6 protein surprisingly lacked both luciferase refolding and polyQ-aggregation-suppressing activities, suggesting this protein may have evolved to chaperone specific and critical clients, such as p53, under conditions of severe stress [35].

Hsp70 levels are abnormally high in a wide variety of tumor cell types and contribute to tumorigenesis and resistance to chemotherapy by directly interfering with several key components of the apoptotic signalling pathway [132]. Similarly, an up-regulation of the HspBP1 NEF was observed in various tumor cell types [133-136]. HspBP1 has been shown to modulate Hsp70s activity either negatively [137, 138] (at high HspBP1/Hsp70 molar ratios) or positively (at low HspBP1/Hsp70 molar ratios) [139]. Interestingly, HspBP1 was shown *in vivo* to bind with a much higher affinity to the stress-inducible Hsp70 than to the abundant and constitutively expressed Hsc70. Hence, HspBP1 is present in great excess (~10 fold) compared to its preferential Hsp70 partner in normal cells. However, the HspBP1/Hsp70 molar ratio may be lowered considerably in tumor cells (from ~10 to ~2) because the up-regulation of Hsp70 is greater than that of HspBP1 in these cells [135]. This may contribute to the Hsp70-mediated resistance to chemotherapy, as tumor cells with high HspBP1/Hsp70 molar ratios were much more susceptible to anticancer drugs than were those with a low ratio [135]. In addition, anticancer drugs up-regulated HspBP1 in various cancer cells, while no effect was observed on Hsp70 levels [135]. These findings suggest that HspBP1 antagonizes the prosurvival function of Hsp70 [135]. In support of this hypothesis, RNAi-mediated depletion of HspBP1 markedly reduced the susceptibility of tumor cells to anticancer drugs [135]. Conversely, transient ectopic expression of HspBP1 in tumor cells enhanced the cathepsin-mediated cell death induced by anticancer drugs [135]. This pro-apoptotic function of HspBP1 appeared to be dependent on its ability to bind to Hsp70 and was inhibited by heat-shock induced up-regulation of Hsp70 [135]. These findings exemplify that part of the functional distinction among Hsp70 isoforms may lie in differential interaction with their co-chaperones, thereby forming specific functional networks in each cell or tissue type. Remarkably, both primate and plant Hsc70 were able to support growth of a Δ*ssa1-4* yeast mutant, whereas the corresponding inducible Hsp70s did not [103], suggesting an evolutionary conserved functional distinction among these molecular chaperones.

The HSPA2 protein appears to play an important role in spermatogenesis in humans and mice *via* its essential chaperoning functions for the cyclinB/cdc2 complex during meiosis, and for spermatid DNA-packaging proteins involved in post-meiotic genome reorganization [6, 140-144]. Additionally, HSPA2 is upregulated in some primary and metastatic breast cancers, and was shown to be required for the growth and survival of several human cancer cells [6, 118, 119, 145]. The depletion of Hsp70 or HSPA2 from cancer cells resulted in markedly different phenotypes and gene expression profiles, whereas their concomitant depletion resulted in synergistic antiproliferative effects, suggesting separate non-overlapping functions for these Hsp70s [119].

By virtue of their cellular protective role, Hsp70s are important candidates in gene-longevity association studies [146, 147]. Indeed, the ability to cope with cellular stress by the induction of Hsp70 was shown to decline with age in numerous *in vivo* and *in vitro* models [146, 148-151]. Moreover, the presence of extra-copies of a heat-inducible Hsp70-encoding gene increased the lifespan of transgenic fly models after a transient mild heat-shock [152]. Specifically, polymorphisms positively or negatively associated with longevity were identified in the genes encoding the MHC-III-linked HSPA1A, HSPA1B and HSPA1L proteins, suggesting that unique anti-ageing functions are fulfilled by these isoforms (reviewed in [146, 147]).

CONCLUSIONS AND FUTURE DIRECTIONS: MOLECULAR BASIS OF HSP70 SPECIALIZATION

Despite a high degree of sequence conservation and overlapping chaperoning functions, members of the Hsp70 superfamily appear to have evolved specialized functions for which they cannot replace each other. This functional specificity is most apparent and widely accepted for compartment-specific and ribosome-associated Hsp70s, and also for the Hsp110 family members that evolved to act as Hsp70 NEFs [6, 109]. However, less attention has been paid to the possibility that a similar specialization might occur among the highly conserved cytosolic Hsp70s [35, 39], although it may provide an answer to why some organisms such as humans or *Drosophila melanogaster* need up to 6 or 10 cytosolic Hsp70s, respectively (Table **1**). A recent study demonstrated functional differences among highly conserved ribosomal protein paralogs in yeast, indicating that the co-existence of multiple members of a given protein family is not only the result of gene duplication and mere redundancy, but also significantly contributes to cellular homeostasis [153].

Future efforts will have to identify and characterize the unique functions played by individual Hsp70s in particular cell or tissue types. Several non-mutually exclusive hypotheses can be made, according to our current knowledge of Hsp70 function, to understand the molecular basis of such specialization. A first model would be that Hsp70 isoforms differentially interact with the various cytosolic Hsp70 co-chaperones, as observed for the preferred interaction of HspBP1 with Hsp70 and not with Hsc70 [135]. A second model would be that Hsp70 isoforms bind to different sets of client proteins or to different types of substrates (*e. g.* native client proteins, nascent proteins, misfolded proteins, protein aggregates, *etc.*). In support of this hypothesis, different peptide binding specificities were described for bacterial DnaK, mammalian Hsc70 and BiP [154, 155]. Similarly, the DnaK protein from the archea *Methanosarcina mazei* failed to fully complement a *dnak* mutant in *Escherichia coli*, and was shown to have markedly different peptide binding properties [156, 157]. Importantly, the 10-kDa C-terminal domain of Hsp70s is the less conserved region and could play an important role in modulating Hsp70-substrate interactions [13] (Fig. **1**). Additionally, different Hsp70 isoforms could cooperate in the process of protein folding by binding to different exposed sites on a given substrate. The side-by-side and thorough biochemical characterization of all Hsp70 isoforms from a single species (*H. Sapiens* or model organisms) will certainly help clarifying the functions and regulations of these essential folding enzymes [35]. Because Hsp70s are relevant targets for drug-based treatments for cancers or protein folding disorders, the identification of the molecular determinants that govern their functional specificities in particular cell or tissue types will greatly improve the efficacy of these approaches without unwillingly affecting vital cellular processes.

ACKNOWLEDGEMENTS

This work was supported by the Centre National de la Recherche Scientifique (CNRS). CNM was the beneficiary of a doctoral grant from the French Ministry of Research.

CONFLICT OF INTEREST

The author(s) confirm that this chapter content has no conflict of interest.

DISCLOSURE

This manuscript is an extended and updated version of the paper published in Current Genomics, Volume 9, Number 5, August 2008, pp. 338-348. (http://www.ncbi.nlm.nih.gov/pmc/articles/PMC2685646/?tool=pubmed)

REFERENCES

[1] Mayer, M.P. and Bukau, B. Hsp70 chaperones: cellular functions and molecular mechanism *Cell Mol Life Sci* **2005**, *62*: 670-684.

[2] Hunt, C. and Morimoto, R.I. Conserved features of eukaryotic hsp70 genes revealed by comparison with the nucleotide sequence of human hsp70 *Proc Natl Acad Sci U S A* **1985**, *82*: 6455-6459.

[3] Lindquist, S. and Craig, E.A. The heat-shock proteins *Annu Rev Genet* **1988**, *22*: 631-677.

[4] Boorstein, W.R., Ziegelhoffer, T. and Craig, E.A. Molecular evolution of the HSP70 multigene family *J Mol Evol* **1994**, *38*: 1-17.

[5] Gupta, R.S. and Singh, B. Phylogenetic analysis of 70 kD heat shock protein sequences suggests a chimeric origin for the eukaryotic cell nucleus *Curr Biol* **1994**, *4*: 1104-1114.

[6] Daugaard, M., Rohde, M. and Jaattela, M. The heat shock protein 70 family: Highly homologous proteins with overlapping and distinct functions *FEBS Lett* **2007**, *581*: 3702-3710.

[7] Flaherty, K.M., DeLuca-Flaherty, C. and McKay, D.B. Three-dimensional structure of the ATPase fragment of a 70K heat-shock cognate protein *Nature* **1990**, *346*: 623-628.

[8] Swain, J.F., Dinler, G., Sivendran, R., Montgomery, D.L., Stotz, M. and Gierasch, L.M. Hsp70 chaperone ligands control domain association *via* an allosteric mechanism mediated by the interdomain linker *Mol Cell* **2007**, *26*: 27-39.

[9] Vogel, M., Mayer, M.P. and Bukau, B. Allosteric regulation of Hsp70 chaperones involves a conserved interdomain linker *J Biol Chem* **2006**, *281*: 38705-38711.

[10] Schmid, D., Baici, A., Gehring, H. and Christen, P. Kinetics of molecular chaperone action *Science* **1994**, *263*: 971-973.

[11] McCarty, J.S., Buchberger, A., Reinstein, J. and Bukau, B. The role of ATP in the functional cycle of the DnaK chaperone system *J Mol Biol* **1995**, *249*: 126-137.

[12] Zhu, X., Zhao, X., Burkholder, W.F., Gragerov, A., Ogata, C.M., Gottesman, M.E. and Hendrickson, W.A. Structural analysis of substrate binding by the molecular chaperone DnaK *Science* **1996**, *272*: 1606-1614.

[13] Schlecht, R., Erbse, A.H., Bukau, B. and Mayer, M.P. Mechanics of Hsp70 chaperones enables differential interaction with client proteins *Nat Struct Mol Biol* **2011**, *18*: 345-351.

[14] Cheetham, M.E. and Caplan, A.J. Structure, function and evolution of DnaJ: conservation and adaptation of chaperone function *Cell Stress Chaperones* **1998**, *3*: 28-36.

[15] Kelley, W.L. The J-domain family and the recruitment of chaperone power *Trends Biochem Sci* **1998**, *23*: 222-227.

[16] Walsh, P., Bursac, D., Law, Y.C., Cyr, D. and Lithgow, T. The J-protein family: modulating protein assembly, disassembly and translocation *EMBO Rep* **2004**, *5*: 567-571.

[17] Hennessy, F., Nicoll, W.S., Zimmermann, R., Cheetham, M.E. and Blatch, G.L. Not all J domains are created equal: implications for the specificity of Hsp40-Hsp70 interactions *Protein Sci* **2005**, *14*: 1697-1709.

[18] Qiu, X.B., Shao, Y.M., Miao, S. and Wang, L. The diversity of the DnaJ/Hsp40 family, the crucial partners for Hsp70 chaperones *Cell Mol Life Sci* **2006**, *63*: 2560-2570.

[19] Kabani, M. Structural and functional diversity among eukaryotic Hsp70 nucleotide exchange factors *Protein Pept Lett* **2009**, *16*: 623-660.

[20] Kabani, M., Beckerich, J.M. and Brodsky, J.L. The yeast Sls1p and Fes1p proteins define a new family of Hsp70 nucleotide exchange factors *Current Genomics* **2003**, *4*: 465-473.

[21] Liu, Q. and Hendrickson, W.A. Insights into hsp70 chaperone activity from a crystal structure of the yeast Hsp110 Sse1 *Cell* **2007**, *131*: 106-120.

[22] Shaner, L. and Morano, K.A. All in the family: atypical Hsp70 chaperones are conserved modulators of Hsp70 activity *Cell Stress Chaperones* **2007**, *12*: 1-8.

[23] Oh, H.J., Chen, X. and Subjeck, J.R. Hsp110 protects heat-denatured proteins and confers cellular thermoresistance *J Biol Chem* **1997**, *272*: 31636-31640.

[24] Oh, H.J., Easton, D., Murawski, M., Kaneko, Y. and Subjeck, J.R. The chaperoning activity of hsp110. Identification of functional domains by use of targeted deletions *J Biol Chem* **1999**, *274*: 15712-15718.

[25] Polier, S., Dragovic, Z., Hartl, F.U. and Bracher, A. Structural basis for the cooperation of Hsp70 and Hsp110 chaperones in protein folding *Cell* **2008**, *133*: 1068-1079.

[26] Shaner, L., Wegele, H., Buchner, J. and Morano, K.A. The yeast Hsp110 Sse1 functionally interacts with the Hsp70 chaperones Ssa and Ssb *J Biol Chem* **2005**, *280*: 41262-41269.

[27] Yam, A.Y., Albanese, V., Lin, H.T. and Frydman, J. Hsp110 cooperates with different cytosolic HSP70 systems in a pathway for *de novo* folding *J Biol Chem* **2005**, *280*: 41252-41261.

[28] Yamagishi, N., Ishihara, K. and Hatayama, T. Hsp105alpha suppresses Hsc70 chaperone activity by inhibiting Hsc70 ATPase activity *J Biol Chem* **2004**, *279*: 41727-41733.

[29] Andreasson, C., Fiaux, J., Rampelt, H., Mayer, M.P. and Bukau, B. Hsp110 Is a Nucleotide-activated Exchange Factor for Hsp70 *J Biol Chem* **2008**, *283*: 8877-8884.

[30] Dragovic, Z., Broadley, S.A., Shomura, Y., Bracher, A. and Hartl, F.U. Molecular chaperones of the Hsp110 family act as nucleotide exchange factors of Hsp70s *EMBO J* **2006**, *25*: 2519-2528.

[31] Raviol, H., Sadlish, H., Rodriguez, F., Mayer, M.P. and Bukau, B. Chaperone network in the yeast cytosol: Hsp110 is revealed as an Hsp70 nucleotide exchange factor *EMBO J* **2006**, *25*: 2510-2518.

[32] Schuermann, J.P., Jiang, J., Cuellar, J., Llorca, O., Wang, L., Gimenez, L.E., Jin, S., Taylor, A.B., Demeler, B., Morano, K.A., Hart, P.J., Valpuesta, J.M., Lafer, E.M. and Sousa, R. Structure of the Hsp110:Hsc70 nucleotide exchange machine *Mol Cell* **2008**, *31*: 232-243.

[33] Brocchieri, L., Conway de Macario, E. and Macario, A.J. hsp70 genes in the human genome: Conservation and differentiation patterns predict a wide array of overlapping and specialized functions *BMC Evol Biol* **2008**, *8*: 19.

[34] Gong, W.J. and Golic, K.G. Genomic deletions of the Drosophila melanogaster Hsp70 genes *Genetics* **2004**, *168*: 1467-1476.

[35] Hageman, J., van Waarde-Verhagen, M., Zylicz, A., Walerych, D. and Kampinga, H.H. The diverse members of the mammalian HSP70 machine show distinct chaperone-like activities *Biochem J* **2011**:

[36] Lin, B.L., Wang, J.S., Liu, H.C., Chen, R.W., Meyer, Y., Barakat, A. and Delseny, M. Genomic analysis of the Hsp70 superfamily in Arabidopsis thaliana *Cell Stress Chaperones* **2001**, *6*: 201-208.

[37] Martineau, C.N., Le Dall, M.T., Melki, R., Beckerich, J.M. and Kabani, M. Molecular and functional characterization of the only known hemiascomycete ortholog of the carboxyl terminus of Hsc70-interacting protein CHIP in the yeast Yarrowia lipolytica *Cell Stress Chaperones* **2012**, *17*: 229-241.

[38] Nikolaidis, N. and Nei, M. Concerted and nonconcerted evolution of the Hsp70 gene superfamily in two sibling species of nematodes *Mol Biol Evol* **2004**, *21*: 498-505.

[39] Sharma, D., Martineau, C.N., Le Dall, M.T., Reidy, M., Masison, D.C. and Kabani, M. Function of SSA subfamily of Hsp70 within and across species varies widely in complementing Saccharomyces cerevisiae cell growth and prion propagation *PLoS One* **2009**, *4*: e6644.

[40] Wada, S., Hamada, M. and Satoh, N. A genomewide analysis of genes for the heat shock protein 70 chaperone system in the ascidian Ciona intestinalis *Cell Stress Chaperones* **2006**, *11*: 23-33.

[41] Werner-Washburne, M., Stone, D.E. and Craig, E.A. Complex interactions among members of an essential subfamily of hsp70 genes in Saccharomyces cerevisiae *Mol Cell Biol* **1987**, *7*: 2568-2577.

[42] Pelham, H.R. Hsp70 accelerates the recovery of nucleolar morphology after heat shock *Embo J* **1984**, *3*: 3095-3100.

[43] Li, G.C., Li, L., Liu, R.Y., Rehman, M. and Lee, W.M. Heat shock protein hsp70 protects cells from thermal stress even after deletion of its ATP-binding domain *Proc Natl Acad Sci U S A* **1992**, *89*: 2036-2040.

[44] Plumier, J.C., Ross, B.M., Currie, R.W., Angelidis, C.E., Kazlaris, H., Kollias, G. and Pagoulatos, G.N. Transgenic mice expressing the human heat shock protein 70 have improved post-ischemic myocardial recovery *J Clin Invest* **1995**, *95*: 1854-1860.

[45] Radford, N.B., Fina, M., Benjamin, I.J., Moreadith, R.W., Graves, K.H., Zhao, P., Gavva, S., Wiethoff, A., Sherry, A.D., Malloy, C.R. and Williams, R.S. Cardioprotective effects of 70-kDa heat shock protein in transgenic mice *Proc Natl Acad Sci U S A* **1996**, *93*: 2339-2342.

[46] Boorstein, W.R. and Craig, E.A. Structure and regulation of the SSA4 HSP70 gene of Saccharomyces cerevisiae *J Biol Chem* **1990**, *265*: 18912-18921.

[47] Boorstein, W.R. and Craig, E.A. Transcriptional regulation of SSA3, an HSP70 gene from Saccharomyces cerevisiae *Mol Cell Biol* **1990**, *10*: 3262-3267.

[48] Werner-Washburne, M., Becker, J., Kosic-Smithers, J. and Craig, E.A. Yeast Hsp70 RNA levels vary in response to the physiological status of the cell *J Bacteriol* **1989**, *171*: 2680-2688.

[49] Werner-Washburne, M. and Craig, E.A. Expression of members of the Saccharomyces cerevisiae hsp70 multigene family *Genome* **1989**, *31*: 684-689.

[50] Nelson, R.J., Ziegelhoffer, T., Nicolet, C., Werner-Washburne, M. and Craig, E.A. The translation machinery and 70 kd heat shock protein cooperate in protein synthesis *Cell* **1992**, *71*: 97-105.

[51] Pfund, C., Lopez-Hoyo, N., Ziegelhoffer, T., Schilke, B.A., Lopez-Buesa, P., Walter, W.A., Wiedmann, M. and Craig, E.A. The molecular chaperone Ssb from Saccharomyces cerevisiae is a component of the ribosome-nascent chain complex *EMBO J* **1998**, *17*: 3981-3989.

[52] Michimoto, T., Aoki, T., Toh-e, A. and Kikuchi, Y. Yeast Pdr13p and Zuo1p molecular chaperones are new functional Hsp70 and Hsp40 partners *Gene* **2000**, *257*: 131-137.

[53] Gautschi, M., Lilie, H., Funfschilling, U., Mun, A., Ross, S., Lithgow, T., Rucknagel, P. and Rospert, S. RAC, a stable ribosome-associated complex in yeast formed by the DnaK-DnaJ homologs Ssz1p and zuotin *Proc Natl Acad Sci U S A* **2001**, *98*: 3762-3767.

[54] Yan, W., Schilke, B., Pfund, C., Walter, W., Kim, S. and Craig, E.A. Zuotin, a ribosome-associated DnaJ molecular chaperone *EMBO J* **1998**, *17*: 4809-4817.

[55] Gautschi, M., Mun, A., Ross, S. and Rospert, S. A functional chaperone triad on the yeast ribosome *Proc Natl Acad Sci U S A* **2002**, *99*: 4209-4214.

[56] Lopez-Buesa, P., Pfund, C. and Craig, E.A. The biochemical properties of the ATPase activity of a 70-kDa heat shock protein (Hsp70) are governed by the C-terminal domains *Proc Natl Acad Sci U S A* **1998**, *95*: 15253-15258.

[57] Pfund, C., Huang, P., Lopez-Hoyo, N. and Craig, E.A. Divergent functional properties of the ribosome-associated molecular chaperone Ssb compared with other Hsp70s *Mol Biol Cell* **2001**, *12*: 3773-3782.

[58] Hundley, H., Eisenman, H., Walter, W., Evans, T., Hotokezaka, Y., Wiedmann, M. and Craig, E. The *in vivo* function of the ribosome-associated Hsp70, Ssz1, does not require its putative peptide-binding domain *Proc Natl Acad Sci U S A* **2002**, *99*: 4203-4208.

[59] Conz, C., Otto, H., Peisker, K., Gautschi, M., Wolfle, T., Mayer, M.P. and Rospert, S. Functional characterization of the atypical Hsp70 subunit of yeast ribosome-associated complex *J Biol Chem* **2007**, *282*: 33977-33984.

[60] Huang, P., Gautschi, M., Walter, W., Rospert, S. and Craig, E.A. The Hsp70 Ssz1 modulates the function of the ribosome-associated J-protein Zuo1 *Nat Struct Mol Biol* **2005**, *12*: 497-504.

[61] Jaiswal, H., Conz, C., Otto, H., Wolfle, T., Fitzke, E., Mayer, M.P. and Rospert, S. The chaperone network connected to human ribosome-associated complex *Mol Cell Biol* **2011**, *31*: 1160-1173.

[62] Otto, H., Conz, C., Maier, P., Wolfle, T., Suzuki, C.K., Jeno, P., Rucknagel, P., Stahl, J. and Rospert, S. The chaperones MPP11 and Hsp70L1 form the mammalian ribosome-associated complex *Proc Natl Acad Sci U S A* **2005**, *102*: 10064-10069.

[63] Sahi, C. and Craig, E.A. Network of general and specialty J protein chaperones of the yeast cytosol *Proc Natl Acad Sci U S A* **2007**, *104*: 7163-7168.

[64] Kabani, M., Beckerich, J.M. and Brodsky, J.L. Nucleotide exchange factor for the yeast hsp70 molecular chaperone ssa1p *Mol Cell Biol* **2002**, *22*: 4677-4689.

[65] Dragovic, Z., Shomura, Y., Tzvetkov, N., Hartl, F.U. and Bracher, A. Fes1p acts as a nucleotide exchange factor for the ribosome-associated molecular chaperone Ssb1p *Biol Chem* **2006**, *387*: 1593-1600.

[66] Brodsky, J.L., Hamamoto, S., Feldheim, D. and Schekman, R. Reconstitution of protein translocation from solubilized yeast membranes reveals topologically distinct roles for BiP and cytosolic Hsc70 *J Cell Biol* **1993**, *120*: 95-102.

[67] Brodsky, J.L. Post-translational protein translocation: not all hsc70s are created equal *Trends Biochem Sci* **1996**, *21*: 122-126.

[68] Brodsky, J.L., Bauerle, M., Horst, M. and McClellan, A.J. Mitochondrial Hsp70 cannot replace BiP in driving protein translocation into the yeast endoplasmic reticulum *FEBS Lett* **1998**, *435*: 183-186.

[69] Gao, B.C., Biosca, J., Craig, E.A., Greene, L.E. and Eisenberg, E. Uncoating of coated vesicles by yeast hsp70 proteins *J Biol Chem* **1991**, *266*: 19565-19571.

[70] Eisenberg, E. and Greene, L.E. Multiple roles of auxilin and hsc70 in clathrin-mediated endocytosis *Traffic* **2007**, *8*: 640-646.

[71] Chappell, T.G., Welch, W.J., Schlossman, D.M., Palter, K.B., Schlesinger, M.J. and Rothman, J.E. Uncoating ATPase is a member of the 70 kilodalton family of stress proteins *Cell* **1986**, *45*: 3-13.

[72] Chang, H.C., Newmyer, S.L., Hull, M.J., Ebersold, M., Schmid, S.L. and Mellman, I. Hsc70 is required for endocytosis and clathrin function in Drosophila *J Cell Biol* **2002**, *159*: 477-487.

[73] Greene, L.E. and Eisenberg, E. Dissociation of clathrin from coated vesicles by the uncoating ATPase *J Biol Chem* **1990**, *265*: 6682-6687.

[74] Derkatch, I.L., Bradley, M.E., Hong, J.Y. and Liebman, S.W. Prions affect the appearance of other prions: the story of [PIN(+)] *Cell* **2001**, *106*: 171-182.

[75] Sondheimer, N. and Lindquist, S. Rnq1: an epigenetic modifier of protein function in yeast *Mol Cell* **2000**, *5*: 163-172.

[76] Wickner, R.B. [URE3] as an altered URE2 protein: evidence for a prion analog in Saccharomyces cerevisiae *Science* **1994**, *264*: 566-569.

[77] Paushkin, S.V., Kushnirov, V.V., Smirnov, V.N. and Ter-Avanesyan, M.D. Propagation of the yeast prion-like [psi+] determinant is mediated by oligomerization of the SUP35-encoded polypeptide chain release factor *EMBO J* **1996**, *15*: 3127-3134.

[78] Glover, J.R., Kowal, A.S., Schirmer, E.C., Patino, M.M., Liu, J.J. and Lindquist, S. Self-seeded fibers formed by Sup35, the protein determinant of [PSI+], a heritable prion-like factor of S. cerevisiae *Cell* **1997**, *89*: 811-819.

[79] King, C.Y., Tittmann, P., Gross, H., Gebert, R., Aebi, M. and Wuthrich, K. Prion-inducing domain 2-114 of yeast Sup35 protein transforms *in vitro* into amyloid-like filaments *Proc Natl Acad Sci U S A* **1997**, *94*: 6618-6622.

[80] Taylor, K.L., Cheng, N., Williams, R.W., Steven, A.C. and Wickner, R.B. Prion domain initiation of amyloid formation *in vitro* from native Ure2p *Science* **1999**, *283*: 1339-1343.

[81] Prusiner, S.B. Prions *Proc Natl Acad Sci U S A* **1998**, *95*: 13363-13383.

[82] Tuite, M.F. and Cox, B.S. Propagation of yeast prions *Nat Rev Mol Cell Biol* **2003**, *4*: 878-890.

[83] Chernoff, Y.O. Stress and prions: lessons from the yeast model *FEBS Lett* **2007**, *581*: 3695-3701.

[84] Glover, J.R. and Lindquist, S. Hsp104, Hsp70, and Hsp40: a novel chaperone system that rescues previously aggregated proteins *Cell* **1998**, *94*: 73-82.

[85] Chernoff, Y.O., Lindquist, S.L., Ono, B., Inge-Vechtomov, S.G. and Liebman, S.W. Role of the chaperone protein Hsp104 in propagation of the yeast prion-like factor [psi+] *Science* **1995**, *268*: 880-884.

[86] Moriyama, H., Edskes, H.K. and Wickner, R.B. [URE3] prion propagation in Saccharomyces cerevisiae: requirement for chaperone Hsp104 and curing by overexpressed chaperone Ydj1p *Mol Cell Biol* **2000**, *20*: 8916-8922.

[87] Ferreira, P.C., Ness, F., Edwards, S.R., Cox, B.S. and Tuite, M.F. The elimination of the yeast [PSI+] prion by guanidine hydrochloride is the result of Hsp104 inactivation *Mol Microbiol* **2001**, *40*: 1357-1369.

[88] Jung, G. and Masison, D.C. Guanidine hydrochloride inhibits Hsp104 activity *in vivo*: a possible explanation for its effect in curing yeast prions *Curr Microbiol* **2001**, *43*: 7-10.

[89] Kryndushkin, D.S., Alexandrov, I.M., Ter-Avanesyan, M.D. and Kushnirov, V.V. Yeast [PSI+] prion aggregates are formed by small Sup35 polymers fragmented by Hsp104 *J Biol Chem* **2003**, *278*: 49636-49643.

[90] Wegrzyn, R.D., Bapat, K., Newnam, G.P., Zink, A.D. and Chernoff, Y.O. Mechanism of prion loss after Hsp104 inactivation in yeast *Mol Cell Biol* **2001**, *21*: 4656-4669.

[91] Satpute-Krishnan, P., Langseth, S.X. and Serio, T.R. Hsp104-dependent remodeling of prion complexes mediates protein-only inheritance *PLoS Biol* **2007**, *5*: e24.

[92] Jung, G., Jones, G., Wegrzyn, R.D. and Masison, D.C. A role for cytosolic hsp70 in yeast [PSI(+)] prion propagation and [PSI(+)] as a cellular stress *Genetics* **2000**, *156*: 559-570.

[93] Newnam, G.P., Wegrzyn, R.D., Lindquist, S.L. and Chernoff, Y.O. Antagonistic interactions between yeast chaperones Hsp104 and Hsp70 in prion curing *Mol Cell Biol* **1999**, *19*: 1325-1333.

[94] Schwimmer, C. and Masison, D.C. Antagonistic interactions between yeast [PSI(+)] and [URE3] prions and curing of [URE3] by Hsp70 protein chaperone Ssa1p but not by Ssa2p *Mol Cell Biol* **2002**, *22*: 3590-3598.

[95] Roberts, B.T., Moriyama, H. and Wickner, R.B. [URE3] prion propagation is abolished by a mutation of the primary cytosolic Hsp70 of budding yeast *Yeast* **2004**, *21*: 107-117.

[96] Sondheimer, N., Lopez, N., Craig, E.A. and Lindquist, S. The role of Sis1 in the maintenance of the [RNQ+] prion *EMBO J* **2001**, *20*: 2435-2442.

[97] Jones, G., Song, Y., Chung, S. and Masison, D.C. Propagation of Saccharomyces cerevisiae [PSI+] prion is impaired by factors that regulate Hsp70 substrate binding *Mol Cell Biol* **2004**, *24*: 3928-3937.

[98] Kryndushkin, D. and Wickner, R.B. Nucleotide exchange factors for Hsp70s are required for [URE3] prion propagation in Saccharomyces cerevisiae *Mol Biol Cell* **2007**, *18*: 2149-2154.

[99] Lian, H.Y., Zhang, H., Zhang, Z.R., Loovers, H.M., Jones, G.W., Rowling, P.J., Itzhaki, L.S., Zhou, J.M. and Perrett, S. Hsp40 interacts directly with the native state of the yeast prion protein Ure2 and inhibits formation of amyloid-like fibrils *J Biol Chem* **2007**, *282*: 11931-11940.

[100] Fan, Q., Park, K.W., Du, Z., Morano, K.A. and Li, L. The role of Sse1 in the *de Novo* formation and variant determination of the [PSI+] prion *Genetics* **2007**, *177*: 1583-1593.

[101] Krzewska, J. and Melki, R. Molecular chaperones and the assembly of the prion Sup35p, an *in vitro* study *EMBO J* **2006**, *25*: 822-833.

[102] Shorter, J. and Lindquist, S. Destruction or potentiation of different prions catalyzed by similar Hsp104 remodeling activities *Mol Cell* **2006**, *23*: 425-438.

[103] Tutar, Y., Song, Y. and Masison, D.C. Primate Chaperones Hsc70 (Constitutive) and Hsp70 (Induced) Differ Functionally in Supporting Growth and Prion Propagation in Saccharomyces cerevisiae *Genetics* **2006**, *172*: 851-861.

[104] Martineau, C.N., Beckerich, J.M. and Kabani, M. Flo11p-independent control of "mat" formation by hsp70 molecular chaperones and nucleotide exchange factors in yeast *Genetics* **2007**, *177*: 1679-1689.

[105] Martineau, C.N., Melki, R. and Kabani, M. Swa2p-dependent clathrin dynamics is critical for Flo11p processing and 'Mat' formation in the yeast Saccharomyces cerevisiae *FEBS Lett* **2010**, *584*: 1149-1155.

[106] Shaner, L., Sousa, R. and Morano, K.A. Characterization of Hsp70 binding and nucleotide exchange by the yeast Hsp110 chaperone Sse1 *Biochemistry* **2006**, *45*: 15075-15084.

[107] Kampinga, H.H., Hageman, J., Vos, M.J., Kubota, H., Tanguay, R.M., Bruford, E.A., Cheetham, M.E., Chen, B. and Hightower, L.E. Guidelines for the nomenclature of the human heat shock proteins *Cell Stress Chaperones* **2009**, *14*: 105-111.

[108] Tavaria, M., Gabriele, T., Kola, I. and Anderson, R.L. A hitchhiker's guide to the human Hsp70 family *Cell Stress Chaperones* **1996**, *1*: 23-28.

[109] Bukau, B., Weissman, J. and Horwich, A. Molecular chaperones and protein quality control *Cell* **2006**, *125*: 443-451.

[110] Goate, A.M., Cooper, D.N., Hall, C., Leung, T.K., Solomon, E. and Lim, L. Localization of a human heat-shock HSP 70 gene sequence to chromosome 6 and detection of two other loci by somatic-cell hybrid and restriction fragment length polymorphism analysis *Hum Genet* **1987**, *75*: 123-128.

[111] Milner, C.M. and Campbell, R.D. Structure and expression of the three MHC-linked HSP70 genes *Immunogenetics* **1990**, *32*: 242-251.

[112] Fourie, A.M., Peterson, P.A. and Yang, Y. Characterization and regulation of the major histocompatibility complex-encoded proteins Hsp70-Hom and Hsp70-1/2 *Cell Stress Chaperones* **2001**, *6*: 282-295.

[113] Leung, T.K., Rajendran, M.Y., Monfries, C., Hall, C. and Lim, L. The human heat-shock protein family. Expression of a novel heat-inducible HSP70 (HSP70B') and isolation of its cDNA and genomic DNA *Biochem J* **1990**, *267*: 125-132.

[114] Parsian, A.J., Sheren, J.E., Tao, T.Y., Goswami, P.C., Malyapa, R., Van Rheeden, R., Watson, M.S. and Hunt, C.R. The human Hsp70B gene at the HSPA7 locus of chromosome 1 is transcribed but non-functional *Biochim Biophys Acta* **2000**, *1494*: 201-205.

[115] Goldfarb, S.B., Kashlan, O.B., Watkins, J.N., Suaud, L., Yan, W., Kleyman, T.R. and Rubenstein, R.C. Differential effects of Hsc70 and Hsp70 on the intracellular trafficking and functional expression of epithelial sodium channels *Proc Natl Acad Sci U S A* **2006**, *103*: 5817-5822.

[116] Elefant, F. and Palter, K.B. Tissue-specific expression of dominant negative mutant Drosophila HSC70 causes developmental defects and lethality *Mol Biol Cell* **1999**, *10*: 2101-2117.

[117] Florin, L., Becker, K.A., Sapp, C., Lambert, C., Sirma, H., Muller, M., Streeck, R.E. and Sapp, M. Nuclear translocation of papillomavirus minor capsid protein L2 requires Hsc70 *J Virol* **2004**, *78*: 5546-5553.

[118] Daugaard, M., Jaattela, M. and Rohde, M. Hsp70-2 is required for tumor cell growth and survival *Cell Cycle* **2005**, *4*: 877-880.

[119] Rohde, M., Daugaard, M., Jensen, M.H., Helin, K., Nylandsted, J. and Jaattela, M. Members of the heat-shock protein 70 family promote cancer cell growth by distinct mechanisms *Genes Dev* **2005**, *19*: 570-582.

[120] Lee, S.H., Kim, M., Yoon, B.W., Kim, Y.J., Ma, S.J., Roh, J.K., Lee, J.S. and Seo, J.S. Targeted hsp70.1 disruption increases infarction volume after focal cerebral ischemia in mice *Stroke* **2001**, *32*: 2905-2912.

[121] Van Molle, W., Wielockx, B., Mahieu, T., Takada, M., Taniguchi, T., Sekikawa, K. and Libert, C. HSP70 protects against TNF-induced lethal inflammatory shock *Immunity* **2002**, *16*: 685-695.

[122] Huang, L., Mivechi, N.F. and Moskophidis, D. Insights into regulation and function of the major stress-induced hsp70 molecular chaperone *in vivo*: analysis of mice with targeted gene disruption of the hsp70.1 or hsp70.3 gene *Mol Cell Biol* **2001**, *21*: 8575-8591.

[123] Shim, E.H., Kim, J.I., Bang, E.S., Heo, J.S., Lee, J.S., Kim, E.Y., Lee, J.E., Park, W.Y., Kim, S.H., Kim, H.S., Smithies, O., Jang, J.J., Jin, D.I. and Seo, J.S. Targeted disruption of hsp70.1 sensitizes to osmotic stress *EMBO Rep* **2002**, *3*: 857-861.

[124] Kwon, S.B., Young, C., Kim, D.S., Choi, H.O., Kim, K.H., Chung, J.H., Eun, H.C., Park, K.C., Oh, C.K. and Seo, J.S. Impaired repair ability of hsp70.1 KO mouse after UVB irradiation *J Dermatol Sci* **2002**, *28*: 144-151.

[125] Hampton, C.R., Shimamoto, A., Rothnie, C.L., Griscavage-Ennis, J., Chong, A., Dix, D.J., Verrier, E.D. and Pohlman, T.H. HSP70.1 and -70.3 are required for late-phase protection induced by ischemic preconditioning of mouse hearts *Am J Physiol Heart Circ Physiol* **2003**, *285*: H866-874.

[126] Hunt, C.R., Dix, D.J., Sharma, G.G., Pandita, R.K., Gupta, A., Funk, M. and Pandita, T.K. Genomic instability and enhanced radiosensitivity in Hsp70.1- and Hsp70.3-deficient mice *Mol Cell Biol* **2004**, *24*: 899-911.

[127] Singleton, K.D. and Wischmeyer, P.E. Effects of HSP70.1/3 gene knockout on acute respiratory distress syndrome and the inflammatory response following sepsis *Am J Physiol Lung Cell Mol Physiol* **2006**, *290*: L956-961.

[128] Dix, D.J., Garges, J.B. and Hong, R.L. Inhibition of hsp70-1 and hsp70-3 expression disrupts preimplantation embryogenesis and heightens embryo sensitivity to arsenic *Mol Reprod Dev* **1998**, *51*: 373-380.

[129] Rubenstein, R.C. and Zeitlin, P.L. Sodium 4-phenylbutyrate downregulates Hsc70: implications for intracellular trafficking of DeltaF508-CFTR *Am J Physiol Cell Physiol* **2000**, *278*: C259-267.

[130] Rubenstein, R.C., Egan, M.E. and Zeitlin, P.L. *In vitro* pharmacologic restoration of CFTR-mediated chloride transport with sodium 4-phenylbutyrate in cystic fibrosis epithelial cells containing delta F508-CFTR *J Clin Invest* **1997**, *100*: 2457-2465.

[131] Choo-Kang, L.R. and Zeitlin, P.L. Induction of HSP70 promotes DeltaF508 CFTR trafficking *Am J Physiol Lung Cell Mol Physiol* **2001**, *281*: L58-68.

[132] Garrido, C., Brunet, M., Didelot, C., Zermati, Y., Schmitt, E. and Kroemer, G. Heat shock proteins 27 and 70: anti-apoptotic proteins with tumorigenic properties *Cell Cycle* **2006**, *5*: 2592-2601.

[133] Raynes, D.A., Graner, M.W., Bagatell, R., McLellan, C. and Guerriero, V. Increased expression of the Hsp70 cochaperone HspBP1 in tumors *Tumour Biol* **2003**, *24*: 281-285.

[134] Souza, A.P., Albuquerque, C., Torronteguy, C., Frasson, A., Maito, F., Pereira, L., Duval da Silva, V., Zerwes, F., Raynes, D., Guerriero, V. and Bonorino, C. HspBP1 levels are elevated in breast tumor tissue and inversely related to tumor aggressiveness *Cell Stress Chaperones* **2009**, *14*: 301-310.

[135] Tanimura, S., Hirano, A.I., Hashizume, J., Yasunaga, M., Kawabata, T., Ozaki, K. and Kohno, M. Anticancer drugs up-regulate HspBP1 and thereby antagonize the prosurvival function of Hsp70 in tumor cells *J Biol Chem* **2007**, *282*: 35430-35439.

[136] Yokoyama, Y., Kuramitsu, Y., Takashima, M., Fujimoto, M., Iizuka, N., Terai, S., Okita, K., Sakaida, I., Oka, M., Raynes, D.A., Guerriero, V. and Nakamura, K. Increased expression of heat shock protein-binding protein 1 and heat shock protein 70 in human hepatocellular carcinoma tissues *Mol Med Report* **2008**, *1*: 197-201.

[137] McLellan, C.A., Raynes, D.A. and Guerriero, V. HspBP1, an Hsp70 cochaperone, has two structural domains and is capable of altering the conformation of the Hsp70 ATPase domain *J Biol Chem* **2003**, *278*: 19017-19022.

[138] Raynes, D.A. and Guerriero, V., Jr. Inhibition of Hsp70 ATPase activity and protein renaturation by a novel Hsp70-binding protein *J Biol Chem* **1998**, *273*: 32883-32888.

[139] Shomura, Y., Dragovic, Z., Chang, H.C., Tzvetkov, N., Young, J.C., Brodsky, J.L., Guerriero, V., Hartl, F.U. and Bracher, A. Regulation of Hsp70 function by HspBP1: structural analysis reveals an alternate mechanism for Hsp70 nucleotide exchange *Mol Cell* **2005**, *17*: 367-379.

[140] Son, W.Y., Han, C.T., Hwang, S.H., Lee, J.H., Kim, S. and Kim, Y.C. Repression of hspA2 messenger RNA in human testes with abnormal spermatogenesis *Fertil Steril* **2000**, *73*: 1138-1144.

[141] Govin, J., Caron, C., Escoffier, E., Ferro, M., Kuhn, L., Rousseaux, S., Eddy, E.M., Garin, J. and Khochbin, S. Post-meiotic shifts in HSPA2/HSP70.2 chaperone activity during mouse spermatogenesis *J Biol Chem* **2006**, *281*: 37888-37892.

[142] Dix, D.J., Allen, J.W., Collins, B.W., Mori, C., Nakamura, N., Poorman-Allen, P., Goulding, E.H. and Eddy, E.M. Targeted gene disruption of Hsp70-2 results in failed meiosis, germ cell apoptosis, and male infertility *Proc Natl Acad Sci U S A* **1996**, *93*: 3264-3268.

[143] Dix, D.J., Allen, J.W., Collins, B.W., Poorman-Allen, P., Mori, C., Blizard, D.R., Brown, P.R., Goulding, E.H., Strong, B.D. and Eddy, E.M. HSP70-2 is required for desynapsis of synaptonemal complexes during meiotic prophase in juvenile and adult mouse spermatocytes *Development* **1997**, *124*: 4595-4603.

[144] Zhu, D., Dix, D.J. and Eddy, E.M. HSP70-2 is required for CDC2 kinase activity in meiosis I of mouse spermatocytes *Development* **1997**, *124*: 3007-3014.

[145] Daugaard, M., Kirkegaard-Sorensen, T., Ostenfeld, M.S., Aaboe, M., Hoyer-Hansen, M., Orntoft, T.F., Rohde, M. and Jaattela, M. Lens epithelium-derived growth factor is an Hsp70-2 regulated guardian of lysosomal stability in human cancer *Cancer Res* **2007**, *67*: 2559-2567.

[146] Macario, A.J. and Conway de Macario, E. Sick chaperones, cellular stress, and disease *N Engl J Med* **2005**, *353*: 1489-1501.

[147] Singh, R., Kolvraa, S. and Rattan, S.I. Genetics of human longevity with emphasis on the relevance of HSP70 as candidate genes *Front Biosci* **2007**, *12*: 4504-4513.

[148] Singh, R., Kolvraa, S., Bross, P., Jensen, U.B., Gregersen, N., Tan, Q., Knudsen, C. and Rattan, S.I. Reduced heat shock response in human mononuclear cells during aging and its

association with polymorphisms in HSP70 genes *Cell Stress Chaperones* **2006**, *11*: 208-215.

[149] Njemini, R., Abeele, M.V., Demanet, C., Lambert, M., Vandebosch, S. and Mets, T. Age-related decrease in the inducibility of heat-shock protein 70 in human peripheral blood mononuclear cells *J Clin Immunol* **2002**, *22*: 195-205.

[150] Simar, D., Malatesta, D., Koechlin, C., Cristol, J.P., Vendrell, J.P. and Caillaud, C. Effect of age on Hsp72 expression in leukocytes of healthy active people *Exp Gerontol* **2004**, *39*: 1467-1474.

[151] Visala Rao, D., Boyle, G.M., Parsons, P.G., Watson, K. and Jones, G.L. Influence of ageing, heat shock treatment and *in vivo* total antioxidant status on gene-expression profile and protein synthesis in human peripheral lymphocytes *Mech Ageing Dev* **2003**, *124*: 55-69.

[152] Tatar, M., Khazaeli, A.A. and Curtsinger, J.W. Chaperoning extended life *Nature* **1997**, *390*: 30.

[153] Komili, S., Farny, N.G., Roth, F.P. and Silver, P.A. Functional specificity among ribosomal proteins regulates gene expression *Cell* **2007**, *131*: 557-571.

[154] Fourie, A.M., Sambrook, J.F. and Gething, M.J. Common and divergent peptide binding specificities of hsp70 molecular chaperones *J Biol Chem* **1994**, *269*: 30470-30478.

[155] Gragerov, A. and Gottesman, M.E. Different peptide binding specificities of hsp70 family members *J Mol Biol* **1994**, *241*: 133-135.

[156] Zmijewski, M.A., Macario, A.J. and Lipinska, B. Functional similarities and differences of an archaeal Hsp70(DnaK) stress protein compared with its homologue from the bacterium Escherichia coli *J Mol Biol* **2004**, *336*: 539-549.

[157] Zmijewski, M.A., Skorko-Glonek, J., Tanfani, F., Banecki, B., Kotlarz, A., Macario, A.J. and Lipinska, B. The DnaK chaperones from the archaeon Methanosarcina mazei and the bacterium Escherichia coli have different substrate specificities *Acta Biochim Pol* **2007**, *54*: 509-522.

Send Orders of Reprints at reprints@benthamscince.net

Advances in Genome Science, Vol. 1, **2013, 452-490**

CHAPTER 17

Mitochondrial and Nuclear Genes of Mitochondrial Components in Cancer

E. Kirches*

Department of Neuropathology, Otto-von-Guericke University, Magdeburg, Gemany

Abstract: Although the observation of aerobic glycolysis of tumor cells by Otto v. Warburg had demonstrated abnormalities of mitochondrial energy metabolism in cancer decades ago, there was no clear evidence for a functional role of mutant mitochondrial proteins in cancer development until the early years of the 21st century. In the year 2000 a major breakthrough was achieved by the observation, that several genes coding for subunits of the respiratory chain (ETC) complex II, *succinate dehydrogenase* (SDH), are tumor suppressor genes in heritable paragangliomas, fullfilling Knudson's classical two-hit hypothesis. A functional inactivation of both alleles by germline mutations and chromosomal losses in the tumor tissue was found in the patients. Later, *SDH* mutations were also identified in sporadic paragangliomas and pheochromocytomas. Genes of the mitochondrial ATP-synthase and of mitochondrial iron homeostasis have been implicated in cancer development at the level of cell culture and mouse experiments. In contrast to the well established role of some nuclear *SDH* genes, a functional impact of the mitochondrial genome itself (mtDNA) in cancer development remains unclear. Nevertheless, the extremely high frequency of mtDNA mutations in solid tumors raises the question, whether this small circular genome might be applicable to early cancer detection. This is a meaningful approach, especially in cancers, which tend to spread tumor cells early into bodily fluids or faeces, which can be screened by non-invasive methods.

Keywords: Cancer, mtDNA, mitoChip, paraganglioma, Warburg effect, SDH, fumarate hydratase, hexokinase II, tumor suppressor.

1. INTRODUCTION: ENERGY METABOLISM OF CANCER CELLS

1.1. Otto Warburg's Unproven Hypothesis

More than eighty years ago, Otto Heinrich Warburg began his pioneering work on

***Address correspondence to E. Kirches:** Department of Neuropathology, Otto-von-Guericke University, Magdeburg, Gemany; E-mail: elmar.kirches@med.ovgu.de

Christian Neri (Ed)

tumor cell metabolism, which culminated in the hypothesis that endogenous respiration defects of cancer cells, accompanied by the slow development of an enhanced 'aerobic glycolysis' during carcinogenesis`. would be a metabolic switch, actually defining the root of cancer [1, 2]. This metabolic cancer hypothesis in its strict sense has largely been forgotten, since the the era of genetic carcinogenesis concepts revolutionized science. Although changes in energy metabolism are today not regarded to be the cause of cancer, the idea of enhanced aerobic glycolysis ('*Warburg effect*') beeing a general metabolic footprint of cancer, is still present in the literature. It is even used as a rational for the clearly effective metabolic PET imaging (*^{18}F-deoxy-glucose positron emission tomography*) of solid tumors, which exploits the increased accumulation of a labeled glucose-derivative in the tumor tissue, since enhanced glycolytic flux would require enhanced glucose uptake. However, the stimulation of glycolytic flux can be explained more easily by the common and often very pronounced hypoxia of solid tumors. Other molecular alterations may play an additional role for deoxy-glucose enrichment, such as increased activity of the glucose carrier Glut-1 [3, 4]. Instead of using imported glucose mainly glycolytically, some tumors may convert it to fatty acids, since they contain increased amounts of fatty acid synthase [5].

The first one to critisize Warburg's concept was Weinhouse [6], who already mentioned in the year 1956, that some normal tissues, such as brain, retina, intestinal mucosa and kidney medulla, exhibit similar aerobic lactate production, as compared with Warburg's model, *i.e.* ascites tumor cells. Moreover, he stated that several tumor cells show rates of complete glucose and fatty acid oxidation, which were quite similar to normal cells. Remarkably, this topic remained a matter of debate until now. Warburg had noticed that ascites tumor cells under normoxic conditions gain roughly half of their ATP demand by fermentation of glucose to the end product lactate, while the glycolytic ATP contribution of *e.g.* adult mouse liver and kidney is less than 1% 7. While the observation of a high glycolytic ATP contribution under normoxia applies also to several other tumor cells in culture, it does not apply generally to all tumor cell lines, in which both, glycolytic and oxidative ATP contributions, had been precisely measured [for review see 7]. *Vice versa*, some normal tissues exhibit a high contribution of aerobic glycolysis to fulfill their energy demands [7]. No statistically significant difference was observed between the groups of normal and tumor cells, if a meta-analysis of the literature was performed, which included only publications, in which both ATP generating pathways were measured in parallel.

1.2. Hexokinase II, a Molecular Link to the 'Warburg Effect'

Although 'enhanced aerobic glycolysis' is far away from beeing a general and specific feature of cancer cells, molecular alterations exist in some cancers, which support a '*Warburg effect*'. The best studied among them is a switch in *hexokinase* (HK) *isoenzymes* during tumor development. This enzyme catalyzes the rate-limiting first step of the glycolytic pathway, the ATP-consuming phosphorylation of glucose to glucose-6-phosphate (G-6-P). Although glycolysis takes place in the cytoplasm, some authors had described in various articles since 1959 that a fraction of *hexokinase activity* of tumor cells appeared in a particulate cell fraction, which already pointed towards a mitochonrial association. In 1977, Bustamante and Pederson found a 20-fold increase of *hexokinase activity* in the azo-dye-derived rat hepatoma line H-91 as compared to normal or regenerating rat liver [8]. The enzyme was not significantly inhibited by low concentrations (< 0.6 mM) of its reaction product G-6-P in contrast to results reported earlier from *brain hexokinase*. Half of the activity was found in the mitochondrial cell fraction and the enzyme seemed to be coupled directly to the *oxidative phosphorylation* (OXPHOS), *i.e.* to the oxidative mitochondrial pathway of ATP synthesis.

Meanwhile many more molecular details of this relationship have been elucidated [for review see 9]. Among the four known human isoenzymes, mainly HK II and to lesser degree HK I are upregulated in several human cancers, while the low affinity isoform HK IV is silenced [10-12]. This isoenzyme switch could be followed in detail during liver carcinoenesis. In contrast, most normal tissues express only little HK II [13, 14]. The HK II gene promoter was found to be inactivated by methylation in normal hepatocytes, while this methylation was lost in hepatoma cells during carcinogenesis [15]. The results indicated that epigenetic changes during transformation or tumor progression may underlie the isoenzyme switch. However, also a low-grade intra-chromosomal amplification of the HK II gene had been ovserved during hepato-carcinogenesis [16]. Both mechanisms may participate in the 100-fold increase in HK II mRNAs observed in northern blots [17-19].

A molecular basis for the suggested relationship between OXPHOS and HK II 8 was found in the discovery of a physical interaction of HK II with a pore-forming protein in the outer mitochondrial membrane, the *voltage dependent anion chanel* (VDAC) [20, 21]. ATP generated during OXPHOS in the mitochondrial matrix is transported *via* ANT *(adenosine nucleotide translocator)* through the inner mitochondrial membrane and leaves the mitochondria towards the cytoplasm *via* VDAC. In the HK II overexpressing tumor cells, ATP is thus directly channelled to the enzyme [22]. In this way, oxidatively generated ATP can be utilized to stimulate phosphorylation of glucose to G-6-P in the cytoplasm. The high abundance and high activity (decreased product inhibition) of the HK II *hexokinase isoform* in the tumor cells may be an important trigger of enhanced glucose utilization and lactate production of tumor cells even under normoxic conditions, under which mitochondrial ATP is available as an additional fuel of the *hexokinase reaction*. Besides these metabolic considerations, another role of HK II in creating a more malignant phenotype may be discussed, since VDAC is an essential component of the *mitochondrial permeability transition pore* (mPTP). Since the mPTP is suggested to be a major player in the intrinsic, mitochondrial pathway of apoptosis, a binding partner of VDAC may be involved in apoptosis resistance. This could be demonstrated by some authors [23, 24], while disruption of the HKII-VDAC-interaction enhanced apoptosis induction [25]. Genetic knockdown of the gene in glioma cells enhanced their apoptotic sensitivity and reduced tumor growth in an orthotopic intracranial mouse model, while enforced overexpression had the opposite effect [26]. Increased expression of HK II was shown to be associated with worse overall survival in some human cancers [27].

1.3. Vhl-Gene Inactivation Mimics Hypoxia and Induces Cancer: A Second Link to Warburg

Despite induction of neoangiogenesis, hypoxia is a common phenomenon in solid tumors, to which tumor cells have to adapt their energy metabolism in order to survive and to further allow cell proliferation. This requires enhanced glucose uptake and glycolysis, since ATP cannot be delivered by the generally more efficient OXPHOS in the absence of O_2. It is therefore not surprising that the expression of Glut-1 and at least one isoform of all glycolytic enzymes are increased during the HIF-1 (*hypoxia inducible factor 1*)- mediated hypoxia response. The α-subunit of the heterodimeric transcription factor HIF-1 is instable under normoxic onditions, since an O_2-requiering hydroxylation of two prolyl residues is the prerequisite for the recognition of the α-subunit by the E3-ubiquitin ligase complex, which initiates degradation of the protein. Hypoxia thus stabilizes the protein and allows the building of sufficient amounts of active heterodimeric HIF-1, which targets many known hypoxia response genes, Glut-1 and glycolytic enzymes beeing among them.

Inactivation of the *von-Hippel-Lindau* (*VHL*)- tumor suppressor gene is a well-studied example of a pseudo-hypoxia reaction increasing glycolytic flux *via* HIF-1 in tumor cells [28]. The biallelic loss of this tumor suppressor gene is known to predispose patients to various benign and malignant neoplasms, mainly retinal, cerebellar and spinal hemangioblastomas, but also renal cell carcinoma, pheocromocytoma and pancreatic cancer [29, 30]. The protein of the *VHL* gene (pVHL) is a subunit of the E3-ubiquitine ligase complex, which binds ubiquitin to all three isoforms of HIF-1α. A loss of pVHL-function therefore leads to HIF-1α stabilization, followed by a chain of events, which would normally be elicited by hypoxia. VHL inactivation by germ line mutation and *loss of heterozygosity* (LOH) is therefore a second, although rare, molecular alteration in tumors which enables a 'Warburg effect'. The glycolytic product pyruvate would normaly enter the Krebs cycle *via pyruvate dehydrogenase* (PDH). Since the PDH-phosphorylating (inhibiting) enzyme *pyruvate dehydrogenase kinase 1* (PDK1) is among the HIF-1 targets, VHL-inactivation at the same time inhibits Krebs cycle and OXPHOS. As a consequence most of the pyruvate has to be metabolized to lactate. Taken together, this represents a switch from oxidative to glycolytic ATP generation under normoxia. In the following paragraph it will be shown that the chain of events eliciting a pseudo-hypoxia response can also have its primary origin in a mitochondrial dysfunction. The inhibition of two Krebs cycle enzymes by gene mutation can induce tumor growth, which means that enzymes of the oxidative energy metabolism act as true tumor suppressors at least in some relatively rare tumor diseases.

2. ENZYMES OF THE KREBS CYCLE AS TUMOR SUPPRESSORS

2.1. Succinate Dehydrogenase: Another Mode to Mimic Hypoxia

2.1.1. Familal and Non-Familial Paragangliomas

A major breaktrough in answering the question, whether mutations of mitochondrial constituents, which are involved in intermediate and energy metabolism, can be the primary cause of human tumors was achieved in the year 2000 by Baysal and colleagues [31]. The authors analyzed the genetic basis of a relatively rare and benigne inherited human tumor disease, the *hereditary paraganglioma* (PGL), for which at least three gene loci (PGL-1, -2, -3) had been suspected. Although this tumor can occur at various body sites, it is observed most often in the *carotid body* (CB), a tissue representing the major oxygen sensor in mammals. The CB stimulates the cardiopulmonary system under hypoxic conditions [32] by processes, which are thought to be gouverned at the cellular level by HIF-1 [32-34]. Chronic hypoxa, *e.g.* caused by dwelling in high altitudes or some cardiopulmonary disorders, induces cellular hyperplasia or anaplasia in the CB tissue [35-41]. Similar histologic changes observed in PGL [40, 42] were first hints, that some kind of HIF-1 mediated pseudo-hypoxia response may underlie PGL pathogenesis. This led to the hypothesis that the genetic defect in PGL may inactivate an essential component of the oxygen sensing machinery itself or of the signal transducing mechanisms on the way to HIF-1.

The genetic analysis of PGL-1 families [31] led to the discovery that the *SDHD* gene, encoding the small cytochome b subunit within the mitochondrial *succinate dehydrogenase complex* (SDH) is responsable for PGL-1. Germline missense or nonsense mutations in this gene were detected together with LOH of the second allele in the tumor tissue. The *SDHD*-gene thus behaved like a typical tumor suppressor, since biallelic loss was necessary for tumor development. Later it was however found that LOH not always occurred in the tumor tissue. Somewhat mysterious remained the feature of maternal imprinting, *i.e.* the transmission of a diseased phenotype soleley through the fathers. The authors suggested that maternal imprinting may be related to the fact, that in several PLG-1 families chromosomal losses (LOH) of the *SDHD* region in the tumor tissue occurred selectively in the maternal chromosome, but the reason for this remained unknown.

This work demonstrated for the first time that mutations of enzymes of the oxidative mitochondrial metabolism can be tumorigenic. At this point it was not clear, how SDH dysfunction initiates tumor growth in the CB. SDH has a dual function in mitochondria, since it is an enzyme involved in the Krebs cycle and is part of the mitochondrial *electron transport chain* (ETC), where it is designated as 'ETC complex II'. This means that SDH participates in the intermediary metabolism, as well as in the production of reduced flavin nucleotides, which support the electron flow, used for mitochondrial ATP synthesis. At first sight, it was not obvious, how an inhibition of one of these two functions may lead to a pseudo-hypoxia response with HIF-1 activation, hyperplasia, anaplasia and at least development of a benigne tumor. Meanwhile, a hypothesis has been deduced from cell culture studies, which proposes loss of an enzymatic function required for HIF1α-degradation as the key mediator of SDH dysfunction, thus connecting this novel tumor suppressor with the classical *VHL* gene (Fig. **1**).

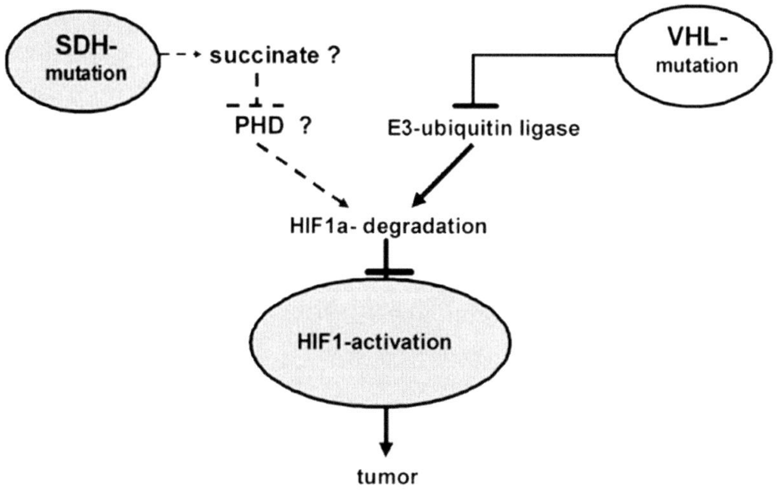

Figure 1: Model derived from cell culture studies to explain the 'pseudohypoxia effect' of succinate dehydrogenase (SDH) inactivation, which may simply inhibit prolyl hydroxylase (PHD) by succinate accumulation. PHD is required to prepare HIF1α to subsequent ubiquitin ligation and degradation. Inactivation of the well known Van-Hippel-Lindau tumor suppressor gene (*VHL*) directly destroys ubiquitin ligase activity.

Studies using *SDHA*-mutant fibroblasts shed some light on the mechanism, although this subunit is currently not known to be especially prone to inactivating mutations in human tumors, while two other subunits are involved in tumorigenesis. In the mutant fibroblasts, SDH-deficiency was correlated with succinate accumulation and HIF-1 nuclear translocation [43]. The latter process was not observed in ATPase-deficient fibroblasts or under cell culture conditions with enhanced production of *reactive oxygen species* (ROS). These results suggested that the accumulation of the SDH substrate succinate in the Krebs cycle may be a better candidate as a starting point of enhanced HIF-1 signaling than a decrease in mitochondrial ATP or an increase in ROS production by the defective enzyme.

Since succinate is a reaction product of the *prolyl hydroxylase* (PHD), which prepares HIF-1α to subsequent degradation, it was suggested, that product-inhibition of PHD may play the key role. This view was strongly supported by the fact that the PHD substrate, α-ketoglurarate (KG), inhibited HIF-1 activation in the mutant fibroblasts. The product/substrate equilibrium of PHD may be critically shifted also by tumorigenic SDH mutations. This may apply at least to some critical mutations of SDHB (see below), a subunit which together with SDHA forms the catalytic core, the inactivation of which should lead to succinate accumulation. In other subunits and probably depending on the location of the affected codons, enhanced ROS production may however be essential [44, 45].

In the same year, when Baysal and colleagues had published their work, the *SDHC* gene was identified as a second PGL locus. Germline mutations caused autosomal dominant paraganglioma type 3 (PGL-3), a form without genetic imprinting [46]. Shortly thereafter mutations of *SDHB* and of *SDHC* / *SDHD*, which anchor the catalytic core to the inner mitochondrial membrane, had been identified as a frequent cause of all three familial PGL-types and as a less frequent cause of non-familial cases [47]. More recently, the list of SDH subunits associated with paranagngliomas has been further extended by addition of SDHA. A first case with germline mutation of this gene and LOH in the tumor tissue has been described by Burnichon and colleagues [48]. The mutation causes a loss of SDH activity and leads to a pseudo-hypoxia response with increased angiogenesis. LOH of this locus was observed in 4.5% of cases in large series of paragangliomas and pheochromocytomas (n = 202). In addition, an accessory protein, needed for the correct incorporation of the cofactor flavin adenine dinucleotide into SDH and for proper function of the enzyme, was found to be associated with paragangliomas. The gene encoding this protein, SDHAF2 (also designated SDH5), was found to be mutated in a Dutch family with hereditary paraganglioma and to co-segregate with the disease [49, 50].

2.1.2. Pheocromocytoma and Other Tumors

Pheocromocytoma, a tumor of the adrenal medulla which may cause secondary hypertension, occurs as a familial disease in about 10% of cases. These cases can be caused by germline mutations of various tumor suppressor genes, *VHL* being among them. This suggested, that altered HIF-1 signaling may play a role in patients with familial pheocromocytoma. *SDHB* mutations were found in two of five families screened and the same gene was occasionally affected in sporadic forms of the disease [51].

Familial *renal cell carcinomas* (RCC) are also among those tumor types, which are sometimes associated with VHL-inactivation. *SDHB* mutations were not only detected in patients suffering from a familial tumor syndrome with early onset RCC and/or PGL [52], but were found also in 4.4% of sporadic RCC cases [53]. Moreover, SDH gertmline mutations were observed in 13.5% of patients with Cowden's syndrome without mutations in the tumor suppressor PTEN, which explains this rare, inherited hamartoma and tumor disease in more han 80% of cases [54].

2.2. Fumarate Hydratase

Fumarate hydratase (FH) deficiency is an encephalopathy inherited as an autosomal recessive trait, and is associated with mostly carboxyterminal germline mutations in the FH gene behind codon 250 [55-58]. Germline mutations clustering in the aminoterminal region before codon 250 were however more recently associated with an autosomal dominant tumor syndrome causing uterine and skin leiomyomata, as well as papillary renal cell cancer [59]. In 84% of the cases, the second allele was inactivated by LOH and the FH gene therefore behaved as a classical tumor suppressor. The tumor suppressor concept was further underlined by the observation, that the leiomyomata showed virtually no

FH enzyme activity. Occasionally, mutations and/or LOH in the chromosomal region, which containes the FH gene, were also observed in non-syndromeuous leiomyomata [60]. FH is an enzyme placed in the Krebs cycle immediately behind SDH, suggesting that tumor development may involve similar mechanisms including an important role for succinate accumulation.

3. OTHER NUCLEAR-ENCODED MITOCHONDRIAL PROTEINS DISCUSSED IN CANCER

3.1. Does Frataxin Play an Inhibiting Role in ROS-Mediated Tumorigenesis?

Friedreich's ataxia is a severe neurodegenerative disease of adulthood, often accompanied by cardial hypertrophy and usually leading to the patient's death within 15 years. It is inherited as a recessive autosomal trait, caused by an intronic GAA trinucleotid expansions in the frataxin gene, which switches off transcription of the affected allele [61]. Frataxin is a mitochondrial protein, which is suggested to be involved in iron homeostasis in the mitochondria, since a reduced amount of the protein in patients with the intragenic trinucleotide expansion leads to intramitochndrial iron deposits. Although the details of frataxin function are currently a matter of debate, these deposits may indicate an insufficient transport of iron to the sites of iron-sulfur cluster biogenesis by frataxin [62-65], which in turn may hamper the incorporation of correctly iron-loaded Fe/S-clusters into various mitochondrial proteins, such as the Fe/S-containing ETC complexes. In cell cultures, frataxin inactivation was shown to result in ETC inhibition, as demonstrated by a diminished mitochondrial membrane potential, decreased O_2-consumption and decreased oxidative ATP synthesis. Frataxin dysfunction may lead to an inhibition of mitochondrial energy metabolism. On the other hand, a dysfunction of ETC complexes, as well as deposits of free iron (by Fenton reaction) can cause enhanced generation of ROS in the mitochondrial matrix and in the intermembrane space. Oxidative stress is thought to further damage ETC complexes and other redox-sensitive proteins, such as the Krebs cycle enzyme aconitase. This oxidative stress component may further inhibit intermediary metabolism and oxidative ATP synthesis.

Shoichet and colleagues analyzed for the frst time a potential tumor suppressing function of frataxin in cell culture and animal models, although patients suffering from Friedreich's ataxia are in no way prone to a higher tumor burden. The idea behind this work was a potential connection between the antioxidative properties of frataxin and ROS-mediated tumorigenesis. The authors analyzed this topic initially in murine 3T3L1 cells, which had been transfected with either a vector expressing human frataxin or a control construct [66]. Both cell clones were exposed to culture conditions with arteficially enhanced ROS production. The cells overexpressing frataxin exhibited a significantly lower number of anchorage-independent foci in the culture dishes and the rate of colony formation in soft agar assays was significantly lower, indicating that frataxin protected the cells from ROS-induced transformation into a tumor phenotype. This was further supported by the observation that only cells from the anchorage-independent foci were able to induce tumor growth when xenografted to nude mice.

While oxidative stress in frataxin-deficient patients with Friedreich:s ataxia may be due to the disturbed synthesis of Fe/S-clusters and mitochondrial iron deposits, it is less obvious how enforced overexpression of frataxin in normal cells may protect them against ROS. The authors explained the protective effect by the observed increase in *glutathione peroxidase* (GPx) activity and in reduced thiols. GPx and the reduced form of glutathione play an important role in the detoxification of H_20_2, whch is built as a product of the SOD- *(superoxide dismutase*) reaction. *Superoxide dismutase 2* (SOD2) is a mitochondrial enzyme, which detoxifies superoxide radicals, released mainly from ETC complexes I and III into the mitochondrial matrix and the intermembrane space.

In a next step, the authors investigated the potential tumor suppressing effect of enforced frataxin expression in the colon cancer lines MIP101, DLD2 and HT29, which lack endogeneous expression of the protein. Mitochondrial oxidative metabolism was enhanced in the transfected cells, as could be shown by an increase of mitochondrial membrane potential, cellular respiration and ATP content, as well as aconitase activity. Increased aconitase activity may be explained in part by decreased oxidative stress. Again the frataxin cells exhibited a lower colony formation rate in soft agar essays and a lower rate of tumor growth after xenotransplantation to nude mice [67]. The most direct evidence for a role of frataxin in carcinogenesis was reported by the same group using targeted hepatic disruption of frataxin expression in mice. The animals had reduced life spans and developed multiple hepatic tumors, in which high apoptotic and mitotic (Ki-67) indices were observed [68]. The liver specimen showed elevated levels of *thiobarbituric-acid reactive substances* (TBARS), a marker of lipid peroxidation, and elevated levels of oxidized glutathione, a classical oxidative stress marker. Activities of those mitochondrial enzymes, which contain Fe/S-moieties, *i.e.* aconitase and ETC complexes I, II, III, were reduced. In accordance with an inhibited oxidative metabolism, the ATP content of livers from knock-out animals was decreased.

The authors did not develop a detailed hypothesis regarding the mechanism of liver tumorigenesis, but speculated that an observed reduction in p38-MAP-kinase phosphorylation may play a role, since activation of this type of *mitogen activated protein kinase* had been discussed earlier as a factor suppressing the formation of hepatic tumors [69, 70]. Although the authors did not discuss this issue, it might be speculated, whether the observed inhibition of SDH (complex II) may participate in tumorigenesis.

Our own experiments with frataxin overexpressing U87MG human glioma cells revealed *in vitro* results, which were only partially consistent with the findings depicted above, while our xenograft experiments in nude mice argued against a general role of frataxin as a tumor suppressor [71]. The H_2O_2 release of isolated mitochondria was attenuated as compared to mock transfected control cells, which was in accordance with the suggested anti-oxidative and protective function of high frataxin levels. This protective function was further supported by a small, but significant activity increase of aconitase, a well-known mitochondrial ROS target. Surprisingly, the determination of cytoplasmic ROS levels using the ROS-sensitive dye DCFH revealed a higher increase of DCF fluorescence over time in the frataxin overexpressing cells. This unexpected result may be explained by the observation of reduced activities of the H_2O_2 degrading enzymes catalase and glutathione peroxidase (GPx), associated with reduced concentrations of the GPx co-substrate glutathione in these cells.

It remains thus unclear, which role frataxin precisely plays for mitochondrial function and oxidative stress in tumor cells. Most important, the frataxin transfected U87MG glioma cells exhibited a significantly accelerated subcutaneous tumor growth in nude mice, which was not in accordance with a general role of this mitochondrial protein as a tumor suppressor, as it appeared from the above described experiments regarding liver tumorigenesis. The role of frataxin in tumors currently remains an open question.

3.2. Other Genes Related to Oxidative Stress

SOD2 is the main enzyme to detoxify the superoxide generated by the *ETC*. Since probably all of the superoxide generated by ETC complex I and a part of that produced by complex III is released into the mitochondrial matrix, SOD2 is located in this compartment. Import of the monomeric precursor protein requires a translocation through both mitochondrial membranes, mediated by the *translocators of the outer* (TOM) *and inner membrane* (TIM), for which a mitochondrial targeting sequence at the aminoterminus is needed. A Val/Ala polymorphism in codon 16 of this leader peptide is known, which predicts a profound conformational change between α-helix and ß-sheet. Ala-16 homozygosity has been shown to be associated with poorer transport efficiacy, leading to a dicreased intramitochondrial formation of the homotetrameric mature enzyme, which in turn reduces the antioxidative capacity of mitochondria due to a lower enzyme activity [72]. The lowered SOD2 activity may usually be without major consequences for the cell, but may lead to oxidative stress, when other antioxidative mechanisms are disturbed at the same time. Li and colleagues observed, that low blood selenium levels enhance the prostate cancer risk in individuals homozygeous for the Ala-16 leader sequene [73]. This observation may be interpreted in terms of a synergism between decreased SOD2-activity and lowered activity of *glutathione peroxidase* (GPx), a selenium-dependent enzyme.

Selenoprotein P is the most important selenium binding protein, which accounts for at least 40% of total plasma selenium load [74]. The protein contains a polymorphism in codon 234, the alanine allele being associated with lower plasma selenium oncentrations in men, which affect GPx activity [75]. It was therefore a logical step to investigate a possible synergism beetween Ala234 and the SOD2-leader polymorphism. Indeed, synergistic effects were observed, since men homozygous for the 'wrong allele' of both genes had a significantly enhanced risk of prostate cancer (*odds ratio* (OR) 1.43) or clinically aggressive prostate cancer (OR 1.60). The OR became even more pronounced in men, who were strong smokers [76].

3.3. Mutations Causing ETC Defects by Generating a Fragile Mitochondrial Genome

The precursors of most mitochondrial proteins are encoded by nuclear genes, translated into polypeptides at cytoplasmic ribosomes and imported into the mitochondria, to which they are targeted by mitochondrial leader sequences at their amino terminus. This rule applies also to the vast majority of polypeptides, which constitute the five OXPHOS complexes of the inner mitochondrial membrane, *i.e.* ETC complexes I to IV and the ATP-synthetase (F_0/F_1-ATPase, complex V). However, with the exception of complex II (succinate dehydrogenase), all OXPHOS complexes contain at least one polypeptide, which is encoded in a small (16.6 kb) circular and intron-free DNA molecule in the mitochondrial matrix space, the mtDNA. Alltogether the mtDNA encodes 13 OXPHOS polypeptides, as well as 2 ribosomal and 22 tRNAs needed for their translation by ribosomes of the mitochondrial matrix.

DNA polymerase γ (POLG) is the enzyme, which replicates this mitochondrial genome. It is necessary to maintain a funtional ETC and OXPHOS. Interestingly, heterozygous missense-, nonsense- or frameshift-mutations in the POLG gene do usually not cause a depletion of mtDNA, but lead to a fragility of the mitochondrial genome, which is prone to multiple large scale deletions in a proportion of mtDNA molecules, which can be detected in Southern blots of skeletal muscle DNA. These deletions usually affect some mtDNA-encoded tRNA genes besides protein-coding sequences and can inhibit ETC complexes I, III and IV as well as the ATP synthetase. The functional defect is translated into a phenotype, which predominantly affects skeletal muscles (muscle weakness and external ophthalmoplegia) and is usually transmitted as an autosomal dominant trait, designated adPEO (*autosomal dominant progressive external ophthalmoplegia*) [77].

The POLG gene contains a polymorphic CAG repeat [78, 79], which is conserved in all species investigated thus far. Since the clearly dominatung repeat length in the human population is 10, selective pressure seems to occur to conserve this length [80]. Men with homozygous deviations from this dominating repeat length were recently shown to have an increased risk of developing testicular germ cell cancer, mainly seminomas [81]. No correlation of repeat length with other clinico-pathological features was observed. Although the fraction of persons with a homozygeous 'non-10' allele was significantly higher among cancer patients (4,9%) as compared to the general population (1.3%), the generally low frequency shows, that the aberrant repeat length can be viewed only as mild risk factor for tumor development.

4. ROLE OF mtDNA-ENCODED MITOCHONDRIAL PROTEINS IN CANCER

4.1. ATPAse Mutation Enhancing Tumorigenicity of Prostate Cancer Cells in Nude Mice

Since more than ten years, a growing number of articles described somatic mutations of the mitochondrial genome (mtDNA) in human tumors, by comparing tumor and adjacent normal tissue or blood usually with sequencing techniques. Sometimes more sophisticated techniques were used, dedicated to the detection of a low burden of mutated mtDNA (heteroplasmy). Often these mutations occurred in the non-coding, so called D-loop region, or were silent base substitutions. Usually, these base exchanges did not include pathogenic mutations known from 'mitochondrial disorders' and could not be predicted to have functional consequences by other criteria, such as introduction of stop-codons, frame-shifts or alteration of conserved tRNA loop-structures. A large proportion of these mutations actually corresponded to known human polymorphisms. They could be suspected to be caused by oxidative stress and to be enriched in the tumor tissue as a consequenvce of the clonal expansion of a single somatic cell.

In the year 2005 however, Petros and colleagues demonstrated a direct causal connection between a known pathogenic mtDNA mutation (T8993G in the ATPase-6 gene) and the tumorigenecity of PC3 prostate cancer cells in nude mice [82]. This mutation was well known to elicit mitochondrial disorders. It can be responsible either for a severe CNS disease leading to symmetrical necroses in the brain and early death (Leigh syndrome) or to the less severe disease NARP *(neuropathy, ataxia and retinitis pigmentosa)*, which seems to depend on the exact proportion of mutant and normal DNA in the CNS. Generally very high burdens of the mutation (> 95% of total mtDNA in the relevant tissues) are thought to be necessary to cause any midrochondrial disorder. This suggests that the ATP synthetase function is significantly hampered only by extremely high mutation loads, but even lower percentages of mutant DNA may cause enhanced ROS production. Petros and colleages detected this transversion as a heteroplasmic germline mutation in a patient with prostate cancer. The authors introduced mtDNA carrying the mutation and in parallel unaffected mtDNA from the same patient into mtDNA-depleted PC3 prostate cancer cells, using the cybrid technique [83]. The PC3 cells with the mitochondrial genotype 8993G developed tumors with a significantly enhanced growth rate in nude mice, as compared to the wild-type. Effects of other variations in the mitochondrial or nuclear genomes were excluded by this sophisticated approach. The mitochondrial haplogroup, as well as other mtDNA positions potentially influencing OXPHOS, were identical in both xenografted tumor cell lines (genotype of the patient). The nuclear background was that one of PC3 tumor cells in both cases. Therefore, the growth difference between the two manipulated tumor cell clones ('transmitochondrial cybrds') could be attributed to the T8993G mutation. Although unproven in their article, the authors discussed a probable role of enhanced mitochondrial ROS production by the mutant cells to support tumor growth, since ROS are known to influence transcription factors [84], signal transducing kinases and regulatory phosphatases and angiogenic factors. The ATP synthetase mutation T8993G is known to induce enhanced ROS production in other cell types, including cybrids based on the osteosarkoma cell line 143B. Increased, but still moderate levels of miochondrial ROS production may be the cause of enhanced tumorigenecity, as suggested by the authors.

4.2. Non-Neoplastic mtDNA Suppresses the Metastatic Potential of Breast Cancer Cells

Recently, Imanishi and colleagues presented convincing results, which demonstrated that the complete replacement of mtDNA in the human breast cancer cell line MDA-MB-231 by mtDNA from a non-neoplastic source (fetal mtDNA) significantly reduced the ability of the breast cancer cells to generate tumor nodes in the lung, following injection of the transmitochondrial cybrids into the tail vein of nude mice [85]. After generating mtDNA-free rho$_0$-cells by long term culture of the tumor cells under sub-lethal concentrations of ethidium bromide, the authors repopulated them with fetal mtDNA. It was already known at that time that the parental tumor cells exhibited a disturbed function of the electron transport chain (complex I dysfunction) and two candidate point mutations in complex I genes to explain this defect had been sequenced. Interestingly, these and various other mtDNA mutations of metastatic MDA-MB-231 cells are not present in the non-metastatic human breast cancer cell line MCF-7. It could be shown that the transfer of fetal mtDNA normalized ETC function, although it could not suppress the enhanced ROS production seen in MDA-MB-231 cells. To reduce the likelihood that mutations induced by ethidium bromide or other influences of the cybridization process may compromise the results of the following *in vivo* experiments, the authors constructed in parallel control cybrids from the same rho_0-cells by re-introducing the original tumor cell mtDNA. A direct comparison of both clones in the nude mouse model revealed a four times higher number of lung tumor nodes ($p < 0.01$) in the controls (MDA mtDNA), as compared to cybrids containing fetal mtDNA (Fig. **2**).

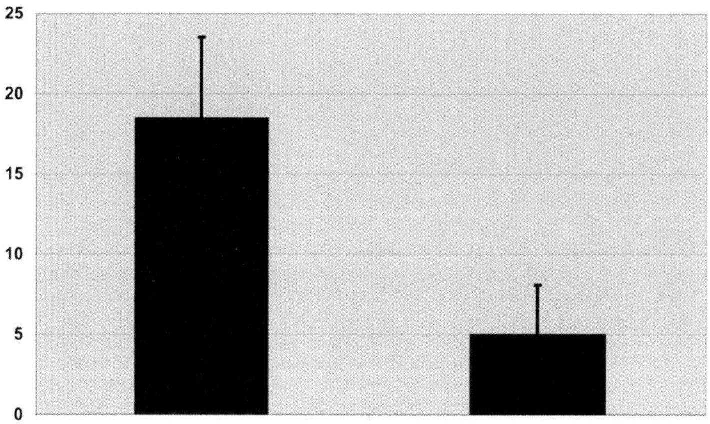

Figure 2: Transmitochondrial cybrids, harbouring the nuclear genome of metatstatic MDA-MB-231 breast cancer cells and either mtDNA of MDA-MB-231 cells (MDA) or mtDNA derived from non-neoplastic fetal fibroblasts (fetal mtDNA), were injected into the tail vein of nude mice. The number of lung nodes observed after 18 to 20 weeks was significantly smaller in cybrids with the non-neoplastic mtDNA ($p < 0.01$). Since cybrids with identical nuclear background had been constructed, the different aggressiveness of both clones could be attributed to their mtDNA (modified according to [85]).

Since ROS levels did not significantly differ between the cybrid clones, the enhanced ROS production of MDA-MB-231 cells could not be attributed to the complex I mutations. Other, ROS-independent mechanisms must be responsible for the suppression of the metastatic potential after recovery of ETC function by mtDNA exchange. Since significant lactate overproduction had been found in the complex I deficient control cybrids and in the parental tumor cells, it was reasonable to speculate about a shift from glygolytic to oxidative energy metabolism as a cause of reduced metastatic potential. However, the authors rejected this hypothesis, since their previous study had already shown that tumor cells harbouring depleted or deleted mitochondrial genomes, although exhibiting higher lactate production, showed actually no metastatic potential [86]. Moreover, the recent study exhibited no metastatic potential of the rho_0-cells themselves, which completely rely on glycolytic energy metabolism and produce the highest lactate levels of all cell clones compared. One conclusion may be, that a too severe disturbance of ETC (mtDNA depletion or deletion) or even its complete inactivation (rho_0-clones) inhibit survival of tumor cells, while moderate ETC dysfunction, *e.g.* by amino acid exchanges in complex I (MDA-MB-231) may contribute to a more metastatic phenotype.

4.3. Mitochondrial DNA Depletion Cooperates with APC-Dependent Intestinal Tumorigenesis

Mice with a heterozygeous knockout of the mitochondrial transcription factor Tfam, which is required for the initiation of mtDNA replication, exhibit a moderate mtDNA depletion, but no obvious phenotype. At least some degree of instability of the mitochondrial genome can be observed. Embryonic fibroblasts, isolated from these mice, showed a dicreased ability to repair H_2O_2-induced oxidative mtDNA damage *in vitro,* as concluded from the reduced recovery of a 10 kb PCR fragment, normalized to a 117 bp fragment to indicate the total amount of mtDNA [87]. Using the same quantitative PCR technique, the authors demonstrated a dramatic increase of mtDNA instability in the intestinal tissue of these mice as compared to their wild type littermates. This increased fragility was interpreated in terms of oxidative mtDNA damage, although the conclusion was not proven by more specific methods.

$Tfam^{+/}$- mice were then crossed with those, harbouring a defective adenomatous polyposis coli (APC) gene ($APC^{Min/+}$), an established mouse model for human intestinal tumorigenesis by loss of function of this tumor suppressor. As compared to $APC^{Min/+}$ mice, the crosses contained a moderately increased number of polyps in the distal part of the small intestine ($p < 0.01$) and a moderately increased total number of medium-sized intestinal polyps ($p < 0.05$) at an age of 18 to 20 weeks [87]. This important result suggested that mtDNA fragility, eventually oxidative damage, in the $Tfam^{+/-}$ mice supports APC-induced intestinal tumorigenesis, while the mitochondrial defect alone was not sufficient to elicit tumorigenesis in any tissue. Although the authors state, that their study does not unequivocally imply a leading role of ROS for the enhanced tumorigenesis of the crossed mice, it is interesting to note that mitochondria isolated from their small intestine exhibited a higher H_2O_2 release, as compared to $APC^{Min/+}$ mice. Since a significant difference between the two genotypes only occurred after inhibition of ETC complex III by antimycin A, it may be concluded that enhanced ROS production *in vivo* occurs only under conditions of a partially inhibited ETC, which may occur in the tumors. Although this argument offers a formal explanation, why the mitochondrial defects supported tumorigensis, but did not elicit the process, the occurrence of natural ETC inhibition in the tumors is not known. Finally, the authors underpinned their hypothesis that ROS play a leading role, by crossing $APC^{Min/+}$ mice with those expressing a mitochondrial targeted catalase, which reduces mitochondrial H_2O_2 release. Indeed, the double transgenic mice developed a moderately lower number of polyps in colon and proximal part of the small intestine, as compared to $APC^{Min/+}$ mice ($p < 0.05$).

The *in vivo* experiments described above (paragraphs 4.1 – 4.3) suggest that at least a small fraction of the non-silent mtDNA point mutations observed in human cancers, and a decrease of mtDNA copy number participate in tumor formation and metastatic potential, instead of representing bystander-effects without any role in tumor biology.

4.4. Spectrum of Human Tumor Diseases, in Which mtDNA Mutations were Observed

Despite Warburg's early suggestion of a causal role of OXPHOS defects in cancer development [1, 2], the mitochondrial genome of cancer cells had not been studied until 1998, when Polyak and colleagues sequenced the mtDNA of ten human colorectal cancer cell lines [88]. The authors detected 12 mutations in 7 of the cell lines (70%). Most base exchanges occurred in rRNA genes or were missense mutations, only two were functional without any doubt (frame shift and chain termination). Although tumor cells often contain a relatively small number of mitochondria with a few mtDNA molecules per mitochondrion, the resulting number of mtDNA copies per cell still garantees, that a new somatic mtDNA mutation initially represents only a minor fraction of total tumor mtDNA, a state called heteroplasmy. Having this situation in mind, it was rather surprising, that most of the mutations were found to be homoplasmic, at least at the sensitivity level of DNA sequencing. Homoplasmic mtDNA mutations were later often observed in human tumor tissues by other authors and do not represent a rare event or a special feature of long cultivated cell lines. As an obvious explanation for homoplasmy, Polyak and colleagues suggested that the mutated mtDNA was dominant at the intracellular and intercellular level in the tumors, from which the cell lines had been derived. This was an important suggestion, since it implied that at least one of the observed mutations per mitochondrial genome should have confered a selective advantage and thus be 'functional'. This hypothesis would suggest a role of mtDNA mutations to promote the proliferation of certain types of tumor cells just in the sense of Warburg's hypothesis. However, selective pressure is not the only explanation for homoplasmy, and not even the most simple one, as will be outlined below (see section 4.5).

In the same year, Habano and colleagues analyzed 45 colorectal tumor samples for 'microsatellite instability' within some homopolymeric sequence tracts of mtDNA [89]. The term 'microsatellite instability', which is still used in current publications, is misleading in a double sense. These sequence tracts do neither fullfill the definition of microsatellites, nor is the low mismatch repair activity in mitochondria the only explanation for their length variability in human tumors, as discussed below (see section 4.5). The term was probably coined to remind nuclear microsatellite instability due to mismatch repair defects in inherited non-polyposis colorectal cancers. The authors detected length changes of the polycytosine tract of the hypervariable region 2 (HVR2) of the mitochondrial D-loop in 20 tumors (44%). This sequence tract, often referred to as D310 in the tumor literature, is meanwhile known to be highly variable in several solid tumor types [90]. It is non-coding (like the whole D-loop) and two different length variants are known to be the dominating constitutive genotypes in normal human tissues, often in a heteroplasmic mixture. Nevertheless, the tract is essential for binding and correct nuclease processing of a RNA primer, which is required for mtDNA replication. In tumors, rare length variants are more frequently observed and even rather long tracts occur, which are never seen as constitutive genotypes in blood samples. In the cancer samples originally analyzed by Habano and colleagues, the authors later described some complex I mutations in 16% of cases in a more detailed analysis [91].

Parrella *et al.* [92] investigated 18 primary breast cancers (invasiv ductal carcinomas) from 17 women by direct sequencing. MtDNA mutations were detected in 7 of them (61%), 42% being length alterations in D310. The remaining were base substitutions in several protein-coding genes. Mutations in the coding and non-coding regions of the mitochondrial genome were detected in half of the investigated head and neck cancers [93, 94], in 64% of bladder and 43% of lung cancer patients [94]. Among patients with multicentric hepatocellular carcinoma, 68% exhibited mtDNA mutations at least in one of the tumors, including a high frequency of D310 length variations (62% of patients).

We and other authors detected a high frequency of D-loop mutations in astrocytic brain tumors [95-97]. We applied D310 length variation to verify clonality in gliomatosis cerebri, an astrocytic tumor disease, in which several brain lobes are diffusely infiltrated without clearly visible tumor margins [98]. The loss of a heteroplasmic D310 genotype in one hemisphere affected by gliomatosis corresponded with a p53 mutation in this hemisphere, thus verifying the clonal expansion of the altered mitochondrial genotype during tumor growth. By screening the mitochondrial genome with time cosuming temperature gradient gel electrophoresis, followed by sequencing, mutations were identified in 40% of medulloblastomas, the most frequent brain tumor of childhood [99].

Studies focusing on the highly susceptible D-loop region or sequencing a few mitochondrial genes in addition, tended to find a lower proportion of tumors harbouring a base exchange or insertions/deletions as compared to studies, which tried to assess as much as possible of the mitochondrial genome. A chip analysis of 94% of the mitochondrial genome in 15 pancreatic cancers detected at least one somatic mutation in every tumor, suggesting that most solid tumors contain mtDNA mutations.

Recent results further supported the existence of mtDNA mutations in a variety of human tumors [100-106], but a large discrepancy occured sometimes regarding mutation frequency within a tumor entity. Wang and co-workers observed mtDNA mutations in only 6 out of 20 (30%) colorectal cancer patients by sequencing the whole mitochondrial genome. Moreover, all mutations were heteroplasmic [100]. This result strongly contrasted with 70% of colorectal tumor cell lines harbouring mostly homoplasmic mutations observed in initial studies. Differences in mutation frequency may be caused by the genetic background of the patients, including the mitochondrial haplogroup, or by differences in the distribution of tumor stages between various studies. On the other hand, a strict quality control of sequence data may have participated in lowering the number of observed "true" mutations, as suggested by Wang and colleagues.

Quality control should include a phylogenetic tree analysis, which detects combinations of base exchanges, which are associated with haplogroups, *i.e.* with branches of the phylogenetic mtDNA tree. Since sample mix up or contaminations would often generate a set of such haplogroup-associated base exchanges in the sequencing data, this type of analysis helps to eliminate fatal errors [107]. A phylogenetic tree approach is an extremely valuable tool for critical re-analysis of sequencing data, allowing the detection of suspicious sets of base exchanges within a single tumor sample. However, it is certainly not meaningful to declare a preferential occurrence of isolated base exchanges at known polymorphic sites in a series of tumor tissues to be suspicious. Without knowledge of the underlying mechanisms, it is not reasonable to expect a priori a random distribution of somatic mtDNA mutations over the whole mitochondrial genome, as briefly disussed in a response included in the above cited paper of Bandelt and Salas [107].

Taken together, the mutation burden of solid tumors is still debated and may be somewhat lower than initially thought, if data were reanalyzed critically using phylogenetic approaches.

4.5. Is there Any Benefit for the Tumors? Positive Selection *Versus* Random Drift

Even polymorphisms of the mtDNA observed in healthy mammals are not necessarily neutral, *e.g.* during the development and aging of organs. This aspect of mitochondrial genetics had already been illuminated in a study creating heteroplasmic mice by genetic engeneering [108] in the year 1997. The mitochondrial genome of two normal healthy mouse strains had been mixed to achieve a balanced artificial heteroplasmy. The two genomes could easily be distinguished by means of PCR. The authors even analyzed heteroplasmic mice with two different nuclear genomes to exclude a potential role of nuclear background. The unexpected result of that study was a tissue-specific up- or downregulation of one of the two mitochondrial genomes during development and adult life of the mice. This result demonstrated that even mitochondrial genomes, which do not carry any known functional mutation, can differ in their replicative fitness in different organs of mammals.

Originally the surprising observation of homoplasmy for a significant proportion of somatic mtDNA mutations in human tumor cells was also interpreted as a replicative advantage of mutated mtDNA and/or a growth or survival advantage for the cells carrying the affected mitochondria [88, 94, 109]. Having in mind the above cited results in heteroplasmic mice, even a proportion of the D-loop, rRNA, tRNA or missense mutations in non-conserved codons, may be interpreted to confer a 'selective advantage'.

However, it is not possible to draw this conclusion simply from the observed frequency of homoplasmic tumors, as could be demonstrated by computer modeling. Coller and colleagues [110] have shown that a sufficient random drift of heteroplasmic mtDNAs theoretically occurs even under conditions of unbiased mtDNA replication and sorting during cell divisons, which can explain the observed frequency of homoplasmy. The computer models were based on some assumptions regarding mtDNA content, mutation rate and number of cell divisions minimally required to generate a detectable solid tumor, which can not be proven in detail. However, even rather conservative estimations of these parameters yielded results, which were in good accordance with the observed rate of homoplasmy. This study forced investigators to accept the notion that homoplasmy of a tumor mutation is no proof of functional relevance. It seems thus likely that a large proportion of all observed mutations are nothing else but indicators of a clonal cell expansion, which may however be exploited as a tool of early cancer detection anyway.

Some sentences may still be dedicated to the usage of the word 'somatic mutation' in the tumor literature, which implies that in all cases a base exchange or deletion/insertion occurred *de novo* in a tumor cell or precursor cell. This assumption seems questionable, at least for some nucleotide positions of mtDNA, which are often heteroplasmic in normal human tissues. The poly-cytosine tract D310, a so called 'hot spot of somatic mutations' in solid tumors, is a good example. The tract occurs in the blood of roughly half of all healthy individuals in one of the two variants C_7TC_6 or C_8TC_6, as determined by sequencing. Especially individuals with the longer tract are usually heteroplasmic, *i.e.* the lymphocytes (and other tissues) contain lower amounts of the shorter variant, as detected by PCR techniques. Even other variants of the tract can occur in normal tissues. It seems thus reasonable to assume that length alterations of this polycytosine stretch in solid tumors, observed during sequencing, are often caused simply by relative shifts between two or more pre-existing heteroplasmic genotypes of the precursor cell. In our laboratory, mtDNA from some glioblastomas had been cloned for extensive analysis of low abundant mtDNA mutations. The 'constitutive genotype' of one patient in this study (normal brain tissue) consisted of two mitchondrial genomes, which could be distinguished by D310 length and two D-loop transitions. One of these genomes was clearly lost in the glioblastoma [95]. This change was attributed to heteroplasmic drift, since three independent events would have been necessary to explain this result by somatic mutations. Unless investigated by specific PCR-assays for a given nucleotide position, one can never be sure that adjacent normal tissue did not contain low amounts of the mtDNA observed as the dominant genotype in a tumor. However, the exact nature of the observed sequence alterations in human tumors plays no role for their potential use in cancer detection, as long as they represent clonal cell expansions. In the following sections the term 'mutation' will be used regardless of the problems associated with it especially regarding D310.

4.6. Are mtDNA Mutations Associated with Patient Survival ?

In some frequent and aggressive human tumor types, such as NSCLC (*non small cell lung cancer*, see below) or breast cancers, an accumulation of somatic mtDNA mutations seemed to be associated with tumor progression, since a higher mutation burden was observed in more advanced cancers. In breast cancers for example, a significant association with histological grade and TNM (*tumor node metastasis*) score was observed [106], which may suggest a role of mtDNA mutations in tumor progression. On the other hand, a pure correlative description can never exclude the possibility that somatic mtDNA mutations may just be a bystander effect of enhanced genotoxic stress in advanced cancers. This argument will even apply to correlations between mutation burden and overall survival of patients (see below).

At least two studies reported preliminary data, suggesting that mutated mtDNA is a marker of poor survival in advanced cancer patients. In a large series of 202 patients suffering from NSCLC of various histological subtypes, mitochondrial rRNA genes and D-loop were analyzed. Mutations were found only in the latter region in 34.7 % of cases [111]. No significant differences in mutation frequency occurred between the histological subtypes (adenocarcinoma, squamous cell carcinoma, bronchoalveolar carcinoma, large cell carcinoma), but mutations accumulated significantly in tumor grades IIIb or higher ($p<0.0001$). Tumor grades had been determined according to the international staging system [112]. The most important result of that study was the significantly decreased survival of advanced cancer patients harbouring mtDNA mutations. This difference was observed within grade IIIb and within grade IV (Fig. **3**). Although a prospective study (fresh material collected prior to cancer treatment), the article did not contain any information regarding chemo- and radiotherapy.

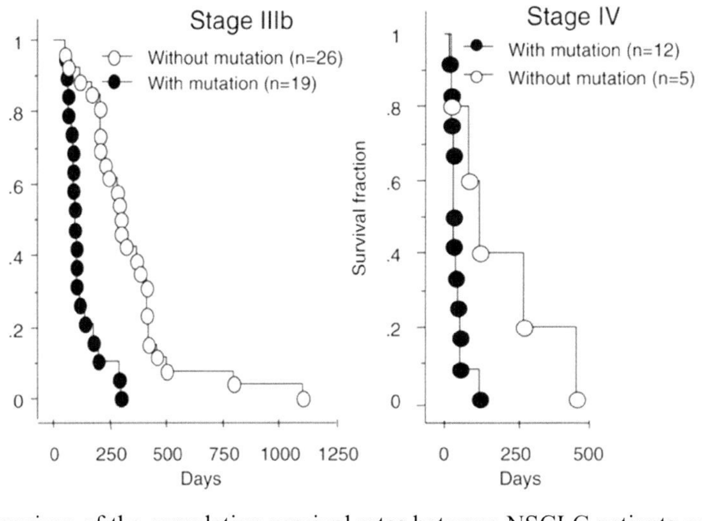

Figure 3: Comparison of the cumulative survival rates between NSCLC patients with and without mitochondrial DNA mutations (modified according to [111]).

Lievre and colleagues [113] retrospectively analyzed the D-loop region of 365 colorectal cancer patients from the Digestive Cancer Registry of Cote d'Or (France). Mutations were detected in 38.3% of the samples. The 3-year survival rate of these patients was 53.5% as compared to 62.1% in the negative cases (p = 0.05). The benefit of a non-mutated status became more pronounced in advanced cancers (Fig. **4**). If all stage III cases were included, the 3-year survival rates of the groups with and without mutation were 47.4% *versus* 70.0% respectively (p<0.02). If fluoro-uracil had been used as adjuvant chemotherapy, the difference was even more pronounced, *i.e.* 45.4% *vs.* 78.3% (p<0.02).

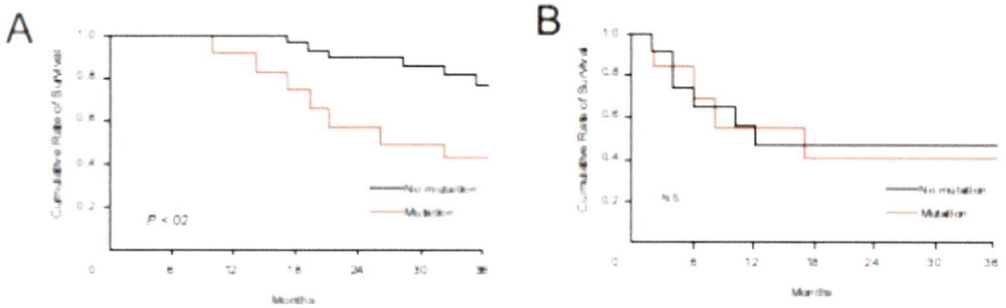

Figure 4: Comparison of cumulative survival rates in patients with stage III colorectal cancer with and without D-loop mutations, either receiving fluorouracil based adjuvant chemotherapy (A) or receiving no adjuvant chemotherapy (B). (modified according to [113]).

Although the results of both studies suggest mtDNA as an additional prognostic marker to refine histology-based prognoses, the suitability of such approaches remains questionable, unless the treatment regimens (chemotherapy and radiation) are more precisely reported and controlled in such analyses. This is not a simple issue in retrospective studies, but could be done by collecting mtDNA from patients, who are dedicated to participate in controlled clinical trials performed for testing new therapy regimens.

5. mtDNA AS A POTENTIAL TOOL OF EARLY CANCER DETECTION

5.1. Are mtDNA Mutations Suitable for Early Cancer Detection?

Although the frequency of mtDNA mutations in solid human tumors is still a matter of debate with sometimes large differences occurring between studies, several studies suggested that D310 length variation and base substitutions in the whole mitochondrial genome are often already detectable in a premalignant stage. In patients with a clinically manifested cancer, these mtDNA alterations can often be detected in paired bodily fluids, probably allowing the development of non-invasive diagnostic assays. Bodily fluids have been analyzed for the first time ten years ago by Fliss and colleagues in cancers of head and neck (n = 13), bladder (n = 14) and lung (n = 14) [94]. In patients harbouring mutated mtDNA in their bladder carcinoma, sequence variants were generally detectable in the paired urine samples. Most paired saliva samples of head and neck cancers were informative too, while the analysis of bronchoalveolar lavage (BAL) of lung cancers was difficult due to the dilution of tumor cell DNA in the relatively large amounts of BAL-fluid. Highly sensitive oligonucleotide mismatch ligation assays, created to specifically amplify a single base exchange, were successful in two cases, thus delivering at least the proof of principle. Since mtDNA usually exhibits a much higher copy number per cell as compared to nuclear single copy genes, this abundance may facilitate PCR-detection in diluted samples. This is well known to the forensic community, since mtDNA can sometimes be used to analyze DNA-traces, for which commonly used markers fail. In the bodily fluids of cancer patients, a second problem occurs, *i.e.* the admixture of tumor cell DNA with DNA of normal epithelial cells, which may also be released into the fluids analyzed. This may be regarded as a severe problem, since it hampers all sequencing strategies, including modern chip technologies (see below), which are not extremely sensitive in discovering heteroplasmic base positions.

Fliss and colleagues thus measured the relative abundance of mutated mtDNA as compared to mutated p53 nuclear DNA in two BAL samples and found the mtDNA sequence to be 19 and 220-fold more abundant. These results suggested that mtDNA may be a more sensitive tool for early cancer detection as compared with nuclear genes. Later, the same group identified mtDNA mutations of three prostate cancer patients in paired urine and, more surprisingly, plasma samles. In one case, the alteration was present in the corresponding PIN lesion (*prostate intraepithelial neoplasm*).

In a large series of 137 premalignant lesions of the head and neck from 93 patients, analyzed selectively for D310 status, 22% of the benigne hyperplasias, but 62% of squamous carcinomas *in situ*, harboured an altered length of the polycytosine tract [114]. The same marker was analyzed in ductal lavage (DL) of the breast in a small series of 14 women with no clinical evidence of breast cancer [115]. Nine of the women were at high-risk of developing cancer, since they harboured germline mutations of the *BRCA1* (*breast cancer 1*) gene, while five women were known to be non-carriers. Only three (30%) of the carriers exhibited detectable length alterations of D310 in DL, which seems to occur in about 30% of breast cancer tissues.

5.2. mitoChip-Technology

The small control-region of mtDNA replication and transcription is usually designated in its total length as 'D-loop' in the tumor literature, although the term *displacement loop* (D-loop) originally characterized only a portion of the control region, able to built a loop structure by binding a short single-stranded primer, displacing the two strands of mtDNA. Although the control region seems to harbour the highest density of point mutations in human tumors, the majority of mutations may be missed be focusing on this region, which comprises only about 6.5% of total mtDNA. Since conventional capillary-sequencing of the complete mitochondial genome (16.6 kb) is laborious and time consuming, parallel sequencing strategies had to be developed. Hybridization of fluorescent DNA samples to oligonucleotide arrays offers the opportunity to sequence the whole mitochondrial genome in a single hybridization experiment. Such arrays (MitoChip) have been constructed on the basis of the Custom Resequencing Microarray platform from Affymetrix (USA). The chip contains around 300.000 regions of ~500 μm^2 (features), each representing an oligonucleotide of 25 bp. Each mtDNA position is represented by four features, which differ only by base #13, being one of the four DNA nucleotides. Fluorescent mtDNA samples, generated by long-PCR and fragmentation, will hybridize most efficiently to the feature representing an exact match, which will thus show the brightest fluorescence. The position of this feature is recognized by a microarray-scanner. This technique showed a nearly perfect reproducibility in repeated hybridizations (>99.99%) and was able to detect mutations in six of nine bodily fluid samples (urine, pancreatic juice) from patients with bladder cancer (n = 5) or pancreatic cancer (n = 4) [116]. While the D-loop was omitted in the first generation chip due to high CG-content, this important region was included in the second version (v2.0), which moreover contained a redundant tiling for 500 of the most common haplotypes [117]. Thus MitoCHip (v2.0) allowed for the first time the examination of the whole mitochondrial genome in a single hybridization experiment.

Sui and colleagues [118] analyzed 14 precancerous lesions of the digestive tract, all of which were found to exhibit at least one mutation. Seven of these samples had been obtained by endoscopic biopsies from Barrett esophagus, showing Barret mucose without concomitant dysplasia (n = 2), low-grade dysplasia (n = 4) or high-grade- dysplasia (n =1), thus demonstrating mtDNA alterations even in the absence of histopathologic evidence of dysplasia. The remaining samples analyzed had been taken from four patients with colorectal adenomas (including one sessile serrated adenoma) and from three patients undergoing colectomy for long-standing idiopathic bowel disease with concomitant dysplasia. None of the analyzed adenomas had shown histopathologic evidence of invasive adenocarcinoma.

The affected sites of mtDNA were distributed over the whole mitochondrial genome (Fig. **5**), again demonstrating the necessity of a global approach. The majority of alterations were heteroplasmic.

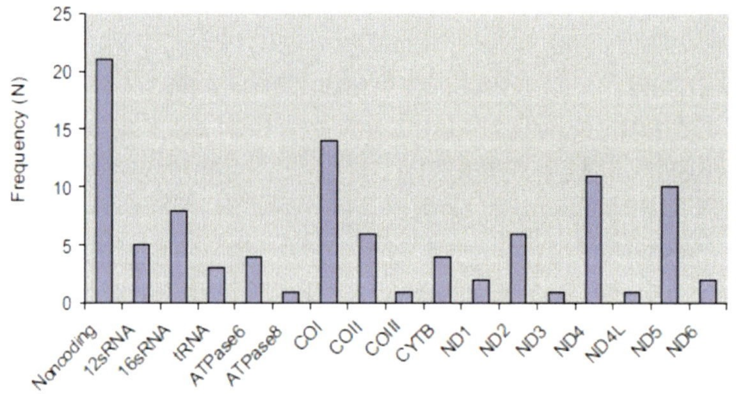

Figure 5: Distribution of mtDNA mutations ovserved with the MitoChip technology in preneoplastic lesions of the gastrointestinal tract (modified according to [118]. Abbreviations = mitochondrial genes according to international notation (see www.MITOMAP.com).

Still using the initial MitoChip version (v1.0), supplemented by capillary sequencing of the D-loop, Jakupciak *et al.* recently analyzed a series of early stage cancers, including bladder (n = 3), kidney (n = 13) and lung (n = 8) and paired bodily fluids, urine or BAL respectively [119]. Mutations of the mtDNA, heteroplasmic in most cases, were detected in all three bladder cancers. They were found in 69% of kidney and 79% of lung cancers. The paired bodily fluids exhibited mutations, *i.e.* their genotype was distinct from the constitutive genotype in blood, in 36% of bodily fluid samples, including 43% of BAL samples.

6. CONCLUDING REMARKS

Several mitochondrial and nuclear genes of mitochondrial components have been associated with cancer. At least for some nuclear genes, which code polypeptides of enzymes involved in the Krebs cycle, a relationship between gene mutation and cancer development in humans is clearly proven. These genes play a role mainly in rare inherited tumors, but screening for somatic mutations is of interest in some sporadic tumor types too.

Despite their high frequency in most solid human tumor types investigated so far, a causal role for a significant portion of mtDNA mutations in cancer cannot be claimed. Although some mtDNA mutations may enhance tumor growth, the enrichment of mutated mtDNA in solid human tumors can be explained by random drift, without need of any selective advantage. Irrespective of their biological role, mtDNA mutations may be a tool for early cancer detection. Studies, using the MitoChip technology, suggest that heteroplasmic and homoplasmic base exchanges are detectable in the majority of solid human tumors, even in early tumor stages or precancerous lesions. Especially in tumors seeding cells into an easily accessable bodily fluid, the screening of such fluids may be a hopeful technique. The MitoChip technology has to be tested in larger series of early cancers to allow an estimation of its practical suitability. However, already at this point two major issues have to be discussed.

First of all, one has to evaluate, what is a meaningful definition of a ‘mutation’, indicative of cancer growth. It is likely that global sequencing approaches will sometimes detect sequence differences between bodily fluid and blood, which are not observed in the tumor tissue of the same patient. *E.g.* in the study of Jakupciak and colleagues, this situation occurred in five patients [119]. Several explanations can be proposed, including oxidative stress in normal epithelial tissues, pre-existing tissue-specific heteroplasmy, or technical reasons. The observed discrepancies between tumor tissues and corresponding fluids imply that global mutation data must be interpreted with care. At least, it will be necessary to learn more about the frequency of mismatches between blood and bodily fluids in unselected individuals without known cancer, premalignant lesions or inflammatory disease in the organ of interest.

A second point requiring discussion is a technical one. As already stated by Sui *et al.* [118], it is important to stress that chip based chemistry is best suited for detection of single base pair alterations, but it is likely that many insertions/deletions, frameshift mutations and large scale deletions are missed by this technique. Generally, array technology should be much less sensitive for detection of heteroplasmic mutations in a huge excess of wild-type mtDNA. *E.g.* one can speculate that early detection of recurrence of a bladder or colon carcinoma, which contains a previously sequenced point mutation (follow-up), should be more meaningful by exploiting specific PCR-assays. However, if the goal is development of techniques for early detection of previously unknown cancers (screening), the whole mitochondrial genome must be analyzed. This can be done efficiently only by array technology, which has already delivered the proof of principal for this purpose. The development of this thechnology is thus connected with great expectations for the future.

ACKNOWLEDGEMENTS

Declared None.

CONFLICT OF INTEREST

The author(s) confirm that this chapter content has no conflict of interest.

DISCOLUSURE

This is an updated report of the earlier review article By Elmar Kirches ‘Mitochondrial and nuclear genes of mitochondrial components in cancer". Current Genomics, 2009 Jun; 10(4):281-293.

REFERENCES

[1] Warburg, O., On the origin of cancer cells. *Science* **1956,** 123, (3191), 309-14.

[2] Warburg, O., On respiratory impairment in cancer cells. *Science* **1956,** 124, (3215), 269-70.

[3] Wachsberger, P. R.; Gressen, E. L.; Bhala, A.; Bobyock, S. B.; Storck, C.; Coss, R. A.; Berd, D.; Leeper, D. B., Variability in glucose transporter-1 levels and hexokinase activity in human melanoma. *Melanoma Res* **2002,** 12, (1), 35-43.

[4] Chang, S.; Lee, S.; Lee, C.; Kim, J. I.; Kim, Y., Expression of the human erythrocyte glucose transporter in transitional cell carcinoma of the bladder. *Urology* **2000,** 55, (3), 448-52.

[5] Thupari, J. N.; Pinn, M. L.; Kuhajda, F. P., Fatty acid synthase inhibition in human breast cancer cells leads to malonyl-CoA-induced inhibition of fatty acid oxidation and cytotoxicity. *Biochem Biophys Res Commun* **2001,** 285, (2), 217-23.

[6] Weinhouse, S., On respiratory impairment in cancer cells. *Science* **1956,** 124, (3215), 267-9.

[7] Zu, X. L.; Guppy, M., Cancer metabolism: facts, fantasy, and fiction. *Biochem Biophys Res Commun* **2004,** 313, (3), 459-65.

[8] Bustamante, E.; Pedersen, P. L., High aerobic glycolysis of rat hepatoma cells in culture: role of mitochondrial hexokinase. *Proc Natl Acad Sci U S A* **1977,** 74, (9), 3735-9.

[9] Mathupala, S. P.; Ko, Y. H.; Pedersen, P. L., Hexokinase II: cancer's double-edged sword acting as both facilitator and gatekeeper of malignancy when bound to mitochondria. *Oncogene* **2006,** 25, (34), 4777-86.

[10] Rempel, A.; Bannasch, P.; Mayer, D., Differences in expression and intracellular distribution of hexokinase isoenzymes in rat liver cells of different transformation stages. *Biochim Biophys Acta* **1994,** 1219, (3), 660-8.

[11] Mayer, D.; Klimek, F.; Rempel, A.; Bannasch, P., Hexokinase expression in liver preneoplasia and neoplasia. *Biochem Soc Trans* **1997,** 25, (1), 122-7.

[12] Pedersen, P. L.; Mathupala, S.; Rempel, A.; Geschwind, J. F.; Ko, Y. H., Mitochondrial bound type II hexokinase: a key player in the growth and survival of many cancers and an ideal prospect for therapeutic intervention. *Biochim Biophys Acta* **2002,** 1555, (1-3), 14-20.

[13] Wilson, J. E., Hexokinases. *Rev Physiol Biochem Pharmacol* **1995,** 126, 65-198.

[14] Wilson, J. E., Isozymes of mammalian hexokinase: structure, subcellular localization and metabolic function. *J Exp Biol* **2003,** 206, (Pt 12), 2049-57.

[15] Goel, A.; Mathupala, S. P.; Pedersen, P. L., Glucose metabolism in cancer. Evidence that demethylation events play a role in activating type II hexokinase gene expression. *J Biol Chem* **2003,** 278, (17), 15333-40.

[16] Rempel, A.; Mathupala, S. P.; Griffin, C. A.; Hawkins, A. L.; Pedersen, P. L., Glucose catabolism in cancer cells: amplification of the gene encoding type II hexokinase. *Cancer Res* **1996,** 56, (11), 2468-71.

[17] Johansson, T.; Berrez, J. M.; Nelson, B. D., Evidence that transcription of the hexokinase gene is increased in a rapidly growing rat hepatoma. *Biochem Biophys Res Commun* **1985,** 133, (2), 608-13.

[18] Rempel, A.; Bannasch, P.; Mayer, D., Microheterogeneity of cytosolic and membrane-bound hexokinase II in Morris hepatoma 3924A. *Biochem J* **1994,** 303 (Pt 1), 269-74.

[19] Mathupala, S. P.; Rempel, A.; Pedersen, P. L., Glucose catabolism in cancer cells. Isolation, sequence, and activity of the promoter for type II hexokinase. *J Biol Chem* **1995,** 270, (28), 16918-25.

[20] Nakashima, R. A.; Mangan, P. S.; Colombini, M.; Pedersen, P. L., Hexokinase receptor complex in hepatoma mitochondria: evidence from N,N'-dicyclohexylcarbodiimide-labeling studies for the involvement of the pore-forming protein VDAC. *Biochemistry* **1986,** 25, (5), 1015-21.

[21] Rosano, C., Molecular model of hexokinase binding to the outer mitochondrial membrane porin (VDAC1): Implication for the design of new cancer therapies. *Mitochondrion* **2011,** 11, (3), 513-9.

[22] Arora, K. K.; Pedersen, P. L., Functional significance of mitochondrial bound hexokinase in tumor cell metabolism. Evidence for preferential phosphorylation of glucose by intramitochondrially generated ATP. *J Biol Chem* **1988,** 263, (33), 17422-8.

[23] Pastorino, J. G.; Hoek, J. B., Hexokinase II: the integration of energy metabolism and control of apoptosis. *Curr Med Chem* **2003,** 10, (16), 1535-51.

[24] Vyssokikh, M. Y.; Brdiczka, D., The function of complexes between the outer mitochondrial membrane pore (VDAC) and the adenine nucleotide translocase in regulation of energy metabolism and apoptosis. *Acta Biochim Pol* **2003,** 50, (2), 389-404.

[25] Zaid, H.; Abu-Hamad, S.; Israelson, A.; Nathan, I.; Shoshan-Barmatz, V., The voltage-dependent anion channel-1 modulates apoptotic cell death. *Cell Death Differ* **2005,** 12, (7), 751-60.

[26] Wolf, A.; Agnihotri, S.; Micallef, J.; Mukherjee, J.; Sabha, N.; Cairns, R.; Hawkins, C.; Guha, A., Hexokinase 2 is a key mediator of aerobic glycolysis and promotes tumor growth in human glioblastoma multiforme. *J Exp Med* **2011,** 208, (2), 313-26.

[27] Gong, L.; Cui, Z.; Chen, P.; Han, H.; Peng, J.; Leng, X., Reduced survival of patients with hepatocellular carcinoma expressing hexokinase II. *Med Oncol.*

[28] Godinot, C.; de Laplanche, E.; Hervouet, E.; Simonnet, H., Actuality of Warburg's views in our understanding of renal cancer metabolism. *J Bioenerg Biomembr* **2007,** 39, (3), 235-41.

[29] Neumann, H. P.; Wiestler, O. D., Clustering of features of von Hippel-Lindau syndrome: evidence for a complex genetic locus. *Lancet* **1991,** 337, (8749), 1052-4.

[30] Brauch, H.; Kishida, T.; Glavac, D.; Chen, F.; Pausch, F.; Hofler, H.; Latif, F.; Lerman, M. I.; Zbar, B.; Neumann, H. P., Von Hippel-Lindau (VHL) disease with pheochromocytoma in the Black Forest region of Germany: evidence for a founder effect. *Hum Genet* **1995,** 95, (5), 551-6.

[31] Baysal, B. E.; Ferrell, R. E.; Willett-Brozick, J. E.; Lawrence, E. C.; Myssiorek, D.; Bosch, A.; van der Mey, A.; Taschner, P. E.; Rubinstein, W. S.; Myers, E. N.; Richard, C. W., 3rd; Cornelisse, C. J.; Devilee, P.; Devlin, B., Mutations in SDHD, a mitochondrial complex II gene, in hereditary paraganglioma. *Science* **2000,** 287, (5454), 848-51.

[32] Gonzalez, C.; Almaraz, L.; Obeso, A.; Rigual, R., Carotid body chemoreceptors: from natural stimuli to sensory discharges. *Physiol Rev* **1994,** 74, (4), 829-98.

[33] Semenza, G. L., Perspectives on oxygen sensing. *Cell* **1999,** 98, (3), 281-4.

[34] Bunn, H. F.; Poyton, R. O., Oxygen sensing and molecular adaptation to hypoxia. *Physiol Rev* **1996,** 76, (3), 839-85.

[35] Arias-Stella, J.; Bustos, F., Chronic hypoxia and chemodectomas in bovines at high altitudes. *Arch Pathol Lab Med* **1976,** 100, (12), 636-9.

[36] Arias-Stella, J.; Valcarcel, J., Chief cell hyperplasia in the human carotid body at high altitudes; physiologic and pathologic significance. *Hum Pathol* **1976,** 7, (4), 361-73.

[37] Edwards, C.; Heath, D.; Harris, P.; Castillo, Y.; Kruger, H.; Arias-Stella, J., The carotid body in animals at high altitude. *J Pathol* **1971,** 104, (4), 231-8.

[38] Saldana, M. J.; Salem, L. E.; Travezan, R., High altitude hypoxia and chemodectomas. *Hum Pathol* **1973,** 4, (2), 251-63.

[39] Nurse, C. A.; Vollmer, C., Role of basic FGF and oxygen in control of proliferation, survival, and neuronal differentiation in carotid body chromaffin cells. *Dev Biol* **1997,** 184, (2), 197-206.

[40] Lack, E. E.; Cubilla, A. L.; Woodruff, J. M., Paragangliomas of the head and neck region. A pathologic study of tumors from 71 patients. *Hum Pathol* **1979,** 10, (2), 191-218.

[41] Lack, E. E.; Perez-Atayde, A. R.; Young, J. B., Carotid body hyperplasia in cystic fibrosis and cyanotic heart disease. A combined morphometric, ultrastructural, and biochemical study. *Am J Pathol* **1985,** 119, (2), 301-14.

[42] Stiller, D.; Katenkamp, D.; Kuttner, K., Jugular body tumors: hyperplasias or true neoplasms? Light and electron microscopical investigations. *Virchows Arch A Pathol Anat Histol* **1975,** 365, (2), 163-77.

[43] Briere, J. J.; Favier, J.; Benit, P.; El Ghouzzi, V.; Lorenzato, A.; Rabier, D.; Di Renzo, M. F.; Gimenez-Roqueplo, A. P.; Rustin, P., Mitochondrial succinate is instrumental for HIF1alpha nuclear translocation in SDHA-mutant fibroblasts under normoxic conditions. *Hum Mol Genet* **2005,** 14, (21), 3263-9.

[44] Ishii, T.; Yasuda, K.; Akatsuka, A.; Hino, O.; Hartman, P. S.; Ishii, N., A mutation in the SDHC gene of complex II increases oxidative stress, resulting in apoptosis and tumorigenesis. *Cancer Res* **2005,** 65, (1), 203-9.

[45] Slane, B. G.; Aykin-Burns, N.; Smith, B. J.; Kalen, A. L.; Goswami, P. C.; Domann, F. E.; Spitz, D. R., Mutation of succinate dehydrogenase subunit C results in increased O2.-, oxidative stress, and genomic instability. *Cancer Res* **2006,** 66, (15), 7615-20.

[46] Niemann, S.; Muller, U., Mutations in SDHC cause autosomal dominant paraganglioma, type 3. *Nat Genet* **2000,** 26, (3), 268-70.

[47] Baysal, B. E.; Willett-Brozick, J. E.; Lawrence, E. C.; Drovdlic, C. M.; Savul, S. A.; McLeod, D. R.; Yee, H. A.; Brackmann, D. E.; Slattery, W. H., 3rd; Myers, E. N.; Ferrell, R. E.; Rubinstein, W. S., Prevalence of SDHB, SDHC, and SDHD germline mutations in clinic patients with head and neck paragangliomas. *J Med Genet* **2002,** 39, (3), 178-83.

[48] Burnichon, N.; Briere, J. J.; Libe, R.; Vescovo, L.; Riviere, J.; Tissier, F.; Jouanno, E.; Jeunemaitre, X.; Benit, P.; Tzagoloff, A.; Rustin, P.; Bertherat, J.; Favier, J.; Gimenez-Roqueplo, A. P., SDHA is a tumor suppressor gene causing paraganglioma. *Hum Mol Genet* 19, (15), 3011-20.

[49] Hao, H. X.; Khalimonchuk, O.; Schraders, M.; Dephoure, N.; Bayley, J. P.; Kunst, H.; Devilee, P.; Cremers, C. W.; Schiffman, J. D.; Bentz, B. G.; Gygi, S. P.; Winge, D. R.; Kremer, H.; Rutter, J., SDH5, a gene required for flavination of succinate dehydrogenase, is mutated in paraganglioma. *Science* **2009,** 325, (5944), 1139-42.

[50] Kunst, H. P.; Rutten, M. H.; de Monnink, J. P.; Hoefsloot, L. H.; Timmers, H. J.; Marres, H. A.; Jansen, J. C.; Kremer, H.; Bayley, J. P.; Cremers, C. W., SDHAF2 (PGL2-SDH5) and hereditary head and neck paraganglioma. *Clin Cancer Res* 17, (2), 247-54.

[51] Astuti, D.; Latif, F.; Dallol, A.; Dahia, P. L.; Douglas, F.; George, E.; Skoldberg, F.; Husebye, E. S.; Eng, C.; Maher, E. R., Gene mutations in the succinate dehydrogenase subunit SDHB cause susceptibility to familial pheochromocytoma and to familial paraganglioma. *Am J Hum Genet* **2001,** 69, (1), 49-54.

[52] Vanharanta, S.; Buchta, M.; McWhinney, S. R.; Virta, S. K.; Peczkowska, M.; Morrison, C. D.; Lehtonen, R.; Januszewicz, A.; Jarvinen, H.; Juhola, M.; Mecklin, J. P.; Pukkala, E.;

Herva, R.; Kiuru, M.; Nupponen, N. N.; Aaltonen, L. A.; Neumann, H. P.; Eng, C., Early-onset renal cell carcinoma as a novel extraparaganglial component of SDHB-associated heritable paraganglioma. *Am J Hum Genet* **2004,** 74, (1), 153-9.

[53] Ricketts, C.; Woodward, E. R.; Killick, P.; Morris, M. R.; Astuti, D.; Latif, F.; Maher, E. R., Germline SDHB mutations and familial renal cell carcinoma. *J Natl Cancer Inst* **2008,** 100, (17), 1260-2.

[54] Ni, Y.; Zbuk, K. M.; Sadler, T.; Patocs, A.; Lobo, G.; Edelman, E.; Platzer, P.; Orloff, M. S.; Waite, K. A.; Eng, C., Germline mutations and variants in the succinate dehydrogenase genes in Cowden and Cowden-like syndromes. *Am J Hum Genet* **2008,** 83, (2), 261-8.

[55] Gellera, C.; Uziel, G.; Rimoldi, M.; Zeviani, M.; Laverda, A.; Carrara, F.; DiDonato, S., Fumarase deficiency is an autosomal recessive encephalopathy affecting both the mitochondrial and the cytosolic enzymes. *Neurology* **1990,** 40, (3 Pt 1), 495-9.

[56] Liochev, S. I.; Fridovich, I., Fumarase C, the stable fumarase of Escherichia coli, is controlled by the soxRS regulon. *Proc Natl Acad Sci U S A* **1992,** 89, (13), 5892-6.

[57] Bourgeron, T.; Chretien, D.; Poggi-Bach, J.; Doonan, S.; Rabier, D.; Letouze, P.; Munnich, A.; Rotig, A.; Landrieu, P.; Rustin, P., Mutation of the fumarase gene in two siblings with progressive encephalopathy and fumarase deficiency. *J Clin Invest* **1994,** 93, (6), 2514-8.

[58] Weaver, T.; Lees, M.; Banaszak, L., Mutations of fumarase that distinguish between the active site and a nearby dicarboxylic acid binding site. *Protein Sci* **1997,** 6, (4), 834-42.

[59] Tomlinson, I. P.; Alam, N. A.; Rowan, A. J.; Barclay, E.; Jaeger, E. E.; Kelsell, D.; Leigh, I.; Gorman, P.; Lamlum, H.; Rahman, S.; Roylance, R. R.; Olpin, S.; Bevan, S.; Barker, K.; Hearle, N.; Houlston, R. S.; Kiuru, M.; Lehtonen, R.; Karhu, A.; Vilkki, S.; Laiho, P.; Eklund, C.; Vierimaa, O.; Aittomaki, K.; Hietala, M.; Sistonen, P.; Paetau, A.; Salovaara, R.; Herva, R.; Launonen, V.; Aaltonen, L. A., Germline mutations in FH predispose to dominantly inherited uterine fibroids, skin leiomyomata and papillary renal cell cancer. *Nat Genet* **2002,** 30, (4), 406-10.

[60] Lehtonen, R.; Kiuru, M.; Vanharanta, S.; Sjoberg, J.; Aaltonen, L. M.; Aittomaki, K.; Arola, J.; Butzow, R.; Eng, C.; Husgafvel-Pursiainen, K.; Isola, J.; Jarvinen, H.; Koivisto, P.; Mecklin, J. P.; Peltomaki, P.; Salovaara, R.; Wasenius, V. M.; Karhu, A.; Launonen, V.; Nupponen, N. N.; Aaltonen, L. A., Biallelic inactivation of fumarate hydratase (FH) occurs in nonsyndromic uterine leiomyomas but is rare in other tumors. *Am J Pathol* **2004,** 164, (1), 17-22.

[61] Campuzano, V.; Montermini, L.; Molto, M. D.; Pianese, L.; Cossee, M.; Cavalcanti, F.; Monros, E.; Rodius, F.; Duclos, F.; Monticelli, A.; Zara, F.; Canizares, J.; Koutnikova, H.; Bidichandani, S. I.; Gellera, C.; Brice, A.; Trouillas, P.; De Michele, G.; Filla, A.; De Frutos, R.; Palau, F.; Patel, P. I.; Di Donato, S.; Mandel, J. L.; Cocozza, S.; Koenig, M.; Pandolfo, M., Friedreich's ataxia: autosomal recessive disease caused by an intronic GAA triplet repeat expansion. *Science* **1996,** 271, (5254), 1423-7.

[62] Lill, R.; Muhlenhoff, U., Iron-sulfur-protein biogenesis in eukaryotes. *Trends Biochem Sci* **2005,** 30, (3), 133-41.

[63] Rotig, A.; de Lonlay, P.; Chretien, D.; Foury, F.; Koenig, M.; Sidi, D.; Munnich, A.; Rustin, P., Aconitase and mitochondrial iron-sulphur protein deficiency in Friedreich ataxia. *Nat Genet* **1997,** 17, (2), 215-7.

[64] Muhlenhoff, U.; Richhardt, N.; Ristow, M.; Kispal, G.; Lill, R., The yeast frataxin homolog Yfh1p plays a specific role in the maturation of cellular Fe/S proteins. *Hum Mol Genet* **2002,** 11, (17), 2025-36.

[65] Puccio, H.; Simon, D.; Cossee, M.; Criqui-Filipe, P.; Tiziano, F.; Melki, J.; Hindelang, C.; Matyas, R.; Rustin, P.; Koenig, M., Mouse models for Friedreich ataxia exhibit cardiomyopathy, sensory nerve defect and Fe-S enzyme deficiency followed by intramitochondrial iron deposits. *Nat Genet* **2001,** 27, (2), 181-6.

[66] Shoichet, S. A.; Baumer, A. T.; Stamenkovic, D.; Sauer, H.; Pfeiffer, A. F.; Kahn, C. R.; Muller-Wieland, D.; Richter, C.; Ristow, M., Frataxin promotes antioxidant defense in a thiol-dependent manner resulting in diminished malignant transformation *in vitro*. *Hum Mol Genet* **2002,** 11, (7), 815-21.

[67] Schulz, T. J.; Thierbach, R.; Voigt, A.; Drewes, G.; Mietzner, B.; Steinberg, P.; Pfeiffer, A. F.; Ristow, M., Induction of oxidative metabolism by mitochondrial frataxin inhibits cancer growth: Otto Warburg revisited. *J Biol Chem* **2006,** 281, (2), 977-81.

[68] Thierbach, R.; Schulz, T. J.; Isken, F.; Voigt, A.; Mietzner, B.; Drewes, G.; von Kleist-Retzow, J. C.; Wiesner, R. J.; Magnuson, M. A.; Puccio, H.; Pfeiffer, A. F.; Steinberg, P.; Ristow, M., Targeted disruption of hepatic frataxin expression causes impaired mitochondrial function, decreased life span and tumor growth in mice. *Hum Mol Genet* **2005,** 14, (24), 3857-64.

[69] Awad, M. M.; Enslen, H.; Boylan, J. M.; Davis, R. J.; Gruppuso, P. A., Growth regulation *via* p38 mitogen-activated protein kinase in developing liver. *J Biol Chem* **2000,** 275, (49), 38716-21.

[70] Iyoda, K.; Sasaki, Y.; Horimoto, M.; Toyama, T.; Yakushijin, T.; Sakakibara, M.; Takehara, T.; Fujimoto, J.; Hori, M.; Wands, J. R.; Hayashi, N., Involvement of the p38 mitogen-activated protein kinase cascade in hepatocellular carcinoma. *Cancer* **2003,** 97, (12), 3017-26.

[71] Kirches, E.; Andrae, N.; Hoefer, A.; Kehler, B.; Zarse, K.; Leverkus, M.; Keilhoff, G.; Schonfeld, P.; Schneider, T.; Wilisch-Neumann, A.; Mawrin, C., Dual role of the mitochondrial protein frataxin in astrocytic tumors. *Lab Invest* **2011,** 91, (12), 1766-76.

[72] Sutton, A.; Imbert, A.; Igoudjil, A.; Descatoire, V.; Cazanave, S.; Pessayre, D.; Degoul, F., The manganese superoxide dismutase Ala16Val dimorphism modulates both mitochondrial import and mRNA stability. *Pharmacogenet Genomics* **2005,** 15, (5), 311-9.

[73] Li, H.; Kantoff, P. W.; Giovannucci, E.; Leitzmann, M. F.; Gaziano, J. M.; Stampfer, M. J.; Ma, J., Manganese superoxide dismutase polymorphism, prediagnostic antioxidant status, and risk of clinical significant prostate cancer. *Cancer Res* **2005,** 65, (6), 2498-504.

[74] Moschos, M. P., Selenoprotein P. *Cell Mol Life Sci* **2000,** 57, (13-14), 1836-45.

[75] Meplan, C.; Crosley, L. K.; Nicol, F.; Beckett, G. J.; Howie, A. F.; Hill, K. E.; Horgan, G.; Mathers, J. C.; Arthur, J. R.; Hesketh, J. E., Genetic polymorphisms in the human selenoprotein P gene determine the response of selenoprotein markers to selenium supplementation in a gender-specific manner (the SELGEN study). *FASEB J* **2007,** 21, (12), 3063-74.

[76] Cooper, M. L.; Adami, H. O.; Gronberg, H.; Wiklund, F.; Green, F. R.; Rayman, M. P., Interaction between single nucleotide polymorphisms in selenoprotein P and mitochondrial superoxide dismutase determines prostate cancer risk. *Cancer Res* **2008,** 68, (24), 10171-7.

[77] Hirano, M.; DiMauro, S., ANT1, Twinkle, POLG, and TP: new genes open our eyes to ophthalmoplegia. *Neurology* **2001,** 57, (12), 2163-5.

[78] Bolden, A.; Noy, G. P.; Weissbach, A., DNA polymerase of mitochondria is a gamma-polymerase. *J Biol Chem* **1977,** 252, (10), 3351-6.

[79] Ropp, P. A.; Copeland, W. C., Cloning and characterization of the human mitochondrial DNA polymerase, DNA polymerase gamma. *Genomics* **1996,** 36, (3), 449-58.

[80] Rovio, A. T.; Marchington, D. R.; Donat, S.; Schuppe, H. C.; Abel, J.; Fritsche, E.; Elliott, D. J.; Laippala, P.; Ahola, A. L.; McNay, D.; Harrison, R. F.; Hughes, B.; Barrett, T.; Bailey, D. M.; Mehmet, D.; Jequier, A. M.; Hargreave, T. B.; Kao, S. H.; Cummins, J. M.; Barton, D. E.; Cooke, H. J.; Wei, Y. H.; Wichmann, L.; Poulton, J.; Jacobs, H. T., Mutations at the mitochondrial DNA polymerase (POLG) locus associated with male infertility. *Nat Genet* **2001,** 29, (3), 261-2.

[81] Blomberg Jensen, M.; Leffers, H.; Petersen, J. H.; Daugaard, G.; Skakkebaek, N. E.; Rajpert-De Meyts, E., Association of the polymorphism of the CAG repeat in the mitochondrial DNA polymerase gamma gene (POLG) with testicular germ-cell cancer. *Ann Oncol* **2008,** 19, (11), 1910-4.

[82] Petros, J. A.; Baumann, A. K.; Ruiz-Pesini, E.; Amin, M. B.; Sun, C. Q.; Hall, J.; Lim, S.; Issa, M. M.; Flanders, W. D.; Hosseini, S. H.; Marshall, F. F.; Wallace, D. C., mtDNA mutations increase tumorigenicity in prostate cancer. *Proc Natl Acad Sci U S A* **2005,** 102, (3), 719-24.

[83] King, M. P.; Attardi, G., Human cells lacking mtDNA: repopulation with exogenous mitochondria by complementation. *Science* **1989,** 246, (4929), 500-3.

[84] Evans, A. R.; Limp-Foster, M.; Kelley, M. R., Going APE over ref-1. *Mutat Res* **2000,** 461, (2), 83-108.

[85] Imanishi, H.; Hattori, K.; Wada, R.; Ishikawa, K.; Fukuda, S.; Takenaga, K.; Nakada, K.; Hayashi, J., Mitochondrial DNA mutations regulate metastasis of human breast cancer cells. *PLoS One* **2011,** 6, (8), e23401.

[86] Ishikawa, K.; Hashizume, O.; Koshikawa, N.; Fukuda, S.; Nakada, K.; Takenaga, K.; Hayashi, J., Enhanced glycolysis induced by mtDNA mutations does not regulate metastasis. *FEBS Lett* **2008,** 582, (23-24), 3525-30.

[87] Woo, D. K.; Green, P. D.; Santos, J. H.; D'Souza, A. D.; Walther, Z.; Martin, W. D.; Christian, B. E.; Chandel, N. S.; Shadel, G. S., Mitochondrial genome instability and ROS enhance intestinal tumorigenesis in APC(Min/+) mice. *Am J Pathol* 180, (1), 24-31.

[88] Polyak, K.; Li, Y.; Zhu, H.; Lengauer, C.; Willson, J. K.; Markowitz, S. D.; Trush, M. A.; Kinzler, K. W.; Vogelstein, B., Somatic mutations of the mitochondrial genome in human colorectal tumours. *Nat Genet* **1998,** 20, (3), 291-3.

[89] Habano, W.; Nakamura, S.; Sugai, T., Microsatellite instability in the mitochondrial DNA of colorectal carcinomas: evidence for mismatch repair systems in mitochondrial genome. *Oncogene* **1998,** 17, (15), 1931-7.

[90] Sanchez-Cespedes, M.; Parrella, P.; Nomoto, S.; Cohen, D.; Xiao, Y.; Esteller, M.; Jeronimo, C.; Jordan, R. C.; Nicol, T.; Koch, W. M.; Schoenberg, M.; Mazzarelli, P.; Fazio, V. M.; Sidransky, D., Identification of a mononucleotide repeat as a major target for mitochondrial DNA alterations in human tumors. *Cancer Res* **2001,** 61, (19), 7015-9.

[91] Habano, W.; Sugai, T.; Yoshida, T.; Nakamura, S., Mitochondrial gene mutation, but not large-scale deletion, is a feature of colorectal carcinomas with mitochondrial microsatellite instability. *Int J Cancer* **1999,** 83, (5), 625-9.

[92] Parrella, P.; Xiao, Y.; Fliss, M.; Sanchez-Cespedes, M.; Mazzarelli, P.; Rinaldi, M.; Nicol, T.; Gabrielson, E.; Cuomo, C.; Cohen, D.; Pandit, S.; Spencer, M.; Rabitti, C.; Fazio, V. M.; Sidransky, D., Detection of mitochondrial DNA mutations in primary breast cancer and fine-needle aspirates. *Cancer Res* **2001,** 61, (20), 7623-6.

[93] Zhou, S.; Kachhap, S.; Sun, W.; Wu, G.; Chuang, A.; Poeta, L.; Grumbine, L.; Mithani, S. K.; Chatterjee, A.; Koch, W.; Westra, W. H.; Maitra, A.; Glazer, C.; Carducci, M.; Sidransky, D.; McFate, T.; Verma, A.; Califano, J. A., Frequency and phenotypic implications of mitochondrial DNA mutations in human squamous cell cancers of the head and neck. *Proc Natl Acad Sci U S A* **2007,** 104, (18), 7540-5.

[94] Fliss, M. S.; Usadel, H.; Caballero, O. L.; Wu, L.; Buta, M. R.; Eleff, S. M.; Jen, J.; Sidransky, D., Facile detection of mitochondrial DNA mutations in tumors and bodily fluids. *Science* **2000,** 287, (5460), 2017-9.

[95] Kirches, E.; Michael, M.; Woy, C.; Schneider, T.; Warich-Kirches, M.; Schneider-Stock, R.; Winkler, K.; Wittig, H.; Dietzmann, K., Loss of heteroplasmy in the displacement loop of brain mitochondrial DNA in astrocytic tumors. *Genes Chromosomes Cancer* **1999,** 26, (1), 80-3.

[96] Kirches, E.; Krause, G.; Warich-Kirches, M.; Weis, S.; Schneider, T.; Meyer-Puttlitz, B.; Mawrin, C.; Dietzmann, K., High frequency of mitochondrial DNA mutations in glioblastoma multiforme identified by direct sequence comparison to blood samples. *Int J Cancer* **2001,** 93, (4), 534-8.

[97] Montanini, L.; Regna-Gladin, C.; Eoli, M.; Albarosa, R.; Carrara, F.; Zeviani, M.; Bruzzone, M. G.; Broggi, G.; Boiardi, A.; Finocchiaro, G., Instability of mitochondrial DNA and MRI and clinical correlations in malignant gliomas. *J Neurooncol* **2005,** 74, (1), 87-9.

[98] Kirches, E.; Mawrin, C.; Schneider-Stock, R.; Krause, G.; Scherlach, C.; Dietzmann, K., Mitochondrial DNA as a clonal tumor cell marker: gliomatosis cerebri. *J Neurooncol* **2003,** 61, (1), 1-5.

[99] Wong, L. J.; Lueth, M.; Li, X. N.; Lau, C. C.; Vogel, H., Detection of mitochondrial DNA mutations in the tumor and cerebrospinal fluid of medulloblastoma patients. *Cancer Res* **2003,** 63, (14), 3866-71.

[100] Wang, C. Y.; Li, H.; Hao, X. D.; Liu, J.; Wang, J. X.; Wang, W. Z.; Kong, Q. P.; Zhang, Y. P., Uncovering the profile of somatic mtDNA mutations in Chinese colorectal cancer patients. *PLoS One* **2011,** 6, (6), e21613.

[101] Jeong, C. W.; Lee, J. H.; Sohn, S. S.; Ryu, S. W.; Kim, D. K., Mitochondrial microsatellite instability in gastric cancer and gastric epithelial dysplasia as a precancerous lesion. *Cancer Epidemiol* **2010,** 34, (3), 323-7.

[102] Cai, F. F.; Kohler, C.; Zhang, B.; Chen, W. J.; Barekati, Z.; Garritsen, H. S.; Lenner, P.; Toniolo, P.; Zhang, J. J.; Zhong, X. Y., Mutations of mitochondrial DNA as potential biomarkers in breast cancer. *Anticancer Res* **2011,** 31, (12), 4267-71.

[103] Shimomura, T.; Hiyama, T.; Oka, S.; Tanaka, S.; Yoshihara, M.; Shimamoto, F.; Chayama, K., Frequent somatic mutations of mitochondrial DNA in traditional serrated adenomas but not in sessile serrated adenomas of the colorectum. *J Gastroenterol Hepatol* **2011,** 26, (10), 1565-9.

[104] Challen, C.; Brown, H.; Cai, C.; Betts, G.; Paterson, I.; Sloan, P.; West, C.; Birch-Machin, M.; Robinson, M., Mitochondrial DNA mutations in head and neck cancer are infrequent and lack prognostic utility. *Br J Cancer* **2011,** 104, (8), 1319-24.

[105] Kuo, S. J.; Chen, M.; Ma, G. C.; Chen, S. T.; Chang, S. P.; Lin, W. Y.; Chen, Y. C.; Lee, T. H.; Lin, T. T.; Liu, C. S., Number of somatic mutations in the mitochondrial D-loop region indicates poor prognosis in breast cancer, independent of TP53 mutation. *Cancer Genet Cytogenet* **2011,** 201, (2), 94-101.

[106] Tseng, L. M.; Yin, P. H.; Yang, C. W.; Tsai, Y. F.; Hsu, C. Y.; Chi, C. W.; Lee, H. C., Somatic mutations of the mitochondrial genome in human breast cancers. *Genes Chromosomes Cancer* **2ß11,** 50, (10), 800-11.
[107] Bandelt, H. J.; Salas, A., Contamination and sample mix-up can best explain some patterns of mtDNA instabilities in buccal cells and oral squamous cell carcinoma. *BMC Cancer* **2009,** 9, 113.
[108] Jenuth, J. P.; Peterson, A. C.; Shoubridge, E. A., Tissue-specific selection for different mtDNA genotypes in heteroplasmic mice. *Nat Genet* **1997,** 16, (1), 93-5.
[109] Habano, W.; Sugai, T.; Nakamura, S. I.; Uesugi, N.; Yoshida, T.; Sasou, S., Microsatellite instability and mutation of mitochondrial and nuclear DNA in gastric carcinoma. *Gastroenterology* **2000,** 118, (5), 835-41.
[110] Coller, H. A.; Khrapko, K.; Bodyak, N. D.; Nekhaeva, E.; Herrero-Jimenez, P.; Thilly, W. G., High frequency of homoplasmic mitochondrial DNA mutations in human tumors can be explained without selection. *Nat Genet* **2001,** 28, (2), 147-50.
[111] Matsuyama, W.; Nakagawa, M.; Wakimoto, J.; Hirotsu, Y.; Kawabata, M.; Osame, M., Mitochondrial DNA mutation correlates with stage progression and prognosis in non-small cell lung cancer. *Hum Mutat* **2003,** 21, (4), 441-3.
[112] Mountain, C. F., Revisions in the International System for Staging Lung Cancer. *Chest* **1997,** 111, (6), 1710-7.
[113] Lievre, A.; Chapusot, C.; Bouvier, A. M.; Zinzindohoue, F.; Piard, F.; Roignot, P.; Arnould, L.; Beaune, P.; Faivre, J.; Laurent-Puig, P., Clinical value of mitochondrial mutations in colorectal cancer. *J Clin Oncol* **2005,** 23, (15), 3517-25.
[114] Ha, P. K.; Tong, B. C.; Westra, W. H.; Sanchez-Cespedes, M.; Parrella, P.; Zahurak, M.; Sidransky, D.; Califano, J. A., Mitochondrial C-tract alteration in premalignant lesions of the head and neck: a marker for progression and clonal proliferation. *Clin Cancer Res* **2002,** 8, (7), 2260-5.
[115] Isaacs, C.; Cavalli, L. R.; Cohen, Y.; Pennanen, M.; Shankar, L. K.; Freedman, M.; Singh, B.; Liu, M.; Gallagher, A.; Rone, J. D.; Dickson, R. B.; Sidransky, D.; Haddad, B. R., Detection of LOH and mitochondrial DNA alterations in ductal lavage and nipple aspirate fluids from hngh-risk patients. *Breast Cancer Res Treat* **2004,** 84, (2), 99-105.
[116] Maitra, A.; Cohen, Y.; Gillespie, S. E.; Mambo, E.; Fukushima, N.; Hoque, M. O.; Shah, N.; Goggins, M.; Califano, J.; Sidransky, D.; Chakravarti, A., The Human MitoChip: a high-throughput sequencing microarray for mitochondrial mutation detection. *Genome Res* **2004,** 14, (5), 812-9.
[117] Zhou, S.; Kassauei, K.; Cutler, D. J.; Kennedy, G. C.; Sidransky, D.; Maitra, A.; Califano, J., An oligonucleotide microarray for high-throughput sequencing of the mitochondrial genome. *J Mol Diagn* **2006,** 8, (4), 476-82.
[118] Sui, G.; Zhou, S.; Wang, J.; Canto, M.; Lee, E. E.; Eshleman, J. R.; Montgomery, E. A.; Sidransky, D.; Califano, J. A.; Maitra, A., Mitochondrial DNA mutations in preneoplastic lesions of the gastrointestinal tract: a biomarker for the early detection of cancer. *Mol Cancer* **2006,** 5, 73.
[119] Jakupciak, J. P.; Maragh, S.; Markowitz, M. E.; Greenberg, A. K.; Hoque, M. O.; Maitra, A.; Barker, P. E.; Wagner, P. D.; Rom, W. N.; Srivastava, S.; Sidransky, D.; O'Connell, C. D., Performance of mitochondrial DNA mutations detecting early stage cancer. *BMC Cancer* **2008,** 8, 285.

INDEX

3D modelling 169, 180

A

A. thaliana 241-6
ABA hypersensitive phenotype 393, 398-9
ABA treatment 383-5, 387, 391-3, 406, 410
ABF4 382-3, 386
Ability 16, 111, 149, 156, 190-1, 200, 256, 262, 276, 278-9, 317, 331, 386, 435, 438-40
Abiotic stress 377, 380, 385, 389-90, 392, 395-8, 404, 406, 408, 411, 414
Abiotic stress conditions 377, 381-2, 386, 390, 394, 397, 405-6, 410, 414
Abiotic stress response 380, 383, 390, 392, 403, 406
Abiotic stress signaling 377, 384, 393, 411, 416
Abiotic stress tolerance 389, 394, 409, 411, 415-16
Abscisic acid (ABA) 382-6, 390, 392-4, 396-9, 405-12, 414-15
Absolute quantification 261, 263, 265-70, 273, 276, 281
Abundance 262, 267-9, 274, 276, 278, 281, 478
relative 265, 275, 277, 478
Accessions 329, 337-9
Accessory proteins 47, 49
Accumulation, succinate 458-9, 461
Accuracy 267, 269, 279, 281, 316-17
Acetate, cyproterone 55-7
Act 17, 50, 52, 146, 171, 193-4, 216, 220, 223, 225-7, 302, 306-7, 352, 357-8, 363
Activation 16, 18-21, 23, 26, 28, 30-1, 33, 147, 156, 355, 367, 377, 401-3, 410-11, 414-15
temporal 15, 29-30, 35-7
Activation Function 46, 48
Active replication origins 16, 29
Activities, enzymatic 306, 379
ADAMTS9 221, 223
Adaptor protein AP-2 353, 355

Christian Neri (Ed)

Adenomas 7, 67, 71, 76, 78, 84, 87, 147
AF-1 46, 48
AFAP patients 67, 78, 86-7
Agonist 48, 50, 52, 56-7, 353
AHA2 402, 411
AKT1 395-6, 400, 404, 411
Alterations 47, 52-3, 101, 112, 115-17, 120-1, 261, 320, 379, 466, 478, 480
Amino acids 47, 49, 54-6, 68, 242, 244, 256, 264, 278, 301, 367, 402, 429
Amounts 25, 30, 34, 263, 266, 269-73, 275, 327-8, 340, 343, 456
    absolute 265, 268, 270-1, 281
    known 265, 276-8
Analyte 265, 267, 269, 271, 281
Analyzing 169-70, 175, 177-8, 317
Androgen 49-55, 57-8
Androgen receptor (AR) 45-50, 52-4, 56-9
Androgen target genes 50, 59
Animal models 189
Animal species 316-17, 326, 334, 346
Animals 170, 190, 199, 203, 245, 268, 316-21, 323, 334, 336-7, 340, 343, 346, 377-80, 462-3
Antiandrogens 49, 52, 57-8
Antiporter 395-6, 399, 401-2
APC gene 66-8, 71, 75, 78
APC germline mutations 76, 78, 84
APC mutations, somatic 71
$APC^{Min/+}$ mice 469-70
Apical-basal polarity 191, 193, 195
Apoptosis 149, 151-3, 155-6, 196, 200, 203-4, 212, 214-15, 226, 455
Apoptosis of -cells 222
Approaches 24, 87, 169, 183, 241, 250-1, 264, 267, 274, 278-9, 316, 320-1, 326, 329, 346
    candidate gene 218-20
AR mutations 49, 52-4, 59
AR pathway 49, 52, 58

Arabidopsis 321, 325-6, 329-31, 333-4, 343, 377, 379, 381-3, 388, 390, 394-6, 398, 408-10, 412-13
Arabidopsis thaliana 241
Arabidopsis TILLING Project (ATP) 299, 307-8, 330-1, 453-5
ARR-1 364
ARRDCs 352, 360-4, 367
Arrestin clan 360, 366-7
Arrestin clan proteins 352, 361, 363, 367
Arrestin core 364, 366
Arrestin Domian containing proteins (ARRDCs) 352, 3620, 366-7
Arrestin family 353-4
Arrestin Fold 352-3, 355, 357, 359, 361, 363, 365, 367
Arrestin N-domain 356, 364, 366
Arrestin proteins 353
  few 356
Arrestin-related proteins 364, 367
  most 366
Arrestins 352-67
  cone 354, 356
  visual 352-3, 355-6, 363-4, 367
Arrestins interact 352-3
Artefacts 177-9, 181
ARTs 364-5
AtCBK3 413-15
AtCIPKs 395-6, 403-4
AtCPK 382-5, 394
AtCRK3 412-15
ATP binding 253, 299
ATP hydrolysis 299, 428
ATPase 402, 411, 432
ATRX 117-18
Attenuated familial adenomatous polyposis (AFAP) 68, 75, 88
Autosomal 4, 6-7, 67-8, 113-14, 116, 459-60, 465

B

Background mutations 345
Background populations 73-4, 83
Base exchanges 466, 472-3, 475
Base excision repair (BER) 66, 70, 72
Base pairs 17, 50, 307
Beta clamp 293
Biallelic MUTYH germline mutations 71, 78, 84
Binding 18, 28, 46-7, 74, 102, 106, 148, 293, 296, 298-9, 301-3, 305-9, 354-5, 412-13, 428-9
  inhibition of 308
  ligand 48-9, 54, 59
Binding calcium 379
Binding protein 114, 119, 296, 298
Binding sites 297, 304, 306, 353-4, 357
Binding surfaces 294-6, 305
Biogenesis 245, 249
Biosynthesis 244, 247, 252, 256
Block surface 170, 172-3, 175-7, 179-81
Blocks 58, 109, 171-2, 177, 180, 182, 302, 384
Blood 266-7, 466, 475, 481-2
Bodily fluids 452, 478, 481-2
Brain 156, 203, 213-16, 437, 453, 466
  larval 190-1, 202-3
Breast 5, 10, 153, 196, 215, 479
Breast cancer 7-9, 80-1, 119, 153-5, 475, 479
Breast cancer cells 467
BELAYA SMERT' (BSM) 245-6, 253
Budding yeast 15-17, 20-3, 25, 28-9, 31, 33-7

C

C. elegans 102-4, 324, 334, 336, 364, 433
C-domains 353-4, 356-8, 361-2, 366
Calcinuerin B-like (CBLs) 377, 379-80, 395-6, 399-401, 407-11

Calcium 377-81, 384-5, 389, 392-6, 398-9, 403-4, 409, 411-14
Calcium concentration 378-9
Calcium dependent protein kinases (CDPKs) 377, 379-95, 399, 411-13
Calcium ions 380-1, 384
Calcium levels 378-9, 384, 399, 402-3, 411
Calcium regulated kinases 377, 379, 381, 383, 385, 387, 389, 391, 393, 395, 397, 399, 401, 403, 415-16
Calcium sensors 379, 396, 400, 414
Calcium signaling protein kinases 377
Calmodulin 377, 379, 384, 413
CaM kinase 377, 380, 411, 413
Cancer 4-9, 11-12, 33, 36, 51, 118-19, 146, 150, 157, 189, 221, 226, 452-5, 466, 481-2
  advanced 475-7
  double 3-4, 8-10
  epithelial 189
  familial accumulation of 3, 8, 10-11
  gynecological 7, 80
  small intestinal 4, 7
Cancer cells 119, 154, 191-2, 203-4, 440, 452-4, 470
Cancer development 195, 452, 470, 481
Cancer phenotype 10, 119
Cancer stem cells 151, 189
Candidate genes 210-11, 213, 217, 219-21, 224, 226, 228, 319
Capturing 171-2, 177-9, 182
Carcinogenesis 3-6, 119, 146, 149, 157, 204, 453-4, 462
Carcinomas 7, 11, 67, 71, 79, 189-90, 216
Carotid body (CB) 457-8
Carriers, biallelic MUTYH 80, 83, 85
Castrate resistant prostate cancer (CRPC) 45, 52, 58
Castration resistance 45
CBL-CIPK signaling pathway 406-7
CBL proteins 379
CBL1 397, 399-400, 403

CBL4 401-2, 404, 411
CBL9 396-8, 400
CCaMK 380, 411, 413, 415
Cdk1-Clb 5, 6 18-19, 30, 35
CDK4 57
CDLS 113
CDPK genes 384-6, 388-90, 393-4
CDPK-related protein kinase 380, 411
CEL 215, 221, 327, 329, 335, 337-8, 345
Cell competition (CC) 200-1
Cell cycle 16, 19-24, 28, 30, 34-5, 37, 103, 114, 146, 156, 195, 221-2, 268, 270, 299
Cell cycle regulation 214-15
Cell development 212-13, 216
Cell proliferation 5, 146, 148-50, 152, 157, 189, 194-5, 245, 455
Cells 21-3, 34-5, 53, 107-8, 111-12, 154-7, 178-9, 189-94, 197-203, 212, 221-3, 266-7, 362-3, 378-9, 462-4
    bsm-null 245
    cancer-derived 147
    eukaryotic 15-17, 379
    follicle 201-2
    guard 382, 393
    mouse ES 116-17
    polarity-deficient 201, 203
    precursor 475
    sic1 35-6
Channels 226, 329, 384, 393-5, 409
Chaperone network 426
Chemical mutagenesis 320, 322-3, 337, 341, 343
Chemical mutagens 323, 326, 328, 334
CHL1 404, 411
Chloroplasts 241-6, 252, 256
Chromatin 48, 106, 108, 110-11, 114, 116, 309
Chromatin structure 18, 37, 101-2, 119

Chromosomal rearrangements 111, 118
Chromosome territories 107-8, 115
Chromosomes 5, 16-17, 20, 24-9, 31-3, 36, 68, 104, 107, 110-11, 113-15, 147, 155-6, 218, 360-1
    interphase 107-8
Chromosomes IX 31-2
Cipk 397-400, 404, 406
CIPK genes 400, 405-7, 410
CIPK members 396, 407, 410-11
CIPKs (CBL-interacting protein kinase) 377, 380, 395-408, 410-11
Clamp 293-306, 308-9
    cognate 296-7, 303
Clamp loaders 297-300, 302, 308
Classes 225, 297, 325, 356, 395, 402, 411
Clathrin 277, 352-3, 355, 434
Clb5 15, 19, 28, 31
CLB5-dependent regions (CDRs) 31-2
Clb5 mutant 31-3
Cleaved fragments 329, 341-4
Client proteins 426, 430, 441
Clones 199-200, 203, 468
Clusters 101-5, 119-20, 254, 295
Co-chaperones 59, 428, 430, 433, 435, 440
Co-expressed genes 101-5
    clustering of 105, 108, 110
Co-expression 103, 106, 120-1
    transcriptional 103-4, 107
Co-expression domains 105
Coactivators 46, 49-50, 52-4, 56
Codon 10, 72, 459-60, 464
Coffin-Siris syndrome (CSS) 116-7
Cofactor binding 48, 53
Colon 67, 105, 147-9, 214-16, 470
Colonoscopy 66, 87-8

Colorectal adenomas 66-8, 71, 76, 78, 84, 86-7, 480
Colorectal cancer 4-5, 7-10, 66, 147
    hereditary non-polyposis 7, 66-7
Complexes 16, 194, 298, 309, 428, 461-3, 465
Conformational changes 299, 301, 307-8, 353, 356, 428, 464
Contributions 16, 37, 75, 79, 119, 122
Control 4, 30, 35, 74, 101-2, 195, 198-9, 384, 386, 427, 462
Corepressors 49-50, 52, 115
Coverage 267, 279-80
CRC 66-8, 70-1, 74, 76, 79, 84-8, 147-9
    sporadic 70-2
CRC-derived cell lines 147
CRC development 66, 68, 71-2, 79, 86-7
CRC patients 67, 70-2, 77, 79, 82-3, 85, 87-8
Creation 326, 334-5, 342-3
Creation of transgenic material 316, 321-3
CREBBP 114
CRKs 380, 411-15
Crop plants 318, 321-2, 334, 407
CSS 116-17
CTCF 109, 119
Cytoplasm 49, 147, 152, 154, 194, 196, 383-4, 387, 454-5
Cytosolic Hsp70s 428, 430-1, 437, 441

D

D-loop 471, 474, 476, 479-80
D-loop mutations 472, 477
Data generation 172-3, 178-81
Databases 80, 262, 317-19, 364
Degradation 21, 322, 353, 363, 427, 456, 458
Detection 70, 86, 181, 276, 278, 280-1, 321, 323, 327, 339-40, 342, 344, 356, 466, 482
Deterministic 15, 23-5, 29

Development 5-6, 11-12, 58-9, 67-8, 70-1, 149-50, 182-3, 213-17, 221, 226-8, 241, 316, 363-4, 377-8, 380
panicle 388-9
plant 245, 256, 332, 381, 390
Development stages, seed 388
Developmental 377, 414, 416
Developmental stages 199, 256, 271, 280, 377, 389, 397-8, 437
Diabetes 112, 121-2, 210-13, 217-28
monogenic 217-20, 223
monogenic forms of 217-18
Diabetes candidate genes 221-2
Diabetes genes 213, 217, 222, 226
few 217
Diabetes heritability 219
Diabetes management 223-5
Diabetes mellitus, permanent neonatal 217-18
Diabetes risk 226
Diabetes subjects 212
Diagnostic criteria, clinical 3-4, 9-11
Dictyostelium arrestin clan proteins 367
Dictyostelium arrestin-domain proteins 367
Differentiation 118, 120, 189-90, 203-4
Digestion 266-7, 271-2
Digital images 172, 176-8, 180-2
Discovery 4, 219-20, 321, 337, 362-3, 455, 457
Disease resistance 321, 338, 340
Diseases 8, 45, 51-3, 58, 75, 85, 87-8, 101, 110-13, 119-22, 210-12, 217-19, 223-7, 319, 460
cardiovascular 121, 211
Disorder of Chromatin Remodeling and Transcription 116-17
Disorders 110, 121, 210
rare Mendelian 110, 120-1
Disruption 54, 104, 110-12, 193, 244-5, 249, 251-2, 254-6, 364, 455
Diversity of arrestin clan proteins 361

DNA 26, 28, 30, 47, 49-50, 72, 102, 107, 109-10, 293-4, 297-301, 303, 305-8, 326-8, 343-4
double-stranded 301
DNA binding 48, 51, 54, 308
DNA binding domain (DBD) 46-7, 54, 146
DNA helicase 302, 307
DNA ligase 6, 295, 303-6, 308
DNA methylation 101, 117-19, 151, 322
DNA mismatch repair gene 3, 11
DNA polymerases 6, 18, 293, 295, 299-304, 306, 465
DNA polymorphisms 337-9
DNA repair 106-7, 113, 117, 293
DNA replication 5, 15-16, 18-19, 22, 24-5, 27-31, 33, 35, 70, 72, 112, 117, 212, 293-5, 299
initiation of 19, 34
DNA structures 298-9, 305, 308-9
DNA units 26
Domains 105, 107, 111, 247, 294, 296-7, 303, 306, 360, 365-7, 381-2, 395, 415, 428
interchromosomal 108
ligand binding 46, 48
Downstream signaling 352-3, 355, 380
Downstream targets 148, 393-4, 396, 399, 401, 416
Drosophila 103-4, 108-9, 113-14, 119, 189-91, 193, 199, 201, 203, 322, 324, 334-5, 343, 352, 363
Drosophila Lgl 196-7
Drosophila lgl mutant 197
Drought 382, 384-7, 390-1, 393-4, 396-8, 401, 405-6, 408-11
Drought stress 391, 398, 409, 414
Drug targets 210, 223
Drugs, anticancer 7, 269, 439
Ductal lavage (DL) 479
DUSP9 217, 221, 223
Dynamic range 273, 280-1

Dysfunction 461, 468-9

E

E-cadherin expression 154
E. coli 17, 270-1, 295, 298-300, 302-3, 306, 320, 402
Early cancer detection 452, 474, 477-8, 481
Early origins 20, 33
EcoTILLING 316, 329, 337-42, 344-6
EcoTILLING experiments 342, 344
EcoTILLING method 337, 339
ECT 403
ECT proteins 402-3
ECT1 396, 402-3
ECT2 402-3
Efficiency 17-18, 23-6, 29, 34-5, 270-2, 276, 278, 317, 321-2, 339
EFIC 174-5, 180-2
Embedding medium 172, 175-7, 180-1
Embryo abortion 251-2, 254
Embryo development 241, 243-6, 248-9
Embryo morphogenesis 242, 247
Embryogenesis 169, 241-4, 249, 256
    genetic mechanisms driving 169
Embryos 169-71, 174, 176, 179, 181-3, 190, 194, 213, 243, 392, 413
    preglobular 241, 247
Encode 241, 243, 247
Encoding 45-6, 457
Endocytosis 352, 355, 358, 362-3, 365, 434
Endometrial 6-7, 9
Endometrial cancer 3-12, 70, 80-1, 86
    familial 3, 8, 11-12
    pathology of 3
Endometrial cancer development 10, 68
Endometrial cancer patients 4, 8-9, 81
Endoplasmic reticulum (ER) 214, 219, 378, 384-5, 403-4, 411, 433

Endosomes 215, 352, 355-6, 358-9, 362, 366
Entry 20-1, 28, 30, 35, 432
ENU 323-4, 335-6
Envelope proteins 242, 253, 255
Enzymes 70, 116, 216, 245, 293, 295, 297-8, 304-5, 307-8, 338, 345, 380-1, 395, 454-8, 460-1
EP300 114
Epi-3D 171, 174-5, 180-1
Episcopic 169, 171-5, 177-83
Episcopic imaging methods 170-1, 174, 178, 180-1
Episcopic microscopy 169, 172
Epithelial tissues 189-90, 195, 199
Essential genes 244, 249, 335
Eukaryotes 16, 20, 36, 103, 120, 303-4, 306, 317, 429
    higher 15, 33-4
Exons 9-10, 45-6, 54, 68, 84, 86, 149, 255, 335, 344
Expressed-Sequence Tag (ESTs) 102, 104, 404, 437
Expressed tissue 213
Expression 57, 120-1, 146-7, 150-3, 155-7, 195, 220, 247-8, 250-1, 381, 383, 386-92, 398-9, 404-9, 436-7
    altered 118-19
Expression and function of KLF 146, 152
Expression pattern of FtsZ genes 250
Expression patterns 120, 176, 380, 390, 393, 396, 398, 413, 432, 437
    analyzing gene 175, 177, 182
Expression profile 157, 385-6
Extracolonic manifestations 67-8, 79-81, 83

F

Factors 16, 19, 49, 75, 87, 106, 146, 178, 275, 305, 308-9, 377, 463
Familial adenomatous polyposis (FAP) 5, 66-8, 71, 79, 86, 88, 147
Familial tumor 3-7, 9, 11-12
Families 4, 9, 48-9, 77, 86, 88, 146, 215, 300, 359-60, 379, 397, 416, 426-7, 460
Family protein 253, 255

FAP patients 67, 79-80, 87
Fen1 295, 297, 304-5, 308
FES1 435-6
FH gene 460-1
Firing 18, 21-4, 29-30
stochastic 15, 21-3, 25
Firing time 18, 23, 26, 33
FLP 197-9
Formation 16, 18, 106, 109, 198, 356, 364, 435-6, 463
Fragments 329, 340-2, 344
Frataxin 461-4
Frequencies 5-6, 11, 46, 53, 72, 79-80, 83, 86, 193, 225, 278, 335, 378, 477, 482
Fruit fly 317, 324-5, 355, 361, 434
Fumarate hydratase (FH) 452, 460-1
Functional analyses of MUTYH mutation variants 75
Functional consequences 68, 74-5, 466
Functional distinction 433-4, 438, 440
Functional specificity 426, 430, 433, 437, 441
Functionality 30, 109, 248, 251
Functions 34-5, 146-7, 151-7, 219-21, 223-4, 245-7, 249, 307, 340, 366-7, 381-4, 407-9, 414-16, 426, 432
plastid 246, 256
predicted 248
prostate 50-1
protective 463
unknown 249, 396, 398
Functions of candidate genes 219

G

G-protein coupled receptors (GPCRs) 352-3, 355-6, 365
GAA sites 71
Gastric cancer 70, 80-1, 150
Gene clusters 101, 103-5, 107, 109, 111, 113, 115, 117, 119, 121
co-expressed 102, 104-5, 121

Gene expression 15, 18, 45, 49, 59, 101-2, 104-5, 107-13, 115-21, 147-8, 169, 171, 175-6, 217, 379
plastid 241, 243
regulating 110, 414
Gene expression patterns 101, 171-3, 182-3
Gene function 107, 170, 316, 319-21, 327, 332, 334-6, 340, 346
Gene mutation 10, 121, 456, 481
Gene product patterns 169, 171, 176
Gene products 169-71, 176, 213, 217, 223, 226, 251
Generations 4-5, 77, 121, 218, 321, 323, 326, 332, 345
Genes 66-8, 101-6, 111-12, 118-21, 169-71, 216-27, 244, 246-52, 254-6, 319-21, 331-3, 335-8, 360-5, 385-91, 405-8
additional 246-7
adjacent 104, 106
androgen-regulated 57
causative 5, 7
chromomethylase 326, 331
few 217, 249
identified 247, 390
inducible 379, 386
most 114, 219, 226
multiple 170, 385, 416
neighboring 106, 120
non-duplicated 102-3
non-paralogous 104, 106, 120
non-redundant 243, 246
plastid protein-encoding 244
protein-coding 472
responsive 50, 382-3, 406, 411
stress marker 398
unrelated 102, 105
Genes co-habiting 101-2
Genes coding 250, 256, 364, 452
Genes encoding 103-4, 116, 243, 245-7, 321, 440

nuclear 243-4, 246
Genes encoding plastid proteins 247-8
Genes encoding proteins 244, 248-9
Genesis 169
Genesis of malformations 169, 183
Genetic diversity 316
Genetic Insights 210-11, 213, 215, 217, 219, 221, 223, 225, 227
Genetic screening 88, 225, 227
Genetic variants 210, 217, 219-20, 225-7
common 210-11, 218, 225
Genetic variation 219-20
Genetics 33, 66, 84-6, 121, 169, 210, 219, 223-5, 318-19, 323-5, 330, 335, 395, 403-4
reverse 316, 320, 326, 415-16
Genome 15-18, 21, 24, 27, 32, 34, 101-2, 120, 241-2, 317-20, 330, 332, 338-41, 344, 474-5
human 103, 105, 120, 436
nuclear 242, 246, 467-8, 474
plant 318, 323
sequenced 316-17, 319
Genome size 316-17, 323, 346
Genomic instability 15-16, 33-5
Genomic neighborhoods 101-4, 106, 110, 114, 120-2
transcriptional 101, 112
Genomic organization 110, 118, 120, 122
nuclear 107-8
Genomic regions, large 110, 113-14
Genomics 27, 33, 316-17, 430
functional 316, 320, 330-1, 334, 336
Genotype 78, 82, 85-6, 218, 320-2, 330, 345-6, 467, 470, 481
Genotype-phenotype correlations 66-7, 82-3, 88
Germline mutations 3, 5, 9-10, 66-8, 73, 75, 78, 85, 88, 452, 459-60
detected 67
Germline MUTYH Mutations 66, 71, 74, 78

Germplasm 316, 340
Gibberellic acid (GA) 387, 390, 392, 413, 415
Globular 241-2, 247, 249, 251
Globular embryos 241, 247
Globular stage 241, 243, 245, 251-2, 256
Glucose 31, 212, 214, 406, 453-5
Glycoprotein 215-16
GPCRs, phosphorylated 355, 357
GPx 462-4
Green fluorescent protein (GFP) 198, 382
Gross chromosomal rearrangements (GCRs) 35
Group II 247-8
Groups 9, 72, 78, 84, 102, 111, 201-2, 221, 247, 271, 279, 352, 432, 454, 477-8
Growth 51, 150, 157, 190, 213-16, 219, 221, 223, 225, 245, 256, 264, 377, 386, 432
Growth advantage 45, 49, 53-4

H

HCC 151-2
HCC-derived cells 151
Heart 214-15, 242, 247, 249, 251
Heat-shock proteins 48, 427
Heating 180-1, 327
Helices 48, 55, 57
Helix 48-9, 55, 57, 299, 365, 367, 464
Hereditary non-polyposis colorectal cancer (HNPCC) 3-11, 66-8, 79, 85-6, 88
Heteroduplexes 326-7, 329, 337, 340, 344-5
Heteroplasmic 473, 475, 480-1
Heterotrimeric G-proteins 272, 353, 358
Heterozygous mutations 341, 344
High resolution episcopic microscopy (HREM) 172, 174, 176, 180-2
HIF-1 456-7, 459
Histological sections 172, 179, 181
HK II 454-5

HMLH1 3, 5, 9-11
HMSH2 3, 5, 9-11
HMSH6 3, 5, 9-11
HMSH6 gene 9-11
HMSH6 mutations 10
HNPCC-associated endometrial cancer 5, 10
HNPCC patients 68, 80, 82, 86
Homozygotes 73, 78, 82, 86
Hormone therapy 45
Housekeeping genes 105
Hsc70 434, 436-8, 441
Hsp70-encoding genes 430, 436
Hsp70 isoforms 426, 430, 433, 440-1
Hsp70 proteins 435, 439
Hsp70s 426-30, 432-5, 437-41
HSPA 436-40
HSPA1A 436-9
HSPA1B 436-8
HSPA1L 436-7
Hsu list 246-8
Hugl-1 189
Human arrestin clan proteins 360-1
Human cancers 110, 118, 200, 454-5, 470
Human Cytosolic Hsp70s 427, 436-7
Human diseases 101, 110-11, 121-2, 319, 322
Human gene mutation database (HGMD) 10
Human serum 267-8
Human tumors 457-8, 466, 471, 473, 475, 479
    solid 477, 481
Humans 6, 33, 101-4, 107, 169-70, 190, 196, 203-4, 317, 319, 322, 339, 353, 402, 440-1
Hydroxyflutamide 55-7

I

Identification frequency 263, 273-8

Images 56, 172-4, 177-82, 341
Imaging 169
Imaging methods 169-75, 177-83
Imaging techniques 169, 171-2, 174, 183
Inactivation 151, 434, 456, 458-9
Induced mutations 323, 329, 331, 333, 335-7, 340
  rare 336, 343
Information 3, 23, 25, 170-1, 178, 204, 225, 249, 251, 256, 261, 268, 319, 356, 390
Inheritance 7, 76-7, 217, 319, 434-5
Inhibition 5, 36, 146, 148-50, 204, 225, 308, 358, 384, 389, 396, 456, 458, 461, 470
Initiation time 22, 25
Insulin 212, 215-16, 219, 222, 270
Insulin resistance 212, 218-20, 222-3
Insulin secretion 210, 212, 214-15, 219-21, 223, 225-7
Intensity 266, 272, 274
Interact 49-50, 53, 56, 114, 214, 293, 297-8, 303, 306, 309, 356-8, 362, 377, 400, 428-9
Interacting 48, 213, 217, 295-6, 357, 365, 384, 395-403, 407-10, 413-14
Interacting partners 385, 396, 400, 408
Interactions 47-8, 55, 72, 107-9, 293-5, 297-301, 303-7, 309, 319, 353-7, 362-3, 382, 396-7, 399-403, 408-10
  protein-protein 274, 380
Interest
  gene of 251, 320-2, 326-7, 334-5, 337, 340, 345
  species of 327, 340, 343
International grape genome program (IGGP) 318
Interventions 210, 223, 225-7
Invasion 148, 150, 153, 193, 202, 204, 388
Ion intensities 263, 265, 276
Isoforms 269, 331, 352, 359, 426, 430, 436, 440, 456
Isotope-labeled standards, stable 270-1
Isotope labeling, stable 261-4

J

J-proteins 431-3

K

KAT1 393-5
KCNJ11 215, 219, 221, 223
KCNQ1 220-1, 223
Kidney 213, 215-17, 453, 480-1
Kinase 20, 255, 358, 377, 380, 395, 406, 412-16
    calcium signaling 401, 414, 416
    calcium-signaling 380, 416
    regulated 380, 411, 413-14
Kinase activities 18, 20-1, 365, 381, 396, 412
Kinase domain 381-2, 400
KLF family members 146, 149, 153, 155-6
KLF4 146-8, 150-1, 153-7
KLF4 expression 147-8, 150, 153, 155
KLF4 protein 147
KLF5 148-50, 153, 155-7
KLF6 150-2, 154-7
    enforced expression of 150-1
KLF6 expression 149, 151-2, 156
KLF6 mutations 149, 156
Knockout 244, 246, 249, 251, 415
Knockout mutants 241, 243-4
KRAS 86
Krebs cycle 456-9, 461, 481
Kruppel like-factors (KLFs) 146-7, 149, 151, 153-5, 157, 214, 222-3
Kyoto Encyclopedia of Genes and Genomes (KEGG) 103

L

Label-free 261-3, 267, 273-6, 281
Lamins 106-7, 109, 111
LC, reverse-phase 262
LC-MS/MS 262, 266, 272

Length 16, 25, 47, 174, 337, 465-6, 475
Level of single genes 387, 397
Levels 52, 75, 102, 110, 118-19, 147, 189, 295, 336, 379, 387, 391-2, 406-7, 414, 437
    undetectable 437
Lgl mutant clones 200, 202
LI-COR DNA Analyzer 329, 333, 342
Ligand binding domain (LBD) 46, 48, 53, 55, 57
Ligand specificity 49, 55
Ligands 45, 48-9, 52-3, 55-7, 59, 224, 293, 295-6, 298, 302, 304, 357
Light 15, 26, 29, 177, 180, 262, 272, 353-4, 378-9, 381, 385, 403, 405, 458
Liquid chromatography (LC) 262-3, 275
Liver 105, 212, 214, 216-17, 463
Liver tumorigenesis 463-4
LOH 146-7, 151, 155, 199, 456-7, 459-61
Loss 55, 106, 112, 115-16, 119, 147, 150-2, 190-1, 195, 197, 201, 378, 383, 402, 458-9
Low abundance proteins 267, 276, 280-1
Lower Uterine Segment (LUS) 11
Lung 119, 196, 214, 216-17, 467, 478, 480
Lung KLF (LKLF) 154-5
Lynch syndrome 3, 11, 67

M

Maintenance 35, 59, 112, 117, 318, 334, 343
Maize 318, 321, 323, 325, 331, 334, 407-8, 412-13
Mammalian Arrestin-Domain Containing Proteins 361
Mammalian cells 15, 20, 24, 34, 110, 269, 275, 278, 430, 439
MAP patients 66-7, 71, 73-82, 84-8
    examined 79-80
    female 80, 82
    studied 77, 79
Mass spectrometry (MS) 261-3, 265, 267, 269, 271, 273, 275-7, 279-81
Mass spectrometer 262-3, 265, 278

Matrix, nuclear 101, 109, 111
Maturity Onset Diabetes of the Young (MODY) 217-18, 221, 225
MDA-MB-231 cells 468
Mechanisms 4, 6, 21, 23, 35, 45, 49, 101-2, 189-90, 220, 222-4, 227-8, 298-300, 308, 319
    regulatory 108, 295-6
MECP2 115
Medical histories, familial 4, 8-10
Medication 213, 221-4, 226
Membrane 196, 355, 385, 391, 395, 404, 408, 410
Mental retardation 114-15, 117, 121
Metastases 153-5, 157, 189, 191-3, 204, 476
Metastatic 189, 192, 203, 469-70
Methods, label-free 262-3, 273-4, 281
Mice 101-2, 107, 117, 146-7, 438, 440, 462, 469-70, 474
Micosatellite instability (MSI) 5, 7-8, 11
Microbes 316-18, 320, 407
Microsatellite instability 5-7, 10, 471
Minute mutations 200
Mismatch repair 6, 11, 66-8, 72, 306-7
Mismatches 6, 306, 327, 329, 340, 344-5, 482
Missense 316, 322, 330, 332, 335-6, 340
Missense mutations 149, 325, 331-2, 471, 474
    somatic 53, 150
MitoChip 452
Mitochondria 378, 430, 455, 458, 461, 464-5, 470-1, 474
Mitochondrial 462, 464, 467, 470, 481-2
Mitochondrial genome 452, 465-6, 469-74, 478-80, 482
Mitochondrial matrix 455, 461-2, 464-5
Mitogen associated protein kinases (MAPK) 150, 416
MMR 3, 6-7, 66-8, 72
MMR gene aberration 9-11
MMR genes 3, 5, 9-10
MMR mutations 72, 85

Model organisms 169-71, 176, 336, 352, 441
Model plant Arabidopsis thaliana 241-2, 318, 330
Models 24-5, 27, 29-30, 101, 106-7, 111, 120-1, 147, 151, 173, 175, 189, 298, 300, 362
deterministic 26-9, 31, 33
incidental expression 106
three-dimensional space 106, 110
Modification, post-translational 265, 267-8, 270, 355
Molecular carcinogenesis 3
Molecular chaperones 246-7, 276, 426, 428, 430, 432, 434, 438, 440
Molecular evolution and disease 101, 111, 113, 115, 117, 119
Molecular weight 277-8, 380
Monoallelic MUTYH mutation carriers 72, 83
Morphogenesis 190, 203, 242
Morphology 78, 175-6, 183
Motifs 48, 50, 56-7, 296, 386
Mouse 104, 108-9, 113, 119-20, 196, 199, 264, 318, 438
MS/MS spectral 274
MSI analysis 8, 10
MSI-positive endometrial cancer 8, 10
MtDNA 452, 465-71, 473, 475, 477-81
fetal 467-8
mutated 466, 471, 474, 476, 478, 481
MtDNA mutations 452, 468, 470-3, 475-7, 480-1
advanced cancer patients harbouring 476
somatic 473-6
Multiple Hsp70 Isoforms 426-7, 429-31, 433, 435, 437, 439, 441
Mutant AR 55, 57-8
Mutant cells 190, 198, 200, 204, 467
Mutant lines 329, 383, 406
Mutant populations 316, 326, 328, 332-4, 336, 342-3
Mutant receptor 56-7
Mutants 35, 49, 53-6, 151, 172, 199, 246-7, 322-3, 326-7, 330-1, 335-6, 342-3, 383-5, 397-8, 400-2

induced 340-1
sos3 401-2
Mutation carriers 73, 77, 81
biallelic MUTYH germline 76, 84
monoallelic MUTYH germline 77, 84
Mutation detection 327, 333, 343
sensitivity of 327, 343
Mutation frequency 323, 473, 476
Mutational density 330, 332
Mutations 7-10, 45, 53-5, 71-3, 75, 112-18, 189-90, 218-20, 322-3, 329-37, 434-6, 457-61, 466-73, 475-7, 479-82
allelic series of 330-1, 340
biallelic 66, 76
causative 68, 225
frameshift 3, 482
gain-of-function 202, 322
nonsense 10, 325, 330, 457
observed 471, 474
splice site 316, 325, 332
various 316, 322, 334, 340
Mutations of nuclear lamina proteins 111
MutS 293, 306
MUTYH 68, 70, 72, 75, 78
MUTYH gene 66-8, 71-3, 75-6, 84, 86, 88
MUTYH germline mutations 68, 72, 84-5
MUTYH glycosylase 68, 70-2, 74
MUTYH mutation carriers 73, 86
MUTYH mutation variants 75, 83
MUTYH mutations 69-75, 78, 82-3, 85-6, 88
biallelic 76, 78-9
Myristoylation motif 390-1, 408

N

N-terminal domain 45-6, 53-4, 360, 380

NAD biosynthesis 252
National Institute of Allergy and Infectious Diseases (NIAID) 217
Natural selection 105, 119, 121
NBD 428
Neighborhoods 104, 106, 111
Nematode 324-5, 334, 361
Nematode arrestin clan proteins 364
Neuroblasts 191, 194-5
Normal tissues 453, 475
Normalization 267, 274, 277-9
Normalized spectral abundance factor (NSAF) 277
Notch 148, 202, 213, 221, 357, 363
NSLC 154-5
Nuclear envelope 106, 111
Nuclear genes 242-3, 452-3, 455, 457, 459, 461, 463, 465, 467, 469, 471, 473, 475, 477-9, 481-2
Nuclear genes encoding plastid proteins 244, 246
Nuclear Genes Encoding Plastid Proteins Required for Arabidopsis thaliana Embryogenesis 251
Nuclear lamina 106-9, 111-12
Nuclear organization 101, 112, 117-18
Nucleotide exchange factors (NEFs) 426, 429, 433, 435
Nucleus 19, 35, 49, 105-6, 110-11, 147, 154, 196, 241, 353, 358, 361, 383-4, 391-3, 403
Nude mice 462-3, 466-7
Number
    large 30, 49, 87, 225, 256, 269, 293, 302, 319-20, 335, 343, 378
    origin 23

O

Obesity 4, 212, 216, 222
Observable peptides 267, 274, 277-8
OGG1 and MHT1 genes in Polyposis patients 70
OMIM 67, 111-17

Oocyte 192, 201-2
Optics 172, 177, 179-80
Organelle 241, 243, 249, 317, 426
Organisms 15, 18, 106, 249, 256, 297, 304, 320, 322-3, 325, 377, 426, 434, 441
Organs 6, 8, 147, 170-1, 175, 181, 192, 199-200, 250, 280, 473-4, 482
Origin activation 15, 17-18, 20-5, 27-31, 33
  temporal program of 16, 23, 25, 28-9, 33
  timing of 17-18
Origin activity 17-18, 37
Origin efficiency 22-3, 28-9, 33-4
Origin firing 16-17, 20-3, 25, 28, 30-1, 34, 36
  stochastic 15, 24-5
Origin recognition complex (ORC) 18-19
Origin selection 20-1
Origins 15-26, 28-9, 31, 34-6
  late 20, 22, 30-1, 33
Origins firing 20-2, 24, 34, 36
OsCIPK 405-8
OsCPK 385-9, 391
Ovarian 5, 9-10, 154
Ovarian cancers 7-8, 80, 153, 196
Overexpression 119, 387-8, 398, 403, 406-7, 409, 434-6, 438
Overexpression lincs 383, 387
Overexpression of OsCDPK7 387-8
Overexpression of Sse1p 435
Oxidative stress 385, 390, 461-4, 466
OXPHOS 454-6, 465, 467

P

Pancreas 213-15, 221, 225, 227
Pancreatic cancer 154, 456, 472, 479
Pancreatic islet 214-16, 220
Pandey 377, 397, 416
Paragangliomas 452

Paralogues 103-5
Pathogenesis 119, 151
Pathology 3, 8, 12, 111, 113-15, 169
Pathways 50, 66, 103, 195, 204, 212-13, 217, 219, 221-6, 307, 358, 384
Patient groups 67, 78
Patient Survival 475-6
Patients 4, 7-12, 45, 51-3, 58, 67, 76, 78-80, 84-8, 115-16, 225, 460-1, 467, 472-3, 475-81
PCNA 293-300, 303-6, 308-9
PCS 267, 271-2, 281
Peak intensity 262-3, 267, 273-6, 278
Peak intensity ratio 262-4
PEG 405, 408
Peptide concatemer 265-7, 271
Peptide-binding domain (PBD) 428-9, 432
Peptide fractionations 280
Peptide ion 262-3, 267, 273, 275, 280
Peptide observability 278-9
Peptides 212, 262, 264, 266-70, 272-6, 278-9, 281, 297-8, 308
    bar-code 273
    identified 261, 274-7, 279
    individual 262, 274, 276, 278
    proteotypic 279
Permanent Neonatal Diabetes Mellitus (PNDM) 217-18
Personalized medicine 210
PGL 457, 460
PGME 247-8, 252-6
PGME category 247
Phase 15-24, 26, 28-31, 34-5, 37, 103, 155
    maturation 242-3
Phase progression 17-18, 21
Phenotypes 66, 78, 88, 170, 189, 193, 195, 246-7, 251, 255, 320, 330, 332, 340, 345-6
    diabetic 217, 219

morphological 169-70, 172
Phenotyping 332
Phosphorylation 18, 150, 195, 265, 268-9, 296, 299, 304, 355, 357-8, 363, 381, 383-4, 400-1, 404
Phosphorylation stoichiometry 268-9
Placenta 213-17, 392
Plant embryogenesis 243
Plant Genetic Resources Conservation Unit (PGRCU) 316
Plant species 316, 320, 326, 380, 391, 407, 411-14
Plant viability 242, 246, 249, 378
Plants 316-22, 325-6, 328, 330, 334, 337-8, 343-4, 346, 377-9, 391-5, 399, 402-3, 407-8, 410-11, 413-14
model 330-1, 346
polyploid 334, 344-5
wild type 398
Plasma membrane 195, 353, 357, 362, 366, 385, 391-3, 397, 401-2, 404, 408-9, 411
Plastid protein-coding genes 246
Plastid proteins, additional genes encoding 243
Plastids 241-3, 245, 247, 249, 251, 253, 255-6, 430
functional 245
Pocket, hydrophobic 48, 294, 296-9, 303
Pol III 300-3
Pol IV 296-7, 302-3
Polar core 354, 356, 359, 362, 367
Polymerases 28, 295-6, 299-302
Polymorphic sites 336, 338-9, 341
Polymorphisms 47, 150, 153, 155, 329, 332, 335, 337-9, 440, 464
Polypeptides 465
Polyposis 66-7, 70, 73-6, 79, 85, 87
Polyposis patients 67, 70, 74, 76, 78, 83-5, 88
tested 84
Pools 327-9, 337, 341, 343-4
Populations 22, 73-4, 77, 86, 121, 225, 316, 323, 326-7, 329, 331-3, 337-43

general 80, 83, 86, 88, 466
Potassium 397, 399, 404, 410
Potassium channels 223, 393, 395-6, 401, 407
Potential origins 22, 34
PP2C 400
PPAR 218-19, 224, 226
PPR protein 252, 255
Principle 176-7, 270, 273, 478
Probability 30, 278-9
Problem 177-8, 180, 226, 302, 319, 341, 344, 475
Processivity 293, 295, 299-302
Proliferate 189-90, 192, 200
Proliferation 146, 148, 150, 152, 193, 196, 204, 224, 471
Promoter 35, 101-2, 109, 148, 151, 154, 198, 251, 382
Promyelocytic leukemia zinc finger (PLZF) 156-7
Propensities 21, 276, 278-9
Prostate 50-2, 55, 196
Prostate cancer 45-6, 51, 53-4, 59, 152-3, 221, 226, 464, 467
Prostatic intraepithelial neoplasia (PIN) 51, 434-5
Protein abundance 261-4, 267-8, 274-9, 281
absolute 276-9, 281
estimate 274, 276-7
relative 274-5, 277
Protein abundance index (PAI) 277
Protein complexes 16, 194, 268
Protein components 16, 109, 414
Protein expression 274-5, 413
Protein folding 427, 437, 441
Protein function 295, 330, 335
Protein identification 261, 278-80
Protein import 246, 249
peptide-dependent 242, 247
Protein interactions 265, 268
Protein kinase domain 331, 382

Protein kinases 194, 378, 395, 412-13
activated 380, 395, 463
dependent 379, 392
Protein length 277-8, 390
Protein levels 75, 147, 387
Protein product 219, 223-4, 336
Protein samples 264, 271
Protein separation 280
Protein sequences 262, 276, 327, 391
Protein size 276-7
Protein trafficking (PT) 247-8, 252-3, 256
Protein tyrosine phosphatase (PTP) 215
Proteins 193-4, 241-2, 253-5, 261-2, 264-8, 270-81, 293-7, 303-9, 365-7, 377-9, 387-8, 401-3, 414-15, 439-41, 460-2
adaptor 277, 352
associated 106, 111, 253
calcium sensor 377, 379
genes encode 226
heat shock 59, 254
human 196, 366
identified 277-8
intact 266-7, 270, 272
interacting 103, 255, 379, 387, 402
key 72, 268
known 249, 319
mitochondrial 461, 464
multiple 268, 271
nuclear 216, 406
nuclear matrix 117, 119
particular 273, 276
partner 309
peptide-concatenated artificial 273
plastid-localized 241-7
regulatory 37, 247, 427

ribosomal 200, 256
single 271, 276, 379
sliding clamp 293
target 267, 270, 272, 379, 401, 408
ubiquitinated 269
various 70, 295
wild type 75, 157
zinc finger 384
Proteome 261-2, 264, 279-81, 366
Proteomics 261, 263, 414
Proto-oncogenes 202-3
Pseudo-hypoxia response 456, 458-9
PSI 434-5
PTGS 320, 322

Q

QconCAT 267, 271-2
Quantification 261, 266-71, 273-4, 279, 281
stoichiometric 261, 265, 268, 272, 277, 281
Quantification of low abundance proteins 267, 280
Quantitative proteomics 261, 263
Quantitative traits 101, 121-2

R

Range 76, 78-9, 177, 278, 322-3, 327, 334-6, 340, 356, 379, 381
Rate 8, 10, 17, 34, 300, 330, 338, 453, 462
Ratio 225, 265, 267, 281, 337, 464
RBS 113-14
Reactive oxygen species (ROS) 70-1, 459, 461-2, 467, 470
Receptor activity 49, 54
Receptors 45, 47-56, 59, 224, 352-3, 356-8
nuclear 45, 47-9
steroid 45, 50
wild-type 48, 53-4

Recombination, homologous 107, 320, 335-6, 346
Recruitment 50, 303-4, 308, 352, 354-5, 357-8
Recruitment of accessory proteins 47, 49
Regions 9, 21, 27, 48, 54, 56, 74, 104-7, 109, 117, 148, 197, 295, 304-6, 479
Regulation 20-1, 33, 57, 106, 119, 148, 212-13, 215, 221, 295, 381, 385-8, 399-402, 408-9, 426
Regulation of gene expression 102, 108, 120, 379
Regulation of Interactions 293, 295, 297, 299, 301, 303, 305, 307, 309
Regulators, positive 146, 382-3
Relative change 262-4, 274
Relative quantification 261-3, 265, 273, 275-6
Relatives 4-5, 8, 85, 212
Release 212, 224, 293, 299, 301, 305-6, 463, 470
Renal cell carcinomas (RCC) 460
Repair 72, 75, 112, 293-4, 302, 308-9
Replicates 20, 22-3, 34, 293, 299, 465
Replication 15, 17, 20-4, 26, 28-31, 36-7, 211, 302, 307
Replication fork 22, 26-8, 302, 306-7
Replication kinetics 28-9, 31
Replication origins 15-22, 24-6, 28, 30-1, 33, 35-7
  late 21
  multiple 17, 31
Replication profiles 20, 22-5, 29, 31
Replication time 16, 26, 28-9
Replication timing 15, 18, 22-3, 25, 30, 33, 107
Replisome 301-2, 306-7
Research, functional 227-8
Researching gene function 170
Resistance genes 338
Resolution 173-4, 177-9
Response 47, 51, 57, 148, 215, 222, 227, 249, 269, 378-80, 382-5, 392, 398, 402, 405-6
Response elements 50
Retromer 215, 352, 359, 363-5
Reverse genetic approaches 316-17, 320, 322-3, 330, 334-5, 346

RFC 295, 297, 299, 305-6, 308
Rho_0-cells 468-9
Rice CDPK genes 386
RIDGEs 104-5
Risk 72, 82-3, 87, 211, 226
  increased 47, 80, 82-3, 150, 465
RNA transcription 110, 113-14, 116
RNAi 316, 320, 322, 336, 346, 406
Roots 387-8, 390, 392, 398, 401-2, 453
ROS production, enhanced 459, 462, 467-8, 470
RTS 114

S

S. cerevisiae 17-18, 21, 25, 30, 34, 103, 113-14, 365, 433, 435
Salt 321, 384-6, 390, 392-4, 396-9, 401, 405-6, 408-11, 414-15
Salt stress 384-7, 389-90, 401-3, 405, 408, 413
Salt stress conditions 382-3, 389, 391, 402, 406
Samples 182, 262-5, 270-1, 273, 275, 281, 319, 327, 329, 337, 340-1, 344, 477, 480
SC phocomelia 113-14
Scaffolds 307, 352, 355
Screening 10, 12, 68, 73, 84, 86-8, 195, 225, 275, 281, 320, 472, 481-2
Scrib 190, 193-5, 202-3
SDH 452, 457-8, 460-1, 463
SDH dysfunction 458
SDHB mutations 459-60
SDS-PAGE 266, 270, 280
SECA 244-6, 253, 255
Sectioning 172, 179-81
Sections 178, 182, 201, 300, 471
Seed development 243-4, 251, 255, 388, 406
  proper 243-4
SeedGenes 241, 255
SeedGenes database 255-6

Seeds 326, 328, 330, 333, 343, 434
    null-mutant 255
Selective androgen receptor modulators (SARMs) 59
Sequence candidate amplicons in multiple parallel reactions (SCAMPR) 329
Sequences of human arrestin clan proteins 361
Sequencing 53, 227, 316, 318-19, 329, 337-8, 340, 472-3, 475
    generation 86, 317, 329
Serial analysis of gene expression (SAGE) 102, 105
Serial reconstruction 171, 174-5, 181
Serial transplantations 193
Shoots 402, 405, 408
Sic1 15, 34-6
Sic1 mutant 35-6
Single nucleotide polymorphisms (SNPs) 152, 218, 225, 316, 326-9, 335-41, 344
Signal transduction pathway 378-80, 427
Signaling pathways 212, 353, 377-8, 395, 414, 416
Signals 357, 378-9
SIM1 172, 174, 176, 180-2g

SIM2 172, 174, 176-7, 180-2
Single genes 245, 387, 397, 430
Site-directed mutagenesis 316, 415
Sites 35, 55, 57, 71, 109, 269-70, 297-8, 300, 320, 335, 338, 354-5, 365, 408, 428
SLC30A8 221, 223, 227
Sliding clamp binding protein 308
Sliding clamps 293-5, 297-309
Sliding clamps during DNA replication and repair 293
SMARCAL1 112, 121
SMARCAL1 protein, functional 112
SNPs, rare 339-40
SOD2 462, 464
Solid tumors 452-3, 455, 473, 475
Somatic mutations 66, 71, 86, 146, 149, 151, 155-6, 466, 472, 475, 481
SOS1 395, 399, 401-2, 408, 411

Soybean 318, 321, 325, 333, 407
Species 102-3, 105, 196, 241, 244, 317, 319, 321-2, 335, 337, 343-4, 346, 361, 407, 414-15
Specificity 47, 71, 151, 266-7, 280-1, 378, 430
Specimens 172-7, 180, 182
  large 177
Spectral count 267, 274-5, 277-9
Spermatogenesis 103, 440
Spleen 213-15
Ssa proteins 432-3, 435
Ssa1 426
SSE1 435-6
Stable isotope-labeled peptides 266-7
Stable isotopes 262-4, 266, 270-1, 281
Staining 177, 180-2
Standard peptides 266-7, 269, 271-2
Standard peptides/proteins 271
Standard proteins 270, 275
Standards 268, 273, 279
STAT 358
Stimuli 114, 353, 377-8, 381, 389, 405, 410, 414
Stochastic firing of origins 23
Stomatal closure 379, 383, 385, 393
Strands 352, 354, 356-7, 361, 364, 479
Strategies 15, 223, 246, 251, 261, 264-6, 268, 270-2, 274, 276, 281, 319-20, 326-7, 337, 343-5
  reverse genetic 320
Stress 219, 378, 380, 382, 385-6, 392, 394-5, 399, 402, 405-6, 410-11, 414-15, 432, 437, 482
  biotic 379, 390, 410
  cold 321, 386-7, 390, 399, 405, 408
  environmental 378, 406
  low potassium 400
Stress conditions 377, 380-1, 384, 386, 389, 393, 396, 398, 405, 407, 412, 416

osmotic 397
Stress signaling 380, 383-4, 413
Stress situations 435, 438
Stromal processing peptidase (SPP) 242, 245-7, 249, 252
Structure 59, 121, 178, 191, 197, 296-7, 303, 352-4, 360, 362, 366, 426, 429
Subsets 17, 22, 25, 28-9, 56-7, 59, 147, 189, 360, 395-6, 430
Substitutions 46, 48, 53-5
Substrates 75, 121, 214, 269, 381, 383, 393, 401, 412, 414, 428-9, 435, 441
Subunits 242, 271, 297-9, 301, 304, 359, 362-3, 365, 452, 456-9
Sulfonylurea derivatives 214-15, 223, 225
Sumoylation 55, 295, 303-4
Surface 48, 108, 172-3, 176, 179, 299, 303, 353, 356
Surveillance 66, 86, 88
Switch 28, 151, 300, 302, 454, 456, 461
Synthetic peptide 266-7, 270-1, 301, 381
stable isotope-labeled 265, 268, 271

T

Target genes 5-6, 106, 196, 321-2, 325, 330, 332-3, 339, 344, 388
regulatory regions of 49-50, 52
Target induced local lesions in genomes (TILLING) 316, 323, 325-6, 329-37, 339-42, 344, 346
TAU1 46
TAU5 46
TCF1 216, 221
TCF2 216, 219, 221
Techniques 108, 170-2, 179, 199, 202, 223, 263, 326, 337, 340-2, 346, 389, 466, 479, 481-2
Temporal program 15, 21, 29
Terminal phenotypes 243, 255
Territories, chromosomal 107-8, 118
Testis 52, 104, 213-15, 437
Thaliana embryogenesis 244, 246, 248, 250
Therapies 45, 49, 52-3, 58-9, 157, 227

Therapy resistance 45
Thiazolidinediones 214, 222-4, 226
Thickness 173, 175, 177-8
TIC 242, 246, 249, 251
TILLING method 326, 328, 331
Time 16-17, 20-1, 23, 26-31, 35, 79, 84, 88, 103, 173, 199, 340-2, 345-6, 463-4, 468
Time expense 178, 183
Tissue architecture 169, 176-7, 181-2, 189-91
    context of 170-1, 183
Tissue contrast 176, 180-1
Tissue samples 169-72, 176, 179, 181-3, 264
Tissues 105, 115, 118, 170-3, 175-6, 180-1, 189-91, 203, 211, 217, 220, 265-8, 381, 387-90, 437
    bsm-null 245
    most 437
    stained 176
Tobacco 392, 413-14
Topology 108, 175, 178
Tract 47, 53, 146, 472, 475
Trafficking 215, 352, 364
Transcript levels 386, 388, 391-2, 408-10, 412, 415
Transcription 47, 105-8, 112, 114, 116-19, 121, 148, 150, 196, 247, 353, 358, 406, 461, 479
Transcription activation units (TAUs) 46
Transcription factors 45, 102, 106, 108, 114, 117, 146, 152, 213-17, 379, 383, 386, 395
Transcripts 249, 388-9, 392, 397, 408, 414
Transgenic material 316, 321-3, 340
Transition 18, 30, 34-5, 241, 335
Transplantation 190-1
Transporters 381, 399-400, 402, 411
Transversions 70-2, 467
Treatment 8, 45, 49, 51, 57-9, 86, 88, 169, 224, 326, 330, 333, 335-6, 362, 390

Trico-Peptide repeat (TPR) 428
Tryptic peptides 269, 271-2, 275, 278
Tumor cells 33, 71, 439, 452-6, 463, 467, 469, 471, 475
    ascites 453
    parental 468-9
Tumor growth 154, 456, 458, 462, 472, 481
Tumor suppressor genes 118-19, 148, 151, 157, 452, 456
Tumor suppressors 146, 149, 452, 456-7, 463-4, 470
Tumor tissues 147, 452-3, 457, 459, 466, 473, 481-2
Tumorigenesis 36, 151, 439, 459, 463
Tumors 3, 7, 71, 86, 147, 152-3, 321, 453, 456-7, 460, 464, 466-7, 470-3, 475
Tumour suppressor 189-90, 196, 203
Tumourigenesis 189, 204
    human 189, 197, 204
Tumours 45, 49, 52, 58-9, 193, 200, 203

U

Ubiquitin 269, 303, 456
Ubiquitination 269, 303, 355, 362-3, 365
Up-regulation 149-50, 155, 388, 432, 439
Updated list 243, 246-8, 256
URE3 434-5
Uterus 213-16

V

Vacuole 365, 378, 393, 401-2, 411
Variable expressivity 112, 121
Variable number tandem repeats (VNTRs) 117
Variants 5, 53-4, 68, 73, 75, 82-3, 86, 88, 152, 218, 220, 224-7, 319, 331-2, 475
Variation 6, 53, 277, 337-8, 341, 343
Variations 219, 264, 352-3, 355, 357, 359, 361-3, 365, 367, 467
VDAC (voltage dependent anion chanel) 455
VHL (von-Hippel-Lindau) 456, 458, 460
Visual 352, 356, 365, 367

Vivo 25, 29, 70, 146, 149-50, 155, 268, 298, 300, 396, 401, 410, 432, 435, 439-40
VNTRs (variable number tandem repeats) 117
Voltage dependent anion chanel (VDAC) 455
Volume data 171, 174-5, 177-9, 181
Von-Hippel-Lindau (VHL) 456, 458, 460
Voxel size 169, 177-8
Voxels 173-4, 178
Vps26 352, 359-60, 363-7
Vps26 proteins 360, 364

W

Warburg Effect 452-4, 456
WFS1 214, 219, 221-2
Wheat 322, 324-5, 332, 334, 388-9, 399, 407
Wild type 28, 35-6, 75, 192, 199, 327, 331, 333, 383, 403
Wild-type background 435-6
Wild-type cells, c-Myc-expressing 201
Wounding 388, 391-2, 397-9

Y

Ydj1p overexpression 435
Yeast 17, 33-4, 102, 109, 271, 275, 320, 359, 361-2, 364-5, 395-7, 400-3, 408-10, 426-8, 430
Yeast chromosomes 16, 20, 28, 31-2
Yeast PCNA 303, 305, 307
Yeast proteins 279-80
Yeast proteome 278-80

Z

Zebrafish 324-5, 334-5, 343

Made in United States
Orlando, FL
04 November 2026

90208973R00324